# 建筑经典选读

杰伊·M. 斯坦 | Jay M. Stein
肯特·F. 斯普雷克尔迈耶 | Kent F. Spreckelmeyer 编著
荆宇辰 译
韩学义 审校

CLASSIC
READINGS IN
ARCHITECTURE

Jay M. Stein + Kent F. Spreckelmeyer

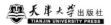

Jay M. Stern, Kent F. Spreckelmeyer
Classic Readings in Architecture
ISBN 0-07-061415-6
Copyright © 1999 by McGraw-Hill Education.

All Rights reserved. No part of this publication may be reproduced or transmitted in any form or by any means, electronic or mechanical, including without limitation photocopying, recording, taping, or any database, information or retrieval system, without the prior written permission of the publisher.

This authorized Bilingual edition is jointly published by McGraw-Hill Education and Tianjin University Press. This edition is authorized for sale in the People's Republic of China only, excluding Hong Kong, Macao SAR and Taiwan.

Copyright © 2016 by McGraw-Hill Education and Tianjin University Press.

版权所有。未经出版人事先书面许可，对本出版物的任何部分不得以任何方式或途径复制或传播，包括但不限于复印、录制、录音，或通过任何数据库、信息或可检索的系统。

本授权双语版由麦格劳—希尔（亚洲）教育出版公司和天津大学出版社合作出版。此版本经授权仅限在中华人民共和国境内（不包括香港特别行政区、澳门特别行政区和台湾地区）销售。

版权 © 2016 由麦格劳—希尔（亚洲）教育出版公司与天津大学出版社所有。

本书封面贴有McGraw-Hill Education公司防伪标签，无标签者不得销售。
天津市版权局著作权合同登记号：02-2014-27

---

图书在版编目（CIP）数据

建筑经典选读 /（美）杰伊·M. 斯坦，（美）肯特·F. 斯普雷克尔迈耶编著；荆宇辰译. —天津：天津大学出版社，2016.9
ISBN 978-7-5618-5648-2

Ⅰ. ①建… Ⅱ. ①杰… ②肯… ③荆… Ⅲ. ①建筑学 Ⅳ. ①TU-0

中国版本图书馆CIP数据核字（2016）第195708号

| 出版发行 | 天津大学出版社 |
|---|---|
| 地　　址 | 天津市卫津路92号天津大学内（邮编：300072） |
| 电　　话 | 发行部：022-27403647 |
| 网　　址 | www.publish.tju.cn |
| 印　　刷 | 廊坊市海涛印刷有限公司 |
| 经　　销 | 全国各地新华书店 |
| 开　　本 | 185mm×240mm |
| 印　　张 | 34.25 |
| 字　　数 | 1120千 |
| 版　　次 | 2016年9月第1版 |
| 印　　次 | 2016年9月第1次 |
| 定　　价 | 95.00元 |

凡购本书，如有缺页、倒页、脱页等质量问题，烦请向我社发行部分联系调换。
版权所有　　侵权必究

# 译序

市场上的建筑专业类图书品种不胜枚举,内容也是纷繁复杂,但稍加考量就会发现图书的品位、价值参差不齐。想要从书海中甄选出对于建筑研习之路有所启迪的好书,无论对于刚刚踏入建筑学门槛的新人,还是执业工作多年的老手,亦或建筑领域的爱好者,都不是一件易事。无论国内,还是国外,传世的精品佳作均属稀缺资源。

为了帮助各个层次的读者在适宜的书籍指引下能更快、更顺畅地步入理想的建筑之路,杰伊·M.斯坦和肯特·F.斯普雷克尔迈耶两位教授走在了最前列,编写出了《建筑经典选读》一书。阅读经编者瘦身浓缩的原作精华,既节省了读者挑选书籍的时间,也减轻了学生通读大部头理论专著的负担。

《建筑经典选读》不仅是斯坦和斯普雷克尔迈耶两位编者的心血,也可算作是美国高等建筑教育的集体结晶。在图书编写过程中,美国建筑学院协会的各成员单位、知名学者和任课教师都为内容的筛选以及篇章的组织等工作予以了大力的支持与协助,这也让该书更具严谨性与权威性。

《建筑经典选读》以美国各高校建筑学专业学生的课程安排和教师的意象调查为基础,兼容编者多年从事建筑学研究和教学的经验心得,囊括文章完全出自名家名作,编排为三个主题,各主题下又设三个分支,共计收录了36篇选读文章。

书稿内容的经典性、编排的权威性、同类书籍的稀缺性更让该书能够独树一帜,在读者中间广泛传播;同时也让译者能够有幸经自己之手将其译为中文,介绍给广大中国读者。鉴于每篇节选均为大师们代表作品中的精髓所在,实难原汁原味地将书中文字及深邃内涵全部翻译呈现为目标语言,译者仅是小心谨慎地将编者前言、名家简介和各篇选读的导读部分译为汉语,节选篇章的原文部分仍保留为英文,而且版式也忠实于英文版的原著。

顺风而呼,其声益彰;顺势而为,其事必成。愿每一位建筑学专业的莘莘学子都能汲取名家大师的睿智灵感,自成一片天地。

荆宇辰

# PREFACE

The purpose of this book is to introduce students of architecture to the important literature of the field. It consists of journal articles and selected book chapters by authors who are recognized as major contributors to architectural history, theory, and practice. The book also will serve as a reference document for architects, design educators, building clients, and others who seek an overview of architecture as an academic discipline as well as a professional pursuit.

The book is a companion work of Jay M. Stein's (1995) *Classic Readings in Urban Planning*. As such, this text is a continuation of the objectives contained in the previous book and is seen as a way of extending introductory material into a field closely allied with planning. The assumption is that one of the best ways to introduce a field is to expose the reader to its great works, the "classics." Webster's (1979: 334) defines *classic* as: "of the highest class; most representative of the excellence of its kind; having recognized worth." Thus, we define classic readings as those that are of superior quality, represent outstanding scholarship, and may have influenced or changed the field.

However, major distinctions between this book and *Classic Readings in Urban Planning* (Stein 1995) are the audience to which each is directed and the nature of the individual readings. The earlier book was written for upper-level undergraduate and graduate-level students of planning, while *Classic Readings in Architecture* will serve as a text for first-year students in architectural curricula. Also, the articles in this book have been selected as much as an introduction to important authors as a review of significant or seminal works. In this respect the term "classic" differs from that in the planning text. The research conducted by the editors for this book revealed that the definitions of "classic readings" in architecture are broader and less focused than the literature of planning. While fairly definitive categories of academic and professional fields can be drawn in planning—transportation, housing, economic development—no such fixed categories exist in architecture. Thus, it is more appropriate to view *Classic Readings in Architecture* as an introduction to significant contributors of architectural knowledge and as a guide to the larger body of architectural literature.

*Classic Readings in Architecture* contains 36 readings organized into three major areas, each with subcategories. Thus, the book is organized as follows: Architecture as Artifact (Architectural History and Theory, Architectural Form, and Architectural Technology); The Context of Architecture (The Urban Environment, The Natural

# 前言

这本《建筑经典选读》意在向建筑学专业的学生推介建筑领域重要的文献专著。本书收录了人们公认的对建筑历史、理论和实践曾做出过突出贡献的重要作者的期刊文章和遴选的专著章节。同时，本书也将为建筑师、讲授设计课程的教师、建筑工程的甲方和其他想了解建筑这一学科的概况以及有志于从事建筑行业的人员提供文献参考。

本书是杰伊·M. 斯坦（1995）编纂的《城市规划经典选读》的姊妹篇。因此，本书延续了前作的编写宗旨，并把介绍性阅读材料的范围扩展到了与规划紧密相联的建筑领域。编者认为引领读者进入某一领域的最佳方式之一便是使其置身于这一领域的伟大作品，即"经典"之中。《韦氏词典》（1979:334）将"经典"一词定义为"处于至高无上的地位；在同类杰出的事物中最具代表性；人们已经认识到它的价值"。鉴于此，我们认为经典选读便是挑选出那些质量上乘的作品选段，它们代表了出色的学识，而且可能已经影响或改变了该学科领域。

但是本书和《城市规划经典选读》（斯坦，1995）相比还是有很大区别的，所面向的读者群和每篇选读的本质属性都不尽相同。《城市规划经典选读》为城市规划专业的高年级本科生和研究生而作，而本书则以建筑学专业一年级新生为主要读者对象。同时，本书在回顾意义重大或对后世影响深远的作品时，除摘录作品中的选段外，也花费笔墨对重要作者进行了介绍。从这方面而言，本书对于"经典"的诠释是不同于《城市规划经典选读》的。在编纂过程中，本书编者揭示出的建筑专业"经典选读"的定义与《城市规划经典选读》相比范围更广，很少聚焦于某一点。尽管在规划中，人们可以列出学术和专业领域相当明确的条目——交通、住房、经济发展——但在建筑学中，这样固定的条目却是不存在的。因此，更适合把本书当作一本导读，让读者了解对建筑学做出过贡献的杰出人物；把本书当作一份指南，引领读者进入庞大的建筑文献的世界。

《建筑经典选读》分为三大部分，每部分下设不同主题，共计36篇选读文章。以此方式，本书编写结构如下：人造艺术的建筑（建筑历史和理论、建筑形式、建筑技术）；建筑的环境（城市环境、自然环境和人文环境）；建筑过程（设计

Environment, and The Human Environment); and The Process of Architecture (The Design Process, The Social Implications of Architecture, and The Architectural Profession). Each reading begins with a brief abstract of the work, and each of the nine category sections concludes with a "suggested readings" list of selected articles and books from that field of study. Faculty, students, and professionals will find the suggested readings to be useful for more extensive study in these areas.

*Classic Readings in Architecture* can be used as a major or supplementary text in introductory courses in architecture, as a reference guide in the various areas of knowledge, and as a companion text in courses that are focused on a specific architectural topic. The book would be valuable, for example, in a course in architectural history as a way of placing an historical concept in the context of a broad survey of architectural theory and practice.

## THE CLASSIC CONCEPT

The objectives of this collection of readings are the same as those outlined in Jay M. Stein's (1995) earlier book on planning, namely that "it is far more interesting and exciting to *directly* experience ideas and read—even struggle—with original writings, than it is to review someone else's synopsis *about* those ideas." In any introductory course in architecture, the student will be exposed to a wide range of architects, buildings, professional standards, and theoretical concepts. We feel that it is important for the student not only to be introduced to these concepts, but to be given the opportunity to hear the voices of the historians, theoreticians, and practitioners who were significant contributors to those ideas. The works contained in *Classic Readings in Architecture* will provide the student with that opportunity.

## ORGANIZING THE SUBJECT AREAS

A primary challenge in assembling these readings has been to create a typology for the architectural literature presented in the book. To develop our typology, we have been guided by the results of a survey conducted in Spring 1996 of introductory architecture courses offered by departments listed in the Association of Collegiate Schools of Architecture (ACSA)'s *Guide to Architecture Schools, 5th ed.* (1994). Schools that responded to the survey are listed at the end of this Preface (we regret any unintended omissions).

The survey of introductory courses served two major purposes. First, it provided the basis for organizing the book into the nine categories by identifying the most frequently covered content areas in introductory courses. Second, the survey provided important input as part of the process of selecting the readings included in the book.

## SELECTING THE READINGS

Similar to the approach used in *Classic Readings in Urban Planning* (Stein 1995), we have used a systematic, four-step process to obtain information from several sources to help identify the "classics" included in this book. The four steps include:

过程、建筑的社会含义和建筑行业）。每篇选读的开头是文章原作的简短摘要，九个主题中每个主题的结尾均附有一份"推荐阅读"书单，列举从相关研究领域选出的文章和图书。教师、学生和建筑行业从业者将会发现这些"推荐阅读"书单对于在这些领域开展更广泛的研究大有裨益。

本书既可以当作建筑学专业入门类课程的主要教材或补充文本使用，也可以当作众多知识领域的参考指南使用，还可以当作关注某一具体建筑主题的课程的同步教材使用。这将会是一本有益的图书，比如在建筑历史课上，本书能够把一个历史概念置于建筑理论和广泛的实践评述之中，加深学生对历史概念的理解。

## 经典概念

这部作品选集的宗旨与杰伊·M. 斯坦（1995）早先推出的关于规划的图书宗旨相同，即"在原汁原味的作品中去体会作者的思想、去阅读，甚至去硬啃原文字句都远比看其他人笔下对这些思想的概述更有趣，也更令人兴奋得多"。建筑学的所有入门课程都会让学生接触到大量的建筑师、建筑物、行业标准和理论概念。我们认为不仅要把这些概念介绍给学生，而且要让他们能够有机会聆听对于提出和发展这些概念曾做出杰出贡献的历史学家、理论家和实践者们的声音，这与介绍概念同等重要。这本《建筑经典选读》中的作品选读就为学生提供了一个聆听大师发声的机会。

## 主题范围的编排

在编纂选读文章的过程中，我们的首要挑战便是对本书收录的建筑文献进行分类归纳。为了推进我们的类型学工作，我们对1996年春季学期一些建筑院系开设的建筑学专业入门课程进行了调查，其调查结果对本书的编排工作具有指导作用。所选建筑院系来自美国建筑学院协会编写的《建筑学院指南（第五版）》（1994），对这次调查做出回应的院系也在前言的最后部分——列出（如因疏忽有所遗漏，深表歉意）。

调查入门类课程有两个主要目的：一是通过识别在入门类课程中最常出现的学习内容，为本书编排的九个主题提供依据；二是为本书中选读内容的筛选提供重要信息。

## 选读内容的遴选

与《城市规划经典选读》（斯坦，1995）使用的方法相似，我们通过系统的四步遴选法从众多的信息源中获取信息，确定本书中收录的"经典"。这四个步骤包括：

1. A survey and analysis of introductory course syllabi.
2. A survey of architecture educators.
3. Identification of award-winning writings.
4. A general review of architecture literature.

The survey and analysis of course syllabi have already been discussed. We also conducted a survey in Fall 1996 of architecture experts listed in the *Guide to Architecture Schools, 5th ed.* (ACSA 1994) representing the now-identified subject areas to be covered in the book. The survey of architecture educators did not follow a purely scientific sampling procedure, but rather is only an attempt to obtain feedback from a diverse group of experts. The questionnaire sent to the respondents identified their field of expertise and asked two simple questions:

1. "Please outline a maximum of ten areas of knowledge that might form chapter headings in this book."
2. "Also please list three articles or book chapters that you consider to be *essential* readings for an introductory course in architecture . . ."

The results of this survey were a major information source in selecting the readings for the book. At the end of this Preface is a list of the architecture educators who were consulted and agreed to the publication of their names (we regret any unintended omissions). Others were consulted but preferred anonymity.

As a third source of input to the classics selection process, we also attempted to identify writings and authors who have received awards or special recognition for their work. Several of the readings selected for the book meet this criterion. Finally, we conducted a major literature review and consulted with several colleagues as to their suggestions for the book.

## CAVEATS

Although surveys of curricula and educators in North American schools of architecture have provided valuable input for organizing the book and selecting the readings, our final selections for this collection were based on two important assumptions. First, the term *architecture* can be defined from at least three distinct perspectives: as a physical artifact; as a part of the larger built, natural, and human environments; and as a process. The readings, therefore, have been classified as contributions to the understanding of architecture within these areas of intellectual and professional knowledge. The second assumption has been that the readings would be by contemporary authors who had contributed an extensive body of knowledge to the literature of architecture. The two exceptions to this rule are the inclusion of works by Vitruvius, on the one hand, and Ernest Boyer and Lee Mitgang, on the other. In the case of the former, we felt strongly that a book of classic readings in architecture should include what must be considered an icon of architectural literature and a starting point of Western architectural thought. In the latter instance, we believe that the most recent study of the field by the American architectural establishment—*Building Community: A New Future for Architecture Education and Practice* (Boyer and Mitgang 1996)—requires the attention of the student of architecture. These two works form the Prologue and Epilogue to the text.

1. 调查和分析入门类课程的教学大纲；
2. 调查建筑学专业的教师；
3. 搜集获奖文章；
4. 对建筑文献进行综述。

我们已经对入门类课程教学大纲的调查和分析进行过讨论。在1996年秋季学期我们又对能够代表书中涉及的当下各明确主题领域的建筑学专家进行了调查。这些建筑学专家名单收录于《建筑学院指南（第五版）》（1994）。对建筑学教师的调查并没有严格遵循科学抽样的程序，这项调查只是从多个组别的专家中获得反馈的一种尝试。调查问卷寄送给每个专业领域明确的被测者，并要求他们回答两个简单的问题。

1. "请最多列出十个可能构成本书章节标题的专业领域。"
2. "请再列出三篇您认为对建筑学入门类课程至关重要的文章或书籍章节……"

调查结果成为本书选读内容的主要信息来源。在前言的最后，我们列出了曾经请教过并同意公开姓名的建筑学教师的名单（如因疏忽有所遗漏，深表歉意）。另有一些我们请教过的教师则希望匿名。

作为遴选经典过程中的第三种信息来源，我们也在努力搜集受到嘉奖的作品和因其作品受到特殊认同的作者。本书中收录的一些选读内容都满足了这一条标准。在筛选过程的最后，我们做出了一份主要文献综述并咨询了诸位同人，听取了他们对本书的建议。

## 附加说明

尽管对北美各建筑学院建筑学课程和教师进行的调查为本书的编排和选读内容的选择提供了极有价值的信息，但是为本书进行的最后编选工作仍是基于我们自己的两个重要设定。设定一，"建筑"一词至少可以从三个不同的角度进行界定：作为物质创造物；作为更大范围内的建成环境、自然环境和人文环境的一部分；作为一个过程。因此，为加深在智力知识和专业知识领域内对建筑的理解，我们把本书中的选读内容按照贡献的不同类别加以分类。设定二，选读内容的作者必须是在广泛的建筑文献知识领域中曾做出贡献的当代人物。此项设定有两个例外，一是维特鲁威（Vitruvius）创作的作品，另一是欧内斯特·勒罗伊·博耶（Ernest Leroy Boyer）和李·D. 米特冈（Lee D.Mitgang）的作品。对于前者，我们有一种强烈的感觉，那就是一本建筑经典选读的书籍应该收录被认为是建筑文献的象征和西方建筑思潮起点的作品。至于后者，我们相信新近关于美国建筑成就研究的《建筑社区：建筑教育和实践的新未来》（博耶和米特冈，1996）应引起建筑学专业学生的关注。这两部作品也构成了本书首尾的序和跋。

Although we followed such a painstaking process to select the readings, the needs of a textbook required several compromises. These include: consideration given to the suitability of the level and complexity of the writings for an introductory course; a desire to include different voices and views; and, finally, the difficulties involved in obtaining reprint permissions in a timely manner and at an affordable rate relative to our constrained budget. Nevertheless, we believe that this book contains an excellent collection of readings and offers a comprehensive introduction to the field of architecture.

*Enjoy the readings and welcome to the adventure of architecture.*

## REFERENCES

Association of Collegiate Schools of Architecture (ACSA). *Guide to Architecture Schools.* 5th ed. Washington, DC, 1994.

Boyer, Ernest L., and Lee D. Mitgang. *Building Community: A New Future for Architecture Education and Practice.* Princeton: The Carnegie Foundation for the Advancement of Teaching, 1996.

Stein, Jay M. *Classic Readings in Urban Planning.* New York: McGraw-Hill, Inc., 1995.

*Webster's New Twentieth Century Dictionary of the English Language Unabridged.* 2nd ed. New York: Simon and Schuster, 1979.

## ACSA COURSE SURVEY RESPONDENTS

Auburn University
Carleton University
Harvard University
Illinois Institute of Technology
Iowa State University
Lawrence Technological University
Morgan State University
North Carolina State University
Notre Dame University
Rensselaer Polytechnic Institute
State University of New York at Buffalo
Tulane University
University of Cincinnati
University of California, Berkeley

California Polytechnic State University, San Luis Obispo
University of Florida
University of Illinois at Urbana-Champaign
University of Illinois at Chicago
University of Kansas
University of Minnesota
University of Nebraska–Lincoln
University of Southwestern Louisiana
University of Virginia
University of Tennessee
University of Wisconsin, Milwaukee
Yale University

## PANEL OF ARCHITECTURE EDUCATORS

Kathryn H. Anthony
Joseph Bilello
Edward J. Cazayoux
Ruth Connell

Michael Crosby
Marleen Davis
Sam Davis
Guido Francescato

尽管我们在甄选的过程中精挑细选，但是为满足其作为一本教科书的需求，我们在挑选标准上也做出了一些折中处理，这些处理包括：考虑到选读内容的难易级别和复杂程度对入门类课程是否适宜；书中要有不同的声音和观点；最后是难以及时获得再版许可，并且由于经费有限难以支付高额版税率的困难。尽管如此，我们仍然相信本书收录了最好的选读内容，提供了建筑领域的全面介绍。

希冀读者享受这些选读内容，以开启你的建筑探索之旅。

## 参考书目

Association of Collegiate Schools of Architecture (ACSA). *Guide to Architecture Schools.* 5th ed. Washington, DC, 1994.

Boyer, Ernest L., and Lee D. Mitgang. *Building Community: A New Future for Architecture Education and Practice.* Princeton: The Carnegie Foundation for the Advancement of Teaching, 1996.

Stein, Jay M. *Classic Readings in Urban Planning.* New York: McGraw-Hill, Inc., 1995.

*Webster's New Twentieth Century Dictionary of the English Language Unabridged.* 2nd ed. New York: Simon and Schuster, 1979.

## 美国建筑学院协会课程调查受访学校

奥本大学
卡尔顿大学
哈佛大学
伊利诺伊理工大学
爱荷华州立大学
劳伦斯理工大学
摩根州立大学
北卡罗莱纳州立大学
圣母大学
伦斯勒理工学院
纽约州立大学布法罗分校
杜兰大学
辛辛那提大学
加利福尼亚大学伯克利分校
加州州立理工大学圣路易斯·奥比斯波分校
佛罗里达大学
伊利诺伊大学厄巴纳—香槟分校
伊利诺伊大学芝加哥分校
堪萨斯大学
明尼苏达大学
内布拉斯加大学林肯分校
西南路易斯安那大学
弗吉尼亚大学
田纳西大学
威斯康星大学密尔沃基分校
耶鲁大学

## 建筑学教师专家小组名单

凯瑟琳·H. 安东尼
约瑟夫·比莱洛
爱德华·J. 卡萨尤克斯
鲁丝·康奈尔
迈克尔·克罗斯比
玛莲·戴维斯
萨姆·戴维
吉多·弗朗西斯卡托

Alan R. Cook
David Cronrath
Don Grant
Sharon Haar
Nancy J. Hubbard
John Klingman
Rochelle Martin
Robert McCarter
Graham Owen

Benjamin Gianni
Stephen Grabow
Gregory S. Palermo
Julia Robinson
Henry Sanoff
Jorge Silvetti
Robert L. Vickery
Donald Watson

## ACKNOWLEDGMENTS

I gratefully acknowledge the ideas and support of my students and colleagues in the Department of Urban and Regional Planning, College of Architecture, at the University of Florida. Research assistants April Alperin and Shenley Neely provided valuable help in the early preparation of the manuscript. Kathryn Younkin, also a research assistant, deserves a special recognition for her hard work, dedication, and attention to details while always managing to have a cheerful disposition. I am most indebted, however, to my two daughters, Danielle and Melissa, for their patience, understanding and cheerfulness. They are always a source of inspiration for me.

*Jay M. Stein*
*Gainesville, Florida*

I am grateful for the suggestions I received during the preparation of the reading lists from colleagues in the School of Architecture and Urban Design at the University of Kansas, especially Dennis Domer, Steve Grabow, Jim Mayo, and Barry Newton. I would like to thank Tim Nielsen for the work he did during the literature surveys, and Cindy Muckey and Barb Seba for technical and editorial assistance. Finally, I would like to thank my wife, Antha, for her support and patience during another project that, once again, took longer than I promised.

*Kent F. Spreckelmeyer*
*Lawrence, Kansas*

We jointly wish to thank our friends and colleagues in the member schools of the Association of Collegiate Schools of Architecture who generously shared their introductory architecture course syllabi and also served as a "panel of experts" in helping to select the "classics" that are this book. Without their contributions, this book would not have been possible. Responsibility for any and all errors are, of course, ours.

We also appreciate the support and professionalism of our first editor, B. J. Clark, who believed in this project from its inception, and his very able successor at McGraw-Hill, Inc., Eric Munson, who helped us to successfully complete the book.

*Jay M. Stein*
*Kent F. Spreckelmeyer*
*June 1998*

| | |
|---|---|
| 艾伦·R.库克 | 本杰明·吉安尼 |
| 大卫·克龙拉特 | 史蒂芬·格拉博 |
| 唐·格兰特 | 格雷戈里·S.巴勒莫 |
| 莎伦·哈尔 | 茱莉亚·罗宾逊 |
| 南希·J.哈伯德 | 亨利·萨诺夫 |
| 约翰·克里格曼 | 乔治·希尔瓦蒂 |
| 罗谢尔·马丁 | 罗伯特·L.维克瑞 |
| 罗伯特·麦卡特 | 唐纳德·沃特森 |
| 格拉汉姆·欧文 | |

**致谢**

我衷心感谢佛罗里达大学建筑学院城市与区域规划系的我的学生及同事对于本书的大力支持和建言献策。我的研究助手阿普里尔·阿尔珀林和雪莉·尼利在原稿的早期准备阶段为我提供了有益的帮助。我的另一个研究助手凯瑟琳·扬金值得特别表扬,她全身心地投入艰苦的工作,专注细节的同时还总是保持乐观的性情。我最要感谢的是我的两个女儿,丹妮尔和梅丽莎,她们对我抱有耐心,理解我的工作,还总是高高兴兴的,让我心生愉快。她们是我灵感的不竭源泉。

杰伊·M.斯坦
盖恩斯维尔,佛罗里达

感谢堪萨斯大学建筑与城市设计学院的同事们在我准备这本选读所收录的书目时给我提供的建议,特别是丹尼斯·多莫、史蒂夫·格拉博、吉姆·梅奥和巴里·牛顿,感谢他们给我的帮助。我同时要感谢蒂姆·尼尔森在文献调查中的出色工作,感谢辛迪·姆奇和巴布·西巴的技术与编辑协助。最后,我要感谢妻子安西娅仍然在支持我的工作,而且对我抱有耐心,尽管我再次投入另一个项目,而且比最初承诺的花费了更长的时间。

肯特·F.斯普雷克尔迈耶
劳伦斯,堪萨斯

我们共同向美国建筑学院协会各成员院校的朋友和同事表达我们的谢意,感谢各成员院校慷慨分享各校的建筑入门类课程教学大纲并作为"专家组"帮助遴选本书中的"经典"选读。没有他们的贡献,本书不可能成形。当然,书中出现任何纰漏错误之处,都应归咎于我们。

我们还要感谢我们的第一位编辑B.J.克拉克从项目伊始便为我们坚持提供支持和专业指导并且感谢他的继任者,麦格劳—希尔公司的编辑艾瑞克·芒森,他能力出众,帮助我们最终圆满完成了本书的编写工作。

杰伊·M.斯坦
肯特·F.斯普雷克尔迈耶
1998年6月

# ABOUT THE EDITORS

JAY M. STEIN is professor and chair of the Department of Urban and Regional Planning, College of Architecture, University of Florida. He previously served as professor and chair, Department of Planning and Design, SUNY-Buffalo; as acting dean, School of Architecture and Planning, SUNY-Buffalo; as visiting professor of Infrastructure Planning and Management, Stanford University; and as a faculty member in the Graduate City Planning Program, College of Architecture, Georgia Institute of Technology. He has also held major positions in national professional organizations and currently serves on four journal editorial boards. Stein is the author of numerous journal articles and four books, including *Classic Readings in Urban Planning* (1995).

KENT F. SPRECKELMEYER is professor of the Architecture Program, School of Architecture and Urban Design, University of Kansas. He has conducted building evaluation and programming studies for a variety of client groups in office and health-care settings. He has taught at the University of Kansas and the Edinburgh College of Art, and he is active in ongoing research projects that investigate the ways that analytic design methods can be integrated into the architectural process. He is the author of a number of books and articles in the fields of design methods and building evaluation.

# 关于编者

杰伊·M. 斯坦，佛罗里达大学建筑学院城市与区域规划系主任、教授。之前曾任纽约州立大学布法罗分校规划和设计系主任、教授；纽约州立大学布法罗分校建筑和规划学院常务院长；斯坦福大学基础设施规划与管理专业访问学者；佐治亚理工学院建筑学院城市规划专业研究生部任课教师。在多家国家级专业机构中也身居要职，现在担任四种期刊编委。斯坦发表了大量期刊文章，并出版了四种图书，其中包括《城市规划经典选读》（1995）。

肯特·F. 斯普雷克尔迈耶，堪萨斯大学建筑与城市设计学院建筑系教授，一直为办公设施和医疗保健设施领域的各类客户做建筑评估和设计研究，现任教于堪萨斯大学和爱丁堡艺术学院，积极投身于持续进行的调查如何将解析设计方法融入建筑过程的研究项目，撰写了大量关于设计方法和建筑评估的专著和文章。

# ABOUT THE CONTRIBUTORS

CHRISTOPHER ALEXANDER is a professor of architecture at the University of California at Berkeley. He is also a practicing architect who has implemented a variety of analytic design tools and ways of design thinking in his practice. Alexander's writings include *Community and Privacy* (with S. Chermayeff), *Notes on the Synthesis of Form*; *A Pattern Language* (with S. Ishikawa and M. Silverstein); and *The Timeless Way of Building*.

REYNER BANHAM was an architectural educator and critic who studied the relationships between modern technology, social processes, and architectural design. He taught at University College, London; the State University of New York at Buffalo; and the University of California at Santa Cruz. His books include *Theory and Design in the First Machine Age*; *The New Brutalism*; *The Architecture of the Well-Tempered Environment*; and *A Concrete Atlantis*.

JUDITH R. BLAU is a Gillian T. Cell University Professor of Sociology at the University of North Carolina, Chapel Hill. She is an expert on the cultural and social dimensions of the architectural, planning, and design professions. Blau has conducted a variety of research projects that focus on the ecology of contemporary cultural organizations. Her books include *Architects and Firms*; *The Shape of Culture*; *Remaking the City*; and *Social Contracts and Economic Markets*.

ERNEST LEROY BOYER served as president of the Carnegie Foundation for the Advancement of Teaching and is co-author (with L. Mitgang) of *Building Community: A New Future for Architectural Education and Practice*. His books have covered education at all levels and include *College: The Undergraduate Experience*; *Scholarship Reconsidered*; *Ready to Learn*; and *The Basic School*. Boyer also served as U.S. Commissioner of Education from 1977–79 and chancellor of the 64-campus State University of New York system from 1970–77.

GEOFFREY BROADBENT is an architectural educator and researcher, was the Head of School, Portsmouth, 1967–88, and retired in 1994. His books include *Design in Architecture*; *Urban Space Design*; and edited compilations on *Signs, Symbols, and Architecture* (with R. Bunt, C. Jencks) and *Meaning and Behavior* (with R. Bunt and T. Llorens). Professor Broadbent is retired, but continues a vigorous lecture schedule, especially in Latin America.

# 关于选读作品的作者

克里斯托弗·亚历山大（1936—），加利福尼亚大学伯克利分校建筑系教授、执业建筑师，在其建筑实践中大量运用解析设计工具和设计思维方法。亚历山大的作品包括《社区与隐私》（同瑟奇·切尔马耶夫（1900—1996，俄裔英国建筑师、工业设计师、作家）合著）、《形式综合论》、《建筑模式语言》（与萨拉·石川佳纯（加利福尼亚大学伯克利分校荣誉建筑教授）和默里·西尔弗斯坦（1943—，美国建筑学者）合著）以及《建筑的永恒之道》。

雷纳·班哈姆（1922—1988），建筑教育家和评论家，主要研究现代技术、社会进程和建筑设计之间的关系。曾在伦敦大学学院、纽约州立大学布法罗分校和加利福尼亚大学圣克鲁兹分校任教。他的著作包括《第一机械时代的理论与设计》、《新粗野主义》、《良好建设环境下的建筑》和《一个真实的亚特兰蒂斯》。

朱迪思·R.布劳（1942—），北卡罗来纳大学教堂山分校吉莉安·T.赛尔社会学教授，是从文化和社会维度研究建筑、规划和设计领域的专家。布劳做过大量专注于当代文化组织的社会生态学方面的研究项目。她的著作包括《建筑师和事务所》、《文化形态》、《重塑城市》和《社会契约与经济市场》。

欧内斯特·勒罗伊·博耶（1928—1995），卡内基教学促进基金会主席，与李·米特冈合著《建筑社区：建筑教育和实践的新未来》一书。他的著作涉及教育的各个阶段，包括《学院：本科的经历》、《反思学术》、《为学习做好准备》和《基础学校》。博耶还于1977—1979年任美国教育专员，1970—1977年任纽约州立大学64个分校区的校监。

杰弗里·勃罗德彭特，建筑教育家和学者，1967—1988年任朴次茅斯大学建筑学院院长，于1994年退休。他的著作包括《建筑中的设计》、《城市空间设计》和汇编作品《符号·象征与建筑》（与理查德·邦特和查尔斯·詹克斯（1939—，美国建筑评论家、园林建筑师和设计师）合集出版）及《意义与行为》（与理查德·邦特和T.洛伦斯合著）。勃罗德彭特教授虽然已经退休，但仍然精力充沛，四处讲学，特别是经常在拉美国家开办讲座。

**DENISE SCOTT BROWN** is a partner with Robert Venturi in the architectural firm of Venturi, Scott Brown and Associates, Inc. As an architect, planner, and urban designer, she is responsible for the firm's urban planning, urban design, campus planning, and architectural and facilities programming projects. Brown's 35 years of interdisciplinary experiences cover building complexes, downtowns, commercial districts, inner-city neighborhoods, recreation areas, university campuses, small towns, and suburbs. She is the 1996 recipient of the ACSA/AIA Topaz Medallion.

**ALAN COLQUHOUN** is an architect and educator who practiced as a partner in the London firm Colquhoun and Miller and also taught at several institutions including the Architectural Association, Princeton, Cornell, and the Polytechnic of Central London. His designs reflected the evolution of his theories of architecture, which are expressed in his books *Essays in Architectural Criticism* and *Modernity and the Classical Tradition*.

**JAMES MARSTON FITCH** is a leading authority in the United States in the field of historic preservation. Dr. Fitch founded Beyer Blinder Belle, an architectural firm in New York City known for its unique design and development approach to historic preservation. He was the founder and director of the graduate programs in historic preservation at both Columbia University and the University of Pennsylvania. He is the recipient of the Louise du Pont Crowninshield Award, The National Trust for Historic Preservation's highest honor for an exemplary career in preservation. His writings include *The American Building: The Historical and Environmental Forces That Shape It*; *Walter Gropius*; *Architecture and the Esthetics of Plenty*; and *Historic Preservation: Curatorial Management of the Built World*.

**KENNETH FRAMPTON** is an architect and architectural historian. He has taught at the Royal College of Art in London, was a fellow of the Institute for Architecture and Urban Studies in New York, and currently is the Ware Professor of Architecture at Columbia University. His books include *Modern Architecture: A Critical History*; *Modern Architecture: 1851–1945*; and *Studies in Tectonic Culture*. He is the 1991 recipient of the ACSA/AIA Topaz Medallion.

**RICHARD BUCKMINSTER FULLER** was an architect, engineer, inventor, and philosopher who is best known in architectural history for his use of the geodesic dome as a structural form. He was a writer and publisher, the principal of the Dymaxion and Geodesic Corporations, and Professor of Design at Southern Illinois University. He authored the books *Critical Path*; *Earth Inc.*; *Operating Manual for Spaceship Earth*; and *Synergetics: Explorations in the Geometry of Thinking*.

**ROBERT GUTMAN** is a sociologist on the faculty of the School of Architecture at Princeton University, where he teaches courses on housing, housing practice, and architectural theory. Dr. Gutman has written extensively on issues of the built environment and the profession. He is an honorary member of the AIA, and has been the recipient of many awards, citations, and fellowships from associations and foundations in architecture and the social sciences. His books include *People and Buildings*; *The Design of American Housing*; *Neighborhood, City and Metropolis*; and *Architectural Practice*.

**EDWARD TWITHER HALL** is a writer, educator, and anthropologist who pioneered the study of the relationships between human culture and the built environment. He has held a variety of positions in higher education, government, and business. His work has bridged the disciplinary boundaries of design, the social sciences, philosophy, and politics. His books include *The Silent Language*; *The Hidden Dimension*; *Beyond Culture*; *The Dance of Life*; *The Fourth Dimension in Architecture*; and *An Anthropology of Everyday Life*.

丹尼斯·斯科特·布朗（1931—），他是罗伯特·文丘里（1925—，美国建筑师）在文丘里与斯科特·布朗建筑事务所的合作伙伴。作为建筑师、城市规划师和城市设计师，布朗负责事务所城市规划、城市设计、校园规划、建筑和设施设计项目。布朗35年的跨领域工作经历内容丰富，涉及综合建筑群、市中心区、商业区、内城区域、娱乐区域、大学校园、小城镇和郊区。她于1996年荣获美国建筑学院协会和美国建筑师协会授予的黄玉奖章。

艾伦·科洪（1921—2012），建筑师和教育家，开办于伦敦的科洪和米勒建筑事务所合伙人，曾在包括建筑协会、普林斯顿大学、康奈尔大学和中央伦敦理工学院在内的众多机构任教。他的建筑设计作品反映了他的建筑理论思想的演变，这些理论思想也被记录在他的著作《建筑评论论文集》和《现代性和经典传统》当中。

詹姆斯·马斯顿·菲奇（1909—2000），美国古建保护领域领军性权威。菲奇博士在纽约市创办了拜尔·布林德·贝尔建筑事务所，该事务所以其独特的设计与开发方法保护文物古建而闻名。他是哥伦比亚大学和宾夕法尼亚大学研究生文物古建保护方向课程的创立者和负责人，荣获路易斯·杜邦·克劳宁希尔德奖，这是国家信托基金为表彰他在文物古建保护方面的突出成就授予的古建保护领域的最高荣誉。他的著作包括《美国建筑：塑造它的历史和环境力量》、《沃尔特·格罗皮乌斯》、《建筑学与多种美学》和《历史保护：管理者对建成世界的管理》。

肯尼斯·弗兰姆普敦（1930—），建筑师和建筑史学家，曾在伦敦皇家艺术学院任教，曾为纽约建筑和城市研究所成员，现为哥伦比亚大学建筑学威尔教授。他的著作包括《现代建筑：一部批判的历史》、《现代建筑：1851—1945》和《建构文化研究》。他荣获了1991年美国建筑学院协会和美国建筑师协会授予的黄玉奖章。

理查德·布克敏斯特·富勒（1895—1983），建筑师、工程师、发明家和哲学家，因将网格穹顶作为建筑结构形式而在建筑历史上为人所熟知；他还是作家和出版商，戴马克松和网格穹顶公司负责人，南伊利诺伊大学设计教授。他留下的著作有《关键路径》、《地球公司》、《地球号太空船操作手册》和《协同论：对思维结构的探索》。

罗伯特·古特曼（1926—2007），普林斯顿大学建筑学院社会学家，讲授住房建造、住房建设和建筑理论课程。古特曼博士写作了大量关于建筑环境和建筑职业的文章和书籍。他是美国建筑师协会的荣誉会员，同时荣获过建筑和社会科学领域许多协会和基金会的奖项、嘉奖和奖金。他的著作包括《人和建筑》、《美国住宅设计》、《社区、城市和大都市》和《建筑实践》。

爱德华·特威切尔·霍尔（1914—2009），作家、教育家和人类学家，研究人类文化和建筑环境相互关系的先驱，在高等学府、政府部门和商业机构担任过许多职务。他在设计、社会科学、哲学和政治学之间搭建起学科沟通的桥梁。他的著作包括《无声的语言》、《隐藏的维度》、《超越文化》、《生命之舞》、《建筑中的第四维》和《生活中的人类学》。

DAVID S. HAVILAND is a professor of architecture and Vice President for Student Life at Rensselaer Polytechnic Institute, Troy, New York. He teaches and conducts research in the building procurement process and practice management, and he has edited a number of practice publications for the American Institute of Architects (AIA), including *Managing Architecture Projects* (the 1988 and 1994 editions) and *The Architect's Handbook of Professional Practice*. Haviland has earned the AIA's Institute Honor for his contributions to practice education, as well as the Architectural Research Centers Consortium's James L. Haecker Award for Distinguished Research Leadership.

DOLORES HAYDEN is professor of architecture, urbanism, and American studies at Yale University. As both a historian and architect, she writes about the social and political history of built environments in the United States and about the politics of design. Her books include *Seven American Utopias: The Architecture of Communitarian Socialism, 1790–1975*; *The Grand Domestic Revolution: A History of Feminist Designs for American Homes, Neighborhoods, and Cities*; *Redesigning the American Dream*; and *The Power of Place: Urban Landscapes as Public History*. She has received numerous awards, including Guggenheim, Rockefeller, and Ford fellowships.

STEVEN IZENOUR is an active participant in all architectural and urban design projects in the architectural firm of Venturi, Scott Brown. He has been the primary designer of several special projects of the firm in exhibition and graphic design, which have won numerous awards and received widespread public recognition. He is a noted lecturer and teacher at schools and universities worldwide.

JOHN BRINCKERHOFF JACKSON was an architectural and landscape critic and writer who popularized the study of vernacular places and structures. He was a professor at both Harvard University and the University of California, Berkeley. His books include *American Space*; *Discovering the Vernacular Landscape*; *The Necessity for Ruins*; and *A Sense of Place, A Sense of Time*.

JANE JACOBS has changed the way that we think about cities and nation states, urban economics, and the value systems within which we live. For over 40 years she has challenged conventional ideas about physical planning and argued the virtues of big, diverse, crowded, and dense cities. Her writings include *The Death and Life of Great American Cities*; *The Economy of Cities*; *Cities and the Wealth of Nations*; and from her Toronto vantage point, *The Question of Separatism*.

KEVIN LYNCH was a professor in the Urban Studies and Planning Program at the Massachusetts Institute of Technology. Lynch's numerous writings—including *The Image of the City*; *A Theory of Good City Form*; *What Time Is This Place?*; and *Site Planning*—all emphasized the importance of people's perceptions in designing the environment. Lynch was also a partner in the environmental design firm of Carr, Lynch Associates, which was involved in many important projects including site planning for the new town of Columbia, Maryland.

DONLYN LYNDON, FAIA, is a professor of architecture at the University of California, Berkeley, and a partner in the architectural firm of Lyndon/Buchanan Associates. Lyndon is the author of numerous journal articles and books, including *The Place of Houses* (with Charles Moore and Gerald Allen); *Chambers for a Memory Palace* (with Charles Moore) and *The City Observed: Boston*. He is the recipient of numerous awards, including the Twenty-Five Year Award given to Moore Lyndon Turnbull Whitaker (MLTW) for Sea Ranch Condominium One and the 1997 ACSA/AIA Topaz Medallion.

大卫·S.哈维兰，纽约州特洛伊市伦斯勒理工学院建筑系教授和主管学生生活的副院长，讲授和研究建筑采购过程和实践管理，为美国建筑师协会编辑了大量指导实践的出版物，包括《建筑项目管理》（1988年版和1994年版）和《建筑师职业实践手册》。哈维兰因其在教育实践上的贡献荣获美国建筑师协会学院大奖和建筑研究中心联盟的詹姆斯·L.哈克突出研究领导奖。

多洛雷斯·海登（1945—，美国建筑师、城市历史学家），耶鲁大学建筑、城市主义和美国研究教授。作为一名历史学家和建筑师，她撰写了关于美国建成环境下社会历史和政治历史的作品以及关于设计政治学的作品。她的著作包括《美国的七个乌托邦：传播社会主义的建筑，1790—1975》、《国内大革命：一部美国女权主义房屋、社区和城市的设计史》、《美国梦的重新设计》和《力量的场所：作为公众历史的城市景观》。她荣获过众多奖项，包括古根海姆、洛克菲勒和福特奖金。

史蒂芬·爱泽努尔（1940—2001，美国建筑师、城市规划专家、理论家），文丘里与斯科特·布朗建筑事务所全部建筑和城市设计项目的积极参与者。他是事务所在展会设计和平面设计中许多特殊项目的首席设计师，赢得众多奖项和公众广泛的认可。他是一位著名的演讲者和教师，在世界各地的大学和学院中举办讲座和授课。

约翰·布林克霍夫·杰克逊（1909—1996），建筑与景观评论家和作家。他使风土建筑研究风靡一时，曾任哈佛大学和加利福尼亚大学伯克利分校教授。他的著作包括《美国空间》、《发现风土景观》、《保护废墟的必要性》和《地域感、时间感》。

简·雅各布斯（1916—2006，美裔加拿大记者、作家和社会活动家），改变了我们对所生活的城市、联邦各州、城市经济学和价值体系的思考方式。40多年来，她一直在挑战城市实体规划的传统做法，并且讨论规模庞大、多样复杂、拥挤稠密的城市具有哪些优点。她的著作包括《美国大城市的死与生》、《城市经济》、《城市和国家财富》和她利用身在加拿大多伦多的优势所写的《分裂主义的问题》。

凯文·林奇（1918—1984，美国城市规划师、作家），麻省理工学院城市研究和规划系教授。林奇著述颇丰，包括《城市意象》、《良好的城市形态理论》、《此地是何时》和《场地规划》，他所有的著作都强调了在环境设计中人的感知的重要性。林奇也是卡尔和林奇环境设计事务所的合伙人，该事务所参与了许多重要项目，包括马里兰州的哥伦比亚新城。

德隆·林顿（1936—），美国建筑师协会会员，加利福尼亚大学伯克利分校建筑系教授，林顿和布坎南建筑事务所合伙人。林顿撰写了大量期刊文章和专著，包括《房屋所在之地》（与查尔斯·摩尔（1925—1993，美国建筑师、教育家、作家）和杰拉德·艾伦（1885—1956，英国学者、英国国教神父、主教）合著）、《记忆宫殿的房间》（与查尔斯·摩尔合著）和《被观察的城市：波士顿》。他曾获得大量奖项，其中包括因设计海岸牧场公寓颁发给摩尔、林顿、特恩布尔与惠特克四人（MLTW）的25周年大奖和1997年美国建筑学院协会和美国建筑师协会授予的黄玉奖章。

CLARE COOPER MARCUS is professor emerita in the Department of Architecture and Landscape Architecture at the University of California, Berkeley. She has contributed numerous articles to design and academic journals and authored four books: *Easter Hill Village: Some Social Implications of Design; Housing as if People Mattered: Site Design Guidelines for Medium-Density Family Housing* (with Wendy Sarkissian); *People Places: Design Guidelines for Urban Open Space* (with Carolyn Francis); and *House as a Mirror of Self: Exploring the Deeper Meaning of Home.* Her consulting firm, People Places, specializes in user-needs/participation approach to design programming, particularly in the area of public housing modernization and public open-space design.

PETER MCCLEARY is a professor at the University of Pennsylvania and is the former chairman of the programs in architecture and historic preservation. He has been involved in private practice in London with Ove Arup and Frank Newby and more recently in Philadelphia. He has published numerous research and interpretive articles on philosophies, history, and concepts of technology. McCleary received the 1994 Association of Collegiate Schools of Architecture (ACSA) Distinguished Professor Medal and the 1992 *Journal of Architectural Education (JAE)* Creative Achievement Award.

IAN LENNOX MCHARG is a landscape architect and educator who has written about and taught in the areas of landscape and environmental design. He has taught at the Universities of Pennsylvania, California, and Washington and his books include *Design with Nature* and *A Quest for Life.* His awards include the Thomas Jefferson Medal from the University of Virginia and the National Medal of Art.

LEE MITGANG is a senior fellow of the Carnegie Foundation for the Advancement of Teaching. He is co-author (with E. Boyer) and principal researcher of the 1996 Carnegie report, *Education and Practice.* Mitgang also authored the 1992 Carnegie report, *School Choice.* Prior to joining Carnegie, Mitgang spent 20 years as a journalist covering education, urban affairs, business, and politics. He has received numerous awards including The John Hancock Award for Excellence on Business and Financial Journalism.

LEWIS MUMFORD was an urban historian who made significant contributions to the literature of architecture, urban design, social history, and philosophy. He is the author of numerous articles and over 30 books, including *The Brown Decades*; *The Culture of Cities*; *Art and Technics*; *The City in History*; and *Sketches from Life.* He taught at numerous universities, including Pennsylvania, California at Berkeley, and MIT. Over a long and highly productive career, Mumford received numerous awards, including honorary membership in the American Institute of Certified Planners and the American Academy of Arts and Letters.

VICTOR PAPANEK was an educator and designer who studied the relationships between built form and human culture. He taught at a number of schools of design and architecture in Europe and North America and was the J. L. Constant Distinguished Professor of Architecture at the University of Kansas. His books include *Design for the Real World*; *Design for Human Scale*; *How Things Don't Work* (with James Hennessey); and *The Green Imperative.*

WOLFGANG F. E. PREISER is a professor of architecture at the University of Cincinnati, School of Architecture and Interior Design. He has held visiting lectureships at a number of universities, and as an international building consultant, Dr. Preiser was co-founder of Architectural Research Consultants, Inc., and Planning Research Institute, Inc. He has edited and written numerous articles and books, including *Post-Occupancy Evaluation* (with H. Rabinowitz and E. T. White); *Design Intervention: Toward a More*

克莱尔·库珀·马库斯，加利福尼亚大学伯克利分校建筑和景观建筑系荣誉教授。她在设计杂志和学术期刊上发表过大量文章，并撰写了四本专著：《复活节山村：设计的某些社会影响》、《重要的住宅：中密度家庭住宅场所设计导则》（与温迪·萨尔基相（澳大利亚咨询师、学者、科廷大学可持续政策研究所兼职副教授、邦德大学可持续发展学院兼职教授）合著）、《人性场所：城市开放空间设计导则》（与卡罗琳·弗朗西斯合著）和《自我写照的住房：探索家的深层含义》。她的咨询事务所"人性场所"专门解决设计规划中使用者需求和参与方法的问题，特别是在公共住房现代化和公共开放空间设计方面。

彼得·麦克利里，宾夕法尼亚大学教授、建筑与历史保护系前系主任。他在伦敦与奥沃·艾拉普（1895—1988，丹麦裔英国结构工程师）和弗兰克·纽比（1926—2001，英国结构工程师）共同完成各种建筑项目，而最近则在费城开展工作。他发表了大量关于哲学、历史和技术概念的研究和说明性文章。麦克利里荣获1994年美国建筑学院协会授予的杰出教授奖章和1992年《建筑教育杂志》颁发的创新成就奖。

伊恩·伦诺克斯·麦克哈格（1920—2001），景观建筑师、教育家，讲授景观和环境设计方面的课程并撰写了相关作品，曾在宾夕法尼亚大学、加利福尼亚大学和华盛顿大学任教。他的著作包括《设计结合自然》和《生命的追求》。他曾获得的奖项包括弗吉尼亚大学托马斯·杰弗逊奖章和国家艺术勋章。

李·D. 米特冈，卡内基教学促进基金会高级研究员，1996年卡内基报告《教育与实践》作者之一（与欧内斯特·勒罗伊·博耶合著）和首席研究员。米特冈还完成了1992年卡内基报告《学校的选择》。在加入卡内基教学促进基金会之前，米特冈从事新闻工作长达20年，报道内容涉及教育、城市民生、商业和政治。他获得了众多荣誉，其中包括约翰·汉考克商业及金融新闻业突出贡献奖。

刘易斯·芒福德（1895—1990），城市历史学家，在建筑文献、城市设计、社会历史和哲学上贡献巨大。他发表了大量文章，并出版了30余本专著，包括《棕色的年代》、《城市文化》、《艺术和技术》、《城市发展史》和《生活速写》。他曾在许多学校任教，包括宾夕法尼亚大学、加利福尼亚大学伯克利分校和麻省理工学院。在其长寿高产的一生中，芒福德获得了大量的荣誉，其中包括美国注册规划师协会和美国艺术暨文学学会荣誉会员。

维克多·帕帕奈克（1923—1998），教育家和设计师，研究建筑形式与人类文化之间的关系，曾在欧洲和北美众多设计和建筑院校授课，为堪萨斯大学永久特聘建筑教授。他的著作包括《为真实世界而设计》、《针对人体尺度的设计》、《事情为何没有进展》（与詹姆斯·亨尼西（美国设计师、荷兰代尔夫特理工大学荣誉教授）合著）和《绿色律令》。

沃尔夫冈·F.E. 普赖泽尔，辛辛那提大学建筑与室内设计学院建筑系教授、多所大学访问学者。作为国际建筑咨询师，普赖泽尔博士帮助建立了建筑研究咨询公司和规划研究所。他撰写了大量文章和专著，包括《使用后评价》（与哈维·拉比诺维茨（1939—，威斯康星大学数学系教授）和爱德华·T. 怀特合著）、《设计干预：朝向人文建筑发展》（与杰奎琳·C. 菲舍

*Humane Architecture* (with J. C. Vischer and E. T. White); *Programming the Built Environment*; *Building Evaluation*; *Professional Practice in Facility Programming*; and *Design Review: Challenging Urban Aesthetic Control* (with B. C. Scheer).

AMOS RAPOPORT is a distinguished professor in the School of Architecture and Urban Planning at the University of Wisconsin–Milwaukee. He has taught previously at the Universities of Melbourne and Sydney (Australia), the University of California at Berkeley, and University College, London. He has held visiting appointments in numerous countries and is one of the founders of the field of Environment-Behavior Studies. Rapoport is the author of numerous papers, chapters, monographs, articles, and the following books: *House Form and Culture*; *Human Aspects of Urban Form*; *The Meaning of the Built Environment*; and *History and Precedent in Environmental Design*.

COLIN ROWE is an architectural educator and historian who is the Andrew Dickson White Professor Emeritus of architecture at Cornell University. He has also served on the faculties of the Universities of Liverpool, Texas, and Cambridge. Rowe is the recipient of numerous awards in recognition of his outstanding contributions to architecture and architecture education, including the 1995 RIBA Gold Medal. His books include *The Mathematics of the Ideal Villa and Other Essays*; *Collage City* (with Fred Koetter); *The Architecture of Good Intentions*; and *As I Was Saying*.

PETER ROWE is the Raymond Garbe Professor of Architecture and Urban Design at Harvard University, where he also serves as dean of the Graduate School of Design. Prior to joining the Harvard faculty in 1985, Rowe was director of the School of Architecture at Rice University and a senior member of several research organizations, including the Rice Center and the Southwest Center for Urban Research. The author of numerous articles, Rowe is also author of the books *Principles for Local Environmental Management*; *Design Thinking*; *Making a Middle Landscape*; *Modernity and Housing*; and *Civic Realism*.

MARIO SALVADORI was a civil engineer, a mathematical physicist, and an architect. He was born and educated in Rome, Italy, and came to the United States in 1939 to escape fascism. He was a professor at Columbia University for 50 years and was the author of 30 books on applied mathematics, structures, and a variety of other topics, translated in 15 languages. Salvadori's achievements have been recognized with many awards, including four honorary university degrees.

DONALD ALAN SCHÖN was Ford Professor Emeritus and senior lecturer at the Department of Urban Studies and Planning at the Massachusetts Institute of Technology. As an educator and organizational consultant, a former government administrator, and director of nonprofit social research organizations, Dr. Schön's research and practice centered on questions of education, professional knowledge, organizational learning, and technological innovation. His books include *Beyond the Stable State*; *Theory in Practice: Increasing Professional Effectiveness* (with Chris Argyris); *The Reflective Practitioner*; *Educating the Reflective Practitioner*; *The Reflective Turn*; *Frame Reflection* (with Martin Rein); and *Organizational Learning II* (with Chris Argyris).

VINCENT JOSEPH SCULLY is the Sterling Professor Emeritus of the History of Art at Yale University and distinguished visiting professor at the University of Miami School of Architecture. His books include *The Shingle Style*; *American Architecture and Urbanism*; *Architecture: The Natural and the Manmade*; *The Earth, the Temple, and the Gods*; *Pueblo: Mountain, Village, Dance*, and *Modern Architecture: The Architecture of Democracy*. He was the recipient of the Tau Sigma Delta Gold Medal in 1996.

尔和爱德华·T. 怀特合著）、《建成环境规划》、《建筑评价》、《设施规划职业训练》和《设计综述：挑战城市美学控制》（与布伦达·卡斯·希尔合著）。

阿摩斯·拉普卜特（1929—），威斯康星大学密尔沃基分校建筑和城市规划学院特聘教授，曾在澳大利亚墨尔本大学和悉尼大学、加利福尼亚大学伯克利分校和伦敦大学学院任教，出访多国的访问学者，环境行为学的创立者之一。拉普卜特撰写了大量论文、篇章、专题著作、文章和下列书籍：《宅形与文化》、《城市形态的人性侧面》、《建成环境的意义》和《环境设计的历史与先例》。

柯林·罗（1920—1999），建筑教育家、建筑历史学家，康奈尔大学安德鲁·迪克森·怀特名誉建筑教授，曾任职于利物浦大学、得克萨斯大学和剑桥大学。罗因其在建筑和建筑教育上的杰出贡献而屡获殊荣，其中包括1995年英国皇家建筑师学会金奖。他的著作包括《理想别墅的数学分析及一些随笔》、《拼贴城市》（与弗瑞德·科特合著）及《善意的建筑》和《正如我曾说过》。

彼得·罗，哈佛大学建筑和城市设计雷蒙德·加布教授，设计研究生院院长。1985年任教于哈佛大学之前，曾任莱斯大学建筑学院院长和许多科研机构包括莱斯研究中心和城市研究西南中心高级研究员。罗发表了大量文章，并出版有《区域性环境管理原则》、《设计思考》、《中等景观设计》、《现代性与住房》和《公民现实主义》。

马里奥·萨瓦多里（1907—1997），市政工程师、数学物理学家和建筑师，生于意大利罗马并在罗马接受教育，1939年为逃避意大利法西斯迫害来到美国。他在哥伦比亚大学任教授并工作了50年，撰写了30部著作，内容涉及应用数学、结构和其他许多主题，被翻译成15种语言。萨瓦多里的学术成就得到广泛认可，荣获许多荣誉，其中包括四所大学授予他的荣誉学位。

唐纳德·艾伦·舍恩（1930—1997），麻省理工学院城市研究与规划系福特荣誉教授，资深讲师，教育家，机构咨询师，前政府官员，公益性社会研究机构负责人。舍恩博士的研究和实践重点在教育问题、职业知识、组织学习和技术创新。他的著作包括《超越稳定状态》、《实践中的理论：提升专业效能》（与克里斯·阿吉里斯（1923—2013，商业理论家、组织学习理论的代表人物）合著）、《反思的实践者》、《反思实践者的教育》、《反思回观》、《结构反思》（与马丁·雷恩合著）和《组织学习2》（与克里斯·阿吉里斯合著）。

文森特·约瑟夫·斯卡利（1920—），耶鲁大学斯特林艺术史荣誉教授，迈阿密大学建筑学院特聘访问学者。他的著作包括《板瓦式风格》、《美国建筑与城市主义》、《建筑：自然和人工》、《大地、庙宇和神灵》、《普韦布洛：高山、村庄和舞蹈》和《现代建筑：民主的建筑》。他荣获1996年陶西格玛德尔塔协会金奖。

ROBERT SOMMER is professor of psychology and chair, Department of Art at the University of California, Davis. He has received a Career Research Award from the Environmental Design Research Association, the Kurt Lewin Award from the Society for the Psychological Study of Social Issues, a research award from the California Alliance for the Mentally Ill, and a Fulbright Fellowship. He is the author of 11 books and numerous articles, many on architecture and design issues.

LOUIS HENRY SULLIVAN was an architect who practiced in Chicago in the late nineteenth and early twentieth centuries. His work defined the principles of skyscraper design, modern ornamentation, and functionalism in the American context. Frank Lloyd Wright, who apprenticed in Sullivan's office, credits Sullivan as the originator of modern American architecture. Sullivan wrote three influential books on architecture: *A System of Architectural Ornament According with a Philosophy of Man's Powers*; *The Autobiography of an Idea*; and *Kindergarten Chats and Other Writings*.

ROBERT VENTURI is a partner in the architectural firm of Venturi, Scott Brown. He is responsible for architectural and urban design. Under his guidance, the firm's achievements in design have been recognized internationally with numerous awards, exhibitions of the firm's work, and special publications. Venturi's book *Complexity and Contradiction in Architecture* has been translated and published in 16 languages. His other books include *Learning from Las Vegas* (with Denise Scott Brown and Steven Izenour) and *A View from the Campidoglio*. He won the Pritzker Prize in 1991.

VITRUVIUS (MARCUS VITRUVIUS POLLIO) was a Roman architect and engineer who lived in the first century B.C. He was the author of the influential architectural treatise *De architectura (The Ten Books on Architecture)*, which served as the text that preserved classical design principles from the Hellenistic to the Renaissance eras. Little is known of Vitruvius's personal or professional life except what we read in the *Ten Books*, although it is probable that he served as an architect and historian in the court of Augustus.

WILLIAM HOLLINGSWORTH WHYTE is a writer, activist, and consultant whose work has focused on the relationship between human culture and urban form. His publications include *The Last Landscape*; *The Organization Man*; *The Social Life of Small Urban Spaces*; and *City: Rediscovering the Center*. He served on the President's Task Force on Natural Beauty and is also the recipient of numerous awards, including the Benjamin Franklin Magazine Writing Award, the Liberty and Justice Book Award, and the Natural Resources Council Award.

FRANK LLOYD WRIGHT was one of the most influential and prolific architects of the twentieth century. He is known primarily for his contributions in defining the principles of "organic" architecture. His designs were executed over a period of 60 years, and his buildings were instrumental in establishing the qualities of modern American architecture. A prolific writer, Wright's books include *Modern Architecture*; *An Autobiography*; *An Organic Architecture*; and *The Future of Architecture*.

BRUNO ZEVI is an architect and architectural historian. He has taught at the Universities of Venice and Rome, and has been awarded honorary degrees from the University of Buenos Aires, Haifa Technion, and the University of Michigan. Zevi has consulted on design projects in Europe and North America. His books include *Storia dell'Architettura Moderna*; *Il Linguaggio Moderno dell'Architettura*; and monographs on Wright, Mendelsohn, and Terrogni. He is an Honorary Fellow of the AIA.

关于选读作品的作者　xxix

罗伯特·萨默（1929—，环境心理学家），加利福尼亚大学戴维斯分校艺术系主任，心理学教授，曾荣获环境设计研究协会职业研究奖、社会问题心理研究学会库尔特·勒温奖、加利福尼亚精神疾病联盟研究奖和富布莱特（1905—1995，美国参议员）奖。他撰写了11本著作和大量文章，大多围绕建筑和设计等主题。

路易斯·亨利·沙利文（1856—1924，美国建筑师，被称为"摩天大楼之父""现代主义之父"），19世纪末和20世纪初在芝加哥执业的建筑师，确定了美国环境中摩天大楼设计、现代装饰和功能主义的原则。曾在沙利文事务所学习的弗兰克·劳埃德·赖特（1867—1959，美国建筑师、室内设计师、作家、教育家）认为沙利文是美国现代建筑的创始人。沙利文撰写了三部影响深远的建筑著作，分别是《人类权力哲学下的建筑装饰体系》、《思想自传》和《幼儿园对话录及其他文章》。

罗伯特·文丘里（1925—，美国建筑师），文丘里与斯科特·布朗建筑事务所的合作伙伴，负责建筑和城市设计。在他的指导下，事务所的设计成就在国际上得到广泛认可，各种奖项、作品展、作品专项出版蜂拥而至。文丘里的《建筑的复杂性与矛盾性》一书被翻译成16种语言出版。他的其他著作包括《向拉斯维加斯学习》（与丹尼斯·斯科特·布朗和史蒂芬·爱泽努尔合著）和《卡比托利欧观察》。他荣获了1991年的普利兹克奖。

维特鲁威（马尔库斯·维特鲁威·波利奥），古罗马建筑师和工程师，生活在公元前1世纪，写下了对后世影响深远的建筑专著《建筑十书》。此书成为从希腊时期到文艺复兴时期保留下来的古典设计原则的范本。虽然维特鲁威很有可能在奥古斯都（公元前63年—公元14年，罗马帝国皇帝）宫廷担任了建筑师和历史学家的职务，但除去我们可以从这本书中读到的，无人知晓维特鲁威的私人生活和职业生涯是什么样子的。

威廉·霍林斯沃斯·怀特（1917—1999），作家、社会活动家和咨询师，关注人类文化和城市形态之间的关系。他的著作包括《最后的景观》、《组织人》、《小型城市空间的社会生活》和《城市：对中心的重新发现》。他曾在总统保护自然美景工作小组中任职，同时荣获众多奖项，其中包括本杰明·富兰克林期刊文章大奖、自由公正图书奖和国家资源委员会奖。

弗兰克·劳埃德·赖特（1867—1959），20世纪最具影响和最多产的建筑师之一，主要因其提出"有机"建筑的原则为世人所熟知。他的设计生涯长达60年，其建筑作品对于确立美国现代建筑的品质发挥了作用。作为一位多产作家，赖特的著作包括《现代建筑》、《一部自传》、《有机建筑》和《建筑的未来》。

布鲁诺·赛维（1918—2000），建筑师和建筑历史学家，任教于威尼斯大学和罗马大学，被布宜诺斯艾利斯大学、海法理工学院和密歇根大学授予荣誉学位。赛维在欧洲和北美开展设计项目咨询。他的著作包括《现代建筑史》、《现代建筑语言》以及关于赖特、门德尔松（1887—1953，犹太裔德国建筑师）、特拉尼（1904—1943，意大利建筑师）的专著。他是美国建筑师协会的荣誉会员。

# CONTENTS

PROLOGUE  1

    READING 1    *Vitruvius (Marcus Vitruvius Pollio)*
                        "Book I"
                        *The Ten Books on Architecture*  2

PART ONE    ARCHITECTURE AS ARTIFACT  21

    1    Architectural History and Theory  23

        READING 2    *Kenneth Frampton*
                        "Cultural Transformations" and "Territorial Transformations"
                        *Modern Architecture: A Critical History*  23
        READING 3    *Colin Rowe*
                        "The Architecture of Utopia"
                        *The Mathematics of the Ideal Villa and Other Essays*  47
        READING 4    *Vincent Joseph Scully*
                        "The Architecture of Community"
                        *The New Urbanism*  62
        READING 5    *Robert Venturi, Denise Scott Brown, and Steven Izenour*
                        "Historical and Other Precedents: Towards an Old Architecture"
                        *Learning from Las Vegas*  72

    2    Architectural Form  96

        READING 6    *Geoffrey Broadbent*
                        "Architects and Their Symbols"
                        *Built Environment*  96

# 目录

序言　1

　　选读1　维特鲁威（马尔库斯·维特鲁威·波利奥）
　　　　　　"第一书"
　　　　　　《建筑十书》　2

## 第一部分　人造艺术的建筑　21

　1　建筑历史和理论　23

　　选读2　肯尼斯·弗兰姆普敦
　　　　　　"文化的变革"与"领土的变革"
　　　　　　《现代建筑：一部批判的历史》　23

　　选读3　柯林·罗
　　　　　　"乌托邦建筑"
　　　　　　《理想别墅的数学分析及一些随笔》　47

　　选读4　文森特·约瑟夫·斯卡利
　　　　　　"社区建筑"
　　　　　　《新城市主义》　61

　　选读5　罗伯特·文丘里，丹尼斯·斯科特·布朗与史蒂芬·爱泽努尔
　　　　　　"历史先例和其他先例：面向老建筑"
　　　　　　《向拉斯维加斯学习》　72

　2　建筑形式　96

　　选读6　杰弗里·勃罗德彭特
　　　　　　"建筑师与他们的象征"
　　　　　　《建成环境》　96

READING 7   Alan Colquhoun
            "Historicism and the Limits of Semiology"
            *Essays in Architectural Criticism*   120
READING 8   James Marston Fitch
            "Experiential Context of the Aesthetic Process"
            *Journal of Architectural Education*   131
READING 9   Bruno Zevi
            "Listing as Design Methodology" and "Asymmetry and Dissonance"
            *The Modern Language of Architecture*   141

3   Architectural Technology   154

    READING 10   Reyner Banham
                 "A Breath of Intelligence"
                 *The Architecture of the Well-Tempered Environment*   154
    READING 11   Peter McCleary
                 "Some Characteristics of a New Concept of Technology"
                 *Journal of Architectural Education*   169
    READING 12   Mario Salvadori
                 "Form-Resistant Structures"
                 *Why Buildings Stand Up*   180
    READING 13   Louis Henry Sullivan
                 "The Tall Office Building Artistically Considered"
                 *Kindergarten Chats and Other Writings*   202

PART TWO   THE CONTEXT OF ARCHITECTURE   209

4   The Urban Environment   211

    READING 14   Dolores Hayden
                 "The Power of Place: A Proposal for Los Angeles"
                 *The Public Historian*   211
    READING 15   Jane Jacobs
                 "The Need for Aged Buildings"
                 *The Death and Life of Great American Cities*   222
    READING 16   Lewis Mumford
                 "Retrospect and Prospect"
                 *The City in History: Its Origins, Its Transformations, and Its Prospects*   230
    READING 17   William Hollingsworth Whyte
                 "Return to the Agora"
                 *City: Rediscovering the Center*   237

选读 7　艾伦·科洪
　　　　"历史主义与符号学之局限"
　　　　《建筑评论论文集》　120

选读 8　詹姆斯·马斯顿·菲奇
　　　　"美学进程中的经验语境"
　　　　《建筑教育杂志》　131

选读 9　布鲁诺·赛维
　　　　"设计方法论"和"非对称性与不协调性"
　　　　《现代建筑语言》　141

3　建筑技术　154

选读 10　雷纳·班哈姆
　　　　"智慧的呼吸"
　　　　《良好建设环境下的建筑》　154

选读 11　彼得·麦克利里
　　　　"新技术概念的一些特点"
　　　　《建筑教育杂志》　169

选读 12　马里奥·萨瓦多里
　　　　"形态抵抗结构"
　　　　《建筑物如何站起来》　180

选读 13　路易斯·亨利·沙利文
　　　　"高层办公建筑的艺术思考"
　　　　《幼儿园对话录及其他文章》　202

第二部分　建筑的环境　209

4　城市环境　211

选读 14　多洛雷斯·海登
　　　　"力量的场所：对洛杉矶的建议"
　　　　《公众史学家》　211

选读 15　简·雅各布斯
　　　　"老建筑之必要"
　　　　《美国大城市的死与生》　222

选读 16　刘易斯·芒福德
　　　　"回顾与展望"
　　　　《城市发展史——起源、演变和前景》　230

选读 17　威廉·霍林斯沃斯·怀特
　　　　"回到古希腊广场"
　　　　《城市：对中心的重新发现》　237

5   The Natural Environment   247

   READING 18   John Brinckerhoff Jackson
                "The American Public Space"
                *The Public Interest*   247
   READING 19   Kevin Lynch
                "The Waste of Place"
                *Wasting Away*   258
   READING 20   Ian Lennox McHarg
                "On Values"
                *Design with Nature*   284

6   The Human Environment   299

   READING 21   Clare Cooper Marcus
                "The House as Symbol of the Self"
                *Designing for Human Behavior*   299
   READING 22   Edward Twitchell Hall
                "The Anthropology of Space: An Organizing Model"
                *The Hidden Dimension*   321
   READING 23   Amos Rapoport
                "On the Cultural Responsiveness of Architecture"
                *Journal of Architectural Education*   329
   READING 24   Robert Sommer
                "Space-Time"
                *Design Awareness*   339

# PART THREE   THE PROCESS OF ARCHITECTURE   351

7   The Design Process   353

   READING 25   Christopher Alexander
                "Goodness of Fit"
                *Notes on the Synthesis of Form*   353
   READING 26   Peter Rowe
                "A Priori Knowledge and Heuristic Reasoning in Architectural Design"
                *Journal of Architectural Education*   363
   READING 27   Donald Alan Schön
                "Toward a Marriage of Artistry and Applied Science in the Architectural Design Studio"
                *Journal of Architectural Education*   374

## 5 自然环境 247

选读 18　约翰·布林克霍夫·杰克逊
　　　　　"美国的公共空间"
　　　　　《公共利益》 247

选读 19　凯文·林奇
　　　　　"场所的浪费"
　　　　　《正在浪费的城市场所》 257

选读 20　伊恩·伦诺克斯·麦克哈格
　　　　　"价值观"
　　　　　《设计结合自然》 284

## 6 人文环境 299

选读 21　克莱尔·库珀·马库斯
　　　　　"作为自我象征的房子"
　　　　　《人类行为设计》 299

选读 22　爱德华·特威切尔·霍尔
　　　　　"空间人类学：一种组织模式"
　　　　　《隐藏的维度》 321

选读 23　阿摩斯·拉普卜特
　　　　　"论建筑的文化反应"
　　　　　《建筑教育杂志》 329

选读 24　罗伯特·萨默
　　　　　"空间—时间"
　　　　　《设计意识》 339

# 第三部分　建筑过程 351

## 7 设计过程 353

选读 25　克里斯托弗·亚历山大
　　　　　"良好的吻合度"
　　　　　《形式综合论》 353

选读 26　彼得·罗
　　　　　"建筑设计中的先验知识和启发性推理"
　　　　　《建筑教育杂志》 362

选读 27　唐纳德·艾伦·舍恩
　　　　　"建筑设计工作室中艺术与应用科学的结合"
　　　　　《建筑教育杂志》 374

READING 28  Frank Lloyd Wright
"The Cardboard House"
*The Future of Architecture* 387

8  The Social Implications of Architecture  399

READING 29  Richard Buckminster Fuller
"Accommodating Human Unsettlement"
*Town Planning Review* 399

READING 30  Donlyn Lyndon
"Design: Inquiry and Implication"
*Journal of Architectural Education* 411

READING 31  Victor Papanek
"Design Responsibility: Five Myths and Six Directions"
*Design for the Real World: Human Ecology and Social Change* 422

9  The Architectural Profession  439

READING 32  Judith R. Blau
"Architecture and the Daedalean Risk"
*Architects and Firms: A Sociological Perspective on Architectural Practice* 439

READING 33  Robert Gutman
"Challenges to Architecture"
*Architectural Practice: A Critical View* 449

READING 34  David S. Haviland
"Some Shifts in Building Design and Their Implications for Design Practices and Management"
*Journal of Architectural and Planning Research* 461

READING 35  Wolfgang F. E. Preiser
"Built Environment Evaluation: Conceptual Basis, Benefits and Uses"
*Journal of Architectural and Planning Research* 474

# EPILOGUE  491

READING 36  Ernest LeRoy Boyer and Lee D. Mitgang
"A Profession in Perspective"
*Building Community: A New Future for Architecture Education and Practice* 492

选读 28　弗兰克·劳埃德·赖特
　　　　　　"纸板屋"
　　　　　　《建筑的未来》　387

8　建筑的社会含义　399

　　　选读 29　理查德·布克敏斯特·富勒
　　　　　　"为无居可归者提供居所"
　　　　　　《城镇规划评论》　399
　　　选读 30　德隆·林顿
　　　　　　"设计：探究与含义"
　　　　　　《建筑教育杂志》　411
　　　选读 31　维克多·帕帕奈克
　　　　　　"设计的责任：五个神话和六个方向"
　　　　　　《为真实世界而设计：人类生态学与社会变革》　422

9　建筑行业　439

　　　选读 32　朱迪思·R. 布劳
　　　　　　"建筑业与代达罗斯的风险"
　　　　　　《建筑师和事务所：一种建筑实践的社会观点》　439
　　　选读 33　罗伯特·古特曼
　　　　　　"建筑业的挑战"
　　　　　　《建筑实践：一种批评观点》　449
　　　选读 34　大卫·S. 哈维兰
　　　　　　"建筑设计中的一些转变以及它们对设计实践与管理的启示"
　　　　　　《建筑和规划研究杂志》　461
　　　选读 35　沃尔夫冈·F.E. 普赖泽尔
　　　　　　"建成环境评价：概念基础、收益和用途"
　　　　　　《建筑和规划研究杂志》　474

跋　491

　　　选读 36　欧内斯特·勒罗伊·博耶和李·D. 米特冈
　　　　　　"对建筑行业的思考"
　　　　　　《建筑社区：建筑教育与实践的新未来》　492

# PROLOGUE 序言

选读 1
"第一书"
维特鲁威（马尔库斯·维特鲁威·波利奥）

本选读为维特鲁威《建筑十书》第一书。《建筑十书》是记录总结公元前 1 世纪已有建筑知识的纲要性著作，内容主要以古希腊的设计和构建方法为基础，几乎涉及建筑、城市规划和市政工程的所有方面。而在各主题下研究探讨的领域包括城市设计、建筑学、建造方法、建筑类型、水力学、时间测算以及军事防御工事。这里节选的"第一书"则讲述了建筑师教育、建筑设计的基本原则、城市选址、城市防御工事以及城市街道和公共建筑布局。在"第一书"中读者会首先接触到维特鲁威关于建筑的著名论断，即好的建筑一定是由"坚固、实用、美观"三种要素共同构成的。

## READING 1
## Book I
**Vitruvius (Marcus Vitruvius Pollio)**

This selection is the first of Vitruvius's *The Ten Books on Architecture*, a compendium of the essential architectural knowledge that was extant during the first century BC. Vitrivius's work, which covers almost every aspect of architecture, city planning, and civil engineering, is based primarily on classical Greek methods of design and construction. Examples of topics examined include urban design, architecture, building methods, building types, hydraulics, the measurement of time, and military fortifications. "Book I" describes the education of the architect, the fundamental principles of architectural design, the siting of cities, urban fortifications, and the layout of urban streets and public buildings. In "Book I" the reader first encounters Vitruvius's famous dictum that good architecture must be a combination of the three elements of "durability, convenience and beauty."

## PREFACE

1 While your divine intelligence and will, Imperator Caesar, were engaged in acquiring the right to command the world, and while your fellow citizens, when all their enemies had been laid low by your invincible valour, were glorying in your triumph and victory,—while all foreign nations were in subjection awaiting your beck and call, and the Roman people and senate, released from their alarm, were beginning to

*Source: The Ten Books on Architecture* by Marcus Vitruvius Pollio, Dover Edition, 1960, is an unabridged and unaltered republication of the first edition of the English translation by Morris Hicky Morgan, originally published by the Harvard University Press in 1914. Reproduced by permission of Dover Publications, Inc. pp. 3–32.

be guided by your most noble conceptions and policies, I hardly dared, in view of your serious employments, to publish my writings and long considered ideas on architecture, for fear of subjecting myself to your displeasure by an unseasonable interruption.

**2** But when I saw that you were giving your attention not only to the welfare of society in general and to the establishment of public order, but also to the providing of public buildings intended for utilitarian purposes, so that not only should the State have been enriched with provinces by your means, but that the greatness of its power might likewise be attended with distinguished authority in its public buildings, I thought that I ought to take the first opportunity to lay before you my writings on this theme. For in the first place it was this subject which made me known to your father, to whom I was devoted on account of his great qualities. After the council of heaven gave him a place in the dwellings of immortal life and transferred your father's power to your hands, my devotion continuing unchanged as I remembered him inclined me to support you. And so with Marcus Aurelius, Publius Minidius, and Gnaeus Cornelius, I was ready to supply and repair ballistae, scorpiones, and other artillery, and I have received rewards for good service with them. After your first bestowal of these upon me, you continued to renew them on the recommendation of your sister.

**3** Owing to this favour I need have no fear of want to the end of my life, and being thus laid under obligation I began to write this work for you, because I saw that you have built and are now building extensively, and that in future also you will take care that our public and private buildings shall be worthy to go down to posterity by the side of your other splendid achievements. I have drawn up definite rules to enable you, by observing them, to have personal knowledge of the quality both of existing buildings and of those which are yet to be constructed. For in the following books I have disclosed all the principles of the art.

## THE EDUCATION OF THE ARCHITECT

**1** The architect should be equipped with knowledge of many branches of study and varied kinds of learning, for it is by his judgement that all work done by the other arts is put to test. This knowledge is the child of practice and theory. Practice is the continuous and regular exercise of employment where manual work is done with any necessary material according to the design of a drawing. Theory, on the other hand, is the ability to demonstrate and explain the productions of dexterity on the principles of proportion.

**2** It follows, therefore, that architects who have aimed at acquiring manual skill without scholarship have never been able to reach a position of authority to correspond to their pains, while those who relied only upon theories and scholarship were obviously hunting the shadow, not the substance. But those who have a thorough knowledge of both, like men armed at all points, have the sooner attained their object and carried authority with them.

**3** In all matters, but particularly in architecture, there are these two points:—the thing signified, and that which gives it its significance. That which is signified is the subject of which we may be speaking; and that which gives significance is a demonstration on scientific principles. It appears, then, that one who professes himself an

**FIGURE 1–1**   Caryatides [From the edition of Vitruvius by Fra Giocondo, Venice, 1511]

architect should be well versed in both directions. He ought, therefore, to be both naturally gifted and amenable to instruction. Neither natural ability without instruction nor instruction without natural ability can make the perfect artist. Let him be educated, skilful with the pencil, instructed in geometry, know much history, have followed the philosophers with attention, understand music, have some knowledge of medicine, know the opinions of the jurists, and be acquainted with astronomy and the theory of the heavens.

**4** The reasons for all this are as follows. An architect ought to be an educated man so as to leave a more lasting remembrance in his treatises. Secondly, he must have a knowledge of drawing so that he can readily make sketches to show the appearance of the work which he proposes. Geometry, also, is of much assistance in architecture, and in particular it teaches us the use of the rule and compasses, by which especially we acquire readiness in making plans for buildings in their grounds, and rightly apply the square, the level, and the plummet. By means of optics, again, the light in buildings can be drawn from fixed quarters of the sky. It is true that it is by arithmetic that the total cost of buildings is calculated and measurements are computed, but difficult questions involving symmetry are solved by means of geometrical theories and methods.

**5** A wide knowledge of history is requisite because, among the ornamental parts of an architect's design for a work, there are many the underlying idea of whose employment he should be able to explain to inquirers. For instance, suppose him to set up the marble statues of women in long robes, called Caryatides, to take the place of columns, with the mutules and coronas placed directly above their heads, he will give the following explanation to his questioners. Caryae, a state in Peloponnesus, sided with the Persian enemies against Greece; later the Greeks, having gloriously won their freedom by victory in the war, made common cause and declared war against the people of Caryae. They took the town, killed the men, abandoned the State to desolation, and carried off their wives into slavery, without permitting them, however, to lay aside the long robes and other marks of their rank as married women, so that they might be obliged not only to march in the triumph but to appear forever after as a type of slavery, burdened with the weight of their shame and so making atonement for their State. Hence, the architects of the time designed for public buildings statues of these women, placed so as to carry a load, in order that the sin and the punishment of the people of Caryae might be known and handed down even to posterity.

**6** Likewise the Lacedaemonians under the leadership of Pausanias, son of Agesipolis, after conquering the Persian armies, infinite in number, with a small force at the battle of Plataea, celebrated a glorious triumph with the spoils and booty, and with the money obtained from the sale thereof built the Persian Porch, to be a monument to the renown and valour of the people and a trophy of victory for posterity. And there they set effigies of the prisoners arrayed in barbarian costume and holding up the roof, their pride punished by this deserved affront, that enemies might tremble for fear of the effects of their courage, and that their own people, looking upon this ensample of their valour and encouraged by the glory of it, might be ready to defend their independence. So from that time on, many have put up statues of Persians supporting entablatures and their ornaments, and thus from that motive have greatly enriched the diversity of their works. There are other stories of the same kind which architects ought to know.

**7** As for philosophy, it makes an architect high-minded and not self-assuming, but rather renders him courteous, just, and honest without avariciousness. This is very important, for no work can be rightly done without honesty and incorruptibility. Let him not be grasping nor have his mind preoccupied with the idea of receiving perquisites, but let him with dignity keep up his position by cherishing a good reputation. These are among the precepts of philosophy. Furthermore philosophy treats of physics (in Greek φυσιολογία) where a more careful knowledge is required because the problems which come under this head are numerous and of very different kinds; as, for example, in the case of the conducting of water. For at points of intake and at curves, and at places where it is raised to a level, currents of air naturally form in one way or another; and nobody who has not learned the fundamental principles of physics from philosophy will be able to provide against the damage which they do. So the reader of Ctesibius or Archimedes and the other writers of treatises of the same class will not be able to appreciate them unless he has been trained in these subjects by the philosophers.

**8** Music, also, the architect ought to understand so that he may have knowledge of the canonical and mathematical theory, and besides be able to tune ballistae, catapultae, and scorpiones to the proper key. For to the right and left in the beams are the

**FIGURE 1–2** Persians [From the edition of Vitruvius by Fra Giocondo, Venice, 1511]

holes in the frames through which the strings of twisted sinew are stretched by means of windlasses and bars, and these strings must not be clamped and made fast until they give the same correct note to the ear of the skilled workman. For the arms thrust through those stretched strings must, on being let go, strike their blow together at the same moment; but if they are not in unison, they will prevent the course of projectiles from being straight.

**9** In theatres, likewise, there are the bronze vessels (in Greek ηχεια) which are placed in niches under the seats in accordance with the musical intervals on mathematical principles. These vessels are arranged with a view to musical concords or harmony, and apportioned in the compass of the fourth, the fifth, and the octave, and so on up to the double octave, in such a way that when the voice of an actor falls in unison with any of them its power is increased, and it reaches the ears of the audience with greater clearness and sweetness. Water organs, too, and the other instruments which resemble them cannot be made by one who is without the principles of music.

**10** The architect should also have a knowledge of the study of medicine on account of the questions of climates (in Greek κλίματα), air, the healthiness and unhealthiness of sites, and the use of different waters. For without these considerations, the healthiness of a dwelling cannot be assured. And as for principles of law, he should know those which are necessary in the case of buildings having party walls, with regard to water dripping from the eaves, and also the laws about drains, windows, and water supply. And other things of this sort should be known to architects, so that, before they begin upon buildings, they may be careful not to leave disputed points for the householders to settle after the works are finished, and so that in drawing up contracts the interests of both employer and contractor may be wisely safeguarded. For if a contract is skilfully drawn, each may obtain a release from the other without disadvantage. From astronomy we find the east, west, south, and north, as well as the theory of the heavens, the equinox, solstice, and courses of the stars. If one has no knowledge of these matters, he will not be able to have any comprehension of the theory of sundials.

**11** Consequently, since this study is so vast in extent, embellished and enriched as it is with many different kinds of learning, I think that men have no right to profess themselves architects hastily, without having climbed from boyhood the steps of these studies and thus, nursed by the knowledge of many arts and sciences, having reached the heights of the holy ground of architecture.

**12** But perhaps to the inexperienced it will seem a marvel that human nature can comprehend such a great number of studies and keep them in the memory. Still, the observation that all studies have a common bond of union and intercourse with one another, will lead to the belief that this can easily be realized. For a liberal education form, as it were, a single body made up of these members. Those, therefore, who from tender years receive instruction in the various forms of learning, recognize the same stamp on all the arts, and an intercourse between all studies, and so they more readily comprehend them all. This is what led one of the ancient architects, Pytheos, the celebrated builder of the temple of Minerva at Priene, to say in his Commentaries that an architect ought to be able to accomplish much more in all the arts and sciences than the men who, by their own particular kinds of work and the practice of it, have brought each a single subject to the highest perfection. But this is in point of fact not realized.

**13** For an architect ought not to be and cannot be such a philologian as was Aristarchus, although not illiterate; nor a musician like Aristoxenus, though not absolutely ignorant of music; nor a painter like Apelles, though not unskilful in drawing; nor a sculptor such as was Myron or Polyclitus, though not unacquainted with the plastic art; nor again a physician like Hippocrates, though not ignorant of medicine; nor in the other sciences need he excel in each, though he should not be unskilful in them. For, in the midst of all this great variety of subjects, an individual cannot attain to perfection in each, because it is scarcely in his power to take in and comprehend the general theories of them.

**14** Still, it is not architects alone that cannot in all matters reach perfection, but even men who individually practise specialties in the arts do not all attain to the highest point of merit. Therefore, if among artists working each in a single field not all, but only a few in an entire generation acquire fame, and that with difficulty, how can an architect, who has to be skilful in many arts, accomplish not merely the feat—in itself a great marvel—of being deficient in none of them, but also that of surpassing all those artists who have devoted themselves with unremitting industry to single fields?

**15** It appears, then, that Pytheos made a mistake by not observing that the arts are each composed of two things, the actual work and the theory of it. One of these, the doing of the work, is proper to men trained in the individual subject, while the other, the theory, is common to all scholars: for example, to physicians and musicians the rhythmical beat of the pulse and its metrical movement. But if there is a wound to be healed or a sick man to be saved from danger, the musician will not call, for the business will be appropriate to the physician. So in the case of a musical instrument, not the physician but the musician will be the man to tune it so that the ears may find their due pleasure in its strains.

**16** Astronomers likewise have a common ground for discussion with musicians in the harmony of the stars and musical concords in tetrads and triads of the fourth and the fifth, and with geometricians in the subject of vision (in Greek λόγος ὀπτικός); and in all other sciences many points, perhaps all, are common so far as the discussion of them is concerned. But the actual undertaking of works which are brought to perfection by the hand and its manipulation is the function of those who have been specially trained to deal with a single art. It appears, therefore, that he has done enough and to spare who in each subject possesses a fairly good knowledge of those parts, with their principles, which are indispensable for architecture, so that if he is required to pass judgement and to express approval in the case of those things or arts, he may not be found wanting. As for men upon whom nature has bestowed so much ingenuity, acuteness, and memory that they are able to have a thorough knowledge of geometry, astronomy, music, and the other arts, they go beyond the functions of architects and become pure mathematicians. Hence they can readily take up positions against those arts because many are the artistic weapons with which they are armed. Such men, however, are rarely found, but there have been such at times; for example, Aristarchus of Samos, Philolaus and Archytas of Tarentum, Apollonius of Perga, Eratosthenes of Cyrene, and among Syracusans Archimedes and Scopinas, who through mathematics and natural philosophy discovered, expounded, and left to posterity many things in connexion with mechanics and with sundials.

**17** Since, therefore, the possession of such talents due to natural capacity is not vouchsafed at random to entire nations, but only to a few great men; since, moreover, the function of the architect requires a training in all the departments of learning; and finally, since reason, on account of the wide extent of the subject, concedes that he may possess not the highest but not even necessarily a moderate knowledge of the subjects of study, I request, Caesar, both of you and of those who may read the said books, that if anything is set forth with too little regard for grammatical rule, it may be pardoned. For it is not as a very great philosopher, nor as an eloquent rhetorician, nor as a grammarian trained in the highest principles of his art, that I have striven to write this work, but as an architect who has had only a dip into those studies. Still, as regards the efficacy of the art and the theories of it, I promise and expect that in these volumes I shall undoubtedly show myself of very considerable importance not only to builders but also to all scholars.

## THE FUNDAMENTAL PRINCIPLES OF ARCHITECTURE

**1** Architecture depends on Order (in Greek τάξις), Arrangement ( in Greek διάθεσις ), Eurythmy, Symmetry, Propriety, and Economy (in Greek οἰκονομία).

**2** Order gives due measure to the members of a work considered separately, and symmetrical agreement to the proportions of the whole. It is an adjustment according to quantity (in Greek ποσότης). By this I mean the selection of modules from the members of the work itself and, starting from these individual parts of members, constructing the whole work to correspond. Arrangement includes the putting of things in their proper places and the elegance of effect which is due to adjustments appropriate to the character of the work. Its forms of expression (in Greek ἰδέαι) are these: groundplan, elevation, and perspective. A groundplan is made by the proper successive use of compasses and rule, through which we get outlines for the plane surfaces of buildings. An elevation is a picture of the front of a building, set upright and properly drawn in the proportions of the contemplated work. Perspective is the method of sketching a front with the sides withdrawing into the background, the lines all meeting in the centre of a circle. All three come of reflexion and invention. Reflexion is careful and laborious thought, and watchful attention directed to the agreeable effect of one's plan. Invention, on the other hand, is the solving of intricate problems and the discovery of new principles by means of brilliancy and versatility. These are the departments belonging under Arrangement.

**3** Eurythmy is beauty and fitness in the adjustments of the members. This is found when the members of a work are of a height suited to their breadth, of a breadth suited to their length, and, in a word, when they all correspond symmetrically.

**4** Symmetry is a proper agreement between the members of the work itself, and relation between the different parts and the whole general scheme, in accordance with a certain part selected as standard. Thus in the human body there is a kind of symmetrical harmony between forearm, foot, palm, finger, and other small parts; and so it is with perfect buildings. In the case of temples, symmetry may be calculated from the thickness of a column, from a triglyph, or even from a module; in the ballista, from the hole or from what the Greeks call the περίτρητος ; in a ship, from the space between the tholepins (διάπηγμα); and in other things, from various members.

**5** Propriety is that perfection of style which comes when a work is authoritatively constructed on approved principles. It arises from prescription (Greek θεματισμῷ), from usage, or from nature. From prescription, in the case of hypaethral edifices, open to the sky, in honour of Jupiter Lightning, the Heaven, the Sun, or the Moon: for these are gods whose semblances and manifestations we behold before our very eyes in the sky when it is cloudless and bright. The temples of Minerva, Mars, and Hercules, will be Doric, since the virile strength of these gods makes daintiness entirely inappropriate to their houses. In temples to Venus, Flora, Proserpine, Spring-Water, and the Nymphs, the Corinthian order will be found to have peculiar significance, because these are delicate divinities and so its rather slender outlines, its flowers, leaves, and ornamental volutes will lend propriety where it is due. The construction of temples of the Ionic order to Juno, Diana, Father Bacchus, and the other gods of that kind, will be in keeping with the middle position which they hold; for the building of such will be an appropriate combination of the severity of the Doric and the delicacy of the Corinthian.

**6** Propriety arises from usage when buildings having magnificent interiors are provided with elegant entrance-courts to correspond; for there will be no propriety in the spectacle of an elegant interior approached by a low, mean entrance. Or, if dentils be carved in the cornice of the Doric entablature or triglyphs represented in the Ionic entablature over the cushion-shaped capitals of the columns, the effect will be spoilt by the transfer of the peculiarities of the one order of building to the other, the usage in each class having been fixed long ago.

**7** Finally, propriety will be due to natural causes if, for example, in the case of all sacred precincts we select very healthy neighbourhoods with suitable springs of water in the places where the fanes are to be built, particularly in the case of those to Aesculapius and to Health, gods by whose healing powers great numbers of the sick are apparently cured. For when their diseased bodies are transferred from an unhealthy to a healthy spot, and treated with waters from health-giving springs, they will the more speedily grow well. The result will be that the divinity will stand in higher esteem and find his dignity increased, all owing to the nature of his site. There will also be natural propriety in using an eastern light for bedrooms and libraries, a western light in winter for baths and winter apartments, and a northern light for picture galleries and other places in which a steady light is needed; for that quarter of the sky grows neither light nor dark with the course of the sun, but remains steady and unshifting all day long.

**8** Economy denotes the proper management of materials and of site, as well as a thrifty balancing of cost and common sense in the construction of works. This will be observed if, in the first place, the architect does not demand things which cannot be found or made ready without great expense. For example: it is not everywhere that there is plenty of pitsand, rubble, fir, clear fir, and marble, since they are produced in different places and to assemble them is difficult and costly. Where there is no pitsand, we must use the kinds washed up by rivers or by the sea; the lack of fir and clear fir may be evaded by using cypress, poplar, elm, or pine; and other problems we must solve in similar ways.

**9** A second stage in Economy is reached when we have to plan the different kinds of dwellings suitable for ordinary householders, for great wealth, or for the high position of the statesman. A house in town obviously calls for one form of construction; that into which stream the products of country estates requires another; this will not be the same in the case of money-lenders and still different for the opulent and luxurious; for the powers under whose deliberations the commonwealth is guided dwellings are to be provided according to their special needs: and, in a word, the proper form of economy must be observed in building houses for each and every class.

## THE DEPARTMENTS OF ARCHITECTURE

**1** There are three departments of architecture: the art of building, the making of time-pieces, and the construction of machinery. Building is, in its turn, divided into two parts, of which the first is the construction of fortified towns and of works for general use in public places, and the second is the putting up of structures for private individuals. There are three classes of public buildings: the first for defensive, the second for religious, and the third for utilitarian purposes. Under defence comes the planning of walls, towers, and gates, permanent devices for resistance against hostile attacks; under religion, the erection of fanes and temples to the immortal gods; under utility, the provision of meeting places for public use, such as harbours, markets, colonnades, baths, theatres, promenades, and all other similar arrangements in public places.

**2** All these must be built with due reference to durability, convenience, and beauty. Durability will be assured when foundations are carried down to the solid ground and materials wisely and liberally selected; convenience, when the arrangement of the apartments is faultless and presents no hindrance to use, and when each class of building is assigned to its suitable and appropriate exposure; and beauty, when the appearance of the work is pleasing and in good taste, and when its members are in due proportion according to correct principles of symmetry.

## THE SITE OF A CITY

**1** For fortified towns the following general principles are to be observed. First comes the choice of a very healthy site. Such a site will be high, neither misty nor frosty, and in a climate neither hot nor cold, but temperate; further, without marshes in the neighbourhood. For when the morning breezes blow toward the town at sunrise, if they bring with them mists from marshes and, mingled with the mist, the poisonous breath of the creatures of the marshes to be wafted into the bodies of the inhabitants, they will make the site unhealthy. Again, if the town is on the coast with a southern or western exposure, it will not be healthy, because in summer the southern sky grows hot at sunrise and is fiery at noon, while a western exposure grows warm after sunrise, is hot at noon, and at evening all aglow.

**2** These variations in heat and the subsequent cooling off are harmful to the people living on such sites. The same conclusion may be reached in the case of inanimate things. For instance, nobody draws the light for covered wine rooms from the south or west, but rather from the north, since that quarter is never subject to change but is always constant and unshifting. So it is with granaries: grain exposed to the sun's course soon loses its good quality, and provisions and fruit, unless stored in a place unexposed to the sun's course, do not keep long.

**3** For heat is a universal solvent, melting out of things their power of resistance, and sucking away and removing their natural strength with its fiery exhalations so that they grow soft, and hence weak, under its glow. We see this in the case of iron which, however hard it may naturally be, yet when heated thoroughly in a furnace fire can be easily worked into any kind of shape, and still, if cooled while it is soft and white hot, it hardens again with a mere dip into cold water and takes on its former quality.

**4** We may also recognize the truth of this from the fact that in summer the heat makes everybody weak, not only in unhealthy but even in healthy places, and that in winter even the most unhealthy districts are much healthier because they are given a solidity by the cooling off. Similarly, persons removed from cold countries to hot cannot endure it but waste away; whereas those who pass from hot places to the cold regions of the north, not only do not suffer in health from the change of residence but even gain by it.

**5** It appears, then, that in founding towns we must beware of districts from which hot winds can spread abroad over the inhabitants. For while all bodies are composed of the four elements (in Greek στοιχεῖα), that is, of heat, moisture, the earthy, and air, yet there are mixtures according to natural temperament which make up the natures of all the different animals of the world, each after its kind.

**6** Therefore, if one of these elements, heat, becomes predominant in any body whatsoever, it destroys and dissolves all the others with its violence. This defect may be due to violent heat from certain quarters of the sky, pouring into the open pores in too great proportion to admit of a mixture suited to the natural temperament of the body in question. Again, if too much moisture enters the channels of a body, and thus introduces disproportion, the other elements, adulterated by the liquid, are impaired, and the virtues of the mixture dissolved. This defect, in turn, may arise from the cooling properties of moist winds and breezes blowing upon the body. In the same way, increase or diminution of the proportion of air or of the earthy which is natural to the body may enfeeble the other elements; the predominance of the earthy being due to overmuch food, that of air to a heavy atmosphere.

**7** If one wishes a more accurate understanding of all this, he need only consider and observe the natures of birds, fishes, and land animals, and he will thus come to reflect upon distinctions of temperament. One form of mixture is proper to birds, another to fishes, and a far different form to land animals. Winged creatures have less of the earthy, less moisture, heat in moderation, air in large amount. Being made up, therefore, of the lighter elements, they can more readily soar away into the air. Fish, with their aquatic nature, being moderately supplied with heat and made up in great part of air and the earthy, with as little of moisture as possible, can more easily exist in moisture for the very reason that they have less of it than of the other elements in their bodies; and so, when they are drawn to land, they leave life and water at the same moment. Similarly, the land animals, being moderately supplied with the elements of air and heat, and having less of the earthy and a great deal of moisture, cannot long continue alive in the water, because their portion of moisture is already abundant.

**8** Therefore, if all this is as we have explained, our reason showing us that the bodies of animals are made up of the elements, and these bodies, as we believe, giving way and breaking up as a result of excess or deficiency in this or that element, we cannot but believe that we must take great care to select a very temperate climate for the site of our city, since healthfulness is, as we have said, the first requisite.

**9** I cannot too strongly insist upon the need of a return to the method of old times. Our ancestors, when about to build a town or an army post, sacrificed some of the cattle that were wont to feed on the site proposed and examined their livers. If the livers of the first victims were dark-coloured or abnormal, they sacrificed others, to see whether the fault was due to disease or their food. They never began to build defensive works in a place until after they had made many such trials and satisfied themselves that good water and food had made the liver sound and firm. If they continued to find it abnormal, they argued from this that the food and water supply found in such a place would be just as unhealthy for man, and so they moved away and changed to another neighbourhood, healthfulness being their chief object.

**10** That pasturage and food may indicate the healthful qualities of a site is a fact which can be observed and investigated in the case of certain pastures in Crete, on each side of the river Pothereus, which separates the two Cretan states of Gnosus and Gortyna. There are cattle at pasture on the right and left banks of that river, but while the cattle that feed near Gnosus have the usual spleen, those on the other side near Gortyna have no perceptible spleen. On investigating the subject, physicians discovered on this side a kind of herb which the cattle chew and thus make their spleen small. The herb is therefore gathered and used as a medicine for the cure of splenetic people. The Cretans call it ἄσπληνον. From food and water, then, we may learn whether sites are naturally unhealthy or healthy.

**11** If the walled town is built among the marshes themselves, provided they are by the sea, with a northern or north-eastern exposure, and are above the level of the seashore, the site will be reasonable enough. For ditches can be dug to let out the water to the shore, and also in times of storms the sea swells and comes backing up into the marshes, where its bitter blend prevents the reproductions of the usual marsh creatures, while any that swim down from the higher levels to the shore are killed at once by the saltness to which they are unused. An instance of this may be found in the Gallic marshes surrounding Altino, Ravenna, Aquileia, and other towns in places of the kind, close by marshes. They are marvellously healthy, for the reasons which I have given.

**12** But marshes that are stagnant and have no outlets either by rivers or ditches, like the Pomptine marshes, merely putrefy as they stand, emitting heavy, unhealthy vapours. A case of a town built in such a spot was Old Salpia in Apulia, founded by Diomede on his way back from Troy, or, according to some writers, by Elpias of Rhodes. Year after year there was sickness, until finally the suffering inhabitants came with a public petition to Marcus Hostilius and got him to agree to seek and find them a proper place to which to remove their city. Without delay he made the most skilful investigations, and at once purchased an estate near the sea in a healthy place, and asked the Senate and Roman people for permission to remove the town. He constructed the walls and laid out the house lots, granting one to each citizen for a mere trifle. This done, he cut an opening from a lake into the sea, and thus made of the lake a harbour for the town. The result is that now the people of Salpia live on a healthy site and at a distance of only four miles from the old town.

## THE CITY WALLS

**1** After insuring on these principles the healthfulness of the future city, and selecting a neighbourhood that can supply plenty of food stuffs to maintain the community, with good roads or else convenient rivers or seaports affording easy means of transport to the city, the next thing to do is to lay the foundations for the towers and walls. Dig down to solid bottom, if it can be found, and lay them therein, going as deep as the magnitude of the proposed work seems to require. They should be much thicker than the part of the walls that will appear above ground, and their structure should be as solid as it can possibly be laid.

**2** The towers must be projected beyond the line of wall, so that an enemy wishing to approach the wall to carry it by assault may be exposed to the fire of missiles on his open flank from the towers on his right and left. Special pains should be taken that there be no easy avenue by which to storm the wall. The roads should be encompassed at steep points, and planned so as to approach the gates, not in a straight line, but from the right to the left; for as a result of this, the right hand side of the assailants, unprotected by their shields, will be next the wall. Towns should be laid out not as an exact square nor with salient angles, but in circular form, to give a view of the

**FIGURE 1–3**  Construction of city walls [From the edition of Vitruvius by Fra Giocondo, Venice, 1511]

enemy from many points. Defence is difficult where there are salient angles, because the angle protects the enemy rather than the inhabitants.

**3** The thickness of the wall should, in my opinion, be such that armed men meeting on top of it may pass one another without interference. In the thickness there should be set a very close succession of ties made of charred olive wood, binding the two faces of the wall together like pins, to give it lasting endurance. For that is a material which neither decay, nor the weather, nor time can harm, but even though buried in the earth or set in the water it keeps sound and useful forever. And so not only city walls but substructures in general and all walls that require a thickness like that of a city wall, will be long in falling to decay if tied in this manner.

**4** The towers should be set at intervals of not more than a bowshot apart, so that in case of an assault upon any one of them, the enemy may be repulsed with scorpiones and other means of hurling missiles from the towers to the right and left. Opposite the inner side of every tower the wall should be interrupted for a space the width of the tower, and have only a wooden flooring across, leading to the interior of the tower but not firmly nailed. This is to be cut away by the defenders in case the enemy gets possession of any portion of the wall; and if the work is quickly done, the enemy will not be able to make his way to the other towers and the rest of the wall unless he is ready to face a fall.

**5** The towers themselves must be either round or polygonal. Square towers are sooner shattered by military engines, for the battering rams pound their angles to pieces; but in the case of round towers they can do no harm, being engaged, as it were, in driving wedges to their centre. The system of fortification by wall and towers may be made safest by the addition of earthen ramparts, for neither rams, nor mining, nor other engineering devices can do them any harm.

**6** The rampart form of defence, however, is not required in all places, but only where outside the wall there is high ground from which an assault on the fortifications may be made over a level space lying between. In places of this kind we must first make very wide, deep ditches; next sink foundations for a wall in the bed of the ditch and build them thick enough to support an earthwork with ease.

**7** Then within this substructure lay a second foundation, far enough inside the first to leave ample room for cohorts in line of battle to take position on the broad top of the rampart for its defence. Having laid these two foundations at this distance from one another, build cross walls between them, uniting the outer and inner foundation, in a comb-like arrangement, set like the teeth of a saw. With this form of construction, the enormous burden of earth will be distributed into small bodies, and will not lie with all its weight in one crushing mass so as to thrust out the substructures.

**8** With regard to the material of which the actual wall should be constructed or finished, there can be no definite prescription, because we cannot obtain in all places the supplies that we desire. Dimension stone, flint, rubble, burnt or unburnt brick,—use them as you find them. For it is not every neighbourhood or particular locality that can have a wall built of burnt brick like that at Babylon, where there was plenty of asphalt to take the place of lime and sand, and yet possibly each may be provided with materials of equal usefulness so that out of them a faultless wall may be built to last forever.

## THE DIRECTIONS OF THE STREETS; WITH REMARKS ON THE WINDS

**1** The town being fortified, the next step is the apportionment of house lots within the wall and the laying out of streets and alleys with regard to climatic conditions. They will be properly laid out if foresight is employed to exclude the winds from the alleys. Cold winds are disagreeable, hot winds enervating, moist winds unhealthy. We must, therefore, avoid mistakes in this matter and beware of the common experience of many communities. For example, Mytilene in the island of Lesbos is a town built with magnificence and good taste, but its position shows a lack of foresight. In that community when the wind is south, the people fall ill; when it is northwest, it sets them coughing; with a north wind they do indeed recover but cannot stand about in the alleys and streets, owing to the severe cold.

**2** Wind is a flowing wave of air, moving hither and thither indefinitely. It is produced when heat meets moisture, the rush of heat generating a mighty current of air. That this is the fact we may learn from bronze eolipiles, and thus by means of a scientific invention discover a divine truth lurking in the laws of the heavens. Eolipiles are hollow bronze balls, with a very small opening through which water is poured into them. Set before a fire, not a breath issues from them before they get warm; but as soon as they begin to boil, out comes a strong blast due to the fire. Thus from this slight and very short experiment we may understand and judge of the mighty and wonderful laws of the heavens and the nature of winds.

**3** By shutting out the winds from our dwellings, therefore, we shall not only make the place healthful for people who are well, but also in the case of diseases due perhaps to unfavourable situations elsewhere, the patients, who in other healthy places might be cured by a different form of treatment, will here be more quickly cured by the mildness that comes from the shutting out of the winds. The diseases which are hard to cure in neighbourhoods such as those to which I have referred above are catarrh, hoarseness, coughs, pleurisy, consumption, spitting of blood, and all others that are cured not by lowering the system but by building it up. They are hard to cure, first, because they are originally due to chills; secondly, because the patient's system being already exhausted by disease, the air there, which is in constant agitation owing to winds and therefore deteriorated, takes all the sap of life out of their diseased bodies and leaves them more meagre every day. On the other hand, a mild, thick air, without draughts and not constantly blowing back and forth, builds up their frames by its unwavering steadiness, and so strengthens and restores people who are afflicted with these diseases.

**4** Some have held that there are only four winds: Solanus from due east; Auster from the south; Favonius from due west; Septentrio from the north. But more careful investigators tell us that there are eight. Chief among such was Andronicus of Cyrrhus who in proof built the marble octagonal tower in Athens. On the several sides of the octagon he executed reliefs representing the several winds, each facing the point from which it blows; and on top of the tower he set a conical shaped piece of marble and on this a bronze Triton with a rod outstretched in its right hand. It was so contrived as to go round with the wind, always stopping to face the breeze and holding its rod as a pointer directly over the representation of the wind that was blowing.

**5** Thus Eurus is placed to the southeast between Solanus and Auster; Africus to the southwest between Auster and Favonius; Caurus, or, as many call it, Corus, between Favonius and Septentrio; and Aquilo between Septentrio and Solanus. Such, then, appears to have been his device, including the numbers and names of the wind and indicating the directions from which particular winds blow. These facts being thus determined, to find the directions and quarters of the winds your method of procedure should be as follows.

**6** In the middle of the city place a marble amussium, laying it true by the level, or else let the spot be made so true by means of rule and level that no amussium is necessary. In the very centre of that spot set up a bronze gnomon or "shadow tracker" (in Greek σκιαθήρας). At about the fifth hour in the morning, take the end of the shadow cast by this gnomon, and mark it with a point. Then, opening your compasses to this point which marks the length of the gnomon's shadow, describe a circle from the centre. In the afternoon watch the shadow of your gnomon as it lengthens, and when it once more touches the circumference of this circle and the shadow in the afternoon is equal in length to that of the morning, mark it with a point.

**7** From these two points describe with your compasses intersecting arcs, and through their intersection and the centre let a line be drawn to the circumference of the circle to give us the quarters of south and north. Then, using a sixteenth part of the entire circumference of the circle as a diameter, describe a circle with its centre on the line to the south, at the point where it crosses the circumference, and put points to the right and left on the circumference on the south side, repeating the process on the north side. From the four points thus obtained draw lines intersecting the centre from one side of the circumference to the other. Thus we shall have an eighth part of the circumference set out for Auster and another for Septentrio. The rest of the entire circumference is then to be divided into three equal parts on each side, and thus we have designed a figure equally apportioned among the eight winds. Then let the directions of your streets and alleys be laid down on the lines of division between the quarters of two winds.

**8** On this principle of arrangement the disagreeable force of the winds will be shut out from dwellings and lines of houses. For if the streets run full in the face of the winds, their constant blasts rushing in from the open country, and then confined by narrow alleys, will sweep through them with great violence. The lines of houses must therefore be directed away from the quarters from which the winds blow, so that as they come in they may strike against the angles of the blocks and their force thus be broken and dispersed.

**9** Those who know names for very many winds will perhaps be surprised at our setting forth that there are only eight. Remembering, however, that Eratosthenes of Cyrene, employing mathematical theories and geometrical methods, discovered from the course of the sun, the shadows cast by an equinoctial gnomon, and the inclination of the heaven that the circumference of the earth is two hundred and fifty-two thousand stadia, that is, thirty-one million five hundred thousand paces, and observing that an eighth part of this, occupied by a wind, is three million nine hundred and thirty-seven thousand five hundred paces, they should not be surprised to find that a single wind, ranging over so wide a field, is subject to shifts this way and that, leading to a variety of breezes.

**10** So we often have Leuconotus and Altanus blowing respectively to the right and left of Auster; Libonotus and Subvesperus to the right and left of Africus; Argestes, and at certain periods the Etesiae, on either side of Favonius; Circias and Corus on the sides of Caurus; Thracias and Gallicus on either side of Septentrio; Supernas and Caecias to the right and left of Aquilo; Carbas, and at a certain period the Ornithiae, on either side of Solanus; while Eurocircias and Volturnus blow on the flanks of Eurus which is between them. There are also many other names for winds derived from localities or from the squalls which sweep from rivers or down mountains.

**11** Then, too, there are the breezes of early morning; for the sun on emerging from beneath the earth strikes humid air as he returns, and as he goes climbing up the sky he spreads it out before him, extracting breezes from the vapour that was there before the dawn. Those that still blow on after sunrise are classed with Eurus, and hence appears to come the Greek name εὖρος as the child of the breezes, and the word for "to-morrow," αὔριον, named from the early morning breezes. Some people do indeed say that Eratosthenes could not have inferred the true measure of the earth. Whether true or untrue, it cannot affect the truth of what I have written on the fixing of the quarters from which the different winds blow.

**12** If he was wrong, the only result will be that the individual winds may blow, not with the scope expected from his measurement, but with powers either more or less widely extended. For the readier understanding of these topics, since I have treated them with brevity, it has seemed best to me to give two figures, or, as the Greeks say, σχήματα, at the end of this book: one designed to show the precise quarters from

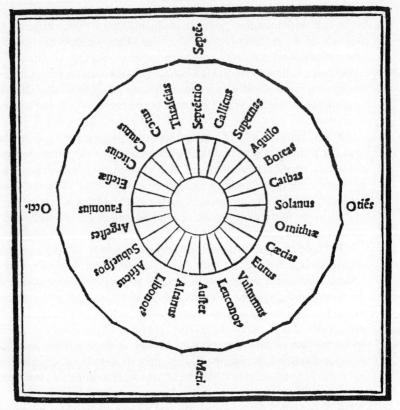

**FIGURE 1–4**  Diagram of the winds [From the edition of Vitruvius by Fra Giocondo, Venice, 1511]

which the winds arise; the other, how by turning the directions of the rows of houses and the streets away from their full force, we may avoid unhealthy blasts. Let A be the centre of a plane surface, and B the point to which the shadow of the gnomon reaches in the morning. Taking A as the centre, open the compasses to the point B, which marks the shadow, and describe a circle. Put the gnomon back where it was before and wait for the shadow to lessen and grow again until in the afternoon it is equal to its length in the morning, touching the circumference at the point C. Then from the points B and C describe with the compasses two arcs intersecting at D. Next draw a line from the point of intersection D through the centre of the circle to the circumference and call it EF. This line will show where the south and north lie.

**13** Then find with the compasses a sixteenth part of the entire circumference; then centre the compasses on the point E where the line to the south touches the circumference, and set off the points G and H to the right and left of E. Likewise on the north side, centre the compasses on the circumference at the point F on the line to the north, and set off the points I and K to the right and left; then draw lines through the centre from G to K and from H to I. Thus the space from G to H will belong to Auster and the south, and the space from I to K will be that of Septentrio. The rest of the circum-

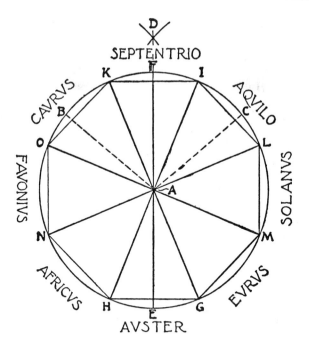

FIGURE 1-5

ference is to be divided equally into three parts on the right and three on the left, those to the east at the points L and M, those to the west at the points N and O. Finally, intersecting lines are to be drawn from M to O and from L to N. Thus we shall have the circumference divided into eight equal spaces for the winds. The figure being finished, we shall have at the eight different divisions, beginning at the south, the letter G between Eurus and Auster, H between Auster and Africus, N between Africus and Favonius, O between Favonius and Caurus, K between Caurus and Septentrio, I between Septentrio and Aquilo, L between Aquilo and Solanus, and M between Solanus and Eurus. This done, apply a gnomon to these eight divisions and thus fix the directions of the different alleys.

## THE SITES FOR PUBLIC BUILDINGS

**1** Having laid out the alleys and determined the streets, we have next to treat of the choice of building sites for temples, the forum, and all other public places, with a view to general convenience and utility. If the city is on the sea, we should choose ground close to the harbour as the place where the forum is to be built; but if inland, in the middle of the town. For the temples, the sites for those of the gods under whose particular protection the state is thought to rest and for Jupiter, Juno, and Minerva, should be on the very highest point commanding a view of the greater part of the city. Mercury should be in the forum, or, like Isis and Serapis, in the emporium: Apollo and Father Bacchus near the theatre; Hercules at the circus in communities which have no gymnasia nor amphitheatres; Mars outside the city but at the training ground, and so Venus, but at the harbour. It is moreover shown by the Etruscan diviners in treatises on their science that the fanes of Venus, Vulcan, and Mars should be situated outside the walls, in order that the young men and married women may not become habituated in the city to the temptations incident to the worship of Venus, and that buildings may be free from the terror of fires through the religious rites and sacrifices which call the power of Vulcan beyond the walls. As for Mars, when that divinity is enshrined outside the walls, the citizens will never take up arms against each other, and he will defend the city from its enemies and save it from danger in war.

**2** Ceres also should be outside the city in a place to which people need never go except for the purpose of sacrifice. That place should be under the protection of religion, purity, and good morals. Proper sites should be set apart for the precincts of the other gods according to the nature of the sacrifices offered to them.

The principle governing the actual construction of temples and their symmetry I shall explain in my third and fourth books. In the second I have thought it best to give an account of the materials used in buildings with their good qualities and advantages, and then in the succeeding books to describe and explain the proportions of buildings, their arrangements, and the different forms of symmetry.

**PROLOGUE—SUGGESTED READINGS**

Alberti, Leon Battista (1404–1472). *The Ten Books of Architecture*. New York: Dover, 1986.

Laugier, Marc Antoine (1711–1769). *An Essay on Architecture*. Los Angeles: Hennessey and Ingalls, 1977.

Palladio, Andrea (1508–1580). *The Four Books of Architecture*. New York: Dover, 1965.

Ruskin, John (1819–1900). *The Seven Lamps of Architecture*. New York: Farrar, Straus, and Giroux, 1981.

# PART ONE 第一部分
## ARCHITECTURE AS ARTIFACT 人造艺术的建筑

# 1

# ARCHITECTURAL HISTORY AND THEORY
# 建筑历史和理论

选读 2
"文化的变革"与"领土的变革"
肯尼斯·弗兰姆普敦

本选读内容包括弗兰姆普敦的《现代建筑》前两章。《现代建筑》是以当代北美和西欧设计理论和实践演进为主题的论文集,它以18世纪中叶到20世纪早期建筑设计的发展和变革为研究脉络,并对20世纪建筑的社会、技术和理论基础做出了简明概述。在"文化的变革"一章中,弗兰姆普敦讲述了促成现代建筑理论与可视化语言的社会和文化因素。在"领土的变革"一章中,他则从交通运输系统、工业化和规划理论的不同角度对19世纪的城市发展进行了分析。

## READING 2

## Cultural Transformations *and* Territorial Transformations

### Kenneth Frampton

This selection contains the first two chapters from Frampton's *Modern Architecture*—a collection of essays on the evolution of contemporary design theories and practices in North America and Western Europe. The book traces the development and transformation of building design from the mid-18th to the early part of the 20th century. It provides a concise overview of the social, technical, and theoretical foundations of 20th century architecture. In "Cultural Transformations," Frampton describes the social and cultural factors that gave rise to the theory and visual language of modern architecture. In "Territorial Transformations," he analyzes the 19th century city from the perspective of transportation systems, industrialization, and planning theories.

*Source: Modern Architecture: A Critical History,* 3rd ed. by Kenneth Frampton. Copyright © 1992 by Thames and Hudson Ltd., London, reprinted by permission of the publishers. pp. 12–28.

**FIGURE 2–1** Soufflot, Ste-Geneviève (now the Panthéon), Paris, 1755–90; crossing piers strengthened by Rondelet in 1806. [Photo used by permission of A. F. Kersting]

## CULTURAL TRANSFORMATIONS: NEO-CLASSICAL ARCHITECTURE
## 1750–1900

> The Baroque system had operated as a kind of double intersection. It had often contrasted with rationalized gardens, building facades decorated with plant motifs. The reign of man and the reign of nature had certainly remained distinct but they had exchanged their characteristics, merging into each other for the sake of ornamentation and prestige. On the other hand the "English style" park, in which man's intervention was supposed to remain invisible, was intended to offer the *purposefulness* of nature, while within, but separate from the actual park, the houses constructed by Morris or Adam manifested the *will* of man, isolating clearly the presence of human reason in the midst of the irrational domains of freely growing vegetation. The Baroque interpenetration of man and nature was now replaced by a separation, thus establishing the distance between man and nature which was a prerequisite for nostalgic contemplation. Now . . . this contemplative separation arose as a compensatory or expiatory reaction against the growing attitude of practical men towards nature. While technical exploitation tended to wage war on nature, houses and parks attempted a reconciliation, a local armistice, introducing the dream of an impossible peace; and to this end man had continued to retain the image of untouched natural surroundings.
>
> Jean Starobinski<br>*L'Invention de la liberté,* 1964

The architecture of Neo-Classicism seems to have emerged out of two different but related developments which radically transformed the relationship between man and nature. The first was a sudden increase in man's capacity to exercise control over nature, which by the mid-17th century had begun to advance beyond the technical frontiers of the Renaissance. The second was a fundamental shift in the nature of human consciousness, in response to major changes taking place in society, which gave birth to a new cultural formation that was equally appropriate to the life styles of the declining aristocracy and the rising bourgeoisie. Whereas technological changes led to a new infrastructure and to the exploitation of an increased productive capacity, the change in human consciousness yielded new categories of knowledge and a historicist mode of thought that was so reflexive as to question its own identity. Where the one, grounded in science, took immediate form in the extensive road and canal works of the 17th and 18th centuries and gave rise to new technical institutions, such as the Ecole des Ponts et Chaussées, founded in 1747, the other led to the emergence of the humanist disciplines of the Enlightenment, including the pioneer works of modern sociology, aesthetics, history and archaeology—Montesquieu's *De l'esprit des lois* (1748), Baumgarten's *Aesthetica* (1750), Voltaire's *Le Siècle de Louis XIV* (1751) and J. J. Winckelmann's *Geschichte der Kunst des Altertum* (History of Ancient Art) of 1764.

The over-elaboration of architectural language in the Rococo interiors of the Ancien Régime and the secularization of Enlightenment thought compelled the architects of the 18th century, by now aware of the emergent and unstable nature of their age, to search for a true style through a precise reappraisal of antiquity. Their motivation was not simply to copy the ancients but to obey the principles on which their work had been based. The archaeological research that arose from this impulse soon led to a major controversy: to which, of four Mediterranean cultures—the Egyptians, the Etruscans, the Greeks and the Romans—should they look for a true style?

One of the first consequences of reassessing the antique world was to extend the itinerary of the traditional Grand Tour beyond the frontiers of Rome, so as to study at its periphery those cultures on which, according to Vitruvius, Roman architecture had been based. The discovery and excavation of Roman cities at Herculaneum and Pompeii, during the first half of the 18th century, encouraged expeditions further afield and visits were soon being made to ancient Greek sites in both Sicily and Greece. The received Vitruvian dictum of the Renaissance—the catechism of Classicism—was now to be checked against the actual ruins. The measured drawings that were published in the 1750s and 1760s, J. D. Le Roy's *Ruines des plus beaux monuments de la Grèce* (1758), James Stuart and Nicholas Revett's *Antiquities of Athens* (1762), and Robert Adam and C. L. Clérisseau's documentation of Diocletian's palace at Split (1764), testify to the intensity with which these studies were pursued. It was Le Roy's promotion of Greek architecture as the origin of the "true style" that raised the chauvinist ire of the Italian architect-engraver Giovanni Battista Piranesi.

Piranesi's *Della Magnificenza ed Architettura de' Romani* of 1761 was a direct attack on Le Roy's polemic: he asserted not only that the Etruscans had antedated the Greeks but that, together with their successors the Romans, they had raised architecture to a higher level of refinement. The only evidence that he could cite in support of his claim was the few Etruscan structures that had survived the ravages of Rome—tombs and engineering works—and these seem to have orientated the remainder of his career in a remarkable way. In one set of etchings after another he represented the dark side of that sensation already classified by Edmund Burke in 1757 as the Sublime, that tranquil terror induced by the contemplation of great size, extreme antiquity and decay. These qualities acquired their full force in Piranesi's work through the infinite grandeur of the images that he portrayed. Such nostalgic Classical images were, however, as Manfredo Tafuri has observed, treated "as a myth to be contested . . . as mere fragments, as deformed symbols, as hallucinating organisms of an 'order' in a state of decay."

Between his *Parere su l'Architettura* of 1765 and his Paestum etchings, published only after his death in 1778, Piranesi abandoned architectural verisimilitude and gave his imagination full rein. In one publication after another, culminating in his extravagantly eclectic work on interior ornamentation of 1769, he indulged in hallucinatory manipulations of historicist form. Indifferent to Winckelmann's pro-Hellenic distinction between innate beauty and gratuitous ornament, his delirious inventions exercised an irresistible attraction on his contemporaries, and the Adam brothers' Graeco-Roman interiors were greatly indebted to his flights of imagination.

In England, where the Rococo had never been fully accepted, the impulse to redeem the excess of the Baroque found its first expression in the Palladianism initiated by the Earl of Burlington, though something of a similar purgative spirit may be detected in the last works of Nicholas Hawksmoor at Castle Howard. By the end of the 1750s, however, the British were already assiduously pursuing instruction in Rome itself where, between 1750 and 1765, the major Neo-Classical proponents could be found in residence, from the pro-Roman and pro-Etruscan Piranesi to the pro-Greek Winckelmann and Le Roy, whose influence had yet to take effect. Among the British contingent were James Stuart, who was to employ the Greek Doric order as early as 1758, and the younger George Dance, who soon after his return to London in 1765 designed Newgate Gaol, a superficially Piranesian structure whose rigorous organization may well have owed something to the Neo-Palladian proportional theories of Robert Morris. The final development of British Neo-Classicism came first in the work of Dance's pupil John Soane, who synthesized to a remarkable degree various influences drawn from Piranesi, Adam, Dance and even from the English Baroque. The Greek Revival cause was then popularized by Thomas Hope, whose *Household Furniture and Interior Decoration* (1807) made available a British version of the Napoleonic "Style Empire," then in the process of being created by Percier and Fontaine.

Nothing could have been further from the British experience than the theoretical development that attended the emergence of Neo-Classicism in France. An early awareness of cultural relativity in the late 17th century prompted Claude Perrault to question the validity of the Vitruvian proportions as these had been received and refined through Classical theory. Instead, he elaborated his thesis of *positive* beauty and *arbitrary* beauty, giving to the former the normative role of standardization and perfection and to the latter such expressive function as may be required by a particular circumstance or character.

This challenge to Vitruvian orthodoxy was codified by the Abbé de Cordemoy in his *Nouveau Traité de toute l'architecture* (1706), where he replaced the Vitruvian attributes of architecture, namely *utilitas, firmitas* and *venustas* (utility, solidity and beauty) by his own trinity of *ordonnance, distribution* and *bienséance*. While the first two of his categories concerned the correct proportioning of the Classical orders and their appropriate disposition, the third introduced the notion of fitness, with which Cordemoy warned against the inappropriate application of Classical or honorific elements to utilitarian or commercial structures. Thus, in addition to being critical of the Baroque, which was the last rhetorical, public manner of the Ancien Régime, Cordemoy's *Traité* anticipated Jacques-François Blondel's preoccupation with appropriate formal expression and with a differentiated physiognomy to accord with the varying social *character* of different building *types*. The age was already having to confront the articulation of a much more complex society.

Apart from insisting on the judicious application of Classical elements, Cordemoy was concerned with their geometrical purity, in reaction against such Baroque devices as irregular columniation, broken pediments and twisted columns. Ornamentation too had to be subject to propriety, and Cordemoy, anticipating Adolf Loos's *Ornament und Verbrechen* (*Ornament and Crime*) by two hundred years, argued that many buildings required no ornament at all. His preference was for astylar masonry and orthogonal structures. For him, the free-standing column was the essence of a pure architecture such as had been made manifest in the Gothic cathedral and the Greek temple.

The Abbé Laugier in his *Essai sur l'architecture* (1753) reinterpreted Cordemoy, to posit a universal "natural" architecture, the primordial "primitive hut" consisting of four tree trunks supporting a rustic pitched roof. After Cordemoy, he asserted this primal form as the basis for a sort of classicized Gothic structure in which there would be neither arches nor pilasters nor pedestals nor any other kind of formal articulation, and where the interstices between the columns would be as fully glazed as possible.

Such a "translucent" structure was realized in Jacques-German Soufflot's church of Ste-Geneviève in Paris, begun in 1755. Soufflot, who in 1750 had been one of the first architects to visit the Doric temples at Paestum, was determined to recreate the lightness, the spaciousness and the proportion of Gothic architecture in Classical (not to say Roman) terms. To this end he adopted a Greek cross plan, the nave and aisles being formed by a system of flat domes and semicircular arches supported on a continuous internal peristyle.

The task of integrating the theory of Cordemoy and the magnum opus of Soufflot into the French academic tradition fell to J. F. Blondel who, after opening his architectural school in Rue de la Harpe in 1743, became the master of that so-called "visionary" generation of architects that included Etienne-Louis Boullée, Jacques Gondoin, Pierre Patte, Marie-Joseph Peyre, Jean-Baptiste Rondelet and, probably the most visionary of all, Claude-Nicolas Ledoux. Blondel set out his main precepts, concerning *composition, type* and *character*, in his *Cours d'architecture*, published from 1750 to 1770. His ideal church design, published in the second volume of his *Cours*, was related to Ste-Geneviève and prominently displayed a representational front, while articulating each internal element as part of a continuous spatial system whose infinite vistas evoked a sense of the Sublime. This church project hints at the simplicity and grandeur that were to inform the work of many of his pupils, most notably Boullée, who after 1772 devoted his life to the projection of buildings so vast as to preclude their realization.

In addition to representing the social character of his creations in accordance with the teachings of Blondel, Boullée evoked the sublime emotions of terror and tranquillity through the grandeur of his conceptions. Influenced by Le Camus de Mézières' *Génie de l'architecture, ou l'analogie de cet art avec nos sensations* (1780), he began to develop his *genre terrible*, in which the immensity of the vista and the unadorned geometrical purity of monumental form are combined in such a way as to promote exhilaration and anxiety. More than any other Enlightenment architect, Boullée was obsessed with the capacity of light to evoke the presence of the divine. This intention is evident in the sunlit diaphanous haze that illuminates the interior of his "Métropole," modelled partly on Ste-Geneviève. A similar light is portrayed in the vast masonry sphere of his projected cenotaph for Isaac Newton, where by night a fire was suspended to represent the sun, while by day it was extinguished to reveal the illusion of the firmament produced by the daylight shining through the sphere's perforated walls.

While Boullée's political sentiments were solidly republican, he remained obsessed with imagining the monuments of some omnipotent state dedicated to the wor-

FIGURE 2–2   Boullée, project for a cenotaph for Isaac Newton, *c.* 1785. Section by "night." [Photo used by permission of Bibliothèque Nationale de France]

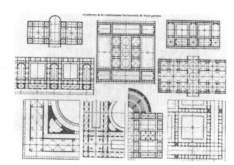

**FIGURE 2–3**  Durand, possible combinations and permutations of plan forms, from his *Précis*, 1802–09.

ship of the Supreme Being. Unlike Ledoux, he was unimpressed by the rural decentralized utopias of Morelly or Jean-Jacques Rousseau. Despite this, his influence in post-Revolutionary Europe was considerable, primarily through the activity of his pupil Jean-Nicolas-Louis Durand, who reduced his extravagant ideas to a normative and economic building typology, set out in the *Précis des leçons données à l'Ecole Polytechnique* (1802–09).

After fifteen years of millennial disarray the Napoleonic era required useful structures of appropriate grandeur and authority, on the condition that they be achieved as cheaply as possible. Durand, the first tutor in architecture at the Ecole Polytechnique, sought to establish a universal building methodology, an architectural counterpart to the Napoleonic Code, by which economic and appropriate structures could be created through the modular permutation of fixed plan types and alternative elevations. Thus Boullée's obsession with vast Platonic volumes was exploited as a means to achieve an appropriate character at a reasonable cost. Durand's criticism of Ste-Geneviève, for example, with its 206 columns and 612 metres (2008 feet) of wall, involved him in making a counter-proposal for a circular temple of comparable area that would require only 112 columns and 248 metres (814 feet) of wall—a considerable economy, with which, according to him, one would have achieved a far more impressive aura.

Ledoux, after his career had been terminated by the Revolution, returned during his imprisonment to develop the scheme of the salt works that he had built for Louis XVI at Arc-et-Senans in 1773–79. He expanded the semicircular form of this complex into the representational core of his ideal city of Chaux, published in 1804 under the title *L'Architecture considérée sous le rapport de l'art, des moeurs et de la législation*. The semicircular salt works itself (which he developed into the oval centre of his city) may be seen as one of the first essays in industrial architecture, inasmuch as it consciously integrated productive units with workers' housing. Each element in this physiocratic complex was rendered according to its character. Thus the salt evaporation sheds on the axis were high-roofed like agricultural buildings and finished in smooth ashlar, with rusticated dressings, while the director's house in the centre was low-roofed and pedimented, rusticated throughout and embellished with Classical porticos. Here and there the walls of the salt sheds and the workers' houses were relieved by grotesque "spouts" of petrified water, which not only symbolized the saline solution on which the enterprise was based but also suggested that the productive system and the labour force had an equally processal status.

**FIGURE 2–4**   Ledoux, ideal city of Chaux, 1804.

In fictitiously developing this limited typology to include all the institutions of his ideal city, Ledoux extended the idea of an architectural "physiognomy" to symbolize the social intention of his otherwise abstract forms. The meanings are established either by conventional symbols, such as the fasces evoking justice and unity on the courthouse, the so-called *Pacifère,* or by isomorphism, as in the case of the *Oikema,* planned in the shape of a penis. This last structure was dedicated to libertinage, whose curious social purpose was to induce virtue through sexual satiety.

A whole world separates Durand's rational permutation of received Classical elements from Ledoux's arbitrary but purgative reconstitution of fragmented Classical parts demonstrated in the toll-gates that he designed for Paris between 1785 and 1789. These *barrières* were just as disconnected from the culture of their time as the idealized institutions of Chaux. With their gradual demolition after 1789 they suffered the same fate as the abstract and unpopular customs boundary that they were intended to administer, the Enceinte des Fermiers Généraux, of which it was said, "Le mur murant Paris rend Paris murmurant."

After the Revolution, the evolution of Neo-Classicism was largely inseparable from the need to accommodate the new institutions of bourgeois society and to represent the emergence of the new republican state. That these forces were initially resolved in the compromise of constitutional monarchy hardly detracted from the role that Neo-Classicism played in the formation of the bourgeois imperialist style. The creation of Napoleon's "Style Empire" in Paris and Frederick II's Francophile "Kulturnation" in Berlin are but separate manifestations of the same cultural tendency. The former made an eclectic use of antique motifs, be they Roman, Greek or

Egyptian, to create the instant heritage of a republican dynasty—a style that revealed itself significantly in the theatrical tented interiors of the Napoleonic campaigns and in the solid Roman embellishments of the capital city, such as Percier and Fontaine's Rue de Rivoli and Arc du Carrousel and Gondoin's Place Vendôme column dedicated to the Grande Armée. In Germany the tendency was first manifested in Carl Gotthard Langhans's Brandenburg Gate, built as the western entry to Berlin in 1793, and in Friedrich Gilly's design for a monument to Frederick the Great, of 1797. Ledoux's primary forms inspired Gilly to emulate the severity of the Doric, thereby echoing the "archaic" power of the Sturm und Drang movement in German literature. Like his contemporary Friedrich Weinbrenner, he projected a spartan Ur-civilization of high moral value, with which to celebrate the myth of the ideal Prussian state. His remarkable monument would have taken the form of an artificial acropolis on the Leipzigerplatz. This temenos would have been entered from Potsdam through a squat triumphal arch capped by a quadriga.

Gilly's colleague and successor, the Prussian architect Karl Friedrich Schinkel, acquired his early enthusiasm for Gothic not from Berlin or Paris, but from his own first-hand experience of Italian cathedrals. Yet after the defeat of Napoleon in 1815, this Romantic taste was largely eclipsed by the need to find an appropriate expression for the triumph of Prussian nationalism. The combination of political idealism and military prowess seems to have demanded a return to the Classic. In any event this was the style that linked Schinkel not only to Gilly but also to Durand, in the creation of his masterpieces in Berlin: his Neue Wache of 1816, his Schauspielhaus of 1821 and his Altes Museum of 1830. While both the guardhouse and the theatre show characteristic features of Schinkel's mature style, the massive corners of the one and the mullioned wings of the other, the influence of Durand is most clearly revealed in the museum, which is a prototypical museum plan taken from the *Précis* and split in half—a transformation in which the central rotunda, peristyle and courtyards are retained and the side wings eliminated . . . While the wide entry steps, the peristyle and the eagles and Dioscuri on the roof symbolized the cultural aspirations of the Prussian state, Schinkel departed from the typological and representational methods of Durand to create a spatial articulation of extraordinary delicacy and power, as the wide peristyle gives way to a narrow portico containing a symmetrical entry stair and its mezzanine (an arrangement which would be remembered by Mies van der Rohe).

The main line of Blondel's Neo-Classicism was continued in the mid-19th century in the career of Henri Labrouste, who had studied at the Ecole des Beaux-Arts (the institution that succeeded the Académie Royale d'Architecture after the Revolution) with A.-L.-T. Vaudoyer, who had been a pupil of Peyre. After winning the Prix de

FIGURE 2–5   Schinkel, Altes Museum, Berlin, 1828–30.
[Photo used by permission of Kunstbibliothale Staatliche Museen Zu Berlin Preudischer Kulturbesitz]

Rome in 1824 Labrouste spent the next five years at the French Academy, devoting much of his time in Italy to a study of the Greek temples at Paestum. Inspired by the work of Jakob-Ignaz Hittorff, Labrouste was among the first to argue that such structures had originally been brightly coloured. This, and his insistence on the primacy of structure and on the derivation of all ornament from construction, brought him into conflict with the authorities after the opening of his own atelier in 1830.

In 1840 Labrouste was named architect of the Bibliothèque Ste-Geneviève in Paris which had been created to house part of the library impounded by the French state in 1789. Based apparently on Boullée's project for a library in the Palais Mazarin, of 1785, Labrouste's design consists of a perimeter wall of books enclosing a rectilinear space and supporting an iron-framed, barrel-vaulted roof which is divided into two halves and further supported in the centre of the space by a line of iron columns.

Such Structural Rationalism was further refined in the main reading room and book stack that Labrouste built for the Bibliothèque Nationale in 1860–68. This complex, inserted into the courtyard of the Palais Mazarin, consists of a reading room covered by an iron and glass roof carried on sixteen cast-iron columns and a multi-storey wrought- and cast-iron book stack. Dispensing with the last trace of historicism, Labrouste designed the latter as a top-lit cage, in which light filters down through iron landings from the roof to the lowest floor. Although this solution was derived from Sydney Smirke's cast-iron reading room and stack built in the courtyard of Robert Smirke's Neo-Classical British Museum in 1854, the precise form of its execution implied a new aesthetic whose potential was not to be realized until the Constructivist work of the 20th century.

The middle of the 19th century saw the Neo-Classical heritage divided between two closely related lines of development: the Structural Classicism of Labrouste and the Romantic Classicism of Schinkel. Both "schools" were confronted by the

FIGURE 2–6   Labrouste, book stack of the Bibliothèque Nationale, Paris, 1860–68. [Photo used by permission of Bibliothèque Nationale de France]

same 19th-century proliferation of new institutions and had to respond equally to the task of creating new building types. They differed largely in the manner in which they achieved these representative qualities: the Structural Classicists tended to emphasize structure—the line of Cordemoy, Laugier and Soufflot; while the Romantic Classicists tended to stress the physiognomic character of the form itself—the line of Ledoux, Boullée and Gilly. Where one "school" seems to have concentrated on such types as prisons, hospitals and railway stations, in the work of men like E.-J. Gilbert and F. A. Duquesney (designer of the Gare de l'Est, Paris, in 1852), the other addressed itself more to representational structures, such as the university museum and library of C. R. Cockerell in England or the more grandiose monuments erected by Leo von Klenze in Germany—above all the latter's highly Romantic Walhalla, completed at Regensburg in 1842.

In terms of theory, Structural Classicism began with Rondelet's *Traité de l'art de bâtir* (1802) and culminated at the end of the century in the writing of the engineer Auguste Choisy, particularly his *Histoire de l'architecture* (1899). For Choisy the essence of architecture is construction, and all stylistic transformations are merely the logical consequence of technical development: "To parade your Art Nouveau is to ignore the whole teaching of history. Not so did the great styles of the past come into being. It was in the suggestion of construction that the architect of the great artistic ages found his truest inspiration." Choisy illustrated the structural determination of his *Histoire* with axonometric projections which revealed the essence of a type of form in a single graphic image, comprising plan, section and elevation. As Reyner Banham has observed, these objective illustrations reduce the architecture that they represent to pure abstraction, and it was this, plus the amount of the information they synthesized, that endeared them to the pioneers of the Modern Movement after the turn of the century.

The emphasis that Choisy's history placed on Greek and Gothic architecture was a late 19th-century rationalization of that Graeco-Gothic ideal which had first been formulated over a century before by Cordemoy. This 18th-century projection of Gothic structure into Classical syntax found its parallel in Choisy's characterization of the Doric as wooden structure transposed into masonry. Just such a transposition was to be practised by Choisy's disciple, Auguste Perret, who insisted on detailing his reinforced-concrete structures after the manner of traditional wood framing.

A Structural Rationalist to the core, Choisy was nonetheless capable of responding to the Romantic sensibility when he wrote of the Acropolis: "The Greeks never visualized a building without the site that framed it and the other buildings that surrounded it . . . each architectural motif, on its own, is symmetrical, but every group is treated like a landscape where the masses alone balance out."

Such a Picturesque notion of partially symmetrical balance would have been as foreign to the teaching of the Beaux-Arts as it was to the polytechnical approach of Durand. Certainly it would have had a limited appeal for Julien Guadet, who sought, in his lecture course *Eléments et théorie de l'architecture* (1902), to establish a normative approach to the composition of structures from technically up-to-date elements, arranged as far as possible according to the tradition of axial composition. It was through Guadet's teaching at the Beaux-Arts, and his influence on his pupils Auguste Perret and Tony Garnier, that the principles of Classical "Elementarist" composition were handed down to the pioneer architects of the 20th century.

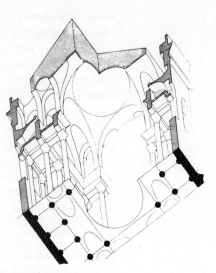

**FIGURE 2–7** Choisy, axonometric projection of part of the Panthéon, Paris (see Figure 2-1), from his *Histoire de l'architecture*, 1899.

## TERRITORIAL TRANSFORMATIONS: URBAN DEVELOPMENTS 1800–1909

> [With] the development of increasingly abstract means of communication, the continuity of rooted communication is replaced by new systems which continue to perfect themselves throughout the 19th century, allowing the population greater mobility and providing information that is more precisely synchronized with the accelerating rhythm of history. Railway, daily press and telegraph will gradually supplant *space* in its previous informative role.
>
> Francoise Choay
> *The Modern City: Planning in the 19th Century,* 1969

The finite city, as it had come into being in Europe over the previous five hundred years, was totally transformed in the space of a century by the interaction of a number of unprecedented technical and socio-economic forces, many of which first emerged in England during the second half of the 18th century. Prominent among them from a technical point of view must be counted such innovations as Abraham Darby's mass-production of cast-iron rails, from 1767, and Jethro Tull's seed-drill cultivation of crops in rows, generally adopted after 1731. Where Darby's invention led to Henry Cort's development of the puddling process for the simplified conversion of cast- to wrought-iron in 1784, Tull's drill was essential to the perfection of Charles Townshend's four-crop rotational system—the principle of "high farming" that became general towards the end of the century.

Such productive innovations had multiple repercussions. In the case of metallurgy, English iron production increased forty-fold between 1750 and 1850 (rising to two million tons a year by 1850); in the case of agriculture, after the Enclosures Act of 1771 inefficient husbandry was replaced by the four-crop system. Where the one was boosted by the Napoleonic Wars, the other was motivated by the need to feed a rapidly growing industrial population.

At the same time the cottage-weaving industry, which had helped to sustain the agrarian economy of the first half of the 18th century, was rapidly changed, first by

James Hargreaves's spinning jenny of 1764, which greatly increased the individual's spinning capacity, and then by Edmund Cartwright's steam-powered loom, first used for factory production in 1784. This last event not only established textile production as a large-scale industry, but also led immediately to the invention of the multi-storey fireproof mill. Thus traditional textile manufacturing was forced to abandon its predominantly rural base and to concentrate both labour and plant, first next to water courses and then, with the advent of steam power, close to coal deposits. With 24,000 power looms in production by 1820, the English mill town was already an established fact.

This process of uprooting—*enracinement,* as Simone Weil has called it—was further accelerated by the use of steam traction for transport. Richard Trevithick first demonstrated the locomotive on cast-iron rails in 1804. The opening of the first public rail service between Stockton and Darlington in 1825 was followed by the rapid development of a completely new infrastructure, Britain having some 10,000 miles of track in place by 1860. The advent of long-distance steam navigation after 1865 greatly increased European migration to the Americas, Africa and Australia. While this migration brought the populations needed to expand the economy of colonial territories and to fill the growing grid-plan cities of the New World, the military, political and economic obsolescence of the traditional European walled city led, after the liberal-national revolutions of 1848, to the wholesale demolition of ramparts and to the extension of the formerly finite city into its already burgeoning suburbs.

These general developments, accompanied by a sudden drop in mortality due to improved standards in nutrition and medical techniques, gave rise to unprecedented urban concentrations, first in England and then, at differing rates of growth, throughout the developing world. Manchester's population grew eight-fold in the course of the century, from 75,000 in 1801 to 600,000 by 1901, as compared to London's six-fold increase over the same period, from around 1 million in 1801 to 6½ million by the turn of the century. Paris grew at a comparable rate but had a more modest beginning, expanding from 500,000 in 1801 to 3 million by 1901. Even these six- to eight-fold increases are modest compared with New York's growth over the same period. New York was first laid out as a gridded city in 1811, in accordance with the Commissioners' Plan of that year, and grew from its 1801 population of 33,000 to 500,000 by 1850 and 3½ million by 1901. Chicago grew at an even more astronomical rate, rising from 300 people at the time of Thompson's grid of 1833 to around 30,000 (of whom something under half had been born in the States) by 1850, and going on to become a city of 2 million by the turn of the century.

The accommodation of such volatile growth led to the transformation of old neighbourhoods into slums, and also to jerry-built new houses and tenements whose main purpose, given the general lack of municipal transport, was to provide as cheaply as possible the maximum amount of rudimentary shelter within walking distance of the centres of production. Naturally such congested developments had inadequate standards of light, ventilation and open space and poor sanitary facilities, such as communal outside lavatories, wash-houses and refuse storage. With primitive drainage and inadequate maintenance, this pattern could lead to the piling up of excrement and garbage and to flooding, and these conditions naturally provoked a high incidence of disease—first tuberculosis and then, more alarmingly for the authorities, a number of outbreaks of cholera in both England and Continental Europe in the 1830s and 1840s.

These epidemics had the effect of precipitating health reform and of bringing about some of the earliest legislation governing the construction and maintenance of dense conurbations. In 1833 the London authorities instructed the Poor Law Commission, headed by Edwin Chadwick, to make enquiries about the origins of a cholera outbreak in Whitechapel. This led to Chadwick's report, *An Inquiry into the Sanitary Conditions of the Labouring Population in Great Britain* (1842), to the Royal Commission on the State of Large Towns and Populous Districts of 1844 and, eventually, to the Public Health Act of 1848. This Act, in addition to others, made local authorities legally responsible for sewerage, refuse collection, water supply, roads, the inspection of slaughterhouses and the burial of the dead. Similar provisions were to occupy Haussmann during the rebuilding of Paris between 1853 and 1870.

The result of this legislation in England was to make society vaguely aware of the need to upgrade working-class housing; but as to the models and means by which this should be achieved there was little initial agreement. Nonetheless, the Chadwick-inspired Society for Improving the Conditions of the Labouring Classes sponsored the erection of the first working-class flats in London in 1844 to the design of the architect Henry Roberts, and followed this resolute beginning with its Streatham Street flats of 1848–50 and a prototypical two-storey worker's cottage containing four flats, again to the design of Roberts, for the Great Exhibition of 1851. This generic model for the stacking of apartments in pairs around a common staircase was to influence the planning of working-class housing for the rest of the century.

The American-backed philanthropic Peabody Trust and various English benevolent societies and local authorities attempted, after 1864, to upgrade the quality of working-class housing, but little of significance was achieved until the slum clearance Acts of 1868 and 1875 and the Housing of the Working Classes Act of 1890, under which local authorities were required to provide public housing. In 1893, when the London County Council (established in 1890) began to build workers' flats under the auspices of this Act, its Architect's Department made a remarkable effort to deinstitutionalize the image of such housing by adapting the Arts and Crafts domestic style . . . to the realization of six-storey blocks of flats. Typical of this development is the Millbank Estate, begun in 1897.

Throughout the 19th century the effort of industry to take care of its own assumed many forms, from the "model" mill, railway and factory towns to projected utopian communities intended as prototypes for some future enlightened state. Among those who manifested an early concern for integrated industrial settlements one must acknowledge Robert Owen, whose New Lanark in Scotland (1815) was designed as a pioneering institution of the co-operative movement, and Sir Titus Salt, whose Saltaire, near Bradford in Yorkshire (founded in 1850), was a paternalistic mill town, complete with traditional urban institutions such as a church, an infirmary, a secondary school, public baths, almshouses and a park.

Neither of these realizations could match in scope and liberating potential the radical vision of Charles Fourier's "new industrial world," as formulated in his essay of that title (*Le Nouveau Monde industriel*) published in 1829. Fourier's non-repressive society was to depend on the establishment of ideal communities or "phalanxes," housed in *phalanstères,* where men were to be related in accordance with Fourier's psychological principle of "passional attraction." Since the phalanstery was projected as being in open country, its economy was to be predominantly agricultural, supple-

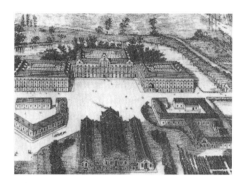

**FIGURE 2–8**  Godin, Familistère, Guise, 1859–70. [Used by permission of Bibliothèque Nationale de France]

mented by light manufacturing. In his earliest writings Fourier outlined the physical attributes of his communal settlement; it was modelled on the layout of Versailles, its central wing being given over to public functions (dining hall, library, winter-garden, etc.), while its side wings were devoted to the workshops and the *caravanseray*. In his *Traité de l'association domestique agricole* (1822) Fourier wrote of the phalanstery as being a miniature town whose streets would have the advantage of not being exposed to the weather. He saw it as a structure whose grandeur, if generally adopted, would replace the petit-bourgeois squalor of the small individual free-standing houses that were, by then, already filling the outer interstices of towns.

Fourier's disciple Victor Considérant, writing in 1838, mixed the metaphor of Versailles with that of the steamship and questioned whether it was "easier to house 1800 men right in the middle of the ocean, six hundred leagues from every shore, . . . than to house in a unitary construction some 1800 good peasants in the heart of Champagne or firmly on the soil of Beauce?" This particular conflation of commune and ship was to be returned to by Le Corbusier, over a century later, in his self-contained commune or Unité d'Habitation, realized with Fourierist overtones at Marseilles in 1952 . . .

The enduring importance of Fourier lies in his radical criticism of industrialized production and social organization, for, despite numerous attempts to create phalansteries in both Europe and America, his new industrial world was fated to remain a dream. Its closest realization was the Familistère, built by the industrialist J.-P. Godin next to his factory at Guise in 1859–70. This complex comprised three residential blocks, a crèche, a kindergarten, a theatre, schools, public baths and a laundry. Each residential block enclosed a top-lit central courtyard which took the place of the elevated corridor streets of the phalanstery. In his book *Solutions sociales* (1870) Godin absorbed the more radical aspects of Fourierism by showing how the system could be adapted to co-operative family living without resorting to the eccentric theories of "passional attraction."

Aside from accommodating the labouring masses, London's 18th-century matrix of streets and squares was extended throughout the 19th century to meet the residetial requirements of a growing urban middle class. No longer satisfied, however, with the scale and texture of the occasional green square—delimited on all sides by streets and continuous terraces—the English Park Movement, founded by the gardener Humphrey Repton, attempted to project the "landscaped country estate" into the city. Repton himself succeeded in demonstrating this, in collaboration with the architect John

Nash, in their layout of Regent's Park in London (1812–27). After the victory over Napoleon in 1815, the proposed development enclosing the park was augmented, under royal patronage, by a continuous "display" façade, penetrating into the existing urban fabric and extending as a more or less uninterrupted ribbon of terraced accommodation from the aristocratic vistas of Regent's Park in the north to the palatial urbanity of St James's Park and Carlton House Terrace in the south.

The squirearchical concept of the Neo-Classical country house set in an irregular landscape (an image derived from the Picturesque work of Capability Brown and Uvedale Price) was thus translated by Nash to the provision of terraced housing on the perimeter of an urban park. This model was first systematically adapted to general use by Sir Joseph Paxton, at Birkenhead Park, built outside Liverpool in 1844. Frederick Law Olmsted's Central Park in New York, inaugurated in 1857, was directly influenced by Paxton's example, even down to its separation of carriage traffic from pedestrians. The concept received its final elaboration in the Parisian parks created by J. C. A. Alphand, where the circulation system totally dictated the manner in which the park was to be used. With Alphand, the park becomes a civilizing influence for the newly urbanized masses.

The irregular lake that Nash created in St James's Park in 1828 out of the rectangular basin that the Mollet brothers had made in 1662 may be taken to symbolize the victory of the English Picturesque over the French Cartesian conception of landscape dating from the 17th century. The French, who had hitherto regarded greenery as another order of architecture and had rendered their avenues as colonnades of trees, were to find the romantic appeal of Repton's irregular landscape irresistible. After the Revolution they remodelled their aristocratic parks into Picturesque sequences.

Yet, for all the power of the Picturesque, the French impulse towards rationality remained, first in the *percements* (wholesale demolition in a straight line to create an entirely new street) of the Artists' Plan for Paris, drawn up in 1793 by a committee of revolutionary artists under the leadership of the painter Jacques-Louis David; and then in Napoleon's arcaded Rue de Rivoli, built after 1806 to the designs of Percier and Fontaine. Where the Rue de Rivoli was to serve as the architectural model not only for Nash's Regent Street but also for the scenographic "façade" of Second Empire Paris, the Artists' Plan demonstrated the instrumental strategy of the *allée,* which was to become the prime tool for the rebuilding of Paris under Napoleon III.

Napoleon III and Baron Georges Haussmann left their indelible mark not only on Paris but also on a number of major cities in France and Central Europe which underwent Haussmann-like regularizations throughout the second half of the century. Their influence is even present in Daniel Burnham's 1909 plan for the gridded city of Chicago, of which Burnham wrote: "The task which Haussmann accomplished for Paris corresponds with the work which must be done for Chicago in order to overcome the intolerable conditions which invariably arise from a rapid growth in population."

In 1853 Haussmann, as the newly appointed Prefect for the Seine, saw these conditions in Paris as being polluted water supply, lack of an adequate sewer system, insufficient open space for both cemeteries and parks, large areas of squalid housing and last, but by no means least, congested circulation. Of these, the first two were undoubtedly the most critical for the everyday welfare of the population. As a consequence of drawing the bulk of its water from the Seine, which also served as the main collector sewer, Paris had suffered two serious outbreaks of cholera in the first half of the century. At the same time, the existing street system was no longer adequate for the administrative centre of an expanding capitalist economy. Under the brief autocracy of Napoleon III, Haussmann's radical solution to the physical aspect of this complex problem was *percement*. His broad purpose was, as Choay has written, "to give unity and transform into an operative whole the 'huge consumer market, the immense workshop' of the Parisian agglomerate." Although the Artists' Plan of 1793 and before that Pierre Patte's plan of 1765 had clearly anticipated the axial and focal structure of Haussmann's Paris, there is, as Choay points out, a discernible shift in the actual location of the axes, from a city organized around traditional *quartiers,* as in the plan made under David, to a metropolis united by the "fever of capitalism."

Saint-Simonian economists and technocrats, mostly from the Ecole Polytechnique, influenced Napoleon III's views as to the economic means and the systematic ends to be adopted in the rebuilding of Paris, emphasizing the importance of rapid and efficient systems of communication. Haussmann converted Paris into a regional metropolis, cutting through its existing fabric with streets whose purpose was to link opposing cardinal points and districts, across the traditional barrier of the Seine. He gave top priority to the creation of more substantial north–south and east–west axes, to the building of the Boulevard de Sébastopol and the easterly extension of the Rue de Rivoli. This basic cross, which served the main railway termini to the north and south, was encircled by a "ring" boulevard which in turn was tied into Haussmann's major traffic distributor, his Etoile complex built around Chalgrin's Arc de Triomphe.

During Haussmann's tenure the city of Paris built some 137 kilometres (85 miles) of new boulevards, which were considerably wider, more thickly lined with trees and better lit than the 536 kilometres (333 miles) of old thoroughfare they replaced. With all this came standard residential plan types and regularized façades, and equally standard systems of street furniture—the *pissoirs,* benches, shelters, kiosks, clocks, lampposts, signs, etc., designed by Haussmann's engineers Eugène Belgrand and Alphand. This entire system was "ventilated" whenever possible by large areas of public open space, such as the Bois de Boulogne and the Bois de Vincennes. In addition, new cemeteries and many small parks, such as the Parc des Buttes Chaumont and the Parc Monceau, were either created or upgraded within the extended boundaries of the city. Above all, there was an adequate sewer system and fresh water piped into the city from the Dhuis valley. In achieving such a comprehensive plan, Haussmann, the apolitical

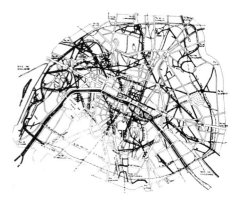

FIGURE 2–9   The regularization of Paris; streets cut by Haussmann are shown in black.

administrator par excellence, refused to accept the political logic of the régime he served. He was finally broken by an ambivalent bourgeoisie, who throughout his tenure supported his "profitable improvements" while at the same time defending their proprietorial rights against his intervention.

Prior to the collapse of the Second Empire, the principle of "regularization" was already being practised outside Paris, particularly in Vienna, where the replacement of demolished fortifications by a display boulevard was taken to its logical extreme in the ostentatious Ringstrasse, built around the old centre between 1858 and 1914. The free-standing monuments of this "open" city expansion, structured around a cranked thoroughfare of enormous width, provoked the critical reaction of the architect Camillo Sitte, who in his influential *Der Stadtebau nach seinen künsterlerischen Grundsatzen* (*City Planning According to Artistic Principles*) of 1889 argued for the enclosure of the major Ringstrasse monuments by buildings and arcades. Sitte's remedial concern cannot be better characterized than in his critical comparison of the traffic-ridden "open" city of the late 19th century with the tranquillity of the medieval or Renaissance urban core:

> During the Middle Ages and Renaissance public squares were often used for practical purposes . . . they formed an entirety with the buildings which enclosed them. Today they serve at best as places for parking vehicles, and they have no relation to the buildings which dominate them . . . In brief, activity is lacking precisely in those places where, in ancient times, it was most intense, near public structures.

Meanwhile, in Barcelona, the regional implications of urban regularization were being developed by the Spanish engineer Ildefonso Cerdá, the inventor of the term *urbanización*. In 1859 Cerdá projected the expansion of Barcelona as a gridded city, some twenty-two blocks deep, bordered by the sea and intersected by two diagonal avenues. Driven by industry and overseas trade, Barcelona filled out this American-scale grid plan by the end of the century. In his *Teoriá general de la urbanización* ("General Theory of Urbanization") of 1867 Cerdá gave priority to a system of circulation and, in particular, to steam traction. For him transit was, in more ways than one, the point of departure for all scientifically-based urban structures. Léon Jaussely's plan for Barcelona of 1902, derived from Cerdá's, incorporated this emphasis on movement into the form of a proto linear city where the separate zones of accommodation and transportation are organized into bands. His design anticipated in certain respects the Russian linear city proposals of the 1920s.

By 1891 intensive exploitation of the city centre was possible, due to two developments essential to the erection of high-rise buildings: the invention in 1853 of the passenger lift, and the perfection in 1890 of the steel frame. With the introduction of the underground railway (1863), the electric tram (1884) and commuter rail transit (1890), the garden suburb emerged as the "natural" unit for future urban expansion. The complementary relationship of these two American forms of urban development—the high-rise downtown and the low-rise garden suburb—was demonstrated in the building boom that followed the great Chicago fire of 1871.

The process of suburbanization had already started around Chicago with the layout in 1869 of the suburb of Riverside, to the picturesque designs of Olmsted. Based in part on the mid-19th-century garden cemetery and in part on the early East Coast suburb, it was linked to downtown Chicago by both a railway and a bridle path.

With the entry into Chicago in 1882 of the steam-powered cable car, the way became open to further expansion. The immediate beneficiary was Chicago's South Side. Yet suburban growth did not really prosper until the 1890s, when, with the introduction of the electric streetcar, suburban transit greatly extended its range, speed and frequency. This led at the turn of the century to the opening up of Chicago's Oak Park suburb, which was to be the proving ground for the early houses of Frank Lloyd Wright. Between 1893 and 1897 an extensive elevated railway was superimposed on the city, encircling its downtown area. All these forms of transit were essential to Chicago's growth. Most important of all for the city's prosperity was the railway, for it brought the first piece of modern agricultural equipment to the prairie—the essential McCormick mechanical reaper invented in 1831—and collected in return both grain and cattle from the great plains, trans-shipping them to the lakeside silos and stockyards which had begun to be built on Chicago's South Side in 1865. It was the railway that redistributed this abundance from the 1880s on, in Gustavus Swift's refrigerated packing cars, and the corresponding growth in trade greatly augmented the extensive passenger traffic centring on Chicago. Thus the last decade of the century saw radical changes in both the methods of town building and the means of urban access—changes which, in conjunction with the grid plan, were soon to transform the traditional city into an ever-expanding metropolitan region where dispersed homestead and concentrated core are linked by continual commuting.

The puritanical entrepreneur George Pullman, who helped to rebuild Chicago after the fire, had been one of the first to appreciate the expanding market in long-distance passenger travel, bringing out his first Pullman sleeping car in 1865. After the achievement of the transcontinental rail link in 1869, Pullman's Palace Car Company prospered, and in the early 1880s he established his ideal industrial town of Pullman, south of Chicago, a settlement that combined workers' residences with a full range of communal facilities, including a theatre and a library as well as schools, parks and playgrounds, all in close proximity to the Pullman factory. This well-ordered complex went

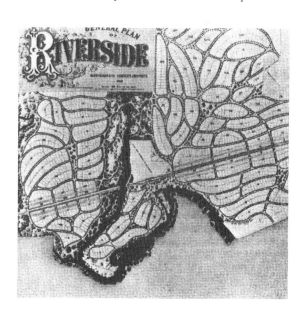

**FIGURE 2–10**   Olmsted, plan of Riverside, Chicago, 1869.

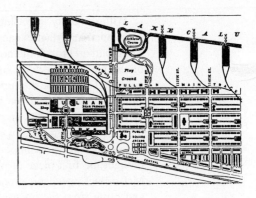

**FIGURE 2–11** S. S. Beman, factory (left) and town of Pullman, Chicago, illustrated in 1885.

far beyond the facilities provided by Godin at Guise some twenty years earlier. It also far exceeds, in its comprehensiveness and clarity, the Picturesque model towns founded in England by the confectioner, George Cadbury at Bournville, Birmingham, in 1879 and by the soap manufacturer W. H. Lever at Port Sunlight, near Liverpool, in 1888. The paternalistic, authoritarian precision of Pullman bears a closer resemblance to Saltaire or to the workers' settlements first established as company policy by Krupp at Essen in the late 1860s.

Rail transit on a much smaller scale, by tram or by train, was to be the main determinant of the two alternative models of the European garden city. One was the axial structure of the Spanish linear garden city, first described by its inventor Arturo Soria y Mata in the early 1880s, and the other was the English concentric garden city, shown as circumnavigated by rail in Ebenezer Howard's *Tomorrow: A Peaceful Path to Real Reform* of 1898. Where Soria y Mata's dynamic interdependent *ciudad lineal* comprised, in his own words of 1882, "A single street of some 500 metres [1640 feet] width and of the length that may be necessary . . . [a city] whose extremities could be Cadiz or St Petersburg or Peking or Brussels," Howard's static yet supposedly independent "Rurisville" was encircled by its rail transit and thereby fixed at an optimum size of between 32,000 and 58,000 people. Where the Spanish model was inherently regional, undetermined and Continental, the English version was self-contained, limited and provincial. Soria y Mata described his "locomotion vertebrae" as incorporating, in addition to transit, the essential services of the 19th-century city—water, gas, electricity and sewerage—compatible with the distribution needs of 19th-century industrial production.

Apart from being an antithesis to the radially planned city, the linear city was a means for building along a triangulated network of pre-existing routes connecting a set of traditional regional centres. While the diagrammatic projection of Howard's city as a satellite town in open country was equally regional, the form of the city itself was less dynamic. On the model of Ruskin's ill-fated St George's Guild of 1871, Howard conceived of his city as an economically self-sufficient mutual aid community, producing little beyond its own needs. The difference between these city models lay finally in the fundamentally different attitudes they adopted to rail transit. Whereas Howard's Rurisville was intended to eliminate the journey to work—the railway being reserved for objects rather than men—the *ciudad lineal* was expressly designed to facilitate communication.

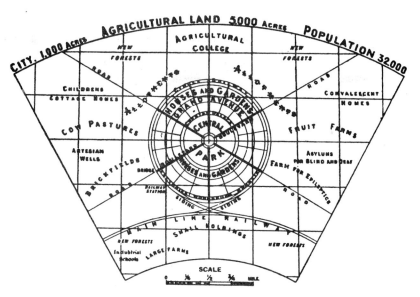

**FIGURE 2–12**  Howard, "Rurisville," schematic garden city from his *Tomorrow,* 1898.

It was, however, the English garden city in its modified form that came to be widely adopted, rather than the linear model sponsored by Soria y Mata's Compañia Madrileña de Urbanización, which only built some 22 kilometres (14 miles) of the 55-kilometre (34-mile) long "necklace" originally projected for the encirclement of Madrid. The failure of this sole example condemned the linear city to a theoretical rather than practical future, and at this level it persisted from the Russian linear cities of the late 1920s to Le Corbusier's ASCORAL planning thesis, first published as *Les Trois Etablissements humains* in 1945.

The radical reinterpretation of Howard's original diagrams, reflected in the layout of the first garden city, Letchworth in Hertfordshire (started in 1903), inaugurated the neo-Sittesque phase in the English garden city movement. That the engineer-planner Raymond Unwin was impressed by Sitte is evident from his highly influential book *Town Planning in Practice,* published in 1909. The preoccupation of Unwin and his colleague Barry Parker with "imaginary irregular towns"—of a kind exemplified for them by such medieval German towns as Nuremberg and Rothenburg-ob-der-Tauber—clearly lies behind their picturesque layout for Hampstead Garden Suburb, designed in 1907. Yet for all his contempt for "bye-law" architecture, Unwin remained as conditioned as any other planner by the constraints imposed by modern standards of hygiene and circulation. Thus, despite the renowned "empirical" success of these pioneering garden cities, the debilitating environment produced subsequently by the English school of town planning stems, at least in part, from Unwin's failure to resolve this implacable dichotomy, that is to reconcile medieval nostalgia with bureaucratic control. The "train-accident" block layouts of the 20th century are among the enduring formal legacies of this failure.

## BIBLIOGRAPHY

Balmer, L.; S. Erni; and U. von Gunten. "Co-operation between Capital and Labour." *Lotus* 12 (September 1976), pp. 59–71.

Banham, R. *Theory and Design in the First Machine Age.* London, 1960. Esp. ch 1–3.

Bartschi, H. P. *Industrialisierung Eisenbahnschlacten und Städtebau.* Stuttgart: ETH/GTA 25, 1983.

Benevolo, L. *The Origins of Modern Town Planning.* Cambridge, MA, 1967.

———. *History of Modern Architecture.* Vol. 1. Cambridge, MA, 1971. Esp. preface and chs. 1–5.

———. *The History of the City.* Cambridge, MA, 1980. Encyclopedic treatment of the History of Western urbanism.

Bentmann, R., and M. Muller. "The Villa as Destination," *9H,* 5, 1983, pp. 104–114 and 7, 1985, pp. 83–104.

Borsi, F. and E. Grodoli. *Vienna 1900.* New York, 1986.

———. *Paris 1900.* New York, 1989.

Boyer, C. *Dreaming of the Rational City: The Myth of American City Planning.* Cambridge, MA, 1983.

Brauman, A. *Le Familistère de Guise ou les équivalents de la richesse.* Brussels, 1976. English text.

Brownlee, D. *Friedrich Weinbrenner, Architect of Karlsruhe.* Philadelphia, 1986.

Buddensieg, T. " 'To Build as One Will . . .' Schinkel's Notions on the Freedom of Building." *Daidalos* 7 (1983), pp. 93–102.

Buder, S. *Pullman: An Experiment in Industrial Order and Community Planning 1880–1930.* New York, 1967.

Burnham, D., and E. H. Bennett. *Plan of Chicago.* Chicago, 1909.

Celik, Z. *Remaking of Istanbul. Portrait of an Ottoman City in the 19th Century.* Seattle, 1986.

Cerdá, I. "A Parliamentary Speech." *AAQ* 9, no. 7 (1977), pp. 23–26.

Choay, F. *L'Urbanisme, utopies et réalités.* Paris, 1965.

———. *The Modern City: Planning in the 19th Century.* New York, 1969. Essential introductory text.

Choisy, A. *Histoire de l'architecture.* Paris, 1899.

Collins, G. "Linear Planning throughout the World." *JSAH* 18 (October 1959), pp. 74–93.

———. "Cities on the Line." *AR,* November 1960, pp. 341–45.

Collins, C. C., and G. R. Collins. *Camillo Sitte and the Birth of Modern City Planning.* London, 1965.

Contal, M. H. "Vittel 1854–1936. Création d'une ville thermale." In *Vittel 1854–1936.* Paris, 1982.

Creese, W. L. *The Legacy of Raymond Unwin.* Cambridge, MA, 1967.

Darley, G. *Villages of Vision.* London, 1976.

Dehio, L. *Friedrich Wilhelm IV von Preussen: Ein Baukunstler der Romantik.* Munich and Berlin, 1961.

Dennis, M. *Court and Garden: From French Hôtel to the City of Modern Architecture.* Cambridge, MA, 1986.

Dickens, A. "The Architect and the Workhouse." *AR,* December 1976, pp. 345–52.

Drexler, A., ed. *The Architecture of the Ecole des Beaux-Arts.* With essays by R. Chafee, N. Levine, and D. van Zanten. New York, 1977.

Duboy, P. *Legueu: Architectural Enigma.* Cambridge, MA, 1986: definitive study with a foreward by Robin Middleton

Etlin, R. A. *The Architecture of Death.* Cambridge, MA, 1984.

Evans, R. "Bentham's Panopticon: An Incident in the Social History of Architecture." *AAQ* 3, no. 2 (April/July 1971), pp. 21–37.

———. "Regulation and Production." *Lotus* 12 (September 1976), pp. 6–14.

Fabos, J.; G. T. Milde; and V. M. Weinmayr. *Frederick Law Olmsted, Sr.* Amherst: University of Massachusetts, 1968.

Fogelson, R. M. *The Fragmented Metropolis: Los Angeles 1850–1930.* Cambridge, MA, 1967.

Forster, K. W. "Sozialer Wohnbau: Geschichte und Gegenwart." *Archithese* 8 (1973), pp. 2–8.

Fortier, B. "Logiques de l'équipement." *AMC* 45 (May 1978), pp. 80–85.

Foster, K. W. "Monument/Memory and the Mortality of Architecture." *Oppositions,* Fall 1982, pp. 2–19.

Fried, A., and P. Sanders. *Socialist Thought.* New York, 1964. Useful for trans. of French utopian socialist texts, Fourier, Saint-Simon, etc.

Gallet, M. *Charles de Wailly 1730–1798.* Paris, 1979.

Geist, J. F., and K. Kurvens. *Das Berliner Miethaus 1740–1862.* Stuttgart, 1982.

Gilmore-Holt, E. *From the Classicists to the Impressionists.* New York, 1966.

Girouard, M. "Neo-Classicism." *AR,* September 1972, pp. 169–80.

Grumbach, A. "The Promenades of Paris." *Oppositions* 8 (Spring 1977).

Guadet, J. *Eléments et théorie de l'architecture.* Paris, 1902.

Hernandez, A. "J. N. L. Durand's Architectural Theory." *Perspecta* 12 (1969).

Herrmann, W. *Laugier and Eighteenth-Century French Theory.* London, 1962.

Hughes, Q. "Neo-Classical Ideas and Practice: St. George's Hall, Liverpool." *AAQ* 5, no. 2 (1973), pp. 37–44.

Jeffery, A. J. "A Future for New Lanark." *AR,* January 1975, pp. 19–28.

Kaufmann, E. *Three Revolutionary Architects, Boullée, Ledoux and Lequeu.* Philadelphia, 1953.

———. *Architecture in the Age of Reason.* New York, 1968.

Kern, S. *The Culture of Time and Space, 1880–1918.* Cambridge, MA, 1983.

Lammert, M. *David Gilly. Ein Baumeister der deutschen Klassizismus.* Berlin, 1981.

Lankheit, K. *Der Tempel der Vernunlt.* Basel and Stuttgart, 1968.

Leatherbarrow, D. "Friedrichstadt—A Symbol of Toleration." *AD,* November–December 1983, pp. 23–31.

Lieb, N., and F. Hufnagl. *Leo von Klenze, Gemälde un Zeichnungen.* Munich, 1979.

Lopez de Aberasturi, A. *Ildefonso Cerdá: la théorie générale de l'urbanisation.* Paris, 1979.

Lowe, D. M. *History of Bourgeois Perception.* Chicago, 1982.

Loyer, F. *Architecture of the Industrial Age.* New York, 1982.

———. *Paris XIXe siècle.* Paris, 1981.

McCormick, T. J. *Charles Louis Clérisseau and the Cremesis of Neoclassicism.* New York, 1990.

Meyer, H., and R. Wade. *Chicago: Growth of a Metropolis.* Chicago, 1969.

Mezzanotte, G. "Edilizia e politica. Appunti sull'edilizia dell'ultimo neoclassicismo." *Casa-bella* 338 (July 1968), pp. 42–53.

Middleton, R. "The Abbé de Cordemoy: The Graeco-Gothic Ideal." *JW&CI,* 1962, p. 1963.

———. "Architects as Engineers: The Iron Reinforcement of Entablatures in 18th-Century France," *AA Files,* 9, Summer, 1985, pp. 54–64.

———. ed. *The Beaux-Arts and Nineteenth Century French Architecture.* Cambridge, MA, 1984.

———. and D. Watkin. *Neoclassical and Nineteenth Century Architecture.* New York, 1987.

Miller, B. "Ildefonso Cerda." *AAQ* 9, no. 7 (1977), pp. 12–22.
Oechslin, W. "Monotonie von Blondel bis Durand." *Werk-Archithese,* January 1977, pp. 29–33.
Oncken, A. *Friedrich Gilly 1772–1800.* Reprint: Berlin, 1981.
Pérez-Gómex, A. *Architecture and the Crisis of Science.* Cambridge, MA, 1983.
Pérouse de Montclos, J. M. *Etienne-Louis Boullée 1728–1799.* Paris, 1969.
Pevsner, N. *Academies of Art, Past and Present.* Cambridge, 1940. Unique study of the evolution of architectural and design education.
———. *Studies in Art, Architecture and Design.* Vol. 1. London, 1968.
———. "Early Working Class Housing." Reprinted in *Studies in Art, Architecture and Design.* Vol. II. London, 1968; reprint, 1982.
Pirrone, G. *Palermo, una capitale.* Milano, 1989.
Posener, J. "Schinkel's Eclecticism and the Architectural." *AD,* November–December 1983, special issue on Berlin, pp. 33–39.
Prey, P. de la Ruffinière du. *John Soane.* Chicago, 1982.
Pundt, H. G. *Schinkel's Berlin.* Cambridge, MA, 1972.
Rella, F. *Il Dispositivo Foucault.* Venice, 1977. With essays by M. Cacciari, M. Tafuri, and G. Teyssot.
Reynolds, J. P. "Thomas Coglan Horsfall and the Town Planning Movement in England." *Town Planning Review* 23 (April 1952), pp. 52–60.
Riemann, G., ed. *Karl Friedrich Schinkel. Reisen nach Italien.* Berlin, 1979.
Rietdorf, A. *Gilly: Wiedergeburt der Architektur.* Berlin, 1943.
Rosenblum, R. *Transformations in Late Eighteenth Century Art.* Princeton, NJ, 1967.
Rowan, A. "Japelli and Cicogarno." *AR,* March 1968, pp. 225–28. On 19th-century Neo-Classical architecture in Padua, etc.
Rykwert, J. *The First Moderns.* Cambridge, MA, 1983.
Saddy, P. "Henri Labrouste: architecte-constructeur." *Les Monuments Historiques de la France,* no. 6 (1975), pp. 10–17.
Service, A. *London 1900.* London and New York, 1979.
Sitte, C. *City Planning According to Artistic Principles.* London, 1965. Trans. of Sitte's text of 1889.
Sola-Morales, M. de. "Towards a Definition: Analysis of Urban Growth in the Nineteenth Century." *Lotus* 19 (June 1978), pp. 28–36.
Starobinski, J. *The Invention of Liberty.* Geneva, 1964. Translated.
———. *The Emblems of Reason.* Charlottesville, VA, 1990.
Stern, R. *New York 1900.* New York, 1984.
Stroud, D. *The Architecture of Sir John Soane.* London, 1961.
———. *George Dance, Architect 1741–1825.* London, 1971.
Sutcliffe, A. *Metropolis 1890–1940.* Chicago, 1984.
———. *Towards the Planned City: Germany, Britain, the United States and France 1780–1914.* New York, 1981.
Szambien, W. *J. N. L. Durand.* Paris, 1984.
Tafuri, M. *Architecture and Utopia: Design and Capitalist Development.* Cambridge, MA, 1976.
Tarn, J. N. "Some Pioneer Suburban Housing Estates," *AR,* May 1968, pp. 367–70.
———. *Working-Class Housing in 19th-Century Britain.* AA Paper no. 7. London, 1971.
Taylor, J. "Charles Fowler: Master of Markets." *AR,* March 1964, pp. 176–82.
Ternois, D., et al. *Soufflot et l'architecture des lumières.* Proceedings of a conference on Soufflot held at the University of Lyons in June 1980. Paris: CNRS, 1980.
Teyssot, G. *Città e utopia nell'illuminismo inglese: George Dance il giovane.* Rome, 1974.

———. "John Soane and the Birth of Style." *Oppositions* 14 (1978), pp. 61–83.

Valdlenaire, A. *Friedrich Weinbrenner.* Karlsruhe, 1919.

Vidler, A. "The Idea of Type: The Transformation of the Academic Ideal 1750–1830." *Oppositions* 8 (Spring 1977). (The same issue contains Quatremère de Quincy's extremely important article on type that appeared in the *Encyclopédie Méthodique* 3, part 2 [Paris, 1825].)

———. *The Writing of the Ways: Architectural Theory in the Late Enlightenment.* Princeton, 1987.

———. *Claude Nicolas Ledoux.* Cambridge, MA, 1990.

Viuari, S. *J.N.L. Durand (1760–1834) Art and Science of Architecture.* New York, 1990.

Watkin, D. *Thomas Hope and the Neo-Classical Idea.* London, 1968.

———. *C. R. Cockerey.* London, 1984.

——— and T. Mellinghoff. *German Architecture and the Classical Ideal.* Milano, 1987.

Wolf, P. "City Structuring and Social Sense in 19th and 20th Century Urbanism." *Perspecta* 13/14 (1971), pp. 220–33.

## 选读 3

### "乌托邦建筑"

**柯林·罗**

柯林·罗在《乌托邦建筑》这篇论文中探讨了几何和数学秩序以何种方式影响我们对建筑形式的看法和观念。罗通过分析文艺复兴时期的乌托邦城市来展现一种理想的物质形态——圆形——是如何事先决定城市的形态而不用顾及人的使用需求、功能需求或是文化差异的。因此，理想的形态成为一种与其相称的人的美德的物质暗喻。罗认为建筑师必须意识到纯粹的几何形式中是包含着权力和专制的，他们也必须准备好允许改变、发展和运动对那些纯粹形式的冲击。从这个角度来讲，他提出了一个建筑师如何从乌托邦式的空想中吸取经验教训的案例，这个案例让建筑师不再迷信物质形态本身能够成为社会健康发展的良方。

## READING 3

# The Architecture of Utopia*

**Colin Rowe**

In the essay "The Architecture of Utopia," Colin Rowe examines the way geometry and mathematical order have influenced our views and concepts of architectural form. Rowe analyzes the Utopian city of the Renaissance to show how an idealized physical form—the circle—can be used to predetermine urban form without regard for human use,

---

*First published in *Granta*, 1959.

*Source: The Mathematics of the Ideal Villa and Other Essays* by Colin Rowe. Copyright © 1976 by The Massachusetts Institute of Technology. Published by MIT Press. Reproduced by permission of MIT Press. pp. 205–23.

functional necessity, or cultural variation. Thus, the ideal form becomes the physical metaphor for the place of corresponding human virtue. Rowe argues that the architect must recognize both the power and the tyranny contained in pure geometric forms, and must be prepared to allow change, growth, and motion to impinge on those forms. In this respect, he presents a case for how architects can absorb the lessons of the Utopian ideal without being trapped into believing that physical form, in itself, can act as a prescription for social health.

> For unto you is paradise opened, the tree of life is planted, the time to come is prepared, plenteousness is made ready, a city is builded, and rest is allowed, yea perfect goodness and wisdom. The root of evil is sealed up from you, weakness and the moth is hid from you, and corruption is fled into hell to be forgotten. Sorrows are passed, and in the end is shewed the treasure of immortality.
>
> 2 Esdras 8: 52–54

Utopia and the image of a city are inseparable. And if, as we might suppose, Utopia does find one of its roots in Jewish millennial thought, the reason is not far to seek. "A city is builded and is set upon a broad field and is full of good things." "Glorious things are spoken of thee, O city of God." "And he carried me away in the spirit to a great and high mountain, and showed me that great city, the holy Jerusalem, descending out of heaven from God." Such biblical references to the felicity of a promised city are too abundant to ignore; but, for all that, the ingredients of the later *Civitas Dei* are as much Platonic as Hebraic and certainly it is a distinctly Platonic deity which presides over the first Utopias to assume specific and architectural form.

Architecture serves practical ends, it is subjected to use; but it is also shaped by ideas and fantasies, and these it can classify, crystallize, and make visible. Very occasionally indeed the architectural crystallization of an idea may even precede the literary one; and thus, if we can consider the Renaissance concept of the ideal city to be Utopian—as surely we may—we find here a notable case of architectural priority. For Thomas More's famous book did not appear until 1516; and by this time the Utopian theme had been well established—in Italian architecture, at least—for almost half a century. Filarete's Sforzinda, the paradigmatic city of so many future Utopian essays, had been projected *c.* 1460; and although Scamozzi's Palma Nova, the first ideal city to achieve concrete form, was not realized until 1593, the themes which it employs were a common architectural currency by 1500.

As illustrated by Sforzinda and Palma Nova (Figures 3-1 and 3-2), the ideal city is usually circular; and, as More said of the towns of his own Utopia, "He that knows one knows them all, they are so alike one another, except where the situation makes some difference."[1] One may withdraw in horror from this calculated elimination of variety; and quite rightly so, for the ideal city, though an entertaining type to inspect, is often a somewhat monotonous environment. But then the ideal city, like Utopia itself, should scarcely be judged in these immediate physical terms. Nor should we evaluate it by either visual or practical criteria, for its rationale is cosmic and metaphysical; and here, of course, lies its peculiar ability to impose itself on the mind.

An architect of the Renaissance, had he felt it necessary to argue about circular, centralized, and radial exercises such as *Palma Nova,* to justify them might possibly have quoted the account in the *Timaeus* where the Demiurge is described as fashioning the universe "in a spherical shape, in which all the radii from the middle are equally distant from the bounding extremities; as this is the most perfect of all figures

**FIGURE 3–1** Plan for an ideal city. From Vincenzo Scamozzi, *L'Idea dell architettura universale* (Venice, 1615). This and the city shown in Figure 3-2 are neither Sforzinda nor Palma Nova. But as Sir Thomas More said of Utopian demonstrations, "They are indeed so alike that he that knows one knows them all." [Used by permission of C. Rowe]

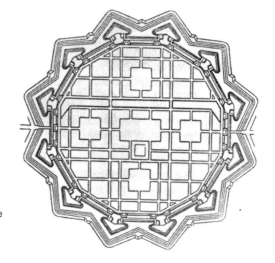

**FIGURE 3–2** Plan for an ideal city. From Buonaiuto Lorini, *Delle Fortificatione Libri Cinque* (Venice, 1592). [Used by permission of C. Rowe]

and the most similar to himself."[2] And thus, the Renaissance architect might have concluded, as an analogy of this divinely created sphere and as an emblem of the artificer who is declared to be immanent within it, the city receives its circular outline.

Now it is by association with the emblematic significance of the sphere that we can understand the persistence of the circular form throughout by far the greater number of architectural Utopias; and so long as our mental inflection is Platonic, as a setting for the headquarters of an ideal state the ideal form of the circle, the mirror of a harmonious cosmic order, follows quite naturally. Thus, on through the seventeenth, eighteenth, and even nineteenth centuries—even as late as 1898 in Ebenezer Howard's prototype for Letchworth Garden City—we still find the circle or conspicuous traces of it.

But there is perhaps another argument involved in this choice of form which is not only Platonic in its bias but also Christian. For possibly the circle is intended both to signify and to assist a redemption of society. It is said to be a natural shape. As obviously, if it is the shape of the universe, it must be. "It is manifest that nature delights principally in round figures, since we find that most things which are generated, made or directed by nature are round."[3] The dictum is from Alberti, one of the first architects to bring a typically Renaissance intellect to bear upon the problems of the city; but the opinion is by no means peculiar to him. Rather, it reflects the characteristic tone of quattrocento humanism; and the argument that the circular city is predominantly "natural" (not at all incongruous with the argument that it is divine), must evidently have introduced a powerful predisposition in its favor. For if a circular city might now be considered to exemplify the laws of nature, how unnatural and therefore in a sense how "fallen" the medieval city must have seemed; while if the medieval city could be thus seen as a counterpart to Babylon, almost as illustration of the working of original sin, the humanistic version of the New Jerusalem could now very well be experienced as the symbol of a regenerated humanity, or a restoration of the injuries of time.

Thus, if nature takes pleasure in the circle (and gives second preference to the square), it is to be expected that these forms should recur elsewhere within the fabric of a Utopian town; and Campanella, for instance, describes, in his *Città del Sole,* a church that is placed in the center of the town and which "is perfectly round, free on all sides, but supported by massive and elegant columns," a church in which "on the altar is nothing but two globes, of which the larger is a celestial, the smaller a terrestrial one, and in the dome are painted the stars of the sky." But, again, the church which is thus specified had been long anticipated by architects. "We cannot doubt that the little temples we make ought to resemble this very great one, which by His immense goodness was perfectly completed with one word of His."[4] This is the opinion of Palladio, and it was also his practice; but Campanella's church and Palladio's advice are already previsioned by the central building in the *Perspective of a Square* (Figure 3-3), variously attributed to Luciano Laurana, Francesco Di Giorgio, or Piero della Francesca and executed *c.* 1470.

Here it might be said that we are shown Utopia infiltrating a preexisting reality; but, nevertheless, this picture can quite well be allowed to serve as a representation of just the city which humanist thought envisaged. "The place where you intend to fix a temple," Alberti recommends, "ought to be noted, famous, and indeed stately, clear from all contagion of secular things";[5] and, following his advice, not only is this particular church round, but the square in which it stands is mathematically proportioned according to Pythagorean principles, while the little palaces about it, grave, regular, and serene, complete the illusion that we have indeed entered into a world where perfect equilibrium is the law.

Fully to understand the revolutionary quality of this space, one should compare it with some medieval square (which, incidentally, one might prefer); and one should compare it then with its innumerable and rather belated progeny. No space of quite this order and regularity came into existence until Michelangelo began his Campidoglio *c.* 1546, and scarcely any space of comparable order was attempted until the seventeenth century. In Paris the Place des Vosges was conceived in 1603. In London the Piazza of Covent Garden dates from the 1630s.[6]

**FIGURE 3–3** Francesco di Giorgio (?), *Perspective of a Square, c.* 1470. [Used by permission of C. Rowe]

And the same may be said of the *Street Scene* attributed to Bramante (Figure 3-4), which provides an urbanistic motif first to be realized in the theater before achieving a more permanent form in building. For it was not until the pontificate of Sixtus V, determined as he was to transform Rome into a modern city, that this kind of straight, serious, and majestic thoroughfare became something which no great city was to be considered complete without. While it is surely equivalent streets which Wren proposed for London, it is again some diluted Napoleonic memory of this avenue of palaces and its closing triumphal arch which persists in the Champs Elysées and in all those innumerable boulevards in Paris and elsewhere that arose to imitate it.

However, the success of the small circular church—which Alberti, in humanist style, insisted on calling a temple—was even greater. Already built by 1502, the diminutive Tempietto of San Pietro in Montorio summarized themes which were then to be monumentally exploited in Michelangelo's St. Peter's, were to be taken up in France, to recur in St. Paul's, to become Hawksmoor's magnificent mausoleum at Castle Howard, and which in Oxford, again set down in the middle of a square, still present themselves for our inspection as the Radcliffe Library.

But we must beware of overstatement; and quite simply to propose all these spaces, all these squares and buildings to be Utopian is to be guilty of obvious overstatement. The Radcliffe Camera, for instance, or the Place des Vosges can scarcely be called Utopian without considerable qualification of this term. For we can scarcely suppose that their respective sponsors, either Oxford or Henri IV, would seriously have been disposed to spend large amounts of capital simply in order to exhibit the plastic corollary to highly abstract political speculation. So much is clear; and yet the Radcliffe Camera and the Place des Vosges, neither of them exactly revolutionary manifestations, are both of them still the products of an architectural culture which had long accepted the once revolutionary *Città ideale* as axiomatic and which habitually derived all the conventional elements of its repertory from this city's accessories. May we not then propose, by way of qualification, that these architectural manifestations, though not in the strict sense Utopian, are still a complement to Utopian thought? That they rose to satisfy emotional needs awakened by humanistic speculation? And that they are products of the same impulse which made of the New Jerusalem an instrument for reshaping the world?

So much may be proposed; but there is still an element of irony which attaches itself to any realization of the Utopian vision. Transcending reality as it does,

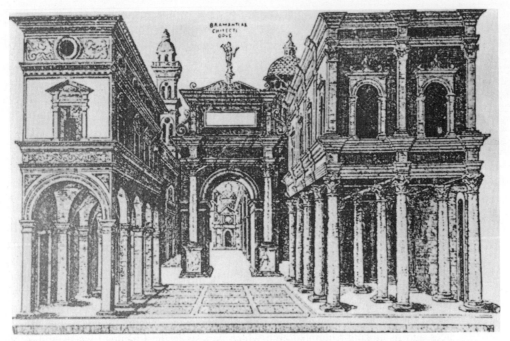

**FIGURE 3-4**  Bramante, *Street Scene*, c. 1500–10 (?). [Used by permission of C. Rowe]

transforming it as it may, Utopia becomes increasingly compromised as it becomes increasingly acceptable. Notoriously, the configuration of the ideal city was "ideal" not only for the philosopher but also for the military engineer; and while the first of such cities, Palma Nova, was established as a Venetian military station towards the Istrian frontier, others, very many with names famous in the history of war, were "idealized" rather through an application to the laws of ballistics than through any devotion to the principles of Plato.[7]

In these and other ways we might demonstrate how a revolutionary idea transforms not only the world but also itself. Penetrating that establishment which it seeks to subvert, permeating it, providing its own color and tone, Utopia ultimately becomes the *ancien régime* against which new demonstrations arise. But the constellation of ideas, partly Christian, partly Platonic, in some senses scientific, which presided over the inception of the late fifteenth century vision could scarcely be expected to occur again, and no subsequent Utopia has ever been able to command an architectural enthusiasm so concentrated and intense as that which Renaissance humanism could draw upon. Thus the successors of the humanists, the philosophers of the Enlightenment, conscious though they might have been of imperfections in the world which they occupied, were only rarely able to enlist the services of their architectural contemporaries.

For these reasons, in the project by Ledoux of 1776 (Figure 3-5)—which we might envisage as typifying the Utopian conditions that the Enlightenment conceived—the circular form seems largely to survive as an intelligible convention. It cannot be wholly sacrificed. Nor can it be wholly convincing. For though the rather deistic turn of mind exhibited by Ledoux might very well be able to accept a circular disposition for cities, for churches, and even for strange cemetery-catacombs—even though a Boullée could find it the supremely appropriate configuration for the Cenotaph for Newton (Figure 3-6)—the concept is no longer quite so "natural" as it once had been; and though a Renaissance idea of "nature" might still provide the mold for revolutionary form, it could no longer wholly absorb the sympathies of that new kind of "natural" man which the impending revolution itself was to evoke.

Romantic individualism and the concept of Utopia were scarcely to be fused; and, as a result, in the nineteenth century the Utopian idea was able to draw on no first-class architectural talent. It persisted, and even proliferated; but it persisted as regards architecture chiefly as a subterranean tradition. Neither the average nor the exceptional nineteenth century architect were ever to be much seduced by Benthamite principles of utility or Positivistic schemes of social reform; so that, in default of the architect's interest, the nineteenth century Utopia retains much of the stamp which the Italians had put upon it between three and four hundred years ago.

But it had in any case become a somewhat provincial idea;[8] and to the better minds of the time, often heavily influenced by subtle hypotheses as to the "organic" nature of society, the *appearance* of Utopia must have come to seem unduly mechanical.[9] Accordingly, deserted by intellect, Utopia now becomes naive; and while its Platonic

FIGURE 3–5    Project, La Saline de Chaux. Claude-Nicolas Ledoux, 1775–79. [Used by permission of C. Rowe]

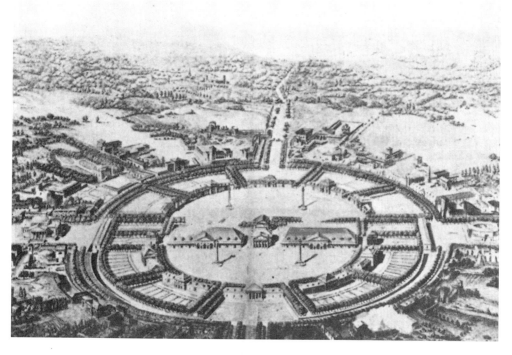

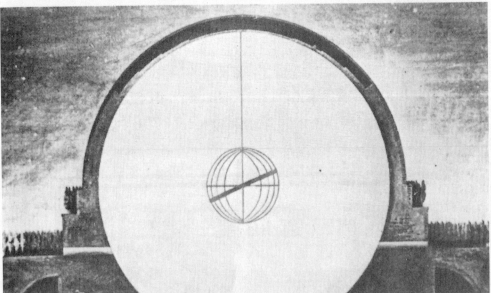

**FIGURE 3–6** Project, Cenotaph for Newton. Elevation and section. Etienne Louis Boullée, 1784. [Used by permission of C. Rowe]

forms persist they are no longer infused with a corresponding content. Also, Utopia seems now to have descended the social scale; for it is apparently no longer concerned with the redemption of society as a whole, but only with the redemption of its lower strata. Therefore the nineteenth century Utopia is apt to wear a look of either strenuous philanthropy or equally strenuous self-help; and the *Happy Colony* "to be built by the working men of Britain in New Zealand" (Figure 3-7) might be taken as an example of the latter look. It is a delightful mid-Victorian proposition; but although in this engaging scheme so many of the former elements of Utopia survive—Platonic solids, delineations of the globe, and even a little central building which has ceased to function as a temple and become instead a model farm—still it is rather to be doubted whether such symbolism any longer had public significance or whether many of those to whom this project was addressed were conscious of its author's transpositions of the traditional iconography.

But, if one kind of Utopia has here received its ultimate formal degradation, can it be said that Utopia has been revived in our own day?

That there is, or rather was, a profound Utopian impulse in modern architecture is indisputable; and, much as one imagines the drawing boards of Renaissance Italy, the drawing boards of the earlier decades of this century certainly seem to have been cluttered with the abstracted images of cities. Two in particular deserve attention: the mechanistic, vitalistic city of the Futurists where dynamism was ceaseless and a life was promised approximating to an absolute orgy of flux; and another city which, though a metropolis, was curiously static and which seems to have been almost as empty of people as the square in the picture at Urbino. Something of the Futurist city has been a component of all subsequent development. But the success of the second city, Le Corbusier's ville radieuse has been inordinate; so that, for the present, one could safely assert that this is the image of the city which controls. Never perhaps to be realized as a whole, its accessories, like those of Sforzinda, have everywhere been adopted. But, if in reality the ville radieuse would be almost as boring as Sforzinda before it, if it has a similar schematic monotony, if, like Sforzinda, it is perhaps one of those general ideas which can never be erased, one of those high abstractions which are empowered to perpetuate themselves, does this make of it a Utopia in the sense which we have so considered Sforzinda?

If Utopias are what Karl Mannheim defines them to be: "orientations transcending reality . . . which, when they pass over into conduct, tend to shatter, either partially or wholly, the order of things prevailing at the time,"[10] then we must surely concede that the ville radieuse is an instrument of some power. But, if we ask with what ideas an "orientation" which "transcends reality" is constructed we are obliged to wonder whether contemporary society can really tolerate such an "orientation."

Judged in terms of results, the most viable Utopia was certainly that evolved *c.* 1500 and which, to borrow a phrase from Jacob Burckhardt, might conveniently be described as an attempt to turn the state into a work of art. Now need we say that the state never can be turned into a work of art? That the attempt to do so is the attempt to bring time to a stop, the impossible attempt to arrest growth and motion? For the work of art (which is also an attempt to bring time to a stop), once it has left its maker, is not subject to change. It enjoys neither growth nor motion. Its mode of existence is not biological. And, though it may instruct, civilize, and even edify the individual who is exposed

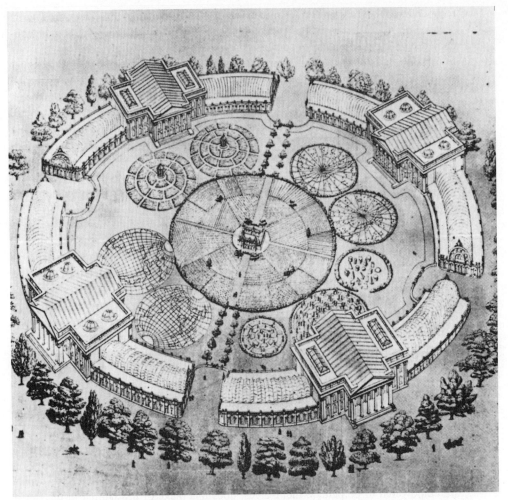

**FIGURE 3–7**  The Happy Colony. From Robert Pemberton, *The Happy Colony* (London, 1854). [Used by permission of C. Rowe]

to it, in itself the work of art will remain constant. For the work of art is not life; and nor, for that matter, is Utopia politics. But in the relation of the individual with the work of art we may still see something comparable to the relation of the state with Utopia. For Utopia too may instruct, civilize, and even edify the political society which is exposed to it. It may do all this; but for all that it cannot, any more than the work of art, become alive. It cannot, that is, *become* the society which it changes; and it cannot therefore change itself.

A crude naturalism experiences great difficulty in recognizing these existential conditions of the Utopian idea; and, to continue the analogy, a crude naturalism approaches Utopia very much indeed as the vulgar are supposed to approach a work of art. It demands an immediate effect. The work of art is *like* life. Utopia is *like* politics. Then let them be so. But the demand that Utopia approximate to a portrait at the Royal Academy or a novel by Arnold Bennett is conceivably more innocent than vulgar. For while the mimetic intention of both art and Utopia requires no emphasis, neither can properly "imitate" life or society except after their own laws—laws which, to all appearance, four hundred and fifty years ago were wonderfully identical. Indeed, at a much later date, Sir Joshua Reynolds' outrageous pronouncement that "the whole beauty and grandeur of art consists in being able to get above all singular forms, local customs, particularities, and details of every kind,"[11] might very well be a definition of "the whole beauty and grandeur" of Utopia. But it is also a statement of classical artistic doctrine, and it introduces the problem of whether we are able to detach Utopia from classicism.

Perhaps with difficulty we may: but only surely if we are prepared to recognize Utopia's limitations; and only too if, while recognizing them, we are also prepared to remember that Utopia is defined as an "orientation" which "transcends reality." Then, and armed with this definition, we might even conclude that, when "reality" is primarily attributed to motion, growth, change, and history, then the preeminent "reality-transcending orientation," the obvious Utopia, might evidently be something not too far removed from that persistent image which has so often been the instrument of change—the classical image of changelessness.

## ADDENDUM 1973

And the classical image of changelessness, that impossible image which has yet been responsible for more changes than one would wish to think about, is so important to this discussion that it might be useful to introduce a rather more strict definition of Utopia than Karl Mannheim's.

A Utopian conception in its fully developed form might be defined as a unified vision which includes:

1 a carefully considered artistic theory or attitude towards art integrated with
2 a fully developed political and social structure conceived of as extant in
3 a locus independent of time, place, history or accident.[12]

Now, for obvious reasons, no Utopian speculator—be he philosopher, architect, or absolute despot—was ever able to combine all these themes. But, if we are here in the realm of myth and if to combine all these themes was a patent impossibility for the speculative intellect of the Renaissance and the Enlightenment (which had a high regard for them all), any comparable endeavor is surely an even more patent impossibility for us at the present day.

We may, indeed, hunger and thirst after righteousness and our moral zeal may well be effusive; but whether our moral passion will, or should, overcome our intellectual fastidiousness must be another matter. We may agonize over a supposed prevalent absence of Utopian reference; but, however much it may be often considered to be so, the relation of society to Utopia is not the relation of a donkey to a carrot. Utopia is an optimum and, therefore, an end condition. Such are the terms of its alliance with classicism. A particular Utopia can be subjected to neither alteration, addition, nor subtraction;[13] but, while recognizing this, if we go on to observe the constantly expressed preference of the present day for the dynamic rather than the static, for becoming rather than being, for process rather than product, perhaps for effort rather than achievement, we can only begin to define a situation which is inimical to the idea of Utopia. And if we then go on to observe our interest in the concrete and the specific, in the paradoxical, in things as found, if we notice our preference for toughness, difficulty, and complication, our insistence on the empirical fact, on data collection, our belief in the work of art as an issue of tensions and balances, as a reconciliation of discordances and opposites, as something essentially and absolutely located in time and place, as something which presents and re-presents its temporal and spatial limitations, as something growing from and thereby illustrating its liaison with existing society, as something intimately involved with particular technologies and with ascertainable functions and techniques, then this can only be still further to establish that the range of often contradictory ideas which we habitually entertain are, *if it is possible to take them together,* considerably more than distinctly hostile to any form of Utopian fantasy.

Or so one might have thought. But any such simple and tolerably commonsense conjecture can only be to deny the historical evidence that a mental orientation towards specifics and a proneness towards Utopian speculation, however logically incompatible they may be, have long been able to enjoy an apparently happy coexistence.

> I have a formal difficulty about the concept of Utopia . . . Does the Utopian postulate a static Utopia, that is to say a society so perfect that any further change (improvement) is inconceivable? Or is Utopia dynamic, a state to which we are continually aspiring but never reach because it itself is continually changing, moving ahead of us? . . . I cannot subscribe to the static version, for its assumptions are inherently absurd. . . .[14]

So wrote one of the contributors to the issue of *Granta* in which this essay was first published; and his point of view is completely understandable—at least so long as we do not find Utopia to be a genuine problem or, alternatively, so long as we are willing directly to equate the notions of Utopia and social progress. However, if we approach Utopia with suspicion (Utopia where the citizens cannot fail to be happy because they cannot choose but be good), if we are sceptical of its combination of progressivism and classicism, then we might probably recognize that the imagined possible fusion of an evolutionary sequence with a perfected condition (of endless *becoming* which will, still, always be complete *being*) is one of the more extraordinary fantasies of the present day—a fantasy which is apt to seem always as wholly benign as it is undoubtedly, well intentioned.

The idea that society can approximate the condition of music, that change and order may become one and the same, that the roads leading into the future may now, for the first time, be rendered free of all bumps and impediments is, of course, one of the root fantasies of modern architecture; and it is apt to be one of those presumptions which travel unexamined. Utopia is both to bring time to an end and, simultaneously, it is to inaugurate an era in which the movements of time will be, for the most part, smooth and predictable. The notion of the millennium is to retain all of its old-style cosmic significance and is then to be made further agreeable by a redecoration with all the gloss of rationality and science.

The state of mind which is here implied, unembarrassed, superstitious, and allegedly enlightened, should now be recognized as constituting the major block to any contemporary Utopian formulation—should this be necessary. A state of mind well disposed toward what it supposes Utopia to be but unequipped with sense of metaphysical difficulty or reservation—the state of mind preeminently of the planner and the empirical sociologist—for the most part unaware of Utopia's historical origins and, generally, conceiving these to be irrelevant: what we have here is, very largely, the imposition of Hegel upon Plato and, further, the tacit insistence that the results of this condition are, necessarily, libertarian.

Which may be, rather extravagantly, to generalize and to jump; but which may also be, opportunistically, to identify those coercive attributes of Utopia which Karl Popper has selected to condemn. For Popper's criticisms of Utopia and the closed society,[15] though they were available to—and apparently ignored by—the contributors to *Granta*,[16] must be conceded as establishing a conspicuous obstacle both to the exercise of Utopian fantasy as well as to the deployment of most of the traditional programs/fantasies of modern architecture.

Utopia, because it implies a planned and hermetically sealed society, leads to suppression of diversity, intolerance, often to stasis presenting itself as change, and, ultimately, to violence. Or, more specifically: if Utopia proposes the achievement of abstract goods rather than the eradication of concrete evils then it is apt to be tyrannical: this since there can far more easily be consensus about concrete evils than there can be about abstract goods. Such is the Popperian message, which, supported as it is by a critique of determinism and a developed theory of the nature of investigation, remains hard to refute and which, very largely, continues to be ignored.

But, of course, when all this has been observed and taken to heart (which, in this context, might mean when modern architecture has been cut down to size and its intentions admitted to be poetry rather than prescription), the problem and the predicament of abstract goods in a world of concrete evils—always an acute question for the architect—remains unalleviated and unexamined; and, if one must agree with Popper about the obligation to eradicate concrete evils, then one must still notice that the issue of abstract goods (with all its Platonic, natural law, and modern-architecture overtones) emphatically persists. For how to designate specific evil without at least *some* theory of general good? And is not Popper's position yet another Utopian formulation—and a Utopian formulation which is particularly Germanic?

> The road to progress was not sought in external deeds or in revolutions, but exclusively in the inner constitution of man and its transformations.

For again, Mannheim—with his discrimination of a largely apolitical and Germanic Utopia (*vide* note 8)—might seem to enter the picture, and Popper's ideal of emancipation through self-knowledge—an emancipation both for the individual and society (the Kantian ideal)—might seem to belong to this important, but still intrinsically local, category.

So, the problem of specific evil, the need for emancipation, a theory of general good and then coercive Utopia as a repository—perhaps *the* repository—of ideas of general good: this might be the conflict of interests that confronts us as we abandon both the folklore of modern architecture and the Popperian destruction of allied propositions; and it is a dilemma which seems to leave us only with the alternatives of Utopia and freedom—with freedom dependent upon Utopia and Utopia always acting to limit freedom.

In American terms this is a quandary which can often become the rule of law versus the will of the people—with "law" or "the people" idealized as the purpose of the occasion might seem to require. But, if this pair of opposites presumes an important willingness to argue and no great willingness to insist, it might also be transferable from the area of ostensible politics to the area of criticism in general. So there are no criteria which cannot be faulted, which are not in continuous fluctuation with their opposites. The flat becomes concave. It also becomes convex. The pursuit of an idea presumes its contradiction. The external world and the senses both equivocate; and criticism, however empirical it may sometimes profess to be, depends always upon an act of faith, upon an assumption ("this is a government of laws not men") of impossible realities but plausible abstractions. But, if the possible, the probable, and the plausibly abstract are always in a continuous condition of intersection, it is perhaps in some such area, where myth and reality interfecundate, that we should be willing to place all extreme fantasies both of Utopia and liberty.

The myth of Utopia and the reality of freedom! Alternatively: the reality of Utopia and the myth of freedom! However stated, what we have here are the intimately interwoven presumptions of both authority *and* liberty; and, if they are—both sets—necessary for survival, both of them the necessary components of discourse, then, if we profess any interest in emancipation but are not anxious to propose anarchy, perhaps it should only be said that *some* affirmation of a limited Utopia remains a psychological obligation. Utopia, in any developed form, in its post-enlightenment form, must surely be condemned as a monstrosity; but, while always a flagrant sociological or political nightmare, as a reference (present even in Popper), as a heuristic device, as an imperfect image of the good society, Utopia will persist—but should persist as possible social metaphor rather than probable social prescription.

## NOTES

1 Thomas More, *Utopia,* Book II, Ch. 11.
2 Plato, *Timaeus,* Bollingen series, 1944, p. 117.
3 James Leoni, *The Architecture of Leon Battista Alberti in Ten Books* . . . (London, 1735), Book VII, Ch. IV.
4 Isaac Ware, *The Four Books of Palladio's Architecture* (London, 1738), Book IV, Preface.
5 Leoni, *The Architecture of Leon Battista Alberti,* Book VII, Ch. III.
6 This is not entirely to discount the splendid Bramantesque piazza at Vigevano dating from the 1490s.
7 See Horst de la Croix, "Military Architecture and the Radial City Plan in Sixteenth Century Italy," *Art Bulletin* 42, no. 4 (1960).
8 Though this might seem to be a somewhat perfunctory handling of Saint-Simon, Fourier, Owen, *et al.,* one might still agree with Karl Mannheim that these were "dreaming their Utopias in the older intellectualist style" (Mannheim, *Ideology and Utopia,* 1st English ed., 1936; reprinted, New York, n.d., p. 245).
9 And one might again agree with Mannheim that "Where as in France, . . . the situation matured into a political attack the intellectualistic took on a rational form with decisively sharp contours. [But] where it was not possible to follow in this path, as in Germany, the Utopia was introverted and assumed a subjective tone." In Germany, Mannheim continues, "the road to progress was not sought in external deeds or revolutions,

but exclusively in the inner constitution of man and its transformations." Mannheim, *Ideology and Utopia,* p. 220.
10  Mannheim, *Ideology and Utopia,* p. 192.
11  Sir Joshua Reynolds, *Literary Works* (London, 1835), Vol. I, p. 333. From Discourse III, delivered in 1770.
12  Carroll William Westfall, "Review of Hermann Bauer, *Kunst und Utopie,*" in *Journal of the Society of Architectural Historians* 26, no. 2 (1967).
13  "A harmony of all the parts in whatsoever subject it appears, fitted together with such proportion and connection, that nothing could be added, diminished, or altered but for the worse": Alberti's definition of beauty (Leoni, *The Architecture of Leon Battista Alberti,* Book VI, Ch. 11) might also serve to illustrate the obvious predicament of Utopia with reference to both history and change.
14  Robin Marris, "Utopia and Conviction," *Granta* 63, no. 1187 (1959).
15  See particularly the article, first published in the *Hibbert Journal* (1948), in the collection *Conjectures and Refutations* (New York and London, 1962). But the judgments which Popper here expresses are obviously to be found more extensively developed in his *The Logic of Scientific Discovery* (first published as *Logik der Forschung* [Vienna, 1934]) (London 1958); in *The Open Society and its Enemies* (London, 1945), and in *The Poverty of Historicism* (London, 1957). A valuable and appreciative criticism of Popper's centrality to the construction of any adequate contemporary critical theory is to be found in G. Radnitzky, *Contemporary Schools of Meta-Science* (Chicago, 1973).
16  That Popper could be overlooked by the contributors to a student magazine in 1959 perhaps should not be surprising; but that a similar failure should have characterized the contributors to vol. 94, no. 2, of the *Proceedings of the American Institute of Arts and Sciences* might arouse curiosity. However the issue *Utopia* of Daedalus (Spring 1965) appears nowhere to display a cognizance of his position.

选读 4
"社区建筑"
文森特·约瑟夫·斯卡利

斯卡利，一位建筑历史学家，在文章中探讨了建筑在塑造人文环境中发挥的重要作用，而这种人文环境可以营造一种社区感。他谴责说，汽车是一种"能使人陷入深度幻想的代步工具"，虽然为我们所有人所钟爱，但正是汽车的普及消除了社区的物理结构，很大程度上导致了社区的分崩离析。他认为流行于本世纪中叶的现代建筑和规划思想仅仅满足对于汽车的需求是无法支持社区构建的。相比之下，现在安德烈斯·杜安伊（Andres Duany，1949—，美国建筑师、城市规划师）和伊丽莎白·普拉特－伊贝克（Elizabeth Plater-Zyberk，1950—，美国建筑师、城市规划师）二人以佛罗里达海岸为代表的作品则把城镇作为一个整体进行处理，让建筑能够重拾"作为一种创建城市手段的传统作用"。斯卡利在文中如是说道："……从根本上来说，建筑不是构建独立房屋的问题，而是塑造整个社区的问题，而且正如巴黎、伊拉克乌鲁克、意大利锡耶纳所做的那样，需要借助法律加以实施。"

# READING 4

## The Architecture of Community

**Vincent Joseph Scully**

Scully, an architectural historian, writes here about architecture's important role in shaping the human environment to create a sense of community. He decries the role of the automobile as a "device of deep illusion," that we all love, but that has contributed enormously to the disintegration of community by obliterating the community's physical structure. He argues that Modern architecture and planning prevalent in the middle years of this century by merely accommodating the automobile were unable to provide community. In contrast, the current work by Andres Duany and Elizabeth Plater-Zyberk, as epitomized by Seaside, Florida, deals with the town as a whole and affords architecture the opportunity to regain "its traditional stature as the means by which cities are made." Scully states: ". . . architecture is fundamentally a matter not of individual buildings but of the shaping of community, and that, as in Paris, Uruk, or Siena, is done by the law."

All human culture is intended to protect human beings from nature in one way or another and to mitigate the effect upon them of nature's immutable laws. Architecture is one of humanity's major strategies in that endeavor. It shelters human beings and reassures them. Its purpose is to mediate between the individual and the natural world by creating the physical reality of the human community, by which the individual is linked to the rest of humanity and nature is in part kept out, in part framed, tamed, and itself humanized. So architecture constructs its own model of reality within nature's implacable order. It is within that model that human beings live; they need it badly, and if it breaks down they may well become insane.

That is exactly what is happening today, and not only in America, but the pattern, as so often in contemporary history, can be perceived most clearly here. This is so, in part, because Americans have in fact destroyed so many kinds of communities during the past generation. The process began directly after World War II, when the remaining trolley tracks, the very lifeline of town and suburb alike, were apparently bought and torn up by the automobile interests. Public transit had been declining since 1914 in any event as the number of automobiles rose. The Redevelopment of the 1960s completed the destruction and showed the true shape of the holocaust. The automobile was, and remains, the agent of chaos, the breaker of the city, and Redevelopment tore most American towns apart to allow it free passage through their centers, which were supposed to be revitalized by affluent suburban shoppers thereby. Instead the reverse took place: The automobile created the suburban shopping mall, which sucked the life out of the old city centers everywhere. This is ironic enough, because the existing center-city communities had themselves been destroyed by Redevelopment in

*Source:* "The Architecture of Community" by Vincent Scully, in *The New Urbanism*, by Peter Katz. Copyright © 1994 by Peter Katz. Published by McGraw-Hill. Reproduced by permission of the McGraw-Hill Companies. pp. 221–30.

order to bring the largely mythical suburbanite shoppers in. To follow I-95 and its various connectors from New England to Florida is to watch that evil process at work from Oak Street in New Haven to Overtown in Miami, at the very end of the road. Their communities physically torn apart and given no opportunity to form new ones, many of the inhabitants of center city began to lose their minds, as who would not. Many of them were African-Americans from the rural South who had been lured to the big cities to work in war plants during World War II; then the factories moved away with perfect cynicism, seeking even cheaper labor back in the South once more; then, in financial panic, the big cities redeveloped themselves in the manner already described and there its people were, out of work under the Piranesian piers of the freeway, in a surreal wasteland with homes, churches, stores, and most of all the orienting street grid of the city, all shot to hell.

By contrast, the suburbs, closed off from the urban population by what Frank Lloyd Wright once called "the iron hand of realty," seemed like Paradise, but they were spawning their own neuroses too, fed by endless hours on the road and no connection with much of anything when one got where one was going. Soon fear came to play its part. It rode behind the locked doors of the automobiles and was eminently justified, whether by the nut on the highway, or the sniper on the overpass, or by what happened if the wrong exit was taken off the connector.

Whatever other factors have been involved in this disintegration of community, it is still the automobile—and how much we all love it—which has done the job. It has not only obliterated the community's physical structure but has also made us feel that the community's psychic protection is unnecessary, and that what the car seems to offer in terms of individual freedom is enough. It is a device of deep illusion and may be said to have rendered all of society insane. Indeed, the years to come will soon show us whether the automobile and what we have thought of as civilization can coexist.

Some of us were writing and teaching all this in the 1960s—even then the eventual effects on American society were predictable enough—and some younger people were apparently listening. Peter Calthorpe, a student at Yale in the 1970s, seems to have been one of them. His "transit-oriented development," as assembled at Laguna West, is an attempt to regroup the suburb into a density which makes public transit feasible. It is shaped by avenues that radiate, like those of Versailles, from a center of public buildings and spaces, among them a "village green." One thinks of the 17th-century grid plan of New Haven, Connecticut, with the great Green in the center. As the grid moved west to shape most of the cities of the continent, the Green, the public space, tended to disappear under the pressure of private greed. Calthorpe now tries to bring it back, reflecting the attempt by many organizations over the past 30 years to preserve or restore public space. One recalls the fight under Margaret Flint of the New Haven Trust for Historic Preservation to preserve the scale and amenity of New Haven's Green itself in 1967. Indeed, the combats of that year—when New Haven's post office and city hall were saved from Redevelopment and after which, by extension, the brutal demolition of low-income neighborhoods was brought to a halt by Senator Lowell Weicker—might be regarded as the true beginning of the contemporary preservation movement, through which, for the first time in the modern period, a popular mass movement has discovered the means and the political clout to force architects and civic officials alike to do what the informed public wants them to do.

That movement, now boldly led by the National Trust for Historic Preservation, does seem to reflect the yearning to rebuild community which is felt by most Americans today. It now seems obvious to almost everybody—as it did once before, in the 1870s, when the Colonial Revival began—that community is what America has most conspicuously lost, and community is precisely what the canonical Modern architecture and planning of the middle years of this century were totally unable to provide. This was so for many reasons; foremost among them was the fact that the Modern architects of the heroic period (Wright, Le Corbusier, Mies Van der Rohe, Gropius and their followers) all despised the traditional city—the finest achievement of Western architecture, put together piece by piece over the centuries—and were determined to replace it with their own personal, utopian, idiosyncratic schemes. Le Corbusier's Ville Radieuse was the most influential of them all; and it furnished the basic model for American Redevelopment itself. Even the social structures involved were eerily the same: both were "cités d'affaires," cities of business, from which the poor were to be excluded. The German Modernists had advanced equally catastrophic ideas, based upon their concept of the "zeitgeist," the "spirit of the age," that did not allow anything which had been done before to be done again or even to be preserved. So Hilbersheimer proposed his endless miles of high-rise slabs, his landscapes of hell, out of which the mass housing of the 1950s took shape, much of it to be dynamited as wholly unlivable hardly more than 20 years later.

In all these cataclysmic proposals for the city there was a true hatred for the world as it was socially constituted, but there was also something else, a consuming contempt for it on aesthetic grounds. The Modern architects of the International Style had largely taken abstract painting as their model, and they came to want to be as free from all constraints as those painters were, free from everything which had always shaped and limited architecture before, in part from statics itself (forms must float) and from roofs, windows, trim and so on, but most of all from the restraints of the urban situation as a whole: from the city, from the community. Their buildings were to be free of zoning laws, and from the need to define the street, and from all respect for whatever already existed on and around the site. They were to be free—like Lever House or the Pan Am Building or the Whitney Museum or even the Guggenheim Museum—to rip the old urbanism apart or to outrage it, or, perhaps most truly, to use its order, while it lasted, as a background before which they could cavort. Most of all they had to be abstract; they could not under any circumstances be inflected toward their surroundings by Classical or vernacular details or stylistic references of any kind. Such would have constituted an immoral act. Here was another madness to complement the others. It is still prevalent today among many architects who, baffled by the complexity of reality, still insist that architecture is a purely self-referential game, having to do with formal invention, linked madly enough with linguistics, or literature, but not at all with the city or with human living on any sane terms. Such architects claim to reflect the chaos of modern life and to celebrate it. Some of them pretend to worship the automobile, and the "space–time continuum," like Marinetti before them idolizing violence, speed, war and Fascism in the end. "Whom Zeus wishes to destroy," said Aeschylus more or less, "he hastens on with madness."

Yet it should be said that there is hardly an architect or critic living today who has not been drawn to Modern architecture during his life and does not love thousands of Modern works of art. But the urban issue has to be faced. The International Style built many beautiful buildings, but its urbanistic theory and practice destroyed the city. It wrote bad law. Its theme in the end was individuality; hence its purest creations were suburban villas, like the Villa Savoie and Philip Johnson's Glass House. These celebrated the individual free from history and time. One could not make a community out of them. In the Glass House especially the individual human being seems wholly liberated from the entire human community. The secret is technology, a chancy thing; plugged into its heating and lighting devices the existential mortal man can dispense with everything else that once stood between nature and himself. He enjoys the sensation of being wholly alone in the world. His architecture cannot, will not, deal with community issues.

So Neo-Modern architecture, in its present "Deconstructivist" phase, though popular in the schools—why not, it offers the ideal academic vocabulary, [is] easy to teach as a graphic exercise and [is] compromised and complicated by nothing that exists outside the academic halls—has been failing for a long time in the larger world of the built environment itself. Here it is clear that the most important development of the past three decades or more has been the revival of the Classical and vernacular traditions of architecture, which have always dealt with questions of community and environment, and their reintegration into the mainstream of Modern architecture. That development in fact began in the late 1940s with an historical appreciation of the American domestic architecture of the 19th century—which I tried to call the Stick and Shingle Styles—and it first took new shape in the present with Robert Venturi's shingled Beach House of 1959. Venturi then went on, in the Vanna Venturi House of the early 1960s, to reassess the early buildings of Frank Lloyd Wright, which had themselves grown directly out of the Shingle Style of the 1880s, and he worked his way wholly back into the Shingle Style itself, as in the Trubek and Wislocki houses of 1970. Here Venturi rediscovered a basic vernacular type, very close indeed to the types "remembered" (as he put it) by Aldo Rossi in Italy very soon thereafter. Many other architects then followed that lead, Robert A.M. Stern foremost in time among them. Stern soon learned to abandon his compulsion to invent—his early houses, though based on Shingle Style models, are Proto-Deconstructivist in form—in favor of trying to learn how to design traditional buildings well, and to group them in ways that make sense. The point became not style but type and, by extension, context. Here again Robert Venturi's work led the way. Wu Hall in Princeton, the Institute for Scientific Information in Philadelphia, and the Sainsbury Wing of the National Gallery in London, all inflect what are otherwise clearly Modern buildings toward the particular "styles" which preexist on each site: Tudorish in Princeton, International Style in Philadelphia, Classical on Trafalgar Square. Each new building thus enhances and completes the existing place on its own terms. The city is healed rather than outraged—poignantly so in London, where the site had actually been blitzed and where early schemes to build upon it lent credence to Prince Charles' remark that Modern architecture had done more damage to England than the Luftwaffe. Now the architect gives up his semi-divine pretension to be Destroyer and Creator and to invent new styles like new religions, and aspires instead to the more humane and realistic role of healer, of physician. Venturi was surely encouraged in this new pragmatism by the work of his wife Denise Scott Brown in neighborhood design and advocacy planning. So, with Venturi, the architect abandons the iconoclasm of much of the Modern movement in favor of the idea that he belongs to a long and continuous architectural tradition, through which cities in the past have on the whole been built correctly and in reasonable accordance with human needs.

Out of this view, which is in fact the natural culmination of the vernacular and Classical revival, the work of Andres Duany and Elizabeth Plater-Zyberk derives. It completes the revival by dealing with the town as a whole. It reclaims for architecture, and for architects, a whole realm of environmental shaping that has been usurped in recent generations by hosts of supposed experts, many of whom, like those of the truly sinister Departments of Transportation everywhere, have played major roles in tearing the environment to bits and encouraging its most cancerous aberrations. With these two young architects, and with their students and colleagues at the University of Miami, architecture regains its traditional stature as the means by which cities are made.

I have written elsewhere how, as architecture students at Yale in the early 1970s, Duany and Plater-Zyberk led my seminar into New Haven's vernacular neighborhoods and showed us all not only how intelligently the individual buildings were put together but also how well they were related to each other to make an urban environment—how effectively the lots worked, and the porches related to the street, and the sidewalks with their fences and their rows of trees bound the whole fabric together, and how street parking was better than parking lots and the automobile could be disciplined, and how, most of all, it could all be done again—and fundamentally had to be done again as all of a piece if it was going to be done right—from the turned posts and the frontal gables to the picket fences, the sidewalks and the trees. Everything that the International Style had hated, everything that the "zeitgeist" had so Germanically consigned to death, came alive again. For me, marinated in Modernism, it was the revelation of a new life in everything. There was no reason whatever why the best of everything had to be consigned to the past. Everything was available to be used again; now, as always in architecture, there were models to go by, types to employ.

So it is important to remember that for Duany and Plater-Zyberk the plan as such did not come first. First came the buildings, the architectural vernacular, because it was after all the buildings which had brought the old New Haven grid up into three dimensions to shape a place. Duany and Plater-Zyberk's critics have never really understood this. It is again a question of types which, with their qualifying details and decoration, have shown themselves capable of shaping civilized places and of fitting together in groups to make towns. Leon Krier was also instrumental in helping us see this, and he became one of Duany and Plater-Zyberk's most important mentors and was to build a beautiful house at Seaside.

Terms like "historicism" are not relevant here—the *zeitgeist* mentality is "historicist," not this one—but ancillary concepts like those relative to symbol are relevant indeed, and no excuse need be made for that fact. Human beings experience all works of visual art in two different but inextricably interrelated ways: empathetically and by association. We feel them both in our bodies and in terms of whatever our culture has taught us. Modernism at its purest fundamentally wanted to eliminate the cultural signs if possible—hence abstraction. It was Venturi himself who, in his epoch-making *Complexity and Contradiction in Architecture* (1966) and his *Learning from Las Vegas* (1972), first brought an awareness of the centrality of symbolism back to architecture, and he was the first to use semiotics as an architectural tool. It was he who introduced literary criticism itself, especially Empson's theory of ambiguity, into the contemporary architectural dialogue. Faced with this, the Neo-Modernists would like to divert it by replacing the relevant primary architectural symbols—those having to do with nature, place and community—with secondary and diversionary ones having to do with linguistics or whatever else may be dredged up out of the riot of sign systems in the human mind.

Not Duany and Plater-Zyberk. Their eyes are on the reality of things as they are. That is why Seaside is so moving. Whatever it may be in fact—a resort community, a modern-day Chautauqua—it has beyond that succeeded, more fully than any other work of architecture in our time has done, in creating an image of community, a symbol of human culture's place in nature's vastness. It does this in terms of the densely three-dimensional organization of its building types as they group together, almost huddle together, on the shore of Florida's panhandle, pressed close up to the gleaming white sand, the green and blue sea and the wild skies of the Gulf of Mexico. Therefore, Seaside is not an affair of plan only, not only of two-dimensional geometry, as all too many planned communities in this century have tended to become. True enough, Seaside's plan as such has a distinguished ancestry. It owes a direct debt not only to Versailles and to the whole French classic planning tradition, out of which Washington no less than modern Paris took its shape, but also to the fine American planning profession that flourished before Gropius came to Harvard in the 1930s and destroyed it at its heart. One thinks here especially of John Nolen's work of the 1920s in Florida, so well illustrated by John Hancock in Jean-Francois Lejeune's brilliant publication *The New City: Foundations* (Fall 1991). All the planning shapes at Seaside are in Nolen's plans for Venice and Clewiston, both in Florida: the grid, the broad hemicycles, the diagonal avenues. And Nolen was of course not alone in his time. Planners of the teens and twenties like Frederick Law Olmsted Jr., Frank Williams, Arthur Shurtlief, Arthur Comey, George and James Ford, every one of them with at least one degree from Harvard, also come to mind, as do many others. It is true that one weakness of these planners, a somewhat Jeffersonian preoccupation from early in the century onward with what they called the "congestion" of the cities, was to play into the hands of the Modernist iconoclasts and the automobile freaks after World War II. Otherwise, the New Urbanism, so-called, is in large part a revival of the Classical and vernacular planning tradition as it existed before International-Style Modernism perverted its methods and objectives.

But Duany and Plater-Zyberk differ from Nolen—and so Seaside from Venice—in one fundamental aspect: They write a code that controls the buildings as well as the plan. They therefore ensure that the three-dimensional reality of the town will fulfill the concept adumbrated in its plat—without themselves having to design every building in it. Hence they encourage many other architects and builders to work, as they can do freely enough, within the overall guidelines. Nolen could not normally exert that much control. So his streets are often ill-defined, his axes climaxed by gas stations, the whole inadequately shaped and contained. Calthorpe so far has been in something of the same fix.

But Duany and Plater-Zyberk had learned not only from Nolen but also from George Merrick, the developer of Coral Gables, upon which he worked most directly from 1921 to 1926, when the hurricane of that year wiped him out but failed to kill his town. Merrick is one of the true heroes of American architecture, and an unlikely one. He was a Florida real estate man of the bad old days of the boom when so many of

Florida's lots were resold two or three times in one day and were in fact under water. But not those of Coral Gables, which also has an unusual plan, involving a perimeter of tightly gridded streets that contains and protects a free-flowing English garden within it—its shapes probably suggested by its several golf courses—and all fundamentally at automobile scale: the automobile scale of the 1920s, that is, the scale of "motoring," which is what Coral Gables, though it once had a fine public transportation system, was fundamentally intended for. We can't really blame Merrick for his beautiful renderings of fine boulevards with a few dignified town cars proceeding along them. Who could have foreseen the explosion of the species that was to come? But what Duany and Plater-Zyberk and Robert Davis learned from Coral Gables was not only the general lesson that a fine coherent small city could be made out of suburban elements but also the specific lesson that it took a Draconian building code to do it. That's what Merrick had, a code that shaped first the beginnings of a Spanish or, perhaps better, a Mediterranean-Revival town and then introduced little villages into it that were French or Chinese or South African Cape Dutch or Southern Colonial—all perfectly delightful, especially the Chinese.

With the hurricane of 1926, Merrick went bankrupt and lost control, and the houses, especially after World War II, became more typically suburban—squashed down, spread out, less urbane and less naturally groupable—while the lots became much bigger, so that a certain structure, or scale, was lost. But much of the code held, and a fundamental urban order continued to be maintained. In the end that order was furnished largely by the trees. They shape the streets and cover them over against the sun and are the major architectural elements that make the place special and unified in every way, and disguise the worst of the houses.

Seaside's structure does not derive from its trees. The windy Gulf coast is not sympathetic to them, and the "jungle" will grow up only to the height that is protected by the houses, but the principle of shaping the streets three-dimensionally is written into the code. Streets are as narrow as possible—automobiles can get through them perfectly well but their scale remains pedestrian—and they are closely defined by picket fences and front porches and by building masses brought tightly up to them. There are no carports and few garages. The cars survive well enough and the street facade retains its integrity. So the important place-maker is the code. It is not "fussy" or "escapist" but essential, and at Seaside it may not have been written strictly enough.

It is curious that the houses there which have been most published in the architectural press—though not in the popular press, which understands the issues better—are those which most stridently challenge the code, as if originality were architecture's main virtue and subversion of community its greatest good. The houses by Walter Chatham at Seaside are the most conspicuous in this regard. Each destroys a type; his own intrudes what looks like a primitivistic cabin, something appropriate to a glade in the Everglades, into a civilized street of humanly scaled windows, flat trim and delicate porches, and so barbarizes it, while his row houses at the town's urban core do the two things no rowhouse can do without destroying the group: interrupting the cornice line and dividing the individual house volume vertically down the middle. Yet Robert Davis encourages Chatham (whom everybody likes anyway) and continues to give him buildings to do, perhaps valuing his intransigence as an image of the therapeutic license within the general order—or simply because it is invariably published. To go further, it would be salutary to see what a really fine architect like Frank Gehry—his work almost Deconstructivist but much too genial, accomplished, and untheoretical for that—might be able to do within the code at Seaside. Gehry has shown not only that he knows how to inflect his apparently anarchic buildings toward the places they are intended for but also that he understands and loves American wood frame construction. He might well find a way to reinterpret the tradition and retain civility in ways that Chatham, for example, has not yet been able to do.

Distressing, though, is the tendency of Duany and Plater-Zyberk, when addressing Neo-Modernists, to suggest that they employ the vernacular in their projects only because it is popular with their clients. This buffoonery, genial enough, nevertheless leaves them open to the charge of "pandering" to the public which their opponents are not slow to advance. But the pandering in this case, as in that of Chatham, is to the architectural magazines and the professional club. It makes a joke of everything Duany and Plater-Zyberk have come to stand for, and it denies the historical facts of their rise. That they should seem to need the approval of professional côteries they have far outclassed may be taken as an aberration of success and an indication of the tight hold (like that of the Marine Corps or the Catholic Church) which the architectural profession exerts on anyone who has ever belonged to it. In any case, another generation—some collaborators, others trained by Duany and Plater-Zyberk, others affiliated with them at the University of Miami and therefore much more liberated than they—will surely carry on the work. Some names that come to mind, and there may surely be others (many mentioned elsewhere in this book), are Jorge Hernandez; Teofilo Victoria; Maria de la Guardia; Jorge, Mari and Luis Trelles; Rocco Ceo; Rafael Portuondo; Geoffrey Ferrell; Charles Barrett; Victor Dover; Joseph Kohl; Jaimie Correa; Mark Schimmenti; Eric Valle; Scott Merrill; Jean-Francois Lejeune; Ramon Trias; Maralys Nepomechie; Gary Greenan; Dan Williams; Monica Ponce de Leon; Richard McLaughlin; Armando Montero; Thorn Grafton; Suzanne Martinson; Rolando Llanes; Sonia Chao; Maria Nardi; Frank Martinez; Ernesto Buch; Douglas Duany; Dennis Hector; Joanna Lombard; Thomas Spain; Roberto Behar; and Rosario Marquardt, whose rapt and noble paintings have helped to set a Mediterranean impress on the school.

It is true that Seaside seems so deceptively *ad hoc* that it can take a good deal of disruption. Could Kentlands? Probably not so much. But the point is clear. All human communities involve an intense interplay between the individual and the law. Without the law there is no peace in the community and no freedom for the individual to live without fear. Architecture is the perfect image of that state of affairs. Ambrogio Lorenzetti showed it to us in Siena, in his *Allegory of Good Government.* There is the country all rounded and rich in vineyards and grain. There is the city wall cutting into it, behind which the hard-edged buildings of the town jostle each other and shape public spaces where the citizens dance together. A figure of *Security* floats above the gate and guards it. Alongside this great scene an allegory of the town government is painted on the wall. The *Commune* sits enthroned, a majestic figure surrounded by virtues. Below him all the citizens of the town are gathered, each dressed in his characteristic costume and all grasping a golden cord which depends from the *Commune* itself. The cord is the law which binds them together and which they hold voluntarily, because it makes them free. In the center of the scene the figure of *Peace* reclines at her ease.

Seaside, which in fact resembles Lorenzetti's densely towered town more than a little—as does Battery Park City at another scale—embodies this necessary duality well. It is interesting in that regard to compare its houses and their groupings with those at Laguna West, where Calthorpe has pointed out that he was not able to control the architectural situation so completely. It is even more interesting to visit the towns along the Gulf near Seaside which imitate it. The picket fences are there, and they help quite a lot, as do the gazebos and the vernacular architecture as well. But the roads are all too wide, the lots usually too big; the density is not really present, so that the automobile still seems to be in command and the pressure of the communal law is not really applied. Therefore, these derivations from Seaside are all less convincing as places. They should not be despised for that, because their movement is in the right direction, but the point remains luminous: architecture is fundamentally a matter not of individual buildings but of the shaping of community, and that, as in Paris, Uruk, or Siena, is done by the law.

Still, one cannot help but hope that the lessons of Seaside and of the other new towns now taking shape can be applied to the problem of housing for the poor. That is where community is most needed and where it has been most disastrously destroyed. Center city would truly have to be broken down into its intrinsic neighborhoods if this were to take place within it. Sadly, it would all have been much easier to do before Redevelopment, when the basic structure of the neighborhoods was still there, than it is today. But, whatever the size of the city as a whole, the "five-minute walk" would have to govern distances and the scale of the buildings themselves should respond to the basically low-rise, suburban-sized environment that, for any number of reasons, most Americans seem to want. It is therefore a real question whether "center city" as we know it can *ever* be shaped into the kind of place most Americans want to live in.

The Clinton neighborhood illustrated in this volume is surely a measurable improvement in that regard over the usual development of Manhattan's blocks, but the scale is still enormous, much larger than that of Vienna's great Gemeindebauten of the period 1919–34, which it otherwise somewhat recalls. It is very urbane, with reasonable public spaces. But in America, unlike Europe, only the rich have normally chosen to live in high-rise apartments. The poor have almost always aspired to what they have been told every American family rates: a single-family house in the suburbs. Ideally they want Seaside. And since we are no longer Modern architects who act upon what we think people ought to have rather than what they want, we should try to figure out how they can get it. The building type itself should present no problem, especially if its basic visual qualities and its sense of personal identification can be captured in a narrower, higher, perhaps even multi-family type. In fact, some of Duany and Plater-Zyberk's basic models, and those of Melanie Taylor and Robert Orr at Seaside, were the two- and three-family wooden houses of New Haven's modest neighborhoods, a good three stories tall with porches, bay windows and high frontal gables, a 19th century blue-collar Stick-and-Shingle structure that defined city streets with a compelling presence and some density but at a moderate scale.

Here, too, a half-forgotten contemporary project comes to mind: Robert Stern's Subway Suburb of 1976. Stern proposed that the city services from subways to sewage that still existed in the South Bronx, above which the city lay burned, wasted and unwanted, should be utilized to create what amounted to a suburban community of single and double houses laid out according to the existing street pattern. Some of the details Stern employed, and perhaps the house types themselves, were not close enough to their superior vernacular prototypes to be convincing today, but the idea was there. HUD later built a few single-family houses in the area that were snapped up at once despite their lack of psychological and physical support from a community group, and similar houses elsewhere can be seen standing in otherwise tragically trashed ghettos all over the Northeast, meticulously groomed behind their chain-link fences. There is reason to believe, therefore, that the Seaside type and related vernacular models, easy and economical to construct, might well be adapted for many urban situations. Such is already being done in dribs and drabs by Habitat for Humanity. But could it be funded as a mass program at urban scale?

Seaside, Kentlands, and Laguna West could be built by developers because there was money to be made out of them. Will it ever be possible to make money by building communities for the poor? Ways may yet be found to do so, some combination of private investment with intelligent government subsidies at all levels may do the trick. The federal government itself once spent so much money on Redevelopment, at a time when the architectural profession hardly knew what to do with the city, that we may hope it will reorder its priorities and begin to spend some now when the profession is better prepared to spend it wisely.

Urban organizations like Chicago's Center for Neighborhood Technology, of which Michael Freedberg is Director of Community Planning, are watching the work of Duany and Plater-Zyberk and Calthorpe and the others very carefully to see if there is anything in it they can use. In Chicago they sit, of course, at the heart of a wonderful urban–suburban order of the recent past, with the Loop, the community of work, and Oak Park, the community of home, perfectly connected by the elevated suburban train. But that order too is breaking down, with work moving centrifugally to Chicago's periphery so that the existing east–west transit serves it perfectly no longer. In the long run a version of Calthorpe's " TODs" might be of relevance here.

To say that there is much hope in this or any other present model would be overstating the case. But there is a lot of determination. One drowns in the urban situation but works with what one has. It would be sad if Seaside, for one, were not to inspire imitations far from the Gulf. Seaside itself sometimes seems to be sinking under the weight of its own success. Everybody in the country appears to be coming to look at it, smothering it, ironically enough, in automobiles during the summer. The only thing that can save it, says Duany, is more Seasides, plenty of them, and this is surely true in the largest social sense. The town of Windsor, for example, by Duany and Plater-Zyberk, with two polo fields, is aimed at as rich a clientele as exists. It offers large "estate" houses around a golf course and others along the shore. In the center, however, is a tightly-gridded town, and that is where every client so far has wanted to be. So the rich, who can choose, choose community, or at least its image. How much more must the poor, who must depend upon it for their lives, want community? If Seaside and the others cannot in the end offer viable models for that, they will remain entirely beautiful but rather sad. Perhaps they will in fact do so, because human beings are moved to act by symbols, and the symbol is there. When the great winds rise up out of the Gulf—and the storm clouds roll in thundering upon the little lighted town with its towered houses—then a truth is felt, involving the majesty of nature and, however partial, the brotherhood of mankind.

选读 5

"历史先例和其他先例：面向老建筑"

罗伯特・文丘里，丹尼斯・斯科特・布朗与史蒂芬・爱泽努尔

在这篇出自《向拉斯维加斯学习》的选读中，作者文丘里、斯科特・布朗和爱泽努尔探讨了在流行文化、大众消费和城市扩张背景下现代建筑的品质问题。以文丘里的早期作品《建筑的复杂性与矛盾性》（1966）为基础，三位作者试图将先例与影像的历史脉络联结到一起，作为解释当代设计师必须处置各种情况的依据。他们认为"商业风土建筑作为建筑象征主义的生动源泉"是有用处的。"我们在拉斯维加斯的研究中已经描述过了在长途高速度的冰冷汽车景观中，人们无福消受纯粹城市空间的精妙之处，在这里空间符号比空间形式更胜一筹。"这本《向拉斯维加斯学习》是由三位作者于1968年在耶鲁大学教授的建筑设计研究课程的内容演化而来。

## READING 5

## Historical and Other Precedents: Towards an Old Architecture

Robert Venturi, Denise Scott Brown, and Steven Izenour

In this selection from *Learning from Las Vegas,* Venturi, Scott Brown, and Izenour explore the qualities of modern architecture in the context of popular culture, mass consumption, and urban sprawl. Building on Venturi's earlier work, *Complexity and Contradiction in Architecture* (1966), the authors seek to draw together the historical threads of precedent and iconography to explain the conditions in which contemporary designers must work. They see the usefulness of "commercial vernacular architecture as a vivid source for symbolism in architecture. We have described in the Las Vegas study the victory of symbols-in-space over forms-in-space in the brutal automobile landscape of great distances and high speed, where the subtleties of pure architectural space can no longer be savored." The book evolved from work conducted in an architectual design studies at Yale University that was taught by the authors in 1968.

### HISTORICAL SYMBOLISM AND MODERN ARCHITECTURE

The forms of Modern architecture have been created by architects and analyzed by critics largely in terms of their perceptual qualities and at the expense of their symbolic meanings derived from association. To the extent that the Moderns recognize the systems of symbols that pervade our environment, they tend to refer to the debasement of our symbols. Although largely forgotten by Modern architects, the historical precedent for symbolism in architecture exists, and the complexities of iconography have contin-

*Source: Learning from Las Vegas* by Robert Venturi, Denise Scott Brown, and Steven Izenour. Copyright © 1977, 1972 by The Massachusetts Institute of Technology. Published by MIT Press. Reproduced by permission of MIT Press. pp. 104–27.

ued to be a major part of the discipline of art history. Early Modern architects scorned recollection in architecture. They rejected eclecticism and style as elements of architecture as well as any historicism that minimized the revolutionary over the evolutionary character of their almost exclusively technology-based architecture. A second generation of Modern architects acknowledged only the "constituent facts" of history, as extracted by Sigfried Giedion,[1] who abstracted the historical building and its piazza as pure form and space in light. These architects' preoccupation with space as *the* architectural quality caused them to read the buildings as forms, the piazzas as space, and the graphics and sculpture as color, texture, and scale. The ensemble became an abstract expression in architecture in the decade of abstract expressionism in painting. The iconographic forms and trappings of medieval and Renaissance architecture were reduced to polychromatic texture at the service of space; the symbolic complexities and contradictions of Mannerist architecture were appreciated for their formal complexities and contradictions; Neoclassical architecture was liked, not for its Romantic use of association, but for its formal simplicity. Architects liked the *backs* of nineteenth century railroad stations—literally the sheds—and tolerated the fronts as irrelevant, if amusing, aberrations of historical eclecticism. The symbol systems developed by the commercial artists of Madison Avenue, which constitute the symbolic ambience of urban sprawl, they did not acknowledge.

In the 1950s and 1960s, these "Abstract Expressionists" of Modern architecture acknowledged one dimension of the hill town–piazza complex: its "pedestrian scale" and the "urban life" engendered by its architecture. This view of medieval urbanism encouraged the megastructural (or megasculptural?) fantasies—in this context hill towns with technological trimmings—and reinforced the antiautomobile bias of the Modern architect. But the competition of signs and symbols in the medieval city at various levels of perception and meaning in both building and piazza was lost on the space-oriented architect. Perhaps the symbols, besides being foreign in content, were at a scale and a degree of complexity too subtle for today's bruised sensibilities and impatient pace. This explains, perhaps, the ironical fact that the return to iconography for some of us architects of that generation was via the sensibilities of the Pop artists of the early 1960s and via the duck and the decorated shed on Route 66: from Rome to Las Vegas, but also back again from Las Vegas to Rome.

## THE CATHEDRAL AS DUCK AND SHED

In iconographic terms, the cathedral is a decorated shed *and* a duck.* The Late Byzantine Metropole Cathedral in Athens is absurd as a piece of architecture (Figure 5-1). It is "out of scale": Its small size does not correspond to its complex form—that is, if form must be determined primarily by structure—because the space that the square room encloses could be spanned without the interior supports and the complex roof configuration of dome, drum, and vaults. However, it is not absurd as a duck—as a domed Greek cross, evolved structurally from large buildings in greater cities, but developed symbolically here to mean cathedral. And this duck is itself decorated with an

---

[1] Sigfried Giedion, *Space, Time and Architecture* (Cambridge, MA: Harvard University Press, 1944), Part I.
* Editors Note: Refer to Reading 6 by Geoffrey Broadbent for definitions of Venturi, Scott Brown, and Izenour's use of the terms "shed" and "duck."

**FIGURE 5–1**  Metropole Cathedral, Athens. [Robert Venturi]

appliqué collage of *objets trouvés*—bas-reliefs in masonry—more or less explicitly symbolic in content.

Amiens Cathedral is a billboard with a building behind it (Figure 5-2). Gothic cathedrals have been considered weak in that they did not achieve an "organic unity" between front and side. But this disjunction is a natural reflection of an inherent contradiction in a complex building that, toward the cathedral square, is a relatively two-dimensional screen for propaganda and, in back, is a masonry systems building. This is the reflection of a contradiction between image and function that the decorated shed often accommodates. (The shed behind is also a duck because its shape is that of a cross.)

The facades of the great cathedrals of the Ile de France are two-dimensional planes at the scale of the whole; they were to evolve at the top corners into towers to connect with the surrounding countryside. But in detail these facades are buildings in themselves, simulating an architecture of space in the strongly three-dimensional relief of their sculpture. The niches for statues—as Sir John Summerson has pointed out—are yet another level of architecture within architecture. But the impact of the facade comes from the immensely complex meaning derived from the symbolism and explicit associations of the aedicules and their statues and from their relative positions and sizes in the hierarchic order of the kingdom of heaven on the facades. In this orchestration of messages, connotation as practiced by Modern architects is scarcely important. The shape of the facade, in fact, disguises the silhouette of nave and aisles behind, and

FIGURE 5–2    Amiens Cathedral, west front. [Jean Roubier, Paris]

the doors and the rose windows are the barest reflections of the architectural complex inside.

## SYMBOLIC EVOLUTION IN LAS VEGAS

Just as the architectural evolution of a typical Gothic cathedral may be traced over the decades through stylistic and symbolic changes, a similar evolution—rare in contemporary architecture—may also be followed in the commercial architecture of Las Vegas. However, in Las Vegas, this evolution is compressed into years rather than decades, reflecting the quicker tempo of our times, if not the less eternal message of commercial rather than religious propaganda. Evolution in Las Vegas is consistently toward more and bigger symbolism. The Golden Nugget casino on Fremont Street was an orthodox decorated shed with big signs in the 1950s—essentially Main Street commercial, ugly and ordinary (Figure 5-3). However, by the 1960s it was all sign; there was hardly any building visible (Figure 5-4). The quality of the "electrographics" was made more strident to match the crasser scale and more distracting context of the new decade and to keep up with the competition next door. The freestanding signs on the Strip, like the towers at San Gimignano, get bigger as well. They grow

**FIGURE 5–3**  Golden Nugget, Las Vegas, pre-1964. [*Learning from Las Vegas* studio, Yale University]

either through sequential replacements, as at the Flamingo, the Desert Inn, and the Tropicana, or through enlargement as with the Caesars Palace sign, where a freestanding, pedimented temple facade was extended laterally by one column with a statue on top—a feat never attempted, a problem never solved in the whole evolution of Classical architecture (Figure 5-5).

## THE RENAISSANCE AND THE DECORATED SHED

The iconography of Renaissance architecture is less overtly propagandistic than is that of medieval or Strip architecture, although its ornament, literally based on the Roman, Classical vocabulary, was to be an instrument for the rebirth of classical civilization. However, since most of this ornament depicts structure—it is ornament symbolic of structure—it is less independent of the shed it is attached to than ornament on medieval and Strip architecture (Figure 5-6). The image of the structure and space reinforces rather than contradicts the substance of the structure and space. Pilasters represent modular sinews on the surface of the wall; quoins represent reinforcement at the ends of the wall; vertical moldings, protection at the edges of the wall; rustication, support at the bottom of the wall; drip cornices, protection from rain on the wall; horizontal moldings, the progressive stages in the depth of the wall; and a combination

**FIGURE 5–4**  Golden Nugget, Las Vegas, post-1964. [*Learning from Las Vegas* studio, Yale University]

of many of these ornaments at the edge of a door symbolizes the importance of the door in the face of the wall. Although some of these elements are functional as well—for instance, the drips are, but the pilasters are not—all are explicitly symbolic, associating the glories of Rome with the refinements of building.

But Renaissance iconography is not all structural. The *stemma* above the door is a sign. The Baroque facades of Francesco Borromini, for instance, are rich with symbolism in bas-relief—religious, dynastic, and other. It is significant that Giedion, in his brilliant analysis of the facade of San Carlo alle Quattro Fontane, described the contrapuntal layerings, undulating rhythms, and subtle scales of the forms and surfaces as abstract elements in a composition in relation to the outside space of the street but without reference to the complex layering of symbolic meanings they contain.

The Italian palace is the decorated shed *par excellence.* For two centuries, from Florence to Rome, the plan of rooms *en suite* around a rectangular, arcaded *cortile* with an entrance penetration in the middle of a facade and a three-story elevation with occasional mezzanines was a constant base for a series of stylistic and compositional variations. The architectural scaffolding was the same for the Strozzi Palace with its three stories of diminishing rustication, for the Rucellai with its quasi-frame of three-ordered pilasters, for the Farnese with its quoined corners complementing the focus of the ornamental central bay and its resultant horizontal hierarchy, and for the Odescalchi with its monumental giant order imposing the image of one dominant story on three (Figures 5-7 and 5-8). The basis for the significant evaluation of the

**FIGURE 5–5** Caesars Palace, extended sign. [*Learning from Las Vegas* studio, Yale University]

CHAPTER 1: ARCHITECTURAL HISTORY AND THEORY 79

FIGURE 5–6  Belvedere Court, Vatican. [Museo Vaticano, Rome]

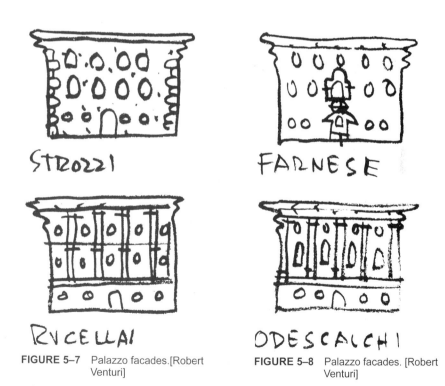

FIGURE 5–7  Palazzo facades. [Robert Venturi]

FIGURE 5–8  Palazzo facades. [Robert Venturi]

**FIGURE 5–9**   Carson Pirie Scott department store, Chicago. [Chicago Architectural Photographing Company]

**FIGURE 5–10**   Howard Johnson's Motor Lodge and Restaurant, Charlottesville, Virginia. [Camera Center, Charlottesville, Virginia]

development of Italian civic architecture from the mid-fifteenth to the mid-seventeenth century lies in the decoration of a shed. Similar ornament adorns subsequent palazzi, commercial and *senza cortili*. The Carson Pirie Scott department store supports at the ground floor a cast-iron cladding of biological patterns in low relief with intricate scale appropriate for sustaining the customers' interest at eye level, while abruptly opposing, in the formal vocabulary above it, the ugly and ordinary symbolism of a conventional loft (Figure 5-9). The conventional shed of a high-rise Howard Johnson motel is more Ville Radieuse slab than palazzo, but the explicit symbolism of its virtually pedimented doorway, a rigid frame in heraldic orange enamel, matches the Classical pediment with feudal crest over the entrance of a patrician palazzo, if we grant the change in scale and the jump in context from urban piazza to Pop sprawl (Figure 5-10).

## NINETEENTH-CENTURY ECLECTICISM

The stylistic eclecticism of the nineteenth century was essentially a symbolism of function, although sometimes a symbolism of nationalism—Henri IV Renaissance in France, Tudor in England, for example. But quite consistently styles correspond to building types. Banks were Classical basilicas to suggest civic responsibility and tradition; commercial buildings looked like burghers' houses; universities copied Gothic rather than Classical colleges at Oxford and Cambridge to make symbols of "embattled learning," as George Howe put it, "tending the torch of humanism through the dark ages of economic determinism,"[2] and a choice between Perpendicular and Decorated for midcentury English churches reflected theological differences between the Oxford and Cambridge Movements. The hamburger-shaped hamburger stand is a current, more literal, attempt to express function via association but for commercial persuasion rather than theological refinement (Figures 5-11–5-13).

Donald Drew Egbert,[3] in an analysis of midcentury submissions for the Prix de Rome at the Ecole des Beaux-Arts—home of the bad guys—called functionalism via

FIGURE 5–11   Eclectic bank. [*Learning from Las Vegas* studio, Yale University]

---

[2]George Howe, "Some Experiences and Observations of an Elderly Architect," *Perspecta 2, The Yale Architectural Journal,* New Haven (1954), p. 4.
[3]Donald Drew Egbert, "Lectures in Modern Architecture" (unpublished), Princeton University, c. 1945.

**FIGURE 5–12** Eclectic church. [*Learning from Las Vegas* studio, Yale University]

**FIGURE 5–13** Hamburger stand, Dallas, Texas. [Spencer Parsons]

association a symbolic manifestation of functionalism that preceded the substantive functionalism that was a basis for the Modern movement: Image preceded substance. Egbert also discussed the balance in the new nineteenth-century building types between expression of function via physiognomy and expression of function via style. For instance, the railroad station was recognizable by its cast-iron shed and big clock. These physiognomic symbols contrasted with the explicit heraldic signing of the Renaissance-eclectic waiting and station spaces up front. Sigfried Giedion called this artful contrast within the same building a gross contradiction—a nineteenth-century "split in feeling"—because he saw architecture as technology and space, excluding the element of symbolic meaning.

## MODERN ORNAMENT

Modern architects began to make the back the front, symbolizing the configurations of the shed to create a vocabulary for their architecture but denying in theory what they were doing in practice. They said one thing and did another. Less may have been more, but the I-section on Mies van der Rohe's fire-resistant columns, for instance, is as complexly ornamental as the applied pilaster on the Renaissance pier or the incised shaft in the Gothic pier. (In fact, less was more work.) Acknowledged or not, Modern ornament has seldom been symbolic of anything nonarchitectural since the Bauhaus vanquished Art Deco and the decorative arts. More specifically, its content is consistently spatial and technological. Like the Renaissance vocabulary of the Classical orders, Mies's structural ornament, although specifically contradictory to the structure it adorns, reinforces the architectural content of the building as a whole. If the Classical orders symbolized "rebirth of the Golden Age of Rome," modern I-beams represent "honest expression of modern technology as space"—or something like that. Note, however, it was "modern" technology of the Industrial Revolution that was symbolized by Mies, and this technology, not current electronic technology, is still the source for Modern architectural symbolism today.

## ORNAMENT AND INTERIOR SPACE

Mies's I-section appliqués represent naked steel-frame construction, and they make the necessarily bulky, enclosed, fire-resistant frame underneath look thinner through their complex articulations. Mies used ornamental marble in his early interiors to define space. The marble and marblelike panels in the Barcelona Pavilion, the House with Three Courts, and other buildings of that period are less symbolic than the later exterior pilasters, although the lush veneering of the marble and its reputation for rarity connote richness (Figure 5-14). Although these "floating" panels can now almost be mistaken for abstract expressionist easel paintings of the 1950s, their purpose was to articulate Flowing Space by directing it within a linear steel frame. Ornament is the servant of Space.

The Kolbe sculpture in this pavilion may have certain symbolic associations, but it too is there primarily to punctuate and direct space; it points up through contrast the machine aesthetic forms around it. A later generation of Modern architects has made these configurations of directional panels and punctuating sculpture the accepted

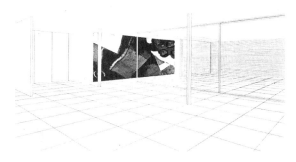

FIGURE 5–14 House with Three Courts; perspective of bedroom wing; Mies van der Rohe. [Mies van der Rohe, Ludwig. House with Three Courts (1934). Interior perspective. Pencil, cut-out reproduction (detail from George Braque: Fruit Dish, Sheet Music and Pitcher, 1926) on illustration board, 30″ × 40″ (76.2 × 101.8 cm). The Museum of Modern Art, New York. Gift of the architect © 1997 The Museum of Modern Art, New York.]

**FIGURE 5–15**  Casino Pio V, Rome. [*Learning from Las Vegas* studio, Yale University]

technique for exhibition and museum display, giving the display elements an informational as well as a space-directing role. Mies's elements were symbolic rather than informational; they contrasted the natural with the machined, demonstrating what Modern architecture was by setting it against what it was not. Neither Mies nor his followers used the forms symbolically to convey other-than-architectural meaning. Social realism in a Mies pavilion would be as unthinkable as a WPA mural in the Petit Trianon (except that the flat roof itself was a symbol of socialism in the 1920s).

In the Renaissance interior too, ornament is used along with plenty of light to direct and punctuate space. But here in contrast with the Mies interiors, it is the constructional elements that are ornamental—the frames, moldings, pilasters, and architraves that reinforce the forms and identify enclosed space—while the surfaces are the neutral context. Inside the Mannerist Casino Pio V, however, pilasters, niches, architraves, and cornices obscure the nature of the space or, rather, make the distinction between wall and vault ambiguous, because these elements, traditionally identified with walls, extend over the vault's surface (Figure 5-15).

In the chapel of the Byzantine Martorama in Sicily there is no question of architectural clarification or of Mannerist ambiguity (Figure 5-16). Instead, representation smothers space, its patterns camouflaging the forms it adorns. The ornamental patterns are almost independent of, and at times contradictory to, walls, piers, soffits, vaults, and dome. These forms are rounded at their edges to accommodate continuous mosaic surfaces, and the gold mosaic background further softens the geometry, while in the obscure light that occasionally highlights significant symbols, space disintegrates into an amorphous glow. The gilded rocaille in the Amalienburg pavilion at Nymphenburg does the same thing with bas-relief (Figure 5-17). Motival bas-relief, splattered like spinach over walls and furniture, hardware and sconces; reflected by mirrors and crystal fixtures; enhanced by generous light yet obscured by

CHAPTER 1: ARCHITECTURAL HISTORY AND THEORY    85

**FIGURE 5–16**    Martorama, Palermo. [Charles Brickbauer]

**FIGURE 5–17**    Amalienburg Pavilion, Nymphenburg. [Charles Brickbauer]

indeterminate curves in plan and section, disintegrates space into an amorphous glitter. Significantly, the Rococo ornament is hardly symbolic and not at all propagandistic. It obscures space, but the ornament is still architectural; in the Byzantine church, propagandistic symbolism overwhelms architecture.

## THE LAS VEGAS STRIP

The Las Vegas Strip at night, like the Martorama interior, is symbolic images in dark, amorphous space; but, like the Amalienburg, it glitters rather than glows (Figure 5-18). Any sense of enclosure or direction comes from lighted signs rather than forms reflected in light (Figure 5-19). The source of light in the Strip is direct; the signs themselves are the source. They do not reflect light from external, sometimes hidden, sources as is the case with most billboards and Modern architecture. The mechanical movement of neon lights is quicker than mosaic glitter, which depends on the passage of the sun and the pace of the observer; and the intensity of light on the Strip as well as the tempo of its movement is greater to accommodate the greater spaces, greater speeds, and greater impacts that our technology permits and our sensibilities respond to. Also, the tempo of our economy encourages that changeable and disposable environmental decoration known as advertising art. The messages are different now, but despite the differences the methods are the same, and architecture is no longer simply the "skillful, accurate, and magnificent play of masses seen in light."

The Strip by day is a different place, no longer Byzantine (Figure 5-20). The forms of the buildings are visible but remain secondary to the signs in visual impact and symbolic content. The space of urban sprawl is not enclosed and directed as in traditional cities. Rather, it is open and indeterminate, identified by points in space and patterns on the ground; these are two-dimensional or sculptural symbols in space rather than buildings in space, complex configurations that are graphic or representational. Acting as symbols, the signs and buildings identify the space by their location and direction, and space is further defined and directed by utility poles and street and

**FIGURE 5–18**   Fremont Street, Las Vegas. [*Learning from Las Vegas* studio, Yale University]

**FIGURE 5–19** Las Vegas Strip at night. [*Learning from Las Vegas* studio, Yale University]

parking patterns. In residential sprawl the orientation of houses toward the street, their stylistic treatment as decorated sheds, and their landscaping and lawn fixtures—wagon wheels, mailboxes on erect chains, colonial lamps, and segments of split-rail fence—substitute for the signs of commercial sprawl as the definers of space (Figures 5-21 and 5-22).

Like the complex architectural accumulations of the Roman Forum, the Strip by day reads as chaos if you perceive only its forms and exclude its symbolic content. The Forum, like the Strip, was a landscape of symbols with layers of meaning evident in the location of roads and buildings, buildings representing earlier buildings, and the sculpture piled all over. Formally the Forum was an awful mess; symbolically it was a rich mix.

The series of triumphal arches in Rome is a prototype of the billboard (*mutatis mutandis* for scale, speed, and content). The architectural ornament, including pilasters, pediments, and coffers, is a kind of bas-relief that makes only a gesture toward architectural form. It is as symbolic as the bas-reliefs of processions and the inscriptions that compete for the surface (Figure 5-23). Along with their function as billboards carrying messages, the triumphal arches in the Roman Forum were spatial markers channeling processional paths within a complex urban landscape. On Route 66 the billboards, set in series at a constant angle toward the oncoming traffic, with a standard distance between themselves and from the roadside, perform a similar formal-spatial function. Often the brightest, cleanest, and best-maintained elements in

**FIGURE 5–20**  Las Vegas Strip by day. [*Learning from Las Vegas* studio, Yale University]

**FIGURE 5–21**  Suburban residential sprawl. [*Learning from Levittown* studio, Yale University]

FIGURE 5–22  Suburban mailbox. [*Learning from Levittown* studio, Yale University]

industrial sprawl, the billboards both cover and beautify that landscape. Like the configurations of sepulchral monuments along the Via Appia (again *mutatis mutandis* for scale), they mark the way through the vast spaces *beyond* urban sprawl. But these spatial characteristics of form, position, and orientation are secondary to their symbolic function. Along the highway, advertising Tanya via graphics and anatomy, like advertising the victories of Constantine via inscriptions and bas-reliefs, is more important than identifying the space (Figure 5-24).

## URBAN SPRAWL AND THE MEGASTRUCTURE

The urban manifestations of ugly and ordinary architecture and the decorated shed are closer to urban sprawl than to the megastructure (Figures 5-25 and 5-26). We have explained how, for us, commercial vernacular architecture was a vivid initial source for symbolism in architecture. We have described in the Las Vegas study the victory of symbols-in-space over forms-in-space in the brutal automobile landscape of great distances and high speed, where the subtleties of pure architectural space can no longer be savored. But the symbolism of urban sprawl lies also in its residential architecture, not only in the strident, roadside communications of the commercial strip (decorated shed or duck). Although the ranch house, split level or otherwise, conforms in its spatial configuration to several set patterns, it is appliquéd with varied though conforming ornament, evoking combinations of Colonial, New Orleans, Regency, Western, French

FIGURE 5–23  Arch of Constantine, Rome. [*Learning from Las Vegas* studio, Yale University]

**FIGURE 5-24**   Tanya billboard, Las Vegas. [*Learning from Las Vegas* studio, Yale University]

**FIGURE 5–25**  Las Vegas Strip. [Denise Scott Brown]

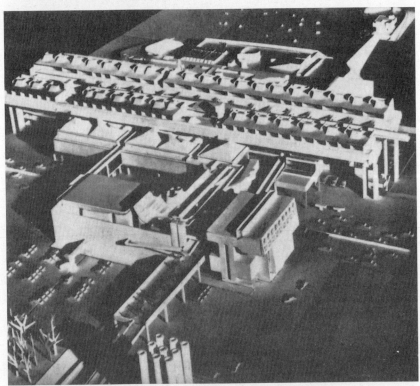

**FIGURE 5–26**  Town center, Cumbernauld, Scotland; Cumbernauld Development Corporation. [Bryan and Shear, Ltd.]

Provincial, Modern, and other styles. Garden apartments—especially those of the Southwest—equally are decorated sheds whose pedestrian courts, like those of motels, are separate from, but close to, the automobile. A comparison of urban sprawl with the megastructure is made in Table 5-1.

Sprawl City's image (Figure 5-27) is a result of process. In this respect it follows the canons of Modern architecture that require form to result from function, structure, and construction methods, that is, from the processes of its making. But for our time the megastructure (Figure 5-28) is a distortion of normal city building process for the

**FIGURE 5–27**   Residential strip. [Denise Scott Brown.]

**FIGURE 5–28**   Habitat, Expo '67. [Moshe Safdie]

TABLE 5–1  COMPARISON OF URBAN SPRAWL WITH MEGASTRUCTURE

| Urban Sprawl | Megastructure |
| --- | --- |
| Ugly and ordinary | Heroic and original |
| Depends on explicit symbolism | Rejects explicit symbolism |
| Symbols in space | Forms in space |
| Image | Form |
| Mixed media | Pure architecture |
| Big signs designed by commercial artists | Little signs (and only if absolutely necessary) designed by "graphic artists" |
| Auto environment | Post- and pre-auto environment |
| Cars | Public transportation |
| Takes the parking lot seriously and pastiches the pedestrian | "Straight" architecture with serious but egocentric aims for the pedestrian; it irresponsibly ignores or tries to "piazzafy" the parking lot |
| Disneyland | Piazzas |
| Promoted by sales staff | Promoted by experts |
| Feasible and being built | Technologically feasible perhaps, but socially and economically unfeasible |
| Popular life-style | "Correct" life-styles |
| Historical styles | Modern style |
| Uses typological models | Uses original creations |
| Process city | Instant city |
| Broadacre city | Ville Radieuse |
| Looks awful | Makes a nice model |
| Architects don't like | Architects like |
| 20th-century communication technology | 19th-century industrial vision |
| Social realism | Science fiction |
| Expedience | Technological indulgence |
| Expedient | Visionary |
| Ambiguous urban image | Traditional urban image |
| Vital mess | "Total design" (and design review boards) |
| Building for markets | Building for Man |
| This year's problems | The old architectural revolution |
| Heterogeneous images | The image of the middle-class intelligentsia |
| The difficult image | The easy image |
| The difficult whole | The easy whole |

sake *inter alia* of image. Modern architects contradict themselves when they support functionalism *and* the megastructure. They do not recognize the image of the process city when they see it on the Strip, because it is both too familiar and too different from what they have been trained to accept.

## HISTORY AND THEORY OF ARCHITECTURE—
## SUGGESTED READINGS

Banham, Reyner. *Theory and Design in the First Machine Age.* Cambridge, MA: MIT Press, 1980.

Colquhoun, Alan. *Modernity and the Classical Tradition.* Cambridge, MA: MIT Press, 1989.

Le Corbusier. (Charles-Edouard Jeanneret) *Towards a New Architecture.* New York: Holt, Rinehart and Winston, 1976.

Frampton, Kenneth. *Studies in Tectonic Culture.* Cambridge, MA: MIT Press, 1995.

Ghirardo, Diane. *Architecture after Modernism.* New York: Thames and Hudson, 1996.

Giedion, Sigfried. *Space, Time, and Architecture: The Growth of a New Tradition.* Cambridge, MA: Harvard University Press, 1967.

Herdeg, Klaus. *The Decorated Diagram: Harvard Architecture and the Failure of the Bauhaus Legacy.* Cambridge, MA: MIT Press, 1997.

Hitchcock, Henry Russell, and Philip Johnson. *The International Style.* New York: Norton, 1996.

Jencks, Charles. *The Language of Post-Modern Architecture.* New York: Rizzoli, 1991.

Kahn, Louis. *What Will Be Has Always Been.* New York: Rizzoli, 1986.

Kostaf, Spiro. *A History of Architecture.* New York: Oxford University Press, 1995.

Moore, Charles; Gerald Allen; and Donlyn Lyndon. *The Place of Houses.* New York: Holt, Rinehart and Winston, 1974.

Nesbitt, Kate, ed. *Theorizing a New Agenda for Architecture.* New York: Princeton Architectural Press, 1996.

Pevsner, Nikolaus. *A History of Building Types.* Princeton: Princeton University, 1976.

Rowe, Colin, and Fred Koetter. *Collage City.* Cambridge, MA: MIT Press, 1978.

Scully, Vincent. *Architecture: The Natural and the Manmade.* New York: St. Martin's Press, 1991.

Summerson, John. *The Classical Language of Architecture.* London: Thames and Hudson, 1980.

Venturi, Robert. *Complexity and Contradiction in Architecture.* New York: Museum of Modern Art, 1977.

Vidler, Anthony. *The Architectural Uncanny.* Cambridge, MA: MIT Press, 1992.

Wittkower, Rudolf. *Architectural Principles in the Age of Humanism.* New York: St. Martin's Press, 1988.

# 2

# ARCHITECTURAL FORM
# 建筑形式

选读 6
"建筑师与他们的象征"
杰弗里·勃罗德彭特

建筑形式的主要特征之一是它可以作为功能、人类文化、政治权力或任何意义的符号或指示物。在形式中有所体验的人会从这些符号或指示物中推断出各种含义。杰弗里·勃罗德彭特根据不断变化的建筑理论与进程必将如何应对视觉可读性和含义清晰性的挑战来对这一现象进行调查。有一个主题在他这篇节选自《建成环境》的文章中得到阐述，即建筑象征了一种视觉语言，这种视觉语言包含着其他交流方式的所有要素。他认为建筑师对与他们的作品息息相关的人负有特殊的责任，应当用措辞准确的语言和对周边环境大文化有意义的方式与人们进行沟通。他通过展现建筑语言的变化如何遮掩、提升或是混淆建筑形式的含义来对这一概念做出分析，但是那些穿越时间和历史长河依然能够为人所解读的最基本符号却是永恒不变的。

建筑范式的转变一直在持续——从19世纪的折中主义到现代运动的功能主义，再到廉价粗俗、以次充好的世界——尽管它们的语言可能发生改变，但是符号是永存的。

## READING 6

### Architects and Their Symbols

#### Geoffrey Broadbent

One of the central characteristics of architectural form is the way it acts as a symbol or signifier of function, human culture, political power, or any kind of meaning that can be inferred by the person that experiences form. Geoffrey Broadbent investigates this phenomenon in light of how changing architectural

*Source*: "Architects and Their Symbols" in *Built Environment* by Geoffrey Broadbent. 1980. Reproduced by permission of Alexandrine Press. Volume 6, no. 1, pp. 10–28.

theories and processes must address the challenge of visual legibility and clarity of meaning. The theme that he develops in this article from *Built Environment* is the notion that architecture embodies a visual language that contains all the ingredients of other forms of communication. He argues that architects have a specific responsibility to the people who interact with their work to speak in terms that are precise and in ways that are meaningful to the larger culture that surrounds the environment. He analyzes this concept by showing how shifts in the architectural language can mask, enhance, or confuse meaning in built form, but the fundamental symbols that are read across time and history remain constant.

Architectural paradigms shift continually—19th century eclecticism to Modern Movement functionalism to the world of honky-tonk, crassness, phoniness—but symbols remain though their language may change.

You'd think, to read some of the journals these days, especially the letters columns, that architects, planners and researchers seem to be completely out of touch, not only with each other but also with "the people" whom all three are supposed to be serving. Architects and planners certainly have done things which "the people" most certainly did not want, and the people themselves don't seem to be able to say what they would like if they could have it. Researchers think they know what has gone wrong and they also have ideas on what can be done to put things right again. But most architects and planners don't seem to want to know. They ignore most of the research which is published or dismiss it as incomprehensible, irrelevant, too complicated, too trivial, not focused on their everyday problems. Researchers argue back, correctly, that *if* the professionals would only listen, then they might realize that the problems which they, the researchers, have tackled really are the deep-seated ones; that it is largely because they ignored them that the professionals got so much wrong. They may go on to argue even further that if the gaps were to be closed between researchers and professionals, then the gaps between professionals and people would also be closed!

It sounds like a lot of bloody-minded, not to say hostile, people, and some of them most certainly are. But often they are people who really care; they have the best of intentions, but they are far too busy, too preoccupied, too set in their ways even to *want* to engage in that creative dialogue which might cause all parties to change. They are each working within a "paradigm," to use that awful word which Thomas Kuhn coined in 1962 to describe the set of social pressures acting on a particular group. He was writing about scientists, but the principles apply also to architects, planners and psychologists, indeed to every kind of group, and the paradigm is that which forces them to conform to the norms of their group. If a scientist wants to gain and retain the respect of his colleagues, to get his work published in the reputable journals, to get invited to conferences and so on, he simply *has* to work within that general framework which everyone in his field accepts as "correct" at the time. Kuhn's point is that the "normal" practitioner always works in this way, but there are always a few brave spirits who *know* that the paradigm is wrong. They think about it, work at it and eventually present alternatives which, better though they may be than the going paradigm, meet, at first, with the greatest hostility, especially from those who have made their reputations within the old paradigm. The "normal" scientist knows what to do and jogs along happily doing it, but then a Newton appears on the scene to challenge the established order of things. At first the majority rejects his views but eventually the opposition dies away, and what seemed new and strange at first becomes the new paradigm. But then an Einstein comes along to show where Newton was wrong, a Heisenberg shows what was wrong with Einstein and so on. And that is how science progresses. The same thing obviously happens in other fields, such as architecture and planning, not to mention psychology, sociology and so on.

Paul Feyerabend goes into greater detail (1978) as to just why and how those who have made their reputations within a particular paradigm are so highly suspicious of those who have set themselves the task of showing its flaws and deficiencies. In his view such challenges come largely from what he calls the "philosophical component" of the field, that is, the researchers and theoreticians who, having pondered deeply on the problems which beset the going paradigm, present those ideas which challenge the *status quo*. Naturally, the average practitioner finds them threatening, even in the comparatively closed world of science: how much more threatened are the practitioners of architecture and planning, challenged as they are not only by "philosophical components" of their particular fields, but also by the great user-public itself, not to mention their self-appointed and highly vocal spokesmen—the critics and journalists. But what *is* that paradigm, that *status quo,* which is open to attack from so many directions? One of the great advantages of Kuhn's paradigm is that he defines it in something like 21 different ways; among other things he suggests that the "notion of style in the arts" could become much clearer if we saw styles themselves as paradigms. The history of architecture obviously is a catalogue of changing styles, Egyptian to Greek to Roman and so on. Within each of these there are what we might call "sub-paradigms" such as: the work of individual architects, particular building types and so on. The Gothic paradigm obviously had individual architects, such as Henry Yevele. It also had churches, cathedrals and castles, not to mention bridges, houses, monasteries, colleges, universities and other building types. The Modern Movement had its Le Corbusiers, its Gropiuses and its Mieses. It also had tall thin curtain-walled slabs for "filing" people at work, large panel pre-cast concrete ones, somewhat longer and lower, for filing them when they were at home. It had light and dry prefabricated systems designed first of all for building schools and then adapted like a Meccano set for universities, hospitals and so on. All these were sub-paradigms of the Modern Movement and it is hardly surprising that those who made their reputations—not to mention their money—designing them, resent very much the ways in which journalists, on the one hand, and researchers, on the other, tell them with some conviction that they were wrong.

But the fact is that, whether we like it or not, paradigm changes of the kind which Kuhn describes are actually occurring at this moment. And, as with all such changes in the past, those responsible for them have concentrated on precisely those aspects of the Modern Movement paradigm which have been causing the greatest distress: the social problems caused by housing families in high-rise flats; the notorious deficiencies of curtain walling in terms of heat loss, solar heat gain, noise penetration and so on; the sheer bleak sterility of that International Style which, having been built right round the world, has severely diminished, if not yet finally destroyed, the sense of identity at any particular place.

So it is hardly surprising that both public criticism and research have been concentrated in these particular areas. I have been asked to take one—the question of identity, meaning and sense of place—to outline what the researchers have done and to show how far the results of their research have been applied into practice.

These are indeed vital questions, for it was a deep-seated tenet of the Modern Movement in architecture that buildings could, and should, be "Functional." This is to say, they should be the simplest, the most direct and the cheapest solution to a particular design problem and any attempt to make them "look like," say, any building from the past, was viewed with the greatest suspicion. That in itself was a natural—and inevitable—reaction to a previous design paradigm; it was an attempt to curb the excesses of 19th century eclecticism.

Small wonder that the founding fathers of the Modern Movement felt very strongly about it. The first shot in this particular argument was fired by Adolph Loos in a famous essay of 1908 called "Ornament and Crime." Savages, he said, tattoo themselves, so it was hardly surprising that a Papuan, for instance, having tattooed his skin, should also go on to tattoo "his boat, his paddles, in short everything he can lay his hands on." But it was a sign of civilization that modern man did not tattoo himself, for "the modern man who tattoos himself is either a criminal or a degenerate. There are prisons in which eighty percent of the inmates show tattoos . . . If someone who is tattooed dies at liberty, it means he has died a few years before committing a murder." Spurious though it obviously was, Loos's argument struck a sympathetic chord amongst those who, wanting to make *their* contribution to the development of architecture—to change the paradigm—were reacting *against* eclecticism.

So it is hardly surprising that, having read Loos, the young Le Corbusier should declare (1923):

> The styles of Louis XIV, XV, XVI, or Gothic, are to architecture what a feather is on a woman's head; it is sometimes pretty, though not always and never anything more.

Gropius too was extremely critical of the "artistic gentleman architects who turned our charming Tudor mansions." He asked the question (1956): "What constitutes a style?" and answered it:

> The irrepressible urge of critics to classify contemporary movements which are still in a state of flux by putting each neatly into a coffin with a style label on it . . . What we looked for was a new approach, not a new style.

And by 1940 J. M. Richards saw the concept of style as a dead and labeled thing. As he put it:

> Presumably all thinking people now agree it is absurd to put up houses that look like miniature castles, petrol stations that look like medieval barns, department stores that look like the palaces of Renaissance bishops—quite apart from being extremely inefficient.

Pevsner too had his views as to how the socially *responsible* architects would face up to the social responsibilities of the mass society. As he first put it in 1936:

> The warmth and directness with which ages of craft and a more personal relation between architect and client endowed buildings of the past may have gone for good. The architect, to represent this century of ours, must be colder, cold to keep in command of mechanical production, cold to design for the satisfaction of anonymous clients.

So, his architecture must *look* cold. Thus the bleak sterility of much modern architecture was a deliberate *expression* of "social responsibility." How fortunate for such architects and critics that the *appearance* of anonymous neutrality could be precisely achieved by those prefabricated building systems which they also tended to favour. The fact that prefabricated buildings were almost always worse than traditional buildings in terms of environmental performance and flexibility of planning, not to mention capital, running and maintenance costs, really was not the point: they *expressed* the application of technology to the solution of social problems. True practicalities do not loom very large when you are driven by visionary commitment, for it is visionary commitments above all that lead to paradigm shifts. That particular shift was certainly successful, with results which are there to be seen brooding on the skyline of virtually all of our cities.

At the extreme, such architecture was reduced to the simple rectangular box which specifically was not intended to "mean" anything. The ultimate of such "meaningless" architecture, I suppose, was Mies van der Rohe's campus for the Illinois Institute of Technology. Every building there *is* reduced to a simple rectangular box: steel-framed with buff brick infilling, glass infilling, or some combination of the two. Two of the smaller buildings, however, literally are brick boxes. Looking, as they do, so similar they "carry," as it were, the same "meaning;" yet one is the chapel and the other merely a filling station. But what a strange sense of values: that equates the great god motor car to exactly the same status as the great god God!

Yet even Mies's buildings were based on the tacit acknowledgement that buildings do "express" something. Admirer as he was of Schinkel's Neo-Classicism—especially the Altes Museum in Berlin (as he put it in 1960: "I learned everything in architecture from that")—he still would not bring himself to design with classical columns.

The trouble was that once you started to "express" the structure, the visual appearance of a structure became much more important than its actual functioning. (The same has happened since with the services as one sees at the Beaubourg in Paris.) Given the Miesian approach to "expressing" structure there were times when the fire officer insisted that your basic steel frame—that which actually held the building up—had to be clad in concrete. So, committed as you were to *expressing* structure you added a fictitious one outside: steel—in the case of Seagram even bronze—I-sections to serve no other purpose but visual effect.

The crucial point about all this is that an architectural paradigm that the Modern Movement was supposed specifically *not* to symbolize turns out to have been fraught with symbolism. Some building types themselves were used so often that we learned to read them as symbols. Take the curtain-walled city centre slab. This obviously became a deliberate symbol. It is extremely wasteful in terms of capital, running and maintenance costs, but it originated in Park Avenue in New York where it came to symbolize business success in exactly the same way as the Gothic cathedrals symbolized the faith of the Middle Ages. So it is hardly surprising that the successful multinationals use it to symbolize their presence wherever they go in the world. As for the grey, concrete multistory housing slab, Sir Leslie Martin's *Land Use and Built Form Studies* had already proved in the early 1960s that you could actually pack more habitable rooms, with the requisite sun penetration, by putting four-storey courtyards across the same area of land (see March and Trace, 1968). But you have to go into four-storey courtyards to see how life goes on there. City councillors, city planners, city architects, housing managers and so on wanted much more visible

**FIGURE 6–1** Two buildings on the campus of the Illinois Institute of Technology: top, Mies van der Rohe's chapel; bottom, the filling station. [Used by permission of *Built Environment*]

**FIGURE 6–2** The painted pipes of the Beaubourg, "expressing" services. [Used by permission of *Built Environment*]

symbols than that. They could point with pride to their high-rise flats and say, "Look what *we* did in our time *for* the people."

One could point to dozens of other examples in the Modern Movement of building types, methods of construction and so on of which the known deficiencies can *only* be explained by the precedence their designers gave to symbolism, rather than to the realities of practical architecture and practical habitation. The fact that they got away with it by *arguing* a "functional" case for building or, more likely, an economic case—which was not necessarily true—merely serves to show that the arguments in themselves were mere symbols, going through the motions, as it were. For the fact is that, whether we like it or not, buildings *do* symbolize; there is no way we can avoid it. The architect is in much the same position as Molière's Monsieur Jourdain who, on being told by his philosophy master that there were only two ways of putting words together, namely poetry and prose, and that the latter included ordinary, everyday speech, exclaimed, with evident surprise: "Well, blow me, I've been speaking prose for more than forty years and I didn't even know it!"

It is hardly surprising that after all the cant written about Modern Architecture which was supposed to transcend the *idea* of style (but which in practice "expressed" all manner of things) that our present paradigm shifters have grasped the nettle. They have been looking at style and meaning, not merely as things to be accepted with reluctance, but as things to be embraced with enthusiasm as the very stuff of which architecture is made. If science itself is subject to styles—in the form of Kuhnian paradigms—then why try to pretend that architects can avoid it? For even those movements that see it as superficial—those movements concerned with user participation, community involvement, advocacy planning and so on—they too are fashion-like paradigms in the strictest Kuhnian sense!

Those who are looking at style again in its "Post Modern" manifestations include not only historians and theorists but also practising architects, not to mention that much rarer creature, the genuine theorist-practitioner. And there is no doubt that collectively they *are* achieving a paradigm shift.

So instead of reviving the idea—which is what the Modern Movement pretended to do—the paradigm which is emerging simply asks us to lie back and enjoy it.

That attitude has been expressed in his laid-back California way by Charles Moore who defines five points for what Charles Jencks and others (1977) have described as Post Modern Architecture. Moore says (1976):

1 Buildings *can* and *should* speak.

2 Therefore they should have *freedom* to speak. Functionalism suppressed that idea at which point architecture simply stopped being interesting for most people. But once we admit that buildings can speak again we should allow them to be wistful, wise, powerful, gentle, silly, just as people are.

3 Functional buildings, on the whole were bleak and hostile. Those which replace them must be inhabitable in the minds and the bodies of human beings, not to mention their plants, statues and other possessions.

4 We should therefore base the design of physical spaces in and around buildings, not on the abstraction of Cartesian geometry, but on the human body and the way we sense spaces.

5 Whether we like it or not, the spaces and shapes of buildings contain certain psychic qualities. We perceive these and they assist the human memory in restructuring connections in time and space.

There is nothing really new in these basic premises of Moore's, for the central one as we shall see had been used for at least 2,000 years.

Let us start with the practitioners: those who have tried deliberately to build an architecture with meaning. That sounds like "artistic gentlemen architects" of the kind which Gropius disliked, but even at the time when he and others of the *avant-garde* were attacking them, they *were* building the kind of architecture that most people liked. Try counting the number of "modern" houses built in Britain between the wars—Jeremy Gould's book (1977) actually catalogues them all—and then compare this with the number of typical suburban semis, with their half-timbered gables, which were built and bought or rented during the same period. Most of them, it is true, were built without benefit of architects, but they *were* in the direct tradition of the 19th century Arts and Crafts Movement which survived, unabated, right through the heroic days of the Modern Movement.

There were indeed "rogue" architects, such as Clough Williams-Ellis, who continued to build his Portmerion—an Italian hill-village on the coast of Cardigan Bay—which he started in 1928. He was revered as a *person* by the most tight-lipped of prefabricators, many of whom privately enjoyed the delights of Portmerion—its sheer human scale and humanity—even though it was diametrically opposed to everything they (publicly) stood for.

**FIGURE 6–3**  Daniel Spoerri's Port Grimaud. [Used by permission of *Built Environment*]

**FIGURE 6–4**  Xanadu, designed by Ricardo Bofill and his Taller di Arquitectura at Calpe. [Used by permission of *Built Environment*]

**FIGURE 6–5**   La Petite Cathédrale. [Used by permission of *Built Environment*]

Others such as Taylor and Green of Norwich continued to build housing in a (pared down) vernacular, while the vernacular has now become a "movement" thanks to Darbourne and Darke, Murray and Maguire, not to mention the Essex Design Guide (see *RIBA Journal* "Vernacular," 1980). Yet there is still an English modesty about most of this. The most spectacular examples are continental ones, such as Daniel Spoerri's Port Grimaud, a holiday village/marina which he started on the Gulf of Saint Tropez as early as 1966. It is in a sense a "Radburn" layout with pedestrian "fingers" reaching into the water, lined with "vernacular" houses which open on their vehicular sides, not to the motor roads of Radburn but onto the marina itself; where other houses have cars, the Port Grimaud houses have boats. The glossy magazines will show you all this as a piece of picturesque kitsch. But the fact is that in terms of people/vehicular relationships, room size, shape and arrangement for (holiday) domestic purposes, sound insulation, sun shading and so on, this is highly practical architecture for the Gulf of Saint Tropez even—no, especially—in economic terms.

Then there is the work of Ricardo Bofill and his Taller di Arquitectura of Barcelona. They dedicated themselves in the early 1960s to fighting the grey, anonymous "cemetery suburb" in which so many people had to live. To do so they experimented with holiday housing at Calpe on the Spanish Mediterranean coast. The whole complex is now called La Manzanera, but their early experiments included Plexus, holiday villas based on the local vernacular (1964), and Xanadu, a pagoda-like block of holiday apartments (1966). Xanadu draws on the local environment in ways we shall be analyzing later. Then they started to build in earnest, with the Barrio Gaudi at Reus (1964), some of the lowest-cost housing in Europe, in which they still managed to establish a coherent sense of identity and sense of place. Then their ambitions grew with La Petite Cathédrale for Cergy Pontoise, a new town outside Paris (1971)—we shall analyse this later for the way it "carries" meaning—and in 1976 there was the Church of Meritxell in Andorra which draws on the local Romanesque, and so on.

The Bofill group too have been dismissed as "rogues" much like Clough Williams-Ellis and certainly few "serious" architects at the time would admit that in Walt Disney's Disneyland (opened 1955) and Disneyworld (opened 1971) were to be seen anything like serious architecture. They were seen, rather, in Peter Blake's words as "honky-tonk, crassness, phoniness." Yet, in design they had been the subject of more analysis—using the full panoply of design method techniques from the Space Programme—than any other environment in the world. And that analysis obviously extended to what the buildings should look like, or, more particularly, what they should *mean*. You can't "read" Disneyworld unless you know Disney's films, not to mention his life story, so the basic steel-frame, concrete and environmental systems are all covered in clip-on glass-reinforced plastic which looks "like" Cinderella's castle or whatever. Except that, unlike the visible, symbolic "structure" of, say, Mies's Seagram building (bronze I-sections for visual effect), the GRP roofs of Disneyland at least serve the purpose of keeping out the rain!

Yet, as Le Corbusier, Gropius and others of the Modern Movement knew, buildings built don't effect paradigm changes unless someone writes about them (Mies got Johnson to do it for him). Articles in journals and even books go right around the world, whereas buildings stay in one place. The buildings I have mentioned started to become paradigm changers *because* people wrote about them: Alison and Peter Smithson on Port Grimaud (1972), Peter Blake (1972) and Roy Landau (1973) on Disneyworld, and myself on Bofill and his Taller (1973, 1975b).

Naturally, those committed to the old paradigm dismissed both the buildings themselves and our writing about the "honky-tonk, crassness, phoniness." But the ramparts had already been breached by one of the theorist-practitioners: Robert Venturi, who, in addition to teaching at the University of Pennsylvania also ran a practice, Venturi and Rauch. The first of his major publications, *Complexity and Contradiction in Architecture*, had been published by the Museum of Modern Art in 1966. This had been subsidized as a study in architectural history, but it was really a polemic for the Baroque as an architecture of "complexity and ambiguity." It ended, like Venturi's other book, with a selection of his buildings in which he attempted to reinterpret the principles of the Baroque *within* the mainstream of the Modern Movement. Most of them were rather careful exercises in which, having set up a basic symmetry, he then somewhat solemnly broke it with results which were, well, complex and contradictory.

The Venturi team's next book was *Learning from Las Vegas* (1972), a deliberate attempt, obviously, to shock those who saw Las Vegas as the epitome of "honky-tonk, crassness, and phoniness." The book is a bold expression of some rather subtle ideas. Venturi, Scott-Brown, Izenour and their students had gone to Las Vegas, Nevada, where they studied the architecture of "The Strip." The Strip consists, among other things, of hotels, each with a large entrance lobby, stuffed to bursting point with one-armed bandits, tables for poker, blackjack, craps and roulette, Keeno boards and Big Six wheels, through which one has to run the gauntlet before even reaching reception. In an extreme case, such as the Stardust Hotel, all this is housed in an anonymous "shed" behind which are some equally anonymous bedroom wings. But the shed, as Venturi says, is *decorated* with a glittering neon sign the full length of the façade and also a programmed sign at the kerbside which of course is highly visible to motorists as they cruise along The Strip.

**FIGURE 6–6** Blake's poultry-stand on Long Island. [Used by permission of *Built Environment*]

**FIGURE 6–7** Decoration on the Stardust Hotel, Las Vegas. [Used by permission of *Built Environment*]

It was from this and other examples that Venturi coined his term the "decorated shed" which is vital to his architectural philosophy. In practice this means that given any architectural problem he plans the most efficient building he can (and Venturi *is* one of the most efficient planners in the United States). Having thus exercised his responsibilities he then feels free to "decorate" his shed, that is to give it a surface treatment which will give "meaning" to his building, much as the signs give identity to an otherwise anonymous Las Vegas Hotel. Venturi contrasts his "shed" with quite another kind of building which he labels the "duck." This derives from his most extreme example, which he found in Peter Blake's book *God's Own Junkyard* (1974), a homage in reverse, as it were, to the worst of what Blake could find in the "honky-tonk, crassness, phoniness" of the typical American commercial strip. Blake's prize example was a poultry-stand on Long Island, New York, which was shaped to look like a duck. One could thus "read" its function—selling poultry—from the duck-like form of the building itself.

It is precisely because they seem so trivial that Venturi's examples raise some of the most important issues as to *how* buildings actually "carry" their meaning, the analysis of which has been the province of researchers in semiotics, or semiology, and related areas. These derive from the analysis of language, which is hardly surprising, for the whole purpose of language, obviously, is the deliberate transfer of ideas and their meanings from one person's brain to another's. This has been studied in a number of ways. One group of analysts, the Information Theorists such as Weiner (1948), Shannon and Weaver (1949) and so on, were concerned with the *efficiency* with which ideas may be converted while the others, including such philosophers as Charles Sanders Peirce (1960) and Ferdinand de Saussure (1906–11) were concerned with the actual *content* of the message, with how words or other symbols actually "carry" their meanings for us. Peirce and Saussure, in their very different ways, were each looking for what they called a general "Theory of Signs" and a sign, in this sense, is something, anything, which "stands for" or reminds us of something else. So a sign can be a word, written or spoken, a gesture, a diagram, a drawing or a picture, a jacket, a tie or a motor car, and, most certainly, a building. This is not the place to trace a history of the field nor even to expound its major concepts. I have tried to do that elsewhere (1977), so let me just outline one or two points.

First of all, it must be admitted that there are enormous problems of terminology and even disagreement between the founding fathers as to whether the field itself should be called Semiology (Saussure) or Semiotic (Peirce). So let me just pick out a few general points which are obviously of relevance to architecture. Saussure, for instance, distinguishes (1906–11) between *language* and *speech*. A *language,* in his terms, is something we all share, including, as it does, the words in the dictionary and a set of rules for stringing them together. But from the whole of the resources which a language makes available, each of us prefers certain words; we also prefer certain ways of putting them together. Saussure attaches the term *speech* to those personal uses and there are obvious parallels in architecture: in the use an individual architect makes of the going style; the way he uses the components of a system, as in his speech he selects from the available "language."

Saussure goes on to describe two uses we make of words; indeed he likens a word to a (classical) column. This also performs two functions: on the one hand a column helps to hold the building up, plays an essential part in its construction, just as a word plays an essential part in the construction of a sentence: it is a part of the *syntax*. But an Ionic column also reminds us that there are other kinds of column, Doric, Corinthian and so on. We *know* it is Ionic because of the place it occupies in the whole range of column types. And similarly, according to Saussure, we know the meaning of a word because of its position within a whole scheme of words: the word "architect," for instance, reminds us of architecture, architectural, architectonic, of building, construction and structure, of the monumental, the vernacular and so on. We may even hear of a politician described as "architect" of a particular policy, so we clarify for ourselves what "architect" means because of all these other words around it.

One of Saussure's key concepts is at least 2,000 years old, for Vitruvius himself wrote: "in all matters, but particularly architecture, there are these two points: the thing signified and that which gives it significance." Saussure's key point is that the two are linked, inextricably, to form what he calls a "sign," that is, something which "stands for" something else. Saussure's *signifier* is that which conveys an idea; the sounds of a spoken word, the marks on paper which form a written one, a drawing, a diagram, a television picture or even, and most certainly, a building. It is, therefore, that which we see, hear, touch or whatever, that which most of us just call "the sign." But that is not enough for Saussure, for whom a "sign" must also have thoughts linked to that physical form. He uses the word *signified* to describe such thoughts, ideas or concepts. Any dictionary will tell us what thoughts are "attached" to which words; the highway code tells us which thoughts are attached to the various road "signs" and so on. The road "sign" in fact makes the point very well. For in a Saussurean view it is not *just* that disc or that triangle attached to a pole, at the side of the road, it is *also* the thoughts which it "carries."

Others since Saussure, while agreeing with him that a sign is more than the thing you see, the thing you hear and so on, have gone on to argue that most of us, most of the time, are having our ideas about *things*. So two English linguists, Odgen and Richards, decided in 1923 that they should incorporate into the "sign" the thing to which it refers and they decided to call it the "referent." Thus they developed their "semiological triangle," still one of the most useful devices in probing the meanings of things:

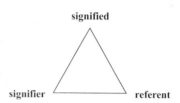

*Where:*

signifier = the spoken or written word, the diagram, drawing, picture, or whatever by which the ideas are being conveyed
signified = the thoughts, ideas, concepts and so on which actually are being conveyed
referent = the object, person or other kind of thing to which those concepts refer.

We can understand a lot more about our buildings if we think of them in these ways. If I write, say, about Ronan Point and illustrate it with a picture, my words and my picture are the "signifiers." They *signify* ideas—the "signifieds" about a block of flats in London. And that block of flats still exists as a thing, that is, a referent, which you can go and see, touch, photograph, or even kick to convince yourself of its physical reality. But it has also come to "stand for" a lot of things; about how unpleasant it is for young families to live in high-rise flats, about the inhumanity of putting real live people into such machine-made and machine-like environments, about the fading of the "great white heat of technology" to a very dull glow indeed. Ronan Point, in other words, is a "signifier" in itself to a highly complex set of ideas.

Obviously we can go into deeper and deeper layers of meaning by using this one device: the three-part sign of Odgen and Richard's semiological triangle. But finally it has its limitations; there comes a point where it can no longer help us because it describes only one type of sign. That is where Charles Sanders Peirce comes in, for he, with his extraordinarily complex mind, once went on record as saying that there were 59,049, that is $3^{10}$, *types* of sign, of which the triangle represents only one. Peirce's language is exceedingly turgid; he would never use an existing word if he could coin a new one, never coin a single-syllable word if he could coin one with three; he changes his mind, and his writing is so ambiguous as to be virtually impenetrable. But he had some amazing insights and one of his many sign-groupings has proved to be exceedingly fruitful.

When you think about it, the crucial feature of Saussure's one type of sign is that we have to *learn* it. Or, rather, we have to learn the relationships between his signifier and his signified. We learn that a particular letter on paper represents a certain sound, that a particular grouping of letters makes a certain word, that this word, according to the dictionary definition, refers to a certain thing, or at least "carries" a certain

**FIGURE 6–8** Roman Point, London. [Used by permission of *Built Environment*]

meaning. And that goes for all the words of a language, which is why we all have difficulties with those foreign languages which we have not had the time or the inclination to learn. Peirce uses the word "symbol" for signs such as this in which relationships between the signifier, the signified and the referent have to be learned. Among other things they include the words of ordinary language, the membership card which admits you to a particular club, the ticket which shows you have paid to ride, the coin or note in your pocket which carries a certain symbolic value and which therefore can be used to buy things. Some buildings obviously are symbols in this sense, such as the church, the cathedral, the temple or the mosque. We have learned by our experience within a particular culture that any large building with a tower, a spire, a pointed roof, pointed windows and so on probably was built as a church. If we go inside and find a nave, aisles, an altar cross and so on we know that it is still used as a church. If, on the other hand, we find that the altar has been taken away, that there is a stage with curtains where the sanctuary used to be, then we know—from our experience—that it has been converted into, say, an arts centre. Conversely, if we see a new and strangely shaped building, the chances are that this will be a church. We have learned from our experience that few buildings except for churches have been built in such strange shapes. And, again, if when we go inside we find that it actually has an altar, a font and so on, then these confirm that it really is a church.

The trouble with symbols is that because they have to be learned, their meanings cannot stay constant. Your membership card runs out unless you renew your subscription, your day return is no good even on the following day, the Bank of England can declare that those "real" old sixpences are no longer legal tender. Words too change their meanings; "prevent" used to mean "go in front of" and "gay" used to mean bright and cheerful. So it's hardly surprising that building forms also change their meanings. A hundred and fifty years ago, for instance, it seemed entirely appropriate to use one of the Greek orders for a museum, an art gallery, an athenaeum, a university, that is to say for any "cultural" building, for Greece was seen as the "cradle" of liberty, of democracy, of philosophy, of mathematics, of sculpture, of everything that was good in civilization, including architecture itself. So what more appropriate for a "cultural" building than the re-use of the Greek orders? But then in the 1930s the Fascist—and Communist—dictators of Europe began to use the Greek orders to express not freedom and democracy, but power—the naked power, that is, of the totalitarian state.

So symbols may be unreliable, which is why some of Peirce's other signs have their value. His "index" is a sign which *indicates,* by some direct physical relationship, the thing to which it is referring. A weather-vane indicates the direction of the wind, a rap on the door indicates that someone is there, a pointing finger indicates a particular book, a drawing, a part of the drawing and so on.

A building too can be an index. It can indicate by steps, canopies and so on where you go in. It can indicate by its plan which way you have to go; art galleries, museums and exhibition pavilions are often planned so as to force you around a set route; some of Le Corbusier's houses were planned as "architectural promenades." The point about an index in the way Peirce uses the term is that you don't have to learn it; you *know,* from its physical form, just what it indicates. You could take anyone, from the most primitive of cultures, put him in a building planned around a set route and he would have to follow the route indicated by the building. Yet put him in front of a

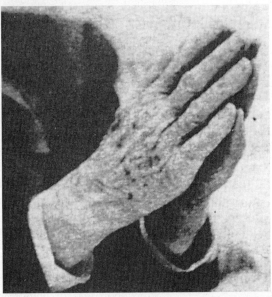

**FIGURE 6–9**  Frank Lloyd Wright's church at Madison and the praying hands it symbolizes [Used by permission of *Built Environment*]

church and, because he had not learned to "read" churches, he would have no idea what is symbolized.

Peirce's icon is, on the face of it, even simpler. It is a sign which is "like" its object. The trouble comes when you start thinking what "like" actually means. One thing can *look* like another—think of Venturi's building as a duck; Saarinen's TWA terminal which looks "like" a bird (or is it a pterodactyl?); the roof of Frank Lloyd Wright's church at Madison which looks "like" his hands at prayer; the roof of Le Corbusier's Ronchamp which *he* says looks "like" the shell of a crab; Kahn's Medical Centre in Philadelphia which looks "like" a Scottish castle and so on.

Sometimes the likeness is appropriate as, say, with Saarinen's Air Terminal, Wright's Church or even Venturi's duck, but sometimes it isn't, as with Le Corbusier's crab shell; such likenesses obviously have to be used with caution. And then there are other kinds of likeness which I have described elsewhere (1977) where the plans of different buildings are derived from the same bubble pattern. The one is "like" the other in terms of planning relationships. Yet they may *look* very different on the ground.

Again, there is much more to it, but as more and more of it gets into print, so the argument is hotting up as to just how relevant all this is for the development of architecture. The Italians started publishing in this field well over 30 years ago but the amount so far published in English really has been very small. Anglo-Saxons tend to be so much more pragmatic. The first, key collection in English, *Meaning in Architecture,* was published in 1969. This was a collection of papers by Reyner Banham, Françoise Choay, Aldo van Eyck, Kenneth Frampton, Alan Colquhoun, Norberg-Schulz, myself and others, edited by Charles Jencks and George Baird. But in some ways it skirted around the subject and none of us even attempted the comprehensive survey which ought to have been included in such a pioneering collection. And yet it made its impact. The discussion in fact started in the margins of the book itself where various authors commented on what the others had written; Janet Daley called it "spitting and snarling" in her review. But still the discussion continued and Robert Venturi was by no means the only practitioner (and teacher) to say he learned something from it. As Venturi himself said:

> Another aspect of this subject [that is "ordinary" architecture] involved meaning and iconography, but the subtleties of these subjects I know little about; they were not part of our education.

Obviously he wished they were, but then he went on to say:

> We studied space. Some English architects are now considering meaning and the whole subject of semeiology [sic] in a way relevant for architecture . . . In my day it was a game in abstraction, without the elements of association and past experience and their effects on perception.

Academic work has proceeded apace and some of it has been published in fairly obscure journals. There have been significant conferences of the International Association for Semiotic Studies (IASS) in Milan (1974) and Vienna (1979), each of which had sessions on architecture, not to mention specialized meetings on architectural semiotics at Castelldefells near Barcelona, Ulm in Germany and so on. The Castelldefells papers came out in Spanish in 1967 and some of them, together with other key papers in the subject, have just been published as *Signs, Symbols and Architecture* (Broadbent, Bunt and Jencks, 1980). The Ulm meeting was organized by Martin Krampen whose *Meaning in the Urban Environment* has just appeared, while another key contributor to the area, Juan Bonta, has also finally published his *Architecture and its Interpretation,* and there is another collection in press: *Meaning and Behaviour in the Built Environment,* this time edited by myself, Dick Bunt and Tomas Llorens.

Yet the field, above all, has been "softened up" by Charles Jencks, especially with his *The Language of Post Modern Architecture* (1977). This, whatever one's doubts about his definitions of "Language," "Post Modern" or even "Architecture," has done more than any other book—with the possible exception of Venturi's—to bring the "meaning" debate into the public arena. And what is more it did so in an extremely readable, well illustrated and provocative way. It is from Charles Jencks's work in particular that the *idea* of meaning in architecture made *Time* magazine, not to mention the Sunday colour supplements. And it is probably from those sources, rather than the work of the researchers, that certain ideas from semiotics have found their way into everyday architectural discourse.

Nathan Silver (1980), for instance, quotes—possibly tongue-in-cheek—some examples of "Architect Talk." They include:

*What happens in the front hall is thematic. I've cut the master bedroom entirely out of my narrative.*

And:

*The upper levels have no conversation with the ground floor.*

Not to mention:

*The building decisively addresses the street.*

Well, of course, he is quoting some pretentious examples. Yet such trivia, widely applied, are the very stuff of which paradigm changes are made. Most extraordinary of all was Richard Rogers who said, in the trailer to a television programme (1980) but not in the programme itself, "We put the components of the building together much as we put together the words in a sentence." Of course, the red, green and blue painted pipes of the Beaubourg "express" the idea of servicing. They *symbolize* the fact that the building has water, drainage, electrical, air-conditioning and other systems while the escalators *indicate* the way up! And the whole thing is an *icon,* if not of Sant Elia's Futurist of *c.* 1914, then at least of Archigram *c.* 1960. So the whole thing is a clear case of application, into a real building, of principles from Peirce's iconicity.

But the liveliest appliers of semiotic into architecture undoubtedly have been such Americans as Robert Venturi himself, Robert Stern and Charles Moore. Take Venturi's decorated shed; this relates, in a curious way, to one of Le Corbusier's ideas. As he put it (1920)

If I show to everyone on Earth—a Frenchman, a Negro, a Laplander—a sphere in the form of a billiard ball (one of the most perfect human materialisations of the sphere), I release in each of these individuals an identical sensation inherent in the spherical form.

The Frenchman might well think of the games to be played with billiard balls, but everyone—according to Le Corbusier—would respond to the *"fixed" sensation released by the primary form.*

The difficulty with such "primary forms," according to Le Corbusier, is that because everyone "reads" them in the same way, they have no particular meaning for anyone. Therefore we must consider what Le Corbusier calls "secondary sensations," which vary from person to person "because they depend upon his cultural capital or hereditary capital." He gives an example of what he means:

If I hold up a primary cubic form, I release in each individual the same primary sensation of the cube; but if I place some black geometric spots on the cube, I immediately release in a civilised man an idea of dice to play with, and a whole series of associations which would follow.

A Papuan would see only an ornament.

In other words one places the primary form into a particular cultural context by the way one treats its surfaces; literally one decorates Venturi's shed.

The question, of course, is how to do it. Robert Stern (1977) suggests three possibilities, each of which he has utilized himself:

**1** *Contextualism* in which forms for the new design are derived from the context into which it is to be placed to which it will then relate in form, colour and scale;

**2** *Allusionism,* based in particular on historical allusions. Stern sees this as far more than mere eclecticism but it may involve the re-use of established and successful types. But above all, it is a matter of trying to recapture an appropriate mood which may involve the incorporation of fragments from the past into the new;

**3** *Ornamentalism.* Stern concerns himself with the physical surfaces which in the case of architecture obviously are walls, while in other kinds of design they would be containers. As he sees it there is no need even to justify the decoration of a visible surface. We need complexity and elaboration and there is no reason why this cannot be applied just for the sake of it.

Such an approach obviously is open to charges of superficiality, triviality and general self-conscious indulgence. Certainly it will be anathema to those who are still living in the Modern Movement paradigm. But isn't that a little dishonest? Obviously it

**FIGURE 6–10** Robert Stern's house for Paul Henry Lang. [Used by permission of *Built Environment*]

is quite impossible to build any building *without* surfaces and the surfaces approved by the Modern Movement were just as susceptible to "honky-tonk, crassness and phoniness" as their Post-Modern successors seem to be.

The white-walled villas of the 1920s were supposed to look "like" machines, Atlantic liners and so on. But buildings are not Atlantic liners, so why should they look like them? And even the alternative surfaces which Le Corbusier and his followers began to use once they had realized that their white-painted plaster streaked, cracked and eventually crumbled—they too certainly had their problems. Rough stone and rough brick were all right in themselves except when they were used and looked like mere infilling, within a concrete frame. So then the idea got around that somehow it was more "honest" to use concrete, exposed as a surface finish. No one could conceive a drabber, greyer, more hostile finish; even when it was cast to perfection—which it seldom was—it would insist on looking drab, grey and hostile, so something had to be done to make it more "human." So what did they do? They used the roughest boards they could find for the formwork and then left the board marks showing: knot holes, joints, grain, splinters, gaps, warts and all. This was supposed to give "human scale" to the concrete, to ensure that it would weather gracefully and so on. But, of course, it did nothing of the kind. The graining of timber is no more appropriate for concrete than sea shells, doggie paw marks or wicker work—and from a distance it looks as drab as ever. For concrete is *not* timber; how inappropriate to make it "look like" timber. Yet board-marked concrete, in its crazy way, did recognize a fundamental fact: that despite Adolph Loos we do need, well, ornament. Pugin, Ruskin and Morris all knew it, while in another kind of paper I might have argued that the eye and brain are so constructed as to *demand* texture and pattern if we are to see surfaces properly, that they need mouldings to reinforce perspective at cornice and skirting and even to ease the transitions from light to dark at window openings and so on. But at the moment our subject is meaning and that is enough on its own. Board-marked concrete obviously gives us texture but beyond that it is literally meaningless. There simply is no reason why concrete should look "like" timber or, to use a photographic analogy, the "negative" of timber. If we are going to ornament, texture, or otherwise decorate, we might as well do so in a way that is meaningful, and Bob Stern, I think, has it right. One of his larger houses was built for Paul Henry Lang, a distinguished music critic and an admirer of classical architecture. The house is an extremely sophisticated "shed"—in the Venturi sense—with a carefully worked out relationship of public and private parts, of master bedroom to guest rooms, odd corners for various family functions, views out, sun shading and so on. But then the façade is "decorated" with a "Palladian" archway and other shapes tacked on in the form of mouldings. They are (just) sufficient to draw the desired allusions (Stern's Allusionism), yet in a way which can be changed almost more easily than wallpaper. So, if the Langs move out and the next owners prefer, say, Gothic, they can have the Palladian mouldings ripped off and others tacked on in their place.

Of course, Stern's Decoration is no more than skin-deep in this case, but his principles can be applied in three-dimensional form. Bofill's Xanadu, for instance, is a splendid example of Contextualism at work, for the basic apartment forms of which it is made could have been put together in many ways. You can group four square apartments to form a larger square, or a cross. You can group eight together to form a double arm cross and, given a steel frame, such basic floor layouts could have been piled vertically to form a straight-up-and-down tower, a pyramid, an inverted pyramid, and so on. But Bofill and his colleagues wanted to *locate* Xanadu at a particular place and nowhere else. So they looked across the bay at the rock in the distance, the Penon de Ifach, and drew the overall form of their building by visual analogy (Peirce would have called it an icon) with the rock. But that gave them a rather blocky building, so they "softened" the outline by drawing on the local vernacular for their pantiled roofs, after which they painted the whole thing in that khaki-green to which the Spanish countryside burns in the summer sun. That is the clearest example I know of Contextualism.

Bofill's group too have used Historical Illusion, in, for instance, their Petite Cathédrale scheme for Cergy Pontoise. Asked to design a whole suburb, with shops, schools, a department store, parking, apartments and so on, they thought immediately of a shopping mall form rather like the great Galleria in Milan. The mall itself would be protected from the weather; it would also encourage social interaction, while the apartments, grouped up and over the top of the gallery, would certainly avoid the bleak sterility of your typical straight-sided slab. But somehow it didn't seem right; why should they build a Milan-type Galleria in the outer suburbs of Paris? But then it occurred to them that the greatest architectural glory of the Ile de France is of course the Gothic cathedral, so what started as a classical Galleria was transformed into a Gothic cathedral!

**FIGURE 6–11** Charles Moore's Piazza d'Italia in New Orleans. [Used by permission of *Built Environment*]

These are spectacular examples but the *principles* have been applied at many scales. Charles Moore's Faculty Club at Santa Barbara, for instance, is an extraordinary example of Contextualism in which elements from Santa Barbara itself, from Spanish Colonial to International Style modern, are drawn into the form of the Club. And perhaps the most controversial of all, his Piazza d'Italia in New Orleans, is intended as a focus for the Italian community. It is centred on a fountain—formed from a relief map of Italy—which in turn is surrounded by classical colonnades redolent of Roman grandeur. As one might expect of Charles Moore, it is wistful and wise, powerful, gentle *and* silly, but already it has become a symbol so that what was once a seedy part of New Orleans—being rapidly taken over by high-rise offices—now has a sense of identity again. The Piazza literally has become the focus for an extensive scheme of urban renewal.

Stern, Moore, Venturi and Bofill *know* they are speaking prose and, just occasionally, writing poetry. But most architects are still like Monsieur Jourdain, whether they are trying to "express" the structure (Mies) or the services (Rogers), drawing allusions to nature by board-marking their concrete, to machines by using Norman Foster surfaces or—and this, of course, is the most popular these days—drawing allusions to the "vernacular." All such architects, whether they like it or not, whether they know it or not, *are* playing the semiotic game. One can play it as an amateur, or one can play it as a professional, but—like all games—you can play it better if you actually know the rules. The rules of this game are being spelled out in the various books I have mentioned and while some, no doubt, will resist, screaming "honky-tonk, crassness, phoniness" until their dying days, the paradigm is shifting, inexorably, against them.

## REFERENCES

Blake, P. "Walt Disney World." *Architectural Forum,* June 1972. Reprinted in P. Finch, *The Art of Walt Disney.* New York: Abrams, 1973.

———. *God's Own Junkyard.* New York: Holt, Reinhart & Winston, 1974.

Bofill, R. *L'Architecture d'un Homme.* Paris: Arthaud, 1978.

Bonta, J. P. *Architecture and Its Interpretation.* London: Lund Humphries, 1979.

Broadbent, G. " 'The Taller of Bofill': Work of the Taller de Arquitectura of Barcelona." *The Architectural Review,* November 1973, pp. 289–97.

———. "The Road to Xanadu and Beyond: Work of the Taller de Arquitectura." *Progressive Architecture,* September 1975a, pp. 68–83.

———. "Taller de Arquitectura." *Architectural Design,* July 1975b, pp. 402–17.

———. "A Plain Man's Guide to the Theory of Signs in Architecture." *Architectural Design,* July/August 1977, pp. 474–83.

Broadbent, G.; R. Bunt; and C. Jencks. *Signs, Symbols and Architecture.* Chichester: John Wiley, 1980.

Broadbent, G.; R. Bunt; and T. Llorens. *Meaning and Behaviour in the Built Environment.* Chichester: John Wiley, in press.

Feyerabend, P. K. *Against Method: Outline of an Anarchistic Theory of Knowledge.* London: New Life Books, 1978.

Gould, J. G. *Modern Housing in Britain 1919–1939.* Architectural Historian Monograph No. 1. London: Society of Architectural Historians of Great Britain, 1977.

Gropius, W. *The Scope of Total Architecture.* London: Allen and Unwin, 1956.

Guttman, L. "An Outline of Some New Methodology for Social Research." *Public Opinion Quarterly* 18 (1954).

Jencks, C. *The Language of Post Modern Architecture.* London: Academy Editions, 1977.
Jencks, C., and G. Baird. *Meaning in Architecture.* New York: George Braziller, 1969.
Krampen, M. *Meaning in the Urban Environment.* London: Pion, 1979.
Kuhn, T. *The Structure of Scientific Revolutions.* Chicago: University of Chicago Press, 1962; enlarged edition, 1970.
Landau, R. "Mickey Mouse: The Great Dictator." *Architectural Design,* September 1973, pp. 591–95.
Le Corbusier. "Le Purisme." *L'Esprit Nouveau,* no. 4 (1920).
———. *Vers Une Architecture.* 1923. Trans. F. Etchells as *Towards a New Architecture.* London: Architectural Press, 1927.
Loos, A. "Ornament and Crime." 1908. Trans. in L. Muntz and G. Kunster, *Adolf Loos, Pioneer of Modern Architecture.* London, 1966.
March, L., and M. Trace. "The Land Use Performance of Selected Arrays of Built Forms." Working Paper No. 2. Land Use and Built Form Studies, Cambridge, 1968.
Mies van der Rohe, M. "Interview with Graham Shankland." London BBC Third Programme, 1960.
Molière. *Le Bourgeois Gentilhomme.* 1670. Ed. Y. Brunswick and P. Ginestier. Paris: Didier, 1968.
Moore, C. "Moore Is More . . . Five Portraits of Charles Moore." *Architecture d'Aujourd'hui,* no. 184 (March/April 1976), pp. 1–64.
Odgen, C. K., and I. A. Richards. *The Meaning of Meaning.* London: Routledge & Kegan Paul, 1923.
Peirce, C. S. *Collected Papers Vols. I to VI.* Ed. C. Hartshorne and P. Weiss. *Vols. VII and VIII.* Ed. A. W. Barker. Cambridge, MA: Harvard University Press, 1960.
Pevsner, N. *Pioneers of the Modern Movement.* London: Faber and Faber, 1936. Revised as *Pioneers of Modern Design.* Harmondsworth: Penguin, 1960.
Richards, J. M. *Modern Architecture.* London: Penguin Books, 1940.
Saussure, F. de. *Course in General Linguistics.* 1906–11. Trans. W. Baskin. London: Peter Owen, 1956.
Shannon, C. E., and W. Weaver. *The Mathematical Theory of Communication.* Urbana: University of Illinois Press, 1949.
Silver, N. "Architect Talk." In *The State of the Language,* ed. L. Michaels and C. Ricks. Berkeley: University of California Press, 1980.
Smithson, A., and P. Smithson. "Signs of Occupancy." *Architectural Design,* February 1972, pp. 91–97.
———. *Without Rhetoric: An Architectural Aesthetic 1955–1972.* London: Latimer, 1973.
Stern, R. "Post Modernism." *Architectural Design* 47, no. 4 (1977), pp. 254–86.
Venturi, R. *Complexity and Contradiction in Architecture.* New York: Museum of Modern Art, 1966.
Venturi, R., D. Scott-Brown, C. Izenour. *Learning from Las Vegas.* Cambridge, MA: MIT Press, 1972.
Venturi, R., and D. Scott-Brown. In *Conversations with Architects,* ed. J. W. Cook and H. Klotz. London: Lund Humphries, 1973.
"Vernacular: The English Disease." *RIBA Journal,* special issue, February 1980.
Weiner, N. *Cybernetics.* 1948. 2nd ed. Cambridge, MA: MIT Press, 1968.

选读 7
"历史主义与符号学之局限"
艾伦·科洪

科洪在这篇论文中探索了存在于建筑形式与蕴藏在人类其他沟通方式中的各种特征之间的关系。他借分析语言学理论来说明，虽然建筑与书面语、口语具有许多相似之处，人们也必须在考虑材料品质、人类的使用情况和物质形态本身的历史意义之后再对建成环境中的符号做出诠释。因此，对建筑的合理诠释取决于它的历史背景。科洪采用这一论据来说明，为了理解如何使用这些形式并证明使用这些形式是正确的，对建筑形式不同风格的描述——哥特式被认为是信念的指示物，新古典主义则被看作是理性的标志——必须从西方宗教思想和启蒙思想的意识形态的角度来看待。

## READING 7

## Historicism and the Limits of Semiology*
### Alan Colquhoun

In this essay, Colquhoun explores the relationships that exist between architectural form and the characteristics embodied in other forms of human communications. He analyzes linguistic theories to demonstrate that while architecture contains many similarities to written and spoken languages, it is also true that symbols in the built environment must be interpreted in ways that account for the material quality, human use, and historical significance of the physical object itself. Thus, architecture is dependent on its historical context for sensible interpretation. Colquhoun uses this argument to show how stylistic descriptions of architectural form—Gothic as a signifier of faith and Neoclassicism as a sign of reason—must be seen in terms of the ideologies of Western religious thought and the Enlightenment in order to understand and justify the uses of these forms.

### HISTORY AND THE ARCHITECTURAL SIGN

Any discussion of architecture considered as a system of signs must come to terms with the fact that semiology is derived from the study of language. Its validity, therefore, depends on the extent to which the signifying component of architecture and other nonlinguistic systems is reducible to something which they have in common with language.

One of the confusing things about semiology is that, because of this basis in linguistics, it has tended to oscillate between the study of language and the study of aesthetics. The studies carried out in Moscow and Prague in the second decade of the twentieth century were concerned with literary criticism as much as with linguistics

*This essay was first published in *op. cit.* September 1972.
Source: *Essays in Architectural Criticism: Modern Architecture and Historical Change* by Alan Colquhoun. Copyright. © 1981 by The Institute for Architecture and Urban Studies and The Massachusetts Institute of Technology. Reproduced by permission of MIT Press. pp. 128–38.

and were therefore normative in aim. Those of Ferdinand de Saussure, on the other hand, were concerned with language, and their aim was descriptive. What unites them, however, is something that might be called the empirical method. The Prague School studied texts, de Saussure studied the spoken language, and both tried to eliminate the normative preconceptions that had dominated literary and linguistic studies in the nineteenth century. From this point of view, semiology can be thought of as coterminous with, and part of, the Modern Movement and therefore intimately bound up with the historicism which, at least in one interpretation, it tries to deny.

It might be interesting to investigate the relation between the Russian and Czech literary criticism of the early 1900s and similar formalistic studies that were carried out in the same period in the plastic arts and architecture in Russia and Germany. But here I shall deal more with the relation of architecture with post-Saussurian semiology—that is, with semiology based on what has come to be known as structural linguistics. I shall try to show the limitations of the linguistic analogy, particularly with regard to such Saussurian concepts as *synchrony* and *diachrony,* when it is applied to aesthetic systems such as architecture. This might clear the air and allow one to look at semiology as a critical tool rather than as an explanatory science. To some extent Roland Barthes has tried to do this, but unfortunately the influence of his *Elements of Semiology* has tended to obscure rather than clarify the relation between aesthetics and linguistics, and semiology is now in danger of becoming an academic study unrelated to practice. What one would hope is that the study of architecture as a sign system could provide a critical tool to bridge the gap between the long-winded trivialities of behavioristic sociology and the formalism which seems to be its only alternative.

The study of language has taught us that we cannot regard the world as a series of facts each of which has its own symbol of representation. By isolating the synchronic aspect of language, de Saussure was able to demonstrate that it is only by operating within the structure of a given language that we gain access to the world of fact. The diachronic or historical study of language, whatever other uses it may have, is not able to reveal how people actually speak language and therefore how they represent the world and communicate about the world with each other. This point of view, which has the characteristic virtues of the empirical method, is based on the assumption that we can only discover the truth by taking observable entities as our object of study. De Saussure does not discuss the ontogenetic or phylogenetic process by which people come to speak; though important, for him these problems lie outside the realm of language. He merely points to the concrete facts of speech and deduces from them a theory of language. At the particular moment in time that he studies a language, he finds a total structure. It is because this structure already exists that the speaker is able to form concepts and to share these concepts within a society. Although de Saussure has to assume that the speaker is unconscious of the logic of the structure he uses, he does not have to assume that such a logic exists a priori as an ideal form. He eliminates this possibility by saying that the structure simply consists of oppositions or analogies between the elements of a code. From the analysis of an infinitely large system, a few basic rules of transformation can be deduced.

From our point of view, however, the interesting thing about de Saussure is that he postulated a general theory of semiotics, which would include the various expressive languages, and of which language itself was in a sense the paradigm. But he does not seem to have been very clear as to the relation in aesthetic systems between the arbitrary sign and the natural sign. He made a basic distinction between the *index*, which is causally related to what it indicates, and *symbols* and *signs*. *Symbols* were motivated (in other words, there was a metaphorical tie between symbols and what they signified), whereas *signs* were arbitrary and conventional. It seems that de Saussure was right in saying that the arbitrary quality of any sign system is necessary if it is to be socially available, because motivated sign systems (the sort of symbols C. G. Jung discusses, and the objects of our dreams) can be purely private. But while it may be true that the signs actually used in aesthetic systems are chosen arbitrarily from all the possible natural signs, this does not mean that signs themselves are arbitrary. And since aesthetic systems are socially available and can be distinguished from language as social phenomena (and not merely as intellectual constructs), their social function as symbols must be taken into account.

It is possible to show that aesthetic systems have properties which do not belong to language as defined by de Saussure. I will give a few examples:

1 In language, change only occurs in one part of the system at a time. In aesthetic systems, change often occurs in the whole system, e.g., the change from Gothic to classical architecture, or from eclecticism to modern.
2 In language, change is always unintentional. In aesthetic systems, change is always intentional (though the intention may not be rationalized).
3 In language, the existence of precise perceptual degrees of difference in the phonic object is relatively unimportant, since it is sufficient for one word to be different from another for differences in meaning to adhere to those two words, or for analogies to exist between words to bring to mind associative meanings. In aesthetic systems, however, precise degrees of difference are important—the difference between the interval of a third and a fifth in music, for example. In music, the ability to distinguish degrees of difference is used to make a structure which is interesting in itself and to create meaning. In language, the ability to distinguish between different degrees of phonetic change is not used in this way, since in language the phonic object is absorbed by its meaning. What are interesting in language are the meanings that are attached to phonic objects, not those objects themselves.
4 De Saussure discusses language as being analogous to economic exchange: "It is not the metal of a piece of money that fixes its value." But in an aesthetic system using metal, it is precisely the intrinsic quality of the metal that is important (though, of course, the semantic properties attributed to this quality will vary from one culture to another).

By aesthetic systems, therefore, I mean systems whose sensible form is interesting in itself. In language, the indissoluble relationship between the signifier and the signified is a function of the arbitrary value of the signifier. In aesthetic systems, on the other hand, the meaning that can be attached to the sign is due to the fact that the sign is itself motivated and the signifier is invested with potential significance. This motivation may take the form of physiognomic properties, or of analogies between the form and the meaning (de Saussure quotes the example of the symbol of Justice carrying a pair of scales). Aesthetic systems in this sense include all the systems traditionally grouped under the fine arts and the applied arts—even though their only, or even their main, purpose may not be a signifying one.

These fundamental differences between language and art mean that in aesthetic systems the study of the diachronic dimension takes on a peculiar importance. Because the changes which occur in aesthetic systems are revolutionary and intentional, these changes are directly related to ideology, and ideology can only be understood in a historical context.

It is true that the historical changes that occur in language may themselves obey certain rules of transformation, and it may therefore be possible to study the dialectical interrelation between the synchronic and diachronic dimensions of language. This idea is suggested by Maurice Merleau-Ponty in his essay "The Phenomenology of Language," where he tries to forge a link between language "as mine" and language as an object of study.[1] What he says is of crucial importance to the whole notion of structure applied to semiology and the human sciences. A structure is by definition a closed system, which is studied from a purely formal point of view. For example, when we study language synchronically, we are not concerned with the semantic element independent of the signifier. The "value" represented by the signifier is taken for granted. It is an indissoluble part of the language we learn, which consists of phonic materials and their corresponding concepts as given.

On the other hand, in any of the nonlinguistic sign systems which are in general use by society, the axioms which form the content of the structure under analysis are guides to conduct and have an immediate bearing on what we think ought to be the case in our social life. We are dealing with values that are ends in themselves and not a means to some other end. It is therefore impossible to avoid trying to get outside this system, and we have to take an attitude toward the values which the system assumes. If this is true in an ontological sense in the case of languages, as Merleau-Ponty suggested, it is even more so in the case of the semiological study of architecture, since this kind of study cannot be separated from criticism and the formation of value judgments.

The historical mutations in language itself are no doubt related to a historical evolution whose laws underlie a confused jumble of accidental facts. But they are not apparent to the user of the language, who does not have to create a metalanguage to explain the language he is using. Hence the need to study language "as ours"—language as the system which exists at any one moment—and hence the distortions which result from studying language purely as a historical object and independently of its diacritical mode of operation, as was done in the nineteenth century.

On the contrary, in aesthetic languages, a metalanguage is a part of the process of creation, whether as a mythological underpinning of practice or as a critical apparatus by which practice can be judged. That is to say, the artist always operates according to procedures which he is conscious of in the form of a set of rules. These rules are normative in the sense that they reflect the attitudes, values, and ideology of a particular society. In Western civilization, architecture, no less than the other arts, has been subject to a rather rapid process of change, and to understand architecture synchronically, in the present or at any other time, it is necessary to study it over a period sufficiently long to observe the changes to which it has been subjected. The period from the eighteenth century to the present is of particular importance in the field of diachronic studies, because it was during this period that historical change itself began to be seen as the field in which values had to be posited. Thus, while art continued to be subject to rules and norms, it began to be impossible to regard these norms as immutable. Throughout this period there is a growing tension between two opposing viewpoints. Both points of view tried to go beyond the obvious conventionality of architecture; but in the first case there was an attempt to find an a priori form of expression that lay behind the conventions, and in the second there was a rejection of the very idea of a priori values and an implicit rejection of aesthetics as anything but a support for ideology.

The first point of view was typical of the eighteenth century, when for the first time the systematic distinction was made between convention and natural law. In linguistics the grammarians made a distinction between speech and the laws of language which were accessible to introspective logic. The numerous inconsistencies and irregularities that appeared in speech were supposed to be the result of historical accidents which had obscured the original language. N. Beauzée, in his *Grammaire Générale* of 1767, gives the following distinction between general grammar and particular grammar: "General grammar is a science because its only object is the reasoned speculation about the immutable and general principles of language . . . particular grammar is an art because it envisages the practical application of the arbitrary and habitual institutions of a particular language to the general principles of language."[2] The distinction which was made between the normative language and a particular language meant that the differences between languages at different times and places had no fundamental importance. Language was not, as it was to be for de Saussure, arbitrary in its very nature but only so to the extent that it had strayed from some supposed ideal norm.

In the study of history the eighteenth century distinguished in a similar way between the accidents of the historical process and the laws which underlay them. "Mankind," said David Hume, "are so much the same in all times and places that history informs us of nothing new or strange in this particular. Its chief use is only to discover the constant and universal principles of human nature."[3] Even Rousseau, concerned as he was to transform society completely, was unaware of anything that could be called process in history. In the famous opening sentence of *The Social Contract*, which begins, "Man is born free; and everywhere he is in chains," he goes on, "How did this change come about? *I do not know.* What can make it legitimate? That question I think I can answer"[4] (my italics). Rousseau was not interested in the historical transformation that has led to society's present state. What interested him was applying to a future society the truths that had been lost in the ramifications of history—truths that possibly, he supposed, were understood in some remote golden age.

In architecture, the Baroque system was radically questioned. What had seemed to be laws suddenly appeared as mere conventions. There was an attempt to reduce architecture to its functional and formal principles, and various formulae were developed to reconcile convention with the principles of reason. The Greco-Roman tradition was subjected to radical adaption and dismemberment, though its fundamental grammar was retained for serious works of architecture.

All these examples indicate that the eighteenth century thought of progress as a *rappel à l'ordre*. If one only looked at the phenomenal world and sought council from one's own natural reason, one could arrive at ethical and aesthetic value judgments that were valid for all time. This led to the notion that there was no difference between the unbiased exercise of reason and the establishment of ideology.

In late eighteenth-century German thought the whole attitude toward history and ideology changed: from Hegel onward history is seen, not as a random process, but as the very field which defines and limits reason. Two consequences follow from this. First, relativity enters into history. The different types of aesthetic language that exist in time and space are no longer seen as so many partial aspects of a universal norm but as the outcome of historical forces that produced them. In linguistics, philology replaces general grammar. The history of language is studied in order to find in it a principle of change and development. A hitherto exotic language, Sanskrit, is discovered to be related to the Romance and Teutonic languages through the postulation of a mother tongue common to them all. In architecture, the ideal Platonic model is no longer sought from the Bible or from the ancient Mediterranean cultures. A study of the age of faith justifies Gothic architecture as the reflection of the ideas and practices of its period.

The historical relativity is accompanied by a growing realization that the nineteenth century possesses no architecture proper to its own age. Boullée had talked about the "character" that different buildings should have. But the deliberate choice of forms signifying particular ideas was not accompanied by a feeling of alienation from the language of classical forms itself. The different forms belonged to the lexicon of language and were part of a given synchronic structure similar to the synchronic structure of language.

These different forms were contained within a single ideology. In the nineteenth century, on the other hand, to choose a form was also to choose between ideologies. If the Gothic style was preferred, it signified that a certain ideology was also preferred; history was being used as a model, not to disclose a universal truth but to measure the ideology of a historical culture—that of the age of faith, in the example I have chosen—against the modern culture. Together with the revival of past styles, a feeling began to develop that, if Gothic was the characteristic style of the age of faith, if Neoclassicism was the characteristic style of the Enlightenment, then the present age should have its own style, rooted in the technical progress that was its own characteristic sign. This growing feeling was the corollary of the fact that relativity was only one aspect of post-Hegelian epistemology. The other aspect was that history was seen as process. History progressed dialectically by transcending itself, each successive period absorbing the previous one and producing a new synthesis. Whether, as in Hegel, this process was seen as teleological—a movement toward the future incarnation of the Ideal that existed outside time—or, as in Marx, it was seen as dialectically working itself out in the class struggle seen according to the Darwinian model, need not concern us. What is important is the idea of history as an intelligible process with a predictable future. Once this idea was established, it was no longer possible to believe that a person in a particular period could, simply by introspection, discover the form of society, the language, and the aesthetic mode that was true for all time. No man could divest himself of his own social conditioning and speak *sub specie aeternitatis*. There were no fixed archetypes to which one could appeal. Instead of trying to discover such Platonic models, man's task was the empirical study of history. As Engels said, paraphrasing Hegel, "the world is not to be comprehended as a complex of ready-made *things* but as a complex of *processes,* in which the things apparently stable . . ., go through an uninterrupted change of coming into being and passing away."[5]

The complex group of philological, sociological, and aesthetic concepts that I have just described I shall call historicism. The word "historicist" is subject to so many interpretations that it is necessary to state clearly the sense in which I am using it here. In architectural history, Nikolaus Pevsner, for example, uses the word "historicism" as synonymous with eclecticism.[6] In my opinion this is misleading. It separates two notions which are different aspects of the same type of thought: eclecticism and the concept of the *Zeitgeist*. Eclecticism is the result of historical relativity. As soon as it is believed that aesthetic codes are the product of particular phases of history, they become laid out before us, as it were. We do not *need* to imitate them, and indeed we should not do so since we believe that their necessity is tied to their own period. But they *can* be used, because they have been studied in and for themselves, as so many equivalent languages. In the classical period one learned Latin and Greek because these were close to the normative language which one was trying to reconstitute. In the nineteenth century one learned Sanskrit because Sanskrit was simply another version of a more general group of languages, none of which had normative priority over another. A culture that was no longer instinctively limited to one type of architectural language, but still lacked a method for developing its own, would be bound to use its new-found historical erudition to search for models in the historical tradition. Creation is impossible in an intellectual void. Until the revulsion from all the existing models reached a certain intensity—until these models had been exhausted of all meaning—it was in accord with artistic tradition to revert to existing models. In a sense the past styles that the nineteenth century used were its own styles; the diachrony of knowledge had become the synchrony of "my language." The same process had, after all, though for different reasons, occurred in the fifteenth century, when architecture went back to the classical model and made it its own and when, in language too, Latin words were introduced into the literary vocabulary. The eclecticism of the nineteenth century was not a unique event in European culture. What were unique were the beginnings of disbelief in the past which the study of the past had generated. This disbelief was reinforced by the other aspect of historicism that I have mentioned, the theory of dialectical development and of history as an irreversible flow.

Two factors seem to have contributed to the overthrow of eclecticism toward the end of the nineteenth century. The first was the rejection of bourgeois culture by the artistic elite. This rejection was not accompanied by any explicit social or political critique. The second was Marxist dialectical materialism and the spread of socialist ideas. The Modern Movement in all the arts can only be understood in relation to these two social factors.

Architecture had a special relation to them. The fact that architecture was one of the useful arts made it hover precariously between the infrastructure and the superstructure in Marxist theory. The operationalism that was at the basis of Marxist thought could be transferred literally to architecture, and function could become the theoretical foundation of a new architectural praxis. "It is no longer a question . . . of inventing interconnections . . . out of our brains, but of discovering them in the facts," Engels had said.[7] What could be more obvious than to consider the facts of statics, structure, new materials, new social programs, and to make out of these facts a new architecture?

What is left out of this equation is the question as to whether an architecture so constituted will still act as a sign. And if it does, what sort of sign is it? Does it become a sign simply by being a fact—an *index* of its physical functions? Or is it a more complex sort of sign involving an idea about those functions and about the mental world of the society which it serves? But in asking these questions I am anticipating problems I shall raise later on.

Some time before socialist/functionalist theory was established in architecture, a movement had been developing in all the arts and in aesthetic theory, which was, as I have said, based more on rejection of bourgeois art than on a systematic critique of social infrastructures. The effect of this on architecture can be shown in a particular example. In the projects of the students working under Otto Wagner in Vienna in the first decade of the twentieth century, all the styles familiar to us in the various arts as Symbolism, Expressionism, and Art Nouveau are displayed side by side.[8] What all these projects have in common is the disintegration and fragmentation of nineteenth-century eclecticism and the attempt to create a new style. It is true that many of these projects refer to existing models of some sort—to Mediterranean, Scottish, or Alpine vernacular architecture, for example, though all these models lie outside the conventional framework of bourgeois architecture. But many of them, particularly those of Karl Maria Kerndle and Josef Hoffmann, do not use any existing architectural model but take over certain Art Nouveau decorative schemes and enlarge them to the scale of the building as a whole. These schemes are rectangular and crystalline and avoid the use of naturalistic motives. We cannot help seeing them as the precursors of the simplified and stereometric forms that were later to emerge in the functionalist architecture of the 1920s. These tendencies in design had their counterpart in theoretical writings on art—particularly those of Theodor Lipps and Wilhelm Worringer, who tried to establish a theory based on psychological constants.

Art, according to Worringer, reveals universal subjective dispositions.[9] If we understand these dispositions, we will arrive at an art that does not use the styles as norms but judges particular manifestations of art according to the fundamental and ageless rules of style itself.

At the beginning of the twentieth century we therefore find two parallel, and in a sense contradictory, theories of architecture, both deriving from post-Hegelian historicism. The first is a theory of moral purpose and objective fact, the second a theory of aesthetics and subjective feeling. But both theories reject the past as a guide to the present and are reductive and essentialist in spirit. The architecture that emerged after the First World War synthesized these theories by establishing functionalism as the dominant strain and calling on the art theory of natural expression and psychological constants whenever functionalism was unable to provide a solution. In linguistic terms, it reduced the architectural sign to an index of function, on the one hand, and the purely natural sign on the other.

Whether we are looking at the social or artistic aspects of this movement, we can see that it had strong deterministic features. By invoking the spirit of the age, modern theorists were presenting the objects of historical and psychological study as facts which were outside the control of the thinking and willing subject and constituted a sort of social or emotional categorical imperative. One aspect of this imperative was that the buildings which would serve the new society were thought of as analogous to biological types established by the laws of evolution. As an element in the range of instruments necessary for society, architecture was seen as belonging to the infrastructure of the historical process. History itself, through the agency of an elite, was determining the characteristic morphology of the new architecture and investing it with a set of *objets-types*. This idea of *objet-type* was closely related to the norm or type that is found in the study of zoological species. However much, in the words of Engels, history can be seen as a Heraclitean flux, it is nonetheless the passing away and coming into being of *types* which enable us to measure its dialectical process from lower to higher forms.

In modern architecture this idea of the biological norm was conflated with the idea of psycho-physical constants. This was an appeal to a determinism of a different kind. For although it reduced the creating of forms to a psychological function beneath consciousness, it could hardly avoid doing this by referring to constant *cultural* values, and in this sense returned to something close to eighteenth-century rationalism. In Le Corbusier these two concepts are clearly stated. On the one hand, the mathematician and the scientist discover the laws of causality; on the other, the artist discovers the laws of harmony between the materials that are given to him.[10]

If we apply the principles of structural linguistics to these phenomena, it is clear that the theories behind modern architecture were opposed to the concept of architecture as a language with a large amount of combinatory freedom or associative meaning. In the sense that architecture was held to be determined by the forces of history, its laws were fixed and there was no place for free play in the ordering of forms. In the other sense of an architecture reflecting universal psychological laws, this freedom was obscured by the old myth of normative beauty. If there was freedom, it could only exist as a sort of symbiosis between the artist and those natural laws which he discovered by intuition.

But the actual practice of the architects of the 1920s, particularly the Russian Constructivists and Le Corbusier, was far from obeying the rules that had been implied by the theory. In fact, their work was still to a large extent based on existing models, even though these models were not necessarily the major styles that had been copied in the nineteenth century. Not only were classical and romantic themes incorporated into the new work but concepts lying outside the field of architecture were transposed. Thus, engineering forms were displayed in the work of the Vesnin brothers in Russia. These forms were a "first articulation" drawn from the visible world and were used, rhetorically, to show that architecture was a matter of construction and not of outworn architectural compositional and decorative schemes. In Le Corbusier the imagery of ships was used to provide a new model for an architectural language and also to establish across history connections, at a deeper level, with age-old human themes—the monastic life of the twelfth century, for instance. This does not mean that modern architecture reverted to a nineteenth-century eclecticism. In the nineteenth century entire semiological schemes were taken over. Modern architecture took fragments of everyday life and fragments found in history. Modern architecture in this sense was essentially constitutive. It broke down the meaning systems into the smallest units that could carry meaning and recombined them, regardless of the entire systems from which they had been extracted.

To what extent does this bring modern architecture into conformity with the linguistic model? In linguistics, by observing how people speak, it is possible to determine laws without destroying freedom. Language exists prior to the individual. But the fact that he has to learn his language does not make him a slave. Indeed, if nature, through the mediation of society, did not present him with a preexistent language, he would not be able to form concepts. He applies the rules of language instinctively, and these rules are simply rules of combination. There is no a priori limit to the concepts that he can form. But in language this freedom of combination is, as I have said, due to the arbitrary nature of the relation between the signifier and signified.

Even in everyday language there exists a number of complex units or syntagmata which it is obligatory to use. Clearly the more of these complex units there are, or the larger each unit is, so the freedom of the speaker will be reduced. Now this is precisely what happens in poetry and literature: literary genres, styles, forms, and types of expression are simply inherited syntagmata. The reason they exist is to give rise to concepts which represent a value in themselves. In *language* the value of the sign is neutral. It is the purpose of *poetry* to turn neutral signs into expressive signs. But although the poet inherits these syntagmata, he is not obliged to use them. Precisely because he has at his disposal a type of language which represents values, he is able to revise these values. This is what Barthes means when he says that poetry is resistant to myth, while language is not. Language never gives full meanings to its concepts: they remain open-ended and ready for recombination. Poetry, on the other hand, being a second-order language, gives a precise emotive significance to its concepts. By dislocating the syntagmata that have been inherited and by arranging words in an order which is different from that of normal speech, it is able to make people aware of the banality and falsity of their concepts. As Barthes says, if myth wants to conquer poetry, it has to swallow it whole and serve it up as a content of its own system. This critical and dismantling activity of poetry results in what Barthes calls a "regressive semiological system."[11]

The same can be said about architecture. Architecture never appears as a neutral combinatory system. The "phonemes" and "morphemes" in architecture (and we can only use these terms metaphorically) are motivated and invested with potential meaning. What could be more motivated than function, structural method, or forms in space? If we are to be as rigorously empirical as de Saussure was about speech, we have to admit that architecture cannot usefully be reduced to a collection of arbitrary elements. Modern architecture tried to reduce its elements to what was essential but not by reducing them to arbitrary units as in linguistic analysis. In the study of language the reduction is purely formal and does not alter the way we speak. In architecture it is reformative and intended to reconstitute architectural meaning.

In this context semiology appears as a magician who reveals by empirical study the complexity and richness of synchronic systems. But it also appears as a possible prescriptive tool and a method of criticism. In this capacity it can easily fall prey to any of the prevailing philosophies of modern architecture, including the "kit of parts" theory that is put forward by certain utopian proponents of systems. This is the result of a confusion which comes from applying the principles of linguistics too literally. This confusion has been demonstrated by Lévi-Strauss in the field of music,[12] and what he says can be applied in general to architecture. Serial music, he points out, eliminates the meaningful structures of the inherited musical language (the relation of tones to the tonic, for example). It therefore no longer operates with two levels of articulation—the one which the composer finds at his disposition and the other in which he is free to create new combinations and meanings. With the apparent absolute freedom which this gives the composer, however, he runs the danger of being rendered inarticulate, unless he is close enough to the existent language to refer to it obliquely or to give a demonstration of its essence by attempting to reduce it to the "degree zero." Igor Stravinsky and Anton Webern might be given as examples of these respective types of reference. The musical "kit of parts" which later serialists propose would only make sense if the language of music were completely natural. This leads to a theory of spontaneous music.[13]

But it also implies that music is like natural language, because natural language is the only sign system where a "kit of parts" provides freedom of combination. But we know that this is only possible in language because the units have already had arbitrary meanings attached to them. Thus we arrive at a nonsense statement: a completely natural system and a completely arbitrary system are the same thing.

We tend to forget that in analyzing language we do not attempt to change it. Structural linguistics is a descriptive, and perhaps explanatory, method. It is concerned with the formal structures underlying language, not with its meaning or value system.

There is a very good reason why semiology in the same way must remain at the formal and descriptive level, and we can see this if we look at structures in a more general sense. Jean Piaget defines a structure as a self-regulating whole with a set of transformational rules and points out that in the field of natural logic, such a structure is always open-ended.[14] Since Kurt Gödel it has been realized that no system of logic can contain its own explanations: a given system is always built upon axioms which are subject to further analysis. The *form* of one system becomes the *content* of the next higher system, and so on. This regress does not concern the mathematician or the physicist, since what matters to him is the self-consistency of the system he has chosen to work within. But it must concern the philosopher, the social scientist, and the designer—and any discussion that deals with the language-like systems in normal use by society. In these systems value judgments have to be made. If a value is relative, it must be relative to *something,* and this something must itself be a value.

Here we come to the root of the problem both of semiology and modern architecture. If a language of any sort is merely the arrangement of minimal structures, these structures must already be full of given meanings, as they are in language. This is the necessary condition of social communication.

**NOTES**

1 Maurice Merleau-Ponty, "On the Phenomenology of Language," *Signs* (Evanston, IL: Northwestern University Press, 1964), pp. 84–97.
2 Quoted and translated from Noam Chomsky, *Cartesian Linguistics: A Chapter in the History of Rationalist Thought* (New York: Harper & Row, 1966), p. 53.
3 David Hume, *An Inquiry Concerning Human Understanding,* ed. Charles W. Hendel (Indianapolis: Bobbs-Merrill Company Inc., 1955), p. 93.
4 Jean-Jacques Rousseau, *The Social Contract,* trans. G. D. H. Cole (New York: E. P. Dutton & Company, Inc., 1950), p. 3.
5 Friedrich Engels, *Ludwig Feuerbach and the Outcome of Classical German Philosophy,* ed. C. P. Dutt (New York: International Publishers, 1935), p. 44.
6 "Modern Architecture and the Historian, or the Return of Historicism," *RIBA Journal* 68 (April 1961), pp. 230–37.
7 Engels, *Ludwig Feuerbach,* p. 59.
8 See Otto Antonia Graf, *Die Vergessene Wagnerschule* (Vienna: Verlag Jugend & Volk, 1969).

9  Wilhelm Worringer, *Abstraction and Empathy: A Contribution to the Psychology of Style,* trans. Michael Bullock (New York: International Universities Press, 1953).
10  Le Corbusier, "In Defense of Architecture," *Oppositions* 4 (October 1974), pp. 93–108.
11  Roland Barthes, "Myth Today," *Mythologies,* trans. Annette Lavers (New York: Hill and Wang, 1972), p. 133.
12  Claude Lévi-Strauss, *The Raw and the Cooked: Introduction to a Science of Mythology: I,* trans. John Weightman and Doreen Weightman (New York: Harper & Row, 1969).
13  This seems to be the point of view of certain semiologists, e.g., Abraham Moles, *Information Theory and Esthetic Perception,* trans. Joel E. Cohen (Urbana: University of Illinois Press, 1966).
14  Jean Piaget, *Structuralism,* trans. and ed. Chaninah Maschler (New York: Basic Books, Inc., 1970).

选读 8

"美学进程中的经验语境"

詹姆斯·马斯顿·菲奇

菲奇在本文中探讨了建筑与人类审美体验之间的关系。他在文中写道:"我的基本论点是建筑的终极任务是发挥支持人类的作用:介入我们和我们所处的自然环境之间,通过这样一种方式来减轻压在我们肩上的总环境负荷。因此,建筑的中心功能是减轻生活压力。"本文架起了菲奇在历史保护领域和建筑评论领域著作间的桥梁——并且对设计过程中建筑审美的不同尺度进行了分析。

## READING 8

## Experiential Context of the Aesthetic Process
### James Marston Fitch

Fitch examines in this article the relationships between architecture and the aesthetic experiences of human beings. He states: "My fundamental thesis is that the ultimate task of architecture is to act in favor of humankind: to interpose itself between us and the natural environment in which we find ourselves, in such a way as to remove the gross environmental load from our shoulders. The central function of architecture is thus to lighten the very stress of life." This article provides a bridge between Fitch's work in the fields of historic preservation and architectural criticism—and analyzes the dimensions of architectural aesthetics in the design process.

*Source:* "Experiential Context of the Aesthetic Process" by James Marston Fitch. *Journal of Architectural Education* 41, no. 2 (Winter 1988). Copyright © 1988 by the Association of Collegiate Schools of Architecture, Inc. (ACSA). Reproduced by permission of MIT Press Journals. pp. 4–9.

## INTRODUCTION

It goes without saying that all architects aspire to the creation of beautiful buildings. But a fundamental weakness in most discussions of architectural aesthetics is a failure to relate it to its matrix of experiential reality. Our whole literature suffers from this conceptual limitation since it tends to divorce the aesthetic process from the rest of experience, as though it were an abstract problem in pure logic. Thus we persist in discussing buildings as though their aesthetic impact upon us were an exclusively visual phenomenon. And this leads immediately to serious misconceptions as to the actual relationship between the building and its human occupants. Our very terminology reveals this misapprehension: we speak of having *seen* such and such a building, of liking or not liking its *looks,* of its *seeming* too large or too small in scale, etc., etc. These are all useful terms, of course, insofar as they convey a part of the whole truth about our relationship to our buildings. But they are also extremely misleading in suggesting that humanity exists in some dimension quite separate and apart from its buildings; that its only relationship with them is one of passive exposure; that this exposure occurs only along the narrow channel of vision; and that the whole experience is quite unaffected by the environment in which it occurs.

The facts are quite otherwise and our modes of thought must be revised to correspond to them. For architecture—like humanity—is totally submerged in the natural external environment. It can never be felt, perceived, experienced, in anything less than multidimensional totality. A change in one aspect or quality of this environment inevitably affects our perception of and response to all the rest. Recognition of this fact is crucial for aesthetic theory, above all for architectural aesthetics. Far from being narrowly based upon any single sense of perception like vision, our response to a building derives from our body's *total* response to and perception of the environmental conditions which that building affords. It is literally impossible to experience architecture in any "simpler" way. *In architecture there are no spectators: there are only protagonists, participants.* The body of critical literature which pretends otherwise is based upon photographs of buildings and not the experience of the actual buildings at all. (It seldom occurs to us to remember that even when we study the pictures of one building in a book or magazine, we always do so while sheltered by another. We could no more enjoy photographs of a beautiful building while seated in a snow-filled meadow than we could respond favorably to a concert in a storm-tossed lifeboat at sea. Most such aesthetic experiences subsume as *sine qua non* the controlled environment of architecture.)

Analogies between architecture and the other forms of art are very common in aesthetic literature. Obviously, architecture does share many formal characteristics with them. Like a painting or sculpture, like a ballet or a symphony, a building may be analyzed from the point of view of proportion, balance, rhythm, color, texture and so on. But such analogies will be misleading unless we constantly bear in mind that our experiential relationship with architecture is fundamentally of a different order from that of the other arts. With architecture, we are *submerged* in the experience, whereas the relationship between us and a painting or a symphony is much more one of simple *exposure*.

Leonardo da Vinci claimed for painting a great advantage over the other forms of art—namely, that the painter had the unique power of fixing, once and for all, not only the vantage point from which his painting was to be viewed but also the internal environment (illumination, atmospheric effects, spatial organization) under which the painted action took place. Such a claim is only partly true for any art form: for architecture it is preposterous. Nevertheless, architects since the Renaissance have accepted without challenge this proposition, thereby obscuring another fundamental difference between the experiencing of works of art and architecture. The first involves a unilinear exposure, a one-way and irreversible sequence of events, while with architecture the experience is polydirectional and random in both time and space.

## SEMIOTICS: NEW FACTOR IN ARCHITECTURAL CRITICISM

Despite the presence of many able persons in the field, the literature of architectural theory and criticism has never stood at a more confused and less productive level than it does today. If the situation were confined to a handful of critics and historians speaking only to each other, it might perhaps be written off as a small, if aberrant, development. But unfortunately this literature has a wide circulation and prestigious institutional backing and its negative impact is affecting both the practice and the teaching of architecture.

The reasons for this qualitative decline are widespread and complex. Many of them no doubt reflect the deepening crisis of our culture as a whole. These may lie beyond our capacity to correct. But some of them are specific to the architectural profession and therefore subject to diagnosis and, one hopes, corrective therapy. One of the principal sources of this malaise in our literature is the recent infusion of two specialized areas of academic philosophy: phenomenology and semiology. Simple dictionary definitions of these terms suggest that, whatever their utility for professional philosophers, they are apt to prove dangerously counter-productive for architectural theorecticians. *Webster's New Collegiate Dictionary* defines phenomenology thus:

> the description of the formal structure of the objects of awareness, and of awareness itself, in abstraction from any claims concerning existence.

*The American Heritage Dictionary* carries the definition a little further into metaphysics:

> the study of all possible appearances in human experience, during which considerations of objective reality and purely subjective response are temporarily left out of account.

Although the earlier (1947) *Webster's Unabridged* does not define phenomenology, it gives this definition of the phenomenist as:

> one who believes only in what he observes in phenomena, having no regard to their causes or consequences; one who does not believe in a priori reasoning or necessary prime principles; one who does not believe in an invariable connection between cause and effect, but holds this generally acknowledged relation to be nothing more than a habitually observed sequence.

*American Heritage* defines semiology as:

> the science dealing with signs or sign language. The use of signs in signalling, as with a semaphore.

while *Webster's New Collegiate* defines semiotics as:

a general philosophical theory of signs and symbols that deals with their function in both artificially constructed and natural (sic!) languages.

But *Webster's Unabridged* has only this definition of semiotics (there is no entry for semiology):

semiotics—pertaining to the signs or symptoms of disease. [!]

These definitions share certain characteristics. Those for semiotics imply that vision is the unique channel of sensory perception, ignoring the fact that most phenomena are experienced through most or all of the others—hearing, olfaction, taste, touch and proprioception; and ignoring the fact that many phenomena—e.g., heat, gravity, the composition and movement of gases—are not visually perceptible at all. Definitions of phenomenology—the field which furnishes the intellectual platform for the architectural semiotician—are more alarming still in their implications for architecture. We are told that "the formal structure of (architecture) can be described in abstraction from the claims of existence." Architectural appearance can thus be evaluated without regard to "objective reality or purely subjective response." The architectural semiologist is at liberty to observe (architectural) phenomena, "having no regard for their causes or consequences." Indeed, he or she is freed from believing in "an invari-effect . . . this generally acknowledged relation (is) nothing more than . . . a habitually observed sequence."

These are conceptual positions which, explicitly or implicitly, permeate the literature of architectural semiologists—and, for that matter, that of the Post-Modernist critics as well. Are we to take this erudite literature literally? What would be the consequences if such a policy were actually applied to the design of structural members? Does the action of gravity on a roof truss or a column display "not an invariable connection between cause and effect . . . but nothing more than . . . a habitually observed sequence"? Is the "formal structure" of architectural space—classroom, theater, laboratory—to be evaluated by a process in which "considerations of objective reality and purely subjective responses are temporarily left out of account"? How can such an approach enable us to specify the air conditioning systems and thermal insulation required to make those spaces habitable?

We should examine carefully the actual circumstances under which phenomenologists and semiologists perform their work. The fact is that neither one deals with experiential reality at all. Instead, each employs words and pictures which purport to describe that reality. While both are essential to communication, both are grievously flawed—one-dimensional facsimiles which, unless we are careful, come to be accepted as substitutes for multidimensional reality itself. The satellite photograph of the sun, for example, will permit the astrophysicist to observe the solar flares on the surface. That observation, detached from "any claims concerning existence" (such as ambient temperatures of 10,000,000°F) does advance our understanding of the universe. But it should never be forgotten, as it too often is, that the physicist's observation is strictly conditioned by the terrestrial temperatures under which *it itself* occurs. A mere 50° change above or below his laboratory's comfortable 70°F would render accurate observations difficult: a 100°F shift would threaten life itself. In real life, there are no circumstances, even "temporary," in which the context of the observation does not modify the observation itself.

## DUAL BASES OF AESTHETIC DECISION

The study, criticism and appreciation of works of art occur in the context of two coexistent levels of experience—cultural and physical, cognitive and sensuous. Thus, when we confront works of art in the museum, our final judgment on the event is the result of a reciprocal process. From the confrontation itself we acquire direct sensuous information by perceptual means—sight, sound, touch and smell. This input is continuously analyzed against a background of cognitive information on that art which we have acquired from our culture. The consequence of this aesthetic process is dialectical: cultural standards modify perception and perceptual input, in turn, modifies our aesthetic response. The end result of this process is a resolution of forces which are at once private and public, personal and societal, sensual and intellectual.

But this process does not take place in a vacuum. It occurs, like life itself, under concrete sets of experiential conditions. Though nominally exogenous, these conditions inevitably modify our judgmental processes. Thus, if the museum is too hot or too cold or too noisy, the aesthetic experience will be negatively affected, no matter what the "objective" value of the art work may be. In addition, of course, a host of internal, endogenous forces are involved in the act of reaching aesthetic decisions. Among them are our physical condition (how tired we are, how much time we have for the viewing, whether we are hungry or thirsty, etc.) as well as our psychic state (whether we know that the art is already esteemed by our peers or as yet unknown to them, whether we view it alone or in company, etc.).

This is the circumambient context of aesthetic decision. And this is the context which is consistently minimized, if not altogether ignored, in the literature of aesthetics and art criticism. The initial error begins with the implicit assumption that, because perception of art is primarily visual, it is exclusively so. And this leads, even if unwittingly, to thinking of the enjoyment of art as a spectator sport. Like people at a tennis match, we are assumed to stand outside the field of action, our only channel of perception being visual. Such an oversimplification ignores the fact that the spectator is submerged in the same set of environmental forces as the art on view. The experience is a first-hand, face-to-face encounter between viewer and viewed, observer and observed, perceiver and perceived. There is no intermediary, no interpreter or interlocutor. At the same time, it is obviously true that our contact with the work of art remains highly tenuous, "disinterested and free" in Kant's definition. When we find the painting or sculpture "bad" (i.e., aesthetically unsatisfactory) or when we "grow tired of looking at" even "good" art work, we can easily terminate the encounter simply by walking away from it.

It is when we try to communicate the aesthetic experience vicariously—that is, by photographic images projected in the lecture hall or printed facsimiles in art journals—that we introduce a whole new set of complicating factors. Not only is the two-dimensional image substituted for the three-dimensional original. Even in purely perceptual terms, without the intervention of the interpreter, this shift is more radical than commonly realized. Vision, here the principal channel of perception, is compromised. The luminous conditions under which the photograph was originally made (color and intensity of light, vantage point of camera, etc.) are imperfectly replicated by even the best photography and printing. And the visual environment depicted inside the image is now enframed in the radically different context of lecture hall or library.

## WORDS AND PICTURES—NECESSARY BUT HAZARDOUS

And now the interlocutor (lecturer, author, critic) interposes himself between the viewer and the viewed: not only does he tell us where the object came from, who made it and why; he also tells us what we ought to think of it. The results of this intervention are literally incalculable. The pictorial information embodied in the artifact is now cocooned in the verbal web of spoken or printed word. The sensuous data of visual perception is now supplemented by cognitive data. Nominally it aims to *supplement:* in reality it tends to supplant it, as the literature of semiotics all too convincingly suggests.

Obviously, the cultural utility of spoken lecture and printed essay is not at issue here. It would, indeed, be hard to imagine the modern consciousness without their ubiquitous presence. What *is* at issue is that their profound influence on aesthetic theory goes largely unremarked. Historically, the art lecture and the art book are very new to human culture. Before printing and photography, aesthetic theory necessarily derived from first-hand contact between user and artifact, viewer and viewed. Early literature—e.g., among the Greeks—might have been profound conceptually: but before the printing press, its propagation was strictly limited in scope and audience.

Architectural aesthetics necessarily differs from that of painting and sculpture, of music, dance and theater in many fundamental respects. Unlike the work of art, architecture is always basically functional and utilitarian, whatever its artistic pretensions. This may also apply to a wide range of other artifacts with both formal and functional properties—from Medieval armor to Minoan drinking cups. But our aesthetic experience with architecture is one of *submergence* rather than passive and contemplative *exposure,* as in the case of art. We *inhabit* the work of architecture; we merely *perceive* the work of art. To fail to understand the qualitative difference between these two aspects of experiential reality may have stultifying effects on the literature of art criticism: on that of architecture, they are nothing less than disastrous.

The literature of architecture, both historiographical and critical, is today heavily dependent upon concepts and terminology borrowed from nearby (and putatively) congruent fields of theoretical activity. One phylum of borrowed aids is verbal/literary; the other pictorial/aesthetic.[1] It well may be that, in lieu of actually experiencing architecture, we have no choice but to employ these borrowed modes of description and analysis. Indeed, actual exposure to architecture as expressed in the cultural tourism of the past few decades, seems only to whet our appetites for more books on architectural history, criticism and travel.

But there are conceptual hazards in the act of writing descriptions and critiques of architecture, just as there are in writing verbal descriptions of such experiences as eating food, drinking wine, or listening to music. In all such cases, we are employing one-dimensional verbal means to describe four-dimensional, multisensory experience—experience of which the aesthetic is only one aspect and where, in any case, sensory perception has sharp limitations (e.g., our inability to perceive botulism in contaminated food or imminent structural failure in the badly-welded roof truss).

If architectural criticism must employ words and pictures to make its points, then it should employ rigorous and precise standards, qualities sadly lacking in much of our current literature. Today, we find the critic saying of the architect that he or she uses "metaphors" to "express" an idea. That the architect employs this or that "grammar," "speaks" such and such "language" of design. Here the critic is describing visual phenomena in verbal/literary terms, attributing to the architect literary qualities which may or may not be implicit in this latter's work. These verbal referents are often hopelessly mixed up with pictorial ones. With the flip of a wrist, the critic can convert any building into an "icon," a "symbol" or an "image." And sometimes the referents are aural: buildings are said to be giving us "signals" or, alternatively, sending us "messages." The most famous among such metaphors is Goethe's dictum: "architecture is frozen music."

Clearly, architects have often used explicit referents in their work. Nineteenth-century historicizing eclecticism was a giant system of literary and pictorial allusion whose "message" could only be comprehended by the literate. The Doric portico was used to suggest the calm, idealized beauty of Periclean democracy. The Gothic style could be used to invoke the virtues of upper class domestic life (in the country house), military prowess (in forts and arsenals), religious piety (in churches and cemeteries). And in the twentieth century, the panelled ceilings of Wright's Prairie houses evoke the Japanese celebration of wooden joinery, just as Le Corbusier's use of pipe railings in the houses of the twenties constitutes a clear allusion to the ocean liners whose efficiency he so admired. But we do not have to hypothecate these referents from a study of engravings or old photographs. They are explicit, well-documented matters of historical record: no semiologist is required to interpret them.

## WHAT ARCHITECTURE "COMMUNICATES" AND HOW

Buildings can be said to "communicate" with their inhabitants in several distinct modes.[2] Perhaps the most literal of these is the way in which buildings (or at least their enclosed volumes) communicate by their acoustical responses to the sounds which we produce in them. These sounds may be either deliberate (as in concerts, lectures, plays) or incidental by-products of other activities (the noise of business machines in an office, children in a schoolroom corridor). In the first instance, the architect manipulates the volume, shape and surfaces of the containing vessel (classroom, theater, concert hall) to transmit the manufactured sound with maximum fidelity. In the second instance, the architect manipulates space and materials to mask or exclude sounds extraneous to the activity itself. But, in any case, it is not really productive to regard the vessel as being any more of an active communicator than the amplifier on a hi-fi set.

Buildings employ accessory means of "communicating" with their users. *Printed signs* ("Exit," "Dade County Court House," "Ladies Room"). *Dedicatory inscriptions* ("here lies the body of . . .," "From this rostrum, on July 4, 1776, spoke . . ."). *Iconographic symbols* (crucifix, Star of David, hammer and sickle). In all these, the means of communication are exclusively visual. But thanks to electronics (amplifiers, public address systems, tapes and recordings) the building also "communicates" by auditory means (ringing bells to mark the end of classes; spoken announcements—"Will the party waiting for Mr. Smith please contact. . ."). It should be noted that all information so transmitted, whether visual or auditory, is cognitive, not sensuous.

One can also say that, across time but strictly within a given culture, building types (Gothic church, porticoed state capitol, multistorey office building) do become metaphors of the institutions which they house, in much the same way that the seated Buddha or the Virgin and Child have become icons of their respective cults. But it ought never to be forgotten that icons *qua* icons have very specific and limited relevances. Except for those building types with explicitly ideological or liturgical functions (churches, tombs, memorials), the iconographic function ranks fairly low in order of architectural priorities.

The semiologist, relying upon photographs or other two-dimensional facsimiles of three-dimensional artifacts, constructs elaborate iconographic systems, replete with "signals" of real or putative significance. But all too often these significances are *inferred* by the critic with little or no objective evidence that the architect consciously and deliberately incorporated them in the design. Nor, in experiential reality, would the spectator be able to isolate the critic's "signals" from all the rich mix of sensuous information with which one is bombarded when one confronts the actual building.

The fact is that semiotics *cannot* deal with the four-dimensional sensuous reality of being actually submerged in architecture. For, under such circumstances, the "signals" (if it can be correct so to describe the sensory stimuli of sight, sound, smell, taste, and touch in such schematic terms) are far too complex to be isolated and evaluated by the observer. Though he or she seldom bothers to tell us so, the semiologist deals only with visually accessible information (photographs, diagrams and words) about the *appearance* of architecture, not at all with its *performance* as an instrument of environmental manipulation.

It must always be remembered that the architectural critic, in preparing a critique, is twice removed from sensory contact with the object of that criticism.

**1** The critic *may* have actually visited the building under review, though often relies upon pictorial and textual materials which are furnished by others.

**2** The critic *always* produces an essay in controlled environmental conditions—usually not the one under discussion. This double separation in time and space isolates the critic from the full experiential impact of actually being an inhabitant of the building under review.

Even more significantly, this isolation acts as a filter through which only visually accessible data are transmitted. This filtration introduces a profound visual bias into all aesthetic judgments. It can perhaps never be altogether eliminated; but it would be minimized if the critic abandoned any conceptual stance as spectator and instead assumed that of participant. The suffocatingly superficial quality of most architectural criticism stems from the critic's unwillingness or inability to understand this, to comprehend the basic fact that architecture is not like painting and sculpture which can only be experienced from the outside.

Failing to understand the difference, the critic stands outside the building, figuratively as well as literally. Most of his or her judgments will be based upon external appearances: interiors nearly always play a secondary role in the architectural critique. This means that what the building does for the passerby is implicitly assumed to be more important than what it does for its inhabitants. This externalizing perspective is consolidated by the fact that most photographs will be daytime shots, if for no other reason than that the camera is a creature of light. Yet half of our experience with architecture around the year occurs in non-daylight hours. And most exterior views will be fair weather exposures, if for no other reason than that rain, sleet and snow complicate the work of the photographer. Yet few American climates outside the Southwest afford the endless sunshine which architectural photography presumes. Thus, even in purely visual terms, the data which the photograph affords us is severely limited in both time and space. It may be an essential part of our apparatus of communication, but we overlook its insuperable limitations only at great risk.

## ART HISTORIAN AS CRITIC: PRISONER OF THE PICTORIAL

The critic with a background in art history approaches architecture from yet another conceptual posture. He or she is trained in the analysis of the formal properties of art works and in the historical process of their evolution across time; taught to look only at the outside of the work because, in a very real sense, *the visually accessible surface is all there is to art.* In painting and sculpture, form and function are one and the same. There is nothing *but* the surface (and in sculpture, of course, the volumes which that surface defines). The cultural function of this surface is to furnish the sensuous basis by which the work pleases, stimulates, informs, exhorts its viewers. Nothing happens inside the work which requires explication—not even the forces which hold it up, hold it together, hold it in shape across time. The fabrication of an oil painting, stone carving, metal casting, serigraph, demands of the artist a working knowledge of complex physical forces. But no comparable hands-on experience with fabrication processes is expected of the art historian. This separation of artistic practice from aesthetic theory is strikingly evident in typical art historical curricula, where the distinction between art history and studio art is rigorously enforced. The young historian is not only not encouraged to learn something of the *metiers* of the objects he or she will analyze but is actively discouraged from doing so. Only the art conservator is visualized as intervening in the life history of the artifact: and this person is normally trained on another and "lower" academic tract.

Given this training, the typical art-historian-turned-architectural critic does what he or she is trained to do: analyzes pictures of buildings in exactly the same way as he or she does paintings themselves. This critic accepts the photograph of the building as a satisfactory surrogate for the four-dimensional experience of the building itself. And here we confront a crucially important divergence in critical method. A good color photograph of a painting, taken from the very viewpoint assumed by the painter (i.e., a few points along an axis vertical to the center of the picture plane) does in fact afford an acceptable stand-in for the painting itself, being a two-dimensional facsimile of a two-dimensional original. There is actually only this locus in space from which the painting can be satisfactorily perceived.

This Leonardian fix in time-space is a great convenience for the art critic, whose whole critical apparatus is based upon it. But, unfortunately, it cannot be transferred to architecture without hazards, for there are thousands of points in time-space from which buildings are commonly observed. When the critic selects—as he or she is compelled to select—a few visual vantage points, he or she risks the commitment conceptually to a tunnel view of the experiential reality of the building.[3]

And this is by no means the only consequence of the critic's commitment. He or she should remember that the volumes delineated in his or her selected photographs are not of *empty forms* but of *inhabited spaces*. This fact will have influenced every aesthetic decision made by the architect while manipulating the visually accessible surfaces of his or her building. How well the architect reconciles the demands of form and function will be a measure of his or her talent; but even the most capricious of architects cannot have wholly ignored the needs of the clients. Nor can they be ignored by the critic in evaluating the architect's accomplishment.

Most critical literature overlooks this crucial consideration; and modern architectural photography makes it all too easy, since it very seldom records actual inhabitants going about their mundane affairs. The reasons for this omission may be innocent enough. Any motion in the field of view is the enemy of the time exposure. So an architectural photograph without people—like the aforementioned photography without bad weather or darkness—is an easy way out for the photographer and, by extension, for the critic. But the cumulative impact of such documentation upon our conceptual construction of the built world is hallucinatory: empty streets, empty portals, empty rooms, all with the enamelled polychromy of a cityscape by DiChirico.

Architecture today needs historians, theoreticians and critics who view it from a conceptual position *inside* the field.[4] It desperately needs persons who understand that, far from being a spectator sport, architecture is our prime instrument of environmental control; that its current inadequacies can only be corrected by a factual analysis of its *performance* and not by mere cosmetic manipulation of its *appearance*. How these spokespeople are trained, what academic degrees they hold, is of secondary importance. One thing is apparent, however: formal problems cannot be fruitfully discussed without a clear understanding of the functional necessities which underlie them. To deny this umbilical connection is to drive our critical literature further into the metaphysical slough which threatens to engulf it.

## NOTES

1 There is, of course, a third phylum of architectural theory which deals with the science and technology of building design, construction and performance. But this vast literature apparently does not qualify for academic attention. It is completely ignored by (may indeed be completely unknown to) the critics of the establishment.
2 To conceive of buildings as "communicating" is, of course, anthropomorphic: it probably produces more confusion at one level of understanding than clarification at another. For, whether the message (or signal, as the semioticians put it) is implicit or explicit, active or passive, objectively present or subjectively read into the experience by the observer, it is ultimately the client who, through the architect, is speaking—not the building itself.
3 C.f. "classic" view of Venturi's Philadelphia home for the aged; down-stream view of FLW's Falling Water; meadow view of Corbu's house for Swiss students at University of Paris.
4 Except for the professional journals—*Architectural Record, Architecture, Inland Architect* and *Progressive Architecture*—criticism is almost wholly in the hands of writers who lack formal academic training for professional status in the field. Among the nation's leading newspapers and weekly journals only two—*Connoisseur* and *The Village Voice*—have regular critics who are licensed architects: Walter McQuade and Michael Sorkin. It is difficult to imagine a comparable situation in other professional areas—e.g., law, medicine, engineering, horticulture, dentistry, or for that matter, art history itself.

选读 9

## "设计方法论"和"非对称性与不协调性"
布鲁诺·赛维

在《现代建筑语言》一书中,布鲁诺·赛维列出了形成 20 世纪建筑基础的各项原则。他认为"现代建筑语言"的定义源于对之前所有关于人类居住目的和形式的假设的质疑。在《现代建筑语言》第一章"设计方法论"中,赛维认为编写建筑功能清单是打破经典秩序和传统惯例桎梏的一种方法。他陈述了功能主义设计方法的基础并且对"归零"原则是如何成为现代建筑的基本准则做出了说明。而在"非对称性与不协调性"一章中,赛维则提出了与经典建筑对称性原则截然相反的现代设计理论,并且阐述了当代设计师抛弃对轴线使用方式的人为限制,从而丰富了建筑语汇的各种方法。

### READING 9

## Listing as Design Methodology *and* Asymmetry and Dissonance

### Bruno Zevi

In *The Modern Language of Architecture,* Bruno Zevi outlines the principles that form the bases of twentieth century architecture. He argues that the definition of a "modern architectural language" is derived by calling into question all previous assumptions about the purposes and forms of human habitation. In "Listing as Design Methodology," the book's first chapter, Zevi states that the process of compiling an inventory of functions that must be accommodated in architecture is a way of breaking the constrictive rules of classical orders and conventions. He presents the foundations for a functionalist design approach and demonstrates how the principle of "going back to the zero degree" has become the fundamental rule of modern architecture. In the chapter "Asymmetry and Dissonance," Zevi places his theory of modern design in opposition to the classical architectural axiom of symmetry and illustrates the ways in which contemporary designers have enriched the architectural vocabulary by rejecting the artificial constraints of the axis.

## LISTING AS DESIGN METHODOLOGY

The list, or inventory, of functions is the generating principle of the modern language in architecture, and it subsumes all other principles. Listing marks the ethical and operational dividing line between those who speak in modern terms and those who chew on dead languages. Every error, every involution, every psychological lapse and

*Source:* "Listing as Design Methodology" and "Asymmetry and Dissonance" by Bruno Zevi in *The Modern Language of Architecture* from University of Washington Press. Copyright © 1978 by University of Washington Press. Reproduced by permission of Bruno Zevi. pp. 7–22.

mental block at the drawing table can be traced back, without exception, to a failure to respect this principle. Therefore it is the basic invariable of the contemporary code.

Implicit in listing, or compiling an inventory of functions, is the dismantling and critical rejection of classical rules, "orders," a priori assumptions, set phrases, and conventions of every type and kind. The inventory springs from an act of cultural annihilation—what Roland Barthes calls "the zero degree of writing"—and leads to a rejection of all traditional norms and canons. It demands a new beginning, as if no linguistic system had ever existed before, as if it were the first time in history that we had to build a house or a city.

The list is an ethical principle even before it becomes an operational one. Indeed, with tremendous effort and immense joy, we must strip away the cultural taboos we have inherited. We must track them down one by one in our minds and desanctify them. For the modern architect, the paralyzing taboos are dogmas, conventions, inertia, all the dead weight accumulated during centuries of classicism. By destroying every institutionalized model, he can break free from idolatry. He can reconstruct and relive the whole process of man's formation and development, realizing that more than once in the course of the millennia, architects have wiped the slate clean and erased every grammatical and syntactical rule. In fact, genuinely creative spirits have always started from scratch. The modern revolution is not unprecedented or apocalyptic. There has been a recurrent struggle against repressive bonds throughout the ages.

Listing, going back to the zero degree, makes you rethink architectural semantics. In the beginning, verbs and conjunctions must be eliminated. Words can no longer be used unless their content and meaning have been analyzed in depth. Some examples will get us to the heart of this methodology of design.

Windows. In the classical tradition a module is selected for the openings of a Renaissance or pseudo-Renaissance building. Then the sequence of modules is examined, along with the relationship between full and empty surface areas. Finally, the horizontal and vertical alignments, that is, the superimposition of the orders, are established. Fortunately the modern architect is free of these formalistic concerns. He is engaged in a more complex and rewarding task of resemanticization. First of all, no repetitive modules. Every window is a word that stands for itself, what it means and what it does. It is not something to be aligned or proportioned. It may be any shape—rectangular, square, round, elliptical, triangular, composite, or free profile. Depending on the room it must light, the window may be anything from a long narrow strip at ceiling or floor level to a cut in the wall or a running band at eye level: whatever may be desired or considered suitable after calculating the specific window's function room by room. There is no reason why every window in a building should be just like the next one and not have a character of its own. Once you get rid of the tyranny of classicism, windows will be all the more effective if they are different and can convey a host of messages.

Classicism breaks the façade into vertical and horizontal sections. But eliminating the juxtaposition and superimposition of modules will make the façade whole again. What is far more important, the façade will become *unfinished.* When the openings—high or low, straight or crooked—are no longer regulated by axial relationships, the façade will cease to be closed and aloof, an end in itself, and begin talking to its surroundings. It will stop being extraneous and hostile and start taking an active part in the city—or landscape.

CHAPTER 2: ARCHITECTURAL FORM 143

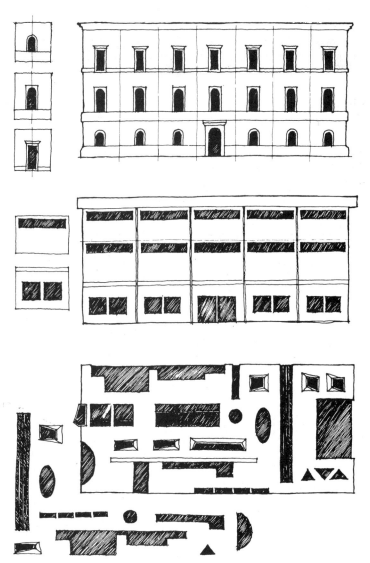

**FIGURE 9–1** The methodology of listing functions, applied to windows. Classicism, whether old classicism (*above*) or the pseudomodern (*center*), is concerned with the module, its repetition, the relationship between full and empty spaces, and alignment. It is concerned with everything except windows. Listing gives back to every element its specific meaning (*below*) and then assembles the various elements. [Bruno Zevi]

Windows are not an appropriate example in discussing modern architecture because, as we will see later, the principle of functional listing precludes the very idea of "façade." Nevertheless, when an architect works in urban fabrics conceived according to preordained schemes and volumes, he is forced to design facades. But that is no reason to give up the modern language. The minute he differentiates windows by form and position, he has done away with the traditional façade and its classical connotations. Indeed, he can inject new life into it by making some windows protrude and others recede, by playing with the thickness of the wall to create a frame of shadow around the glass or, on the contrary, to bring the glass forward into the blaze of light. And why not slant the windows to the surface of the façade? One window can tilt down, focusing on a square, a tree, or a doorway across the street. Another can turn up, framing a piece of sky. A window can be slanted left or right to catch panoramic views, a section of street, a monument, or the sea. Windows can be conceived with a wealth of angles, so that their surfaces are never parallel to the building front.

Even when limited to the detail of windows, the principle of functional listing challenges the classical approach to the façade, takes away its "finished" look, and breaks its square frame by fragmenting the corners of the building and maybe the line between top floor and roof. A double aim is achieved: alternate lighting solutions in the interiors and heightened expressive qualities on the outside.

I can imagine two objections, one of simple dismay and the other of ideological alibis masking dismay. The first objection is that a frightening amount of work is involved in this procedure: if the outline and position of every window have to be thought out separately, the design of a ten-window façade is going to take too much time and energy, far out of proportion to the rewards. The second objection is that such a method may lead to an "academy of misrule," to the triumph of arbitrariness.

The answer to the first objection is that it is largely true. The only correct way to design a window is to study the space it lights, for the perceptual and behavioral value of any space depends on how it is lighted. The fact is that spaces and volumes, the whole building, have to be planned before it is decided what shapes of window to choose. Is modern architecture hard? Probably, but it is splendid because every element, every word of it, is related to a social content. If it were easy, most of the buildings put up today would be truly modern. Suffice it to look at their windows to realize that they are quite often the product of academic irresponsibility.

As to the second objection, that the modern language of architecture tends to be arbitrary: on the contrary, classicism is totally arbitrary, in so far as it gives mythical value to abstract orders that repress freedom and social behavior. Does functional listing lead to disorder? Yes, to sacrosanct disorder that drives out idolatrous order and the taboos imposed by standardized and alienating mass production. The listing method rejects the products of neo-capitalist industry, just as William Morris rejected paleocapitalist products in the second half of the nineteenth century. Industry too often promotes sameness; it categorizes, standardizes, and classicizes. Recent skyscrapers with their curtain walls are more static, boxy, and monolithic than those built fifty years ago. You can see it from the windows as well.

The two objections betray troubled psychological origins. The modern language increases the possibilities of choice, while classical architecture reduces them. Choice creates anguish, a neurotic "anxiety for certainty." What is to be done? There are no tranquilizers for this ailment. But are there in other areas? Does not abstract painting arouse a similar anguish? What about dodecaphonic and aleatory or accidental music? And conceptual art? Is it not anguishing to look at oneself in the mirror for the first time and recognize oneself in an image outside oneself, or to learn that the earth rotates even though it seems to be standing still? Fear of freedom and horror of irrational impulses are at the bottom of this anguish. Let us suppose for a moment that, in a given building, windows could be alike or different without altering their function in any way. The modern language says, let them be different, let there be more choices. The classical code dictates that they all be alike, they must be orderly—like corpses. But the hypothesis that they may be equally functional is absurd, really arbitrary. This merely confirms an established fact, but one that is very hard to instill in the minds of architects: what seems rational and logical, because it is regulated and ordered, is humanly and socially foolish; it makes sense only in terms of despotic power. What is presumed irrational, on the other hand, is generally the result of thinking things through and courageously granting the imagination its rights. Classicism is fine for cemeteries, not for life. Only death can resolve the "anxiety of certainty."

What has been said about windows should be repeated for every aspect of design on any scale: volumes and spaces, their interrelationships, urban complexes, and regional planning. The invariable is always the functional list. Why should a room be cubical or prismatic, instead of free form and harmonious with its uses? Why should a group of rooms form a simple box? Why must a building be conceived as the wrapping for a lot of small boxes packed inside a larger box? Why should it be closed in on itself, making a sharp distinction between the architectural cavities and the urban or natural landscape? Why must all the rooms in an apartment be the same height? And so on. The invariable of modern language consists in whys and what-fors, in not submitting to a priori laws, in rethinking every conventional statement, and in the systematic development and verification of new hypotheses. A will to be free of idolatrous precepts is the mainspring of modern architecture, beginning with Le Corbusier's famous five principles: the "free" plan, the "free" façade, the pilotis that leave the ground "free" under the building, the roof garden that implies the "free" use of the top of the building, and even the strip window, in so far as it offers further evidence that the façade is "free" of structural elements.

The list approach continually makes a clean start. It verifies and challenges even the five principles, as Le Corbusier himself did in his later years, from Ronchamp on. Indeed, his earlier "purism" imposed heavy design restrictions, because the plan was "free" only within the perimeter of a "pure" geometric figure. Why should we sanctify geometry, or straight lines, or right angles? The functional list says *no* to these prescriptions as well. It affects content and form, individual ethics and collective life, just as language does.

The following chapters examine other applications of this invariable. There is no modern architecture outside the list process. The rest is fraud, classicist or pseudomodern. It is a crime, when there is a proper language of architecture to speak.

## ASYMMETRY AND DISSONANCE

Where then? *Anywhere else.* When you criticize something for being symmetrically

FIGURE 9–2  The methodology of listing functions, applied to volumes. Old and pseudomodern classicism boxes man's activities, ignoring their specific differences. Then it sets the boxes above and beside each other to form a larger box (*left*). Listing gives meaning back to volumes, groups them, but preserving their individuality (*right*). [Bruno Zevi]

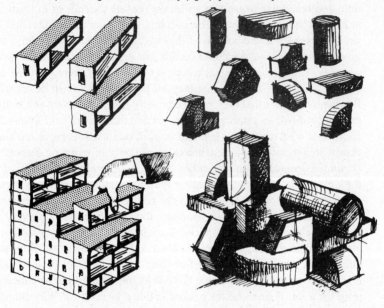

arranged, and you are asked where else to put it, your answer should be: *anywhere else.* There is only one place that is radically wrong, the place that is selected "spontaneously," dredging up all the atavistic conventions of the subconscious.

We can take an even simpler example than the window to demonstrate this, a picture. Here is a wall. Where shall we hang the picture? In the center, of course. No, *anywhere else.* To the right or left, higher up or lower down, anywhere but there. If you hang the picture in the middle, it splits the wall into two equal parts. It reduces the visual dimensions and makes them trivial. The picture seems to be framed and isolated by the wall, when it could open up the room and give it breathing space.

Symmetry is one of the invariables of classicism. Therefore asymmetry is an invariable of the modern language. Once you get rid of the fetish of symmetry, you will have taken a giant step on the road to a democratic architecture.

Symmetry = economic waste + intellectual cynicism. Any time you see a house consisting of a central core with two symmetrical lateral extensions you can reject it out of hand. What is in the left wing? The living room, perhaps. And in the right one? Bedrooms and bathrooms. Is there any conceivable reason why the two enveloping volumes should be identical? The architect wasted space by enlarging the living room to make it the same size as the bedrooms. Or else he restricted essential functions of the sleeping area to keep it the same size as the living room. And look at the height of the ceilings. Why should a vast living room have a low ceiling? On the other hand, if the bedroom ceiling is too high, the space seems visually cramped and suffocating. It

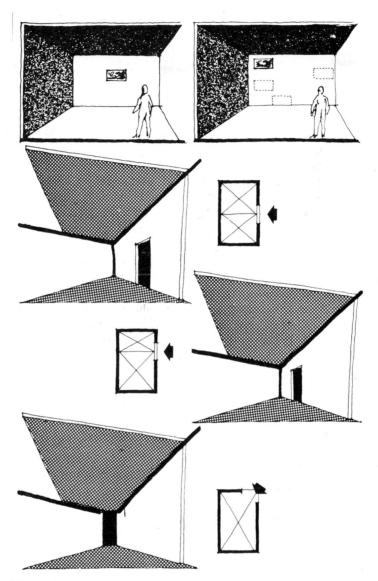

**FIGURE 9–3** Where to hang a picture. Anywhere except in the middle of a wall (*above*). Where to put a door. Anywhere except in the middle (*center*). The farther the door is from the middle, the deeper the room will look (*center, below*). The corner door is the ideal: it enhances the diagonal (*below*). [Bruno Zevi]

is a flagrant waste, both economically and esthetically; a double injury and a double sacrifice. On the altar of what taboo is this sacrifice laid? On the altar of symmetry.

Symmetry = a spasmodic need for security, fear of flexibility, indetermination, relativity, and growth—in short, fear of living. The schizophrenic cannot bear the temporal aspect of living. To keep his anguish under control, he requires immobility. Classicism is the architecture of conformist schizophrenia. Symmetry = passivity or, in Freudian terms, homosexuality. This is explained by psychoanalysts. *Homo*logous parts instead of *hetero*nymous parts. It is infantile fear of the father—the academy, in this case, is a father figure, protective of the cowardly child—who will castrate you if you attack a heteronymous figure, the woman, the mother. As soon as one becomes passive and accepts symmetry, the anguish seems to subside, because the father no longer threatens, he possesses.

Perhaps the whole history of architecture could be reread in terms of symmetry neurosis. Certainly that of Western architecture could be. It is no accident, for example, that Italy was the first country to revive the worship of this idol during the Renaissance, while other countries continued to develop the Gothic style. The economy of the Italian peninsula was going through a severe crisis which the dominant classes tried to conceal behind a classicist mask. They evoked the Greco-Roman past in a mythical key in order to camouflage the instability of the present. They assumed a courtly, forbidding, or Olympian air to hide the desolation of society. It has always been like that: symmetry is the façade of sham power trying to appear invulnerable. The public buildings of Fascism, Nazism, and Stalinist Russia are all symmetrical. Those of South American dictatorships are symmetrical. Those of theocratic institutions are symmetrical; they often have a double symmetry. Can you imagine an asymmetrical Victor Emmanuel Monument in Rome, out of balance, varied in its parts, with an equestrian statue to the left or right rather than in the center? An Italy capable of building that kind of monument would have been another kind of nation, one committed to the creation of a democratic state administration, an efficient service sector, a society balanced between northern and southern regions and based on justice. As a matter of fact, such a country would not have wasted public funds on a marble monstrosity like the Victor Emmanuel Monument. Such a society would not have disfigured the Piazza Venezia with something that made its proportions so trivial, by moving the Palazzetto Venezia and tearing down the Palazzo Torlonia; in short, ruining not only an architectural hub but the whole townscape of Rome. It would have used the money to build lower-class housing, schools, and libraries and to reform agriculture and public health facilities. The Victor Emmanuel Monument reflects the fragility of a backward nation that pretends to be progressive by striking a triumphant, monumental, arrogant, and bombastic attitude. The flame of the Unknown Soldier at the foot of the Arc de Triomphe in Paris and the Cenotaph in London pale in modesty before this horror, whose symmetry rises to titanic heights of wickedness.

There are symmetrical buildings that are not rhetorical, but all rhetorical buildings—symbols of totalitarian power or products of sloth and cynicism—are symmetrical. On closer examination, moreover, nonrehetorical symmetrical buildings prove to be only partially symmetrical, generally only on the main front. This leads to another observation: symmetry has been used in the most obscene way to deform and falsify the arrangement of historic monuments. The most striking example: the Propy-

CHAPTER 2: ARCHITECTURAL FORM  149

FIGURE 9–4   Rome, Piazza Venezia. The old narrow square (*above*) could have accommodated an evocative monument like Le Corbusier's "Open Hand" (*second row, left*). Instead it was blasted open to make room for the pharaonic Victor Emmanuel Monument (*right and third row*). Of course, no asymmetry was allowed (*below*). [Bruno Zevi]

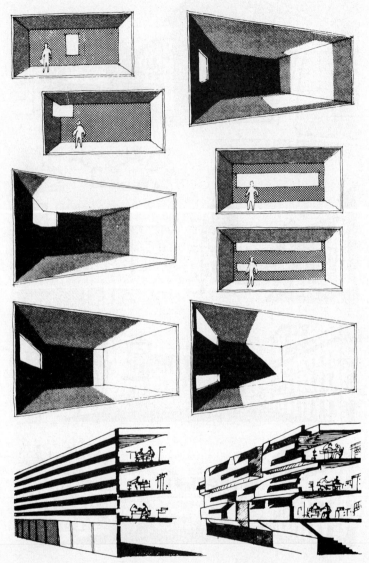

**FIGURE 9–5** How to light a room. Not in the middle (*above*). Any other arrangement would be better: corner window, strip window, double strip (*center*). In the Rome railway station, a double glass strip provides light for the offices (*below, left*), but a greater variety of shapes would have been preferable (*right*). [Bruno Zevi]

laea of the Athenian Acropolis. These have a blasphemously asymmetric plan; but since the Ecole des Beaux-Arts could not admit that such a heretical structure stood at the very entrance to the sanctuary of classicism, Mnesicles' work was displayed as if it were symmetrical. Why? Because in a moment of mental aberration the Greeks had made a mistake, and it had to be corrected. Another example: the Erechtheum, a quite irregular and asymmetrical building, so "modern" that in a way it is a forerunner of Adolf Loos's multileveled Raumplan. What weight did the Erechtheum carry in the Beaux-Arts doctrine? None. It was not symmetrical, so it could serve no purpose.

Take a room, for example. Where should the entrance door be? Anywhere, just so long as it is not in the middle of a wall. That would split the space in two. What "anywhere else" really means is the most conveniently uncentral position, so that the diagonal can be enhanced to create the maximum sense of depth. And to accent the diagonal view, why not detach the entrance door from the wall surface and tilt it? Fine, let us give it a specific meaning, different from the other doors.

The same room. Where should the light come from? Anywhere, as long as it is not in the center of a wall, dividing the room into three sections, an illuminated one between two areas of darkness. Let us give each window new meaning as a specific light carrier in function of the interior space. If there is no view outside, try a strip window at floor level, another one at the ceiling (with a different width to avoid symmetry), and perhaps vertical strips at the corners to light the walls. In the offices of the Rome railway station there are two strips of window per floor, one at desk level and one at the ceiling. This is a satisfactory arrangement, although classicized by too much repetition of the motif. When windows are installed in opposite walls, they must not face each other directly: they will merely light each other and not the room. Take the Room of the Months in the famous Palazzo Schifanoia in Ferrara. Every window faces a full panel on the other side of the room, thus providing magnificent lighting for the marvelous Este frescoes.

Symmetry is a single, though macroscopic, symptom of a tumor whose cells have metastasized everywhere in geometry. The history of cities could be interpreted as the clash between geometry (an invariable of dictatorial or bureaucratic power) and free forms (which are congenial to human life). For hundreds of thousands of years the paleolithic community was ignorant of geometry. But as soon as neolithic settlements began and hunter-cultivators were subjected to a tribal chief, the chessboard made its appearance. Political absolutism imposes geometry, and absolutist governments regiment the urban structure by establishing axes and then more axes, either parallel to each other or intersecting at right angles. Barracks, prisons, and military installations are rigidly geometrical. Citizens are not allowed to make a natural curved turning to the left or the right. They must spring round 90 degrees like marionettes. The plans of new cities are generally laid out on a grillwork. There have been exceptional cases of cities designed on hexagonal or triangular schemes, but they have never left the drawing board. New York is a chessboard, with Broadway the only diagonal. Imperial Paris is based on brutal slashes that sadistically gashed the pre-existing popular fabric of the city. Latin America was colonized with peremptory laws that imposed a priori geometrical forms on cities, whatever their natural topography might have been.

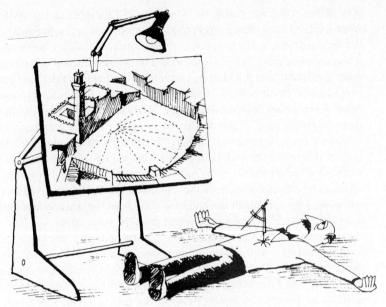

**FIGURE 9–6**  It would be extremely difficult to represent a medieval urban layout (for example, Siena's Piazza del Campo) using T-squares, compasses, and drafting machines. These tools are good only for boxy architecture, which can easily be represented in perspective. [Bruno Zevi]

Cities, and especially capitals, are regular victims of geometrical operations. They survive only because their growth outdistances administrative and political prescriptions. Small towns, on the contrary, and particularly rural towns, are not usually geometrical, but Mafia-run settlements in rural Sicily show mercilessly rigorous geometry.

This age-old cancer, with such illustrious remissions as medieval civilization and country villages, can be extirpated only with an iron will. Architects are so influenced by inhuman and artificial geometry that it seems "natural" and "spontaneous" to them. They know no other language. And this ancestral disease is nourished by the very tools of design: T-squares, compasses, drafting machines. They serve to draw parallel lines, parallel walls, parallel rooms, parallel streets, and right-angled intersections: a world perfectly enclosed in rectangles and prisms, a world easily kept under guard by rifle or machine gun. Coffins package corpses, but being trapezoidal in form they are closer to the shape of their contents. Living men do not even have that concession. They are cynically boxed in abstract and inorganic forms.

At the end of the Middle Ages the taste for freedom from regular geometry, which coincided emblematically with the taste for liberty pure and simple, disappeared. Buildings like the Palazzo Vecchio in Florence and groups of buildings like those in Siena and Perugia look today like something from outer space. Present-day architects could not design them; the language they use will not let them. To re-educate architects, T-squares must be banned, along with compasses and all the equipment that is laid out as a function of the grammar and syntax of classical architecture. Antigeometry and free form, and therefore asymmetry and antiparallelism, are invariables of the modern language of architecture. They mark emancipation through dissonance.

Schoenberg wrote that dissonance should not be considered a piquant seasoning for tasteless sounds. Dissonances are logical components of a new organism that has the same vitality as the prototypes of the past. Schoenberg discovered that music freed from a tonic, or a harmonic center, was fully comprehensible and capable of evoking emotions. Tonality stands for symmetry, proportion, consonance, and geometry. Too many architects have not yet learned this lesson.

## ARCHITECTURAL FORM—SUGGESTED READINGS

Bachelard, Gaston. *The Poetics of Space.* New York: Orion Press, 1964.
Benedikt, Michael. *Deconstructing the Kimbell.* New York: Sites Books, 1991.
Blake, Peter. *Form Follows Fiasco.* Boston: Little, Brown, 1977.
Bloomer, Kent, and Charles Moore. *Body, Memory, and Architecture.* New Haven: Yale University, 1977.
Bonta, Juan. *Architecture and Its Interpretation.* New York: Rizzoli, 1979.
Clark, Roger, and Michael Pause. *Precedents in Architecture.* New York: Van Nostrand Reinhold, 1996.
DeZurko, Edward. *The Origins of Functionalist Theory.* New York: Columbia University Press, 1957.
Evans, Robin. *The Projective Cast: Architecture and Its Three Geometries.* Cambridge, MA: MIT Press, 1995.
Heidegger, Martin. *Poetry, Language and Thought.* New York: Harper and Row, 1971.
Krier, Rob. *Architectural Composition.* New York: Rizzoli, 1988.
Norberg-Schulz, Christian. *Genius Loci: Towards a Phenomenology of Architecture.* New York: Rizzoli, 1979.
Rasmussen, Steen Eiler. *Experiencing Architecture.* Cambridge, MA: MIT Press, 1962.
Rykwert, Joseph. *The Necessity of Artifice.* New York: Rizzoli, 1982.
Thompson, D'Arcy. *On Growth and Form.* New York: Macmillan, 1942.

# 3

# ARCHITECTURAL TECHNOLOGY
# 建筑技术

选读 10
"智慧的呼吸"
雷纳·班哈姆

在《良好建设环境下的建筑》一书中，班哈姆调查了现代环境控制系统对于建筑形式的影响。这部著作的主要观点是设计师应该利用技术提高人类居住的舒适程度，但是不应允许技术凌驾于气候、文化和建筑先例的自然进程之上。在本篇选读，也是这部著作的最后一章里，班哈姆讨论了设计师如何运用高科技玻璃解决各种环境问题。然而，设计师只有认识到在经济实用、文化适宜性、生态进程和物质美学的背景下对技术方法做出判断，才能成功解决各种环境问题。尽管我们已经把这部著作归纳在建筑技术的广泛子类之中，但是读者应该留意本章的最末一句："本书肯定不会再被归在技术门下。"

## READING 10
### A Breath of Intelligence
#### Reyner Banham

In *The Architecture of the Well-Tempered Environment,* Banham investigated the effects of modern environmental control systems on architectural form. The primary theme of this work is that technology should be used by the designer to enhance human comfort, but that it should not be allowed to overpower the natural processes of climate, culture, and architectural precedent. In the selection included here, the final chapter of the book, Banham discusses how high-technology glazing can be employed by the designer to solve myriad environmental problems. However, this can be done successfully only if the designer recognizes that a technical solution will be judged in the context of

*Source:* "A Breath of Intelligence" in *The Architecture of the Well-Tempered Environment* by Reyner Banham. Copyright © 1969, 1984 by The Architectural Press Ltd. Reproduced by permission of The University of Chicago Press.

CHAPTER 3: ARCHITECTURAL TECHNOLOGY  155

**FIGURE 10–1**   Willis, Faber, Dumas offices, Ipswich, England, 1977, by Foster Associates; night view. [Used by permission of John Donat.]

**FIGURE 10–2**   Proposed "Polyvalent Wall," 1981, by Mike Davies; exploded view.

Key

1 Silica weather-skin and deposition substrate
2 Sensor and control logic—external
3 Photoelectric grid
4 Thermal sheet radiator/selective absorber
5 Electro-reflective deposition
6 Micro-pore gas-flow layers
7 Electro-reflective deposition
8 Sensor and control logic—internal
9 Silica deposition substrate and inner skin

[Reproduced with permission from RIBA Journals Ltd. who hold copyright of the illustration and its original accompanying article.]

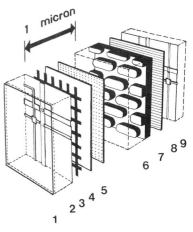

economics, cultural appropriateness, ecological processes, and physical beauty. Although we have placed this work in the broad category of architectural technology, the reader should be mindful of the last sentence of the chapter: "This book must no longer be filed under Technology."

A close student of the rise of environmental technology cannot but note the various and conspicuous roles played by glass throughout history. As the classic selective material, pervious to light but not air or water, it revolutionized the function of holes in massive walls; as a cheap, light, impervious building skin, it made the abolition of those massive walls possible; highly pervious to heat, it made air-conditioning obligatory and high energy consumption unavoidable; selectively pervious to light of different wavelengths, it paradoxically made low energy consumption via the greenhouse effect available. Towards the end of this history, however glass itself has begun to change and diversify; no longer a given, immutable material with, as it were, the single ideal attribute of transparency, glass is now seen as a substance whose behaviour may be specified, self-regulated, or controlled.

It may be specified in various tints, degrees of reflectivity and heat absorption, in double or triple layers with larger or smaller insulating gaps between, and so forth. Mirror and near-mirror glass has already passed into fairly common usage as the preferred cladding for smart office towers and conspicuous downtown hotels, but the aesthetics of its use have rarely risen above the level of a gimmick. One of the rare exceptions is the Willis, Faber, Dumas office building in Ipswich, England, where the architects (Foster Associates) used a three-storey external wall of suspended sheets of dark glass sealed edge-to-edge without glazing-bars to deliver a brilliant display of facetted reflections of the surrounding townscape by day, to reveal an elegantly busy interior by night, and to achieve a measurable economy in environmental power consumption. The use of a growing lawn to insulate the roof shows a nicely judged eclecticism in the choice of appropriate technologies, also.

Self-regulating and controllable glass remains little more than a promise in architecture at present, though most readers will already be familiar with photo-sensitive sunglasses for personal wear that change their tint and/or light-absorption according to the ambient illumination. Controllable glass, whose characteristics change due to molecular response to electrical charges across the pane, is barely out of the laboratory, but its practicality has been demonstrated (at least in the eyes of the technologically convinced).[1] Since the wiring for such glass might be deposited as printed circuitry on the face of the glass itself, it becomes possible to imagine an architecture that does not so much realise the glass paradise of Paul Scheerbart as go far beyond it into the realm of controllable stained glass traceried by its own circuitry in gold or silver, responding to changes in external and internal illumination, and programmed to enhance the daily routines and emotional states of the inhabitants.

"Responsive" environments of this kind have been the stock in trade of futuristic science fiction for some half-century now, but as they edge their way into the realm

---

[1] For a general survey of sensitive/controlled glass technology, and other related matters, see Mike Davies, "A Wall for All Seasons," *RIBA Journal,* February 1981, pp. 55–57.

of the applicable, we have to realise that in practise they will never be cheap. While it was one of the triumphs of Victorian technology to render glass so inexpensive that it could become at certain periods literally the cheapest cladding material in which buildings could be faced, the sophisticated glass technologies of the later twentieth century are increasingly costly to apply in practical uses, either because of the exotic chemistries involved, or because of the elaborate regulating systems required, or contingent costs for special installation procedures involving, say, sealants or gaskets capable of resisting large thermal movements as the panes heat up or cool down.

The new, high-technology glazing will therefore only be employed where it makes good sense economically. While the concept of economic good sense may, occasionally, be stretched to include buildings so important that no expense need be spared, it will normally refer to applications where economic or environmental advantage can be demonstrated objectively—and that will normally involve the calculation of an "energy budget" of some sort. Such budgeting is now not only possible, but quite often reliable, and this may mark the truest of all the gains that mankind has acquired from the past century and a half of environmental engineering.

For most of the history of architecture such calculus was as impossible as it was unnecessary. The occupants of buildings were made comfortable, if at all, by the application of knowledge that was implicit in a kind of craft familiarity with the behaviour of wood, stone, brick, and adobe, understood primarily as structural necessities with inevitable environmental consequences, and of certain other materials, notably textiles, understood primarily as environmental corrective—if you had to build in stone in Northern climates, you hoped to be able to afford to cover your walls with tapestry to reduce the inevitable chill. Knowledge of this sort, though hallowed in tradition, was still unreliable—walls wept moisture, chimneys smoked, palaces were incurably chilly and gloomy, cathedrals collapsed during construction.

In general, however, people and buildings rubbed along well enough, especially since expectations were modest, and miracles were neither promised nor expected. With the rise of the unprecedented materials and technologies of the Industrial Revolution, however, with their concomitant problems and opportunities, the traditional lore of the operation no longer sufficed. Newly applicable information has to be found, understood and applied—the history set out in this book is not only a history of machines and buildings, it is also a history of the application of rational enquiry and creative thinking to environmental management. The rise of solar architecture has reminded us of this afresh; energy budgets, thermal balance sheets, and the like, are essential to its successful deployment; they depend on an increasing body of tabular information derived from experiment and observation, and on calculations of a complexity that might have been too daunting for everyday application were it not for the availability of electronic computation.

Complex mathematics, of course, are not strangers of the history of modern architecture; the triumphs of the *grand constructeurs* of the nineteenth century would not have been thinkable without them, and phrases like "calculation of the resistance of materials" resound through the years after 1900. But this was no more than the replacement of traditional lore (the secrets of the Masons) by experiment and calculation that still referred primarily to the statics of construction and had very little to do with the physiological environment.

Applicable knowledge of an environmental kind, based on observation, however crude, and ordered by mathematics, however simplistic, only began to be readily available in the middle of the nineteenth century with the rise of the heating industries. Specification of hot water systems, and *a fortiori,* hot air systems could not be left entirely to "guess and gas," though successful operation of the systems, as installed, often depended greatly on the accumulated craft of the installers, and the keen senses of the janitors who operated them. But nearly all the useful tables and nomograms developed in this period make sweeping and often unwitting assumptions about the buildings in which the systems are to be installed, supposing them to be of brick, with attics or basements or whatever, and usually careless of the distortions and miscalculations that might be caused by indoor generators of heat (industrial plant, kitchen or laundry equipment) or localised sources of chill (large areas of glass facing away from the sun). Furthermore, these tables and recommended standards could take little note of external climatic conditions, except in a geographically generalised way that perforce ignored local variations.

Yet the kind of rationally holistic vision of building shell, technical equipment, topographical and climatic conditions that went into Frank Lloyd Wright's prairie houses of just before 1910 was matched by one spectacular exercise of comprehensive environmental quantification in those same years—a feat made the more impressive by its author's willingness to back the correctness of his calculations with money. In 1907, for a plant at the Huguet Silk Mills in Wayland, New York, Willis Carrier offered his first performance guarantee. Instead of treating the air-conditioning plant as another branch of structure and offering to guarantee the quality of materials and workmanship, he faced the fact that what his clients were asking from him was to deliver reliably a certain kind of atmosphere, and offered to guarantee the quality of the environment instead. To do this he had to know not only the ability of his plant to handle air and the environmental hazards promoted by the factory's machinery and work force, but also, for the first time, the heat-gain due to the effect of the summer's sun on the building's structure.

Though information sufficiently precise to support exact calculation proved difficult to acquire (throughout these early years of the art, one of Carrier's main labours was in producing sets of standard tables for all kinds of atmospheric calculations) he could at the end of his efforts make statements as precise as

> We guarantee the apparatus we propose to furnish you to be capable of heating your mill to a temperature of 70°F when outside temperature is not lower than 10°F below zero.
>
> We also guarantee you that by means of an adjustable automatic control it will enable you to vary the humidity with varying temperatures and enable you to get any humidity up to 85% with 70°F in the mill in winter.
>
> In summer-time we guarantee that you will be able to obtain 75% humidity in the mill without increasing the temperature above the outside temperature. Or that you may be able to get 85% in the mill with an increase in temperature of approximately 5°F above outside temperature.[2]

---

[2]quoted in full in Ingels, Margaret. *Willis Haviland Carrier, Father of Air Conditioning.* Garden City, NY: Country Life Press, 1952. pp. 31–32.

Though this does not promise absolute control (any humidity at any temperature), it makes a precise promise of sufficient control for the circumstances and states most of the critical tolerances, and in writing it Carrier had performed an almost unprecedented intellectual operation in comprehending the entire environmental complex at the site as a single entity. The work had not originally been so conceived, of course; Carrier had, as usual, been called in to make good a given building proposition which he accepted as found and did not offer to re-design. Nor would most of his contemporaries and successors in his profession; it remains a continuing reproach that specialist environmental engineers have tended to regard themselves as merely the correctors or facilitators of other people's building designs and have rarely taken the initiative in proposing a solution or refusing to implement an ill-conceived one.

Yet a promise was there, however long delayed in practice: the promise that it should be possible to specify an exact environmental desideratum or lay down firm limits of environmental variation and design outwards, as it were, from those requirements, using mechanical and structural methods as appropriate to produce the specified volume and arrangement of the desired environment. The rare examples of this having been done all seem to stem, as usual, from acute industrial need and not—alas—from human delight.

The most conspicuous of such industrial needs at the time of this writing is for "Clean Rooms" in electronics manufacturing plants. In order to avoid the contamination of the silicon "chips" which are the essential components of micro-processors and the like, it is necessary to maintain a fanatically clean atmosphere, free of even the finest dust or microorganisms, while the health and efficiency of the highly skilled work force must be protected by removing as rapidly and thoroughly as possible the noxious by-products of some of the exotic chemistries involved. None of these necessary conditions can be assured by unaided structure, and most would be threatened by "natural" ventilation, so it would seem logical to begin by designing the air-conditioning installation first, then the volumes to be serviced by it, and then the structural shell last.

Logical, but a logic almost never followed in practice. Most such plants consist of available building types—available conceptually (as in *shed, rental unit,* etc.) or in physical fact—modified by the addition of massive air-handling plant, but remaining forever makeshift. Under some other circumstances such *ad hoc* ingenuity might be altogether admirable, but where the needs are as pressing and the tolerances as tight as they are in this ultrahigh technology trade, a radical solution has greater recommendation.

The most radical to date seems to be the "Inmos" plant near Newport in South Wales, where it is claimed that the order of design was to posit the likely bulk of air-handling plant needed to service the known volumes of work space, then to devise a structural system capable of carrying all that equipment at a sufficient height above the work-floor for the duct-work to be distributed above the ceilings of the work spaces, and finally to devise a structural procedure for carrying these ceilings/roofs without unnecessary columns cluttering the working volumes.

The design process was almost certainly less tidy and linear than this, but the final form of the building suggests that the account given above is a fair summary—or allegory—of what happened. The service-packs ride high, and visible, along the spine of the building, above a central corridor, almost as if this were a consciously inverted

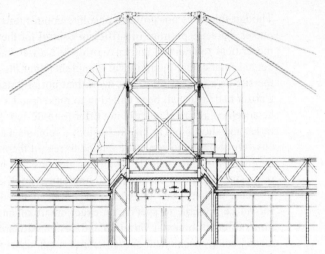

**FIGURE 10–3**   Inmos factory; section. [Used by permission of Richard Rogers Partnership]

version of the Royal Victoria Hospital, where the air-handling plant is *below* a central corridor, and the distributary ducts snake beneath the floor, whereas at Inmos the ductwork sprawls naked across the roofs. The service packs are carried between a long file of H-frames that straddle the corridor, and, from the upper peaks of this frame system, tensile rods extend to carry the long trusses of the roof structure whose outer ends are tied down, not held up, by V-shaped struts. The clean rooms are on one side of the corridor, additionally defended by a ring of semi-clean access space, while less heavily serviced and more sociable spaces are on the other side, and again the echo of the layout of the RVH seems remarkable, however different the external appearance of the two structures.

    The rationality of the solution is equally obvious in both cases, but it may be that the designers of Inmos were as much assisted by the prevailing aesthetic as Henman and Cooper were restricted by theirs in the early nineteen hundreds. Many of the crucial members of the design team from Richard Rogers and Partners who worked on Inmos had already worked on Centre Pompidou, and had thus been long relieved of any inhibitions about the exposure and visual exploitation of environmental installations. The earliest published versions of the design for Inmos show a far greater relish in the rhetorical complexities and baroque exuberances of the air-handling plant and duct-work than real world economics permitted in the built version, suggesting that here was a set of architectural temperaments who, unlike Louis Kahn, did not "hate pipes," but saw them as a modern opportunity to extend the expressive vocabulary of their art.

    The architects may also have been helped by the example of the space suit and its totally regenerative provision for human life support, and the orders of priority that such an environmental device would imply, but there is a crucial difference. Failure of a life support system in the ultimate risk conditions of outer space would normally mean instant physiological disaster; failure of the systems at Inmos would not. As with space systems, there is back-up provision for emergencies; more importantly, the environment outside Inmos is only microscopically different (though the difference is what Inmos is all about) from that within, and the work force could save their lives by walking to a door. Even if a major failure in, say, the acid-fume extract system were to produce immediately injurious conditions, the human physiology contains tolerances large enough for survival—the work force could hold their noses and *sprint* for the door.

These matters are of some consequence over and above this special case, for it is human tolerance, and the relatively small absolute difference between the tempered interior and the outside atmosphere that have long been architecture's license to practice environmentalism. Had the tolerances been finer, the required differences wider, architecture would have had to be radically different to survive. But this license is not absolute, because the tolerances are not infinite. Nor is their range simply fixed or determinable; the environmental needs of the whole living human being are variable in sickness and health, youth and age, education and culture, physical and social circumstance. When British troops in Aden were accused of subtly torturing Arab detainees under interrogation by "deliberately running the air conditioning at 'full cool,' " it may well have been the case that the setting of the air conditioner dial at "full cool" was deliberate, and that the Arabs, as a result, felt subtly tortured, but the motives of the British troops may have been simply to make themselves feel comfortable without possessing the necessary cultural and environmental insight to realise what this might do to persons raised in the local culture and climate. And the same British, or their close cousins, will complain of the "ridiculous way" that Americans run their air-conditioning so cold that one has to remove clothes on leaving the building, without having the cultural and environmental insight to realise that only thus is it possible to wear indoors the mink stoles, etc., which are accepted badges of social rank in Texas and Southern California.

The recognition that there are no absolute environmental standards for human beings has required the environmental sciences to develop methods of assessing performance and needs that depend upon attempts to quantify subjective responses without doing injury to their human validity, to allow for the interaction of what is being assessed with other elements in the environment that are not under study, and to allow for variability in time through fatigue on the one hand, or conscious and unconscious accommodation on the other—faced with a glare of excessive light, one may reduce the amount of illumination, put on dark glasses, screw up the eyes or leave it to the contraction of the iris to compensate.[3] Each of these may be the correct line of action, according to circumstance, and particularly as a function of the length of time to which one is exposed to the glare, for all tolerances seem to be greater where the extreme conditions occur only as occasional peaks in a flow of variables.

This combination of circumstance is fortunate for the shelter industry, since it means that, over time, the spread of tolerances is effectively wider than totally and minutely quantifiable laboratory tests might suggest. Given variation in time, the human body adapts itself to short term changes; the environmental control system

---

[3] The history of physiological environmental studies—of human responses to heat, light and sound—remains to be written. The urgency of the need to get it written while the living memories of some of its pioneers are still available has been recognised by some of the pioneers themselves (e.g. R. G. Hopkinson, late of University College, London), but it will prove to be a formidable task.

**FIGURE 10–4** Space suits, with air-conditioners for use on ground, 1965.

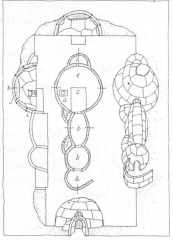

**FIGURE 10–5** An igloo; plan, sections and elevations. Drawn by Jocelyn Bates Helgerson. First appeared in Ian Mackinlay's "The Neglected Hazards of Snow and Cold," *AIA Journal,* February 1983. [Used by permission of Ian Mackinlay, FAIA]

does not have to make instant adaptation to every degree of temperature change in atmosphere or occupant, does not have to anticipate the effects of boiling a kettle or opening the fridge door. In many strictly lethal circumstances the time taken to get up from a chair, walk to a window and open it, is not a life and death consideration, and for less acute situations of vitiation or risk it may not be fatal to wait for someone else to become aware of the problem and open the window for you. It is possible that in the high risk conditions of hard vacuum the instant responses and omni-competence of a space-capsule's life support system are absolutely necessary, and one knows what real life dramas the telemetering of an uncontrolled rise in cabin temperature could occasion during the US moon-flight programme, but here on Earth it will often prove that drawing a blind over a window, or actuating some other equally simple control, is all that is required. In the right circumstances, a truly sophisticated approach to the man/environment system may involve no complex mechanisms at all.

Sophistication is not necessarily the product of highly developed machinery, nor intensive capital investment. It is more a way of using available equipment and resources with cunning and intelligence: the snow-domed igloo of the Eskimo remains a paragon of environmental ingenuity and geometrical sophistication. Its virtues have been rediscovered by high-technology intellects like those of Buckminster Fuller, and then transmitted to Steve Baer and the solar movement.

It should be remembered, however, that the wide acceptance of Fuller's ideas at the beginning of the Sixties was, to a larger extent than could be recognised at the time, a purely fashionable revolt against the forms and usages of the "International Style" that had dominated modern architecture for half a century. In that revolt, many ancient notions and traditional formulae were given new (and uncritical) currency, and environmental wisdom was noisily discovered in vernacular buildings from all over the world.

It had, of course, been the ambition of the International Style Modernists to deliver mankind, all over the world, from the inadequacies of vernacular buildings. Instead, they were to enjoy the benefits of applied science, as in Le Corbusier's proposal to make the ideal temperature of 18℃ available everywhere—from the Poles to the Tropics, "from Moscow to Dakar." Believing this to be ridiculous, dictatorial and inhumane, the opponents of the International Style tended to rush to the other extreme of insisting that all wisdom about the environment is to be found somewhere in vernacular architecture, and that, *therefore,* all vernacular architecture was environmentally wise—which is observably untrue as well as false logic.

Vernaculars (of architecture, language or whatever else) are bodies of culturally transmitted habits which can hold unquestioned sway over the lives of the communities that practise them, even when they have no survival value for those communities. They, obviously, are not likely to be hostile to the survival of those communities, or they would have destroyed them long since, and the long persistence of vernacular practices shows that they have some good generalised congruence with those peoples and the territories they inhabit. Any vernacular—which for the purposes of the present argument could include the Art of Architecture as practised in the Europeanised nations—can be counterproductive when confronted with extreme or aberrant conditions. We all know houses that are "mysteriously" damp or "incurably" draughty within the prevailing lore of the operation, or institutions like the Glasgow School of Art whose heating installations had to be replaced because they "couldn't be made to work;" that is, its plenum system needed different procedures of operation and maintenance from those current in the hot-water-radiator vernacular understood by the city's janitorial culture.

These good-enough-for-general-purposes vernacular procedures may not only fail under extreme conditions, but they may also stretch the limits of physiological tolerance to the point where only a deeply entrenched culture can prevent the resultant human and social inconveniences becoming intolerable. Not only would the level of smell and other atmospheric pollutants inside the otherwise admirable Eskimo igloo be intolerable to the inhabitants of many other cultures, but so would—at another extreme—the way in which the heating arrangements of traditional Japanese houses can immobilise the inhabitants in bulky clothing around a sunken pit containing a miniscule charcoal brazier. Anthropology abounds in examples where cultural rigidity and fixed repertoire of architectural forms are welded into a seemingly permanent deadlock with results that may perhaps preserve a body of ancient wisdom—or an embalmed corpus of ancestral folly.

**FIGURE 10–6** Terrace housing, Sydney, Australia; structural sun-shading of street fronts. [Photo used by permission of Max Dupain and Assoc.]

A textbook example of these vernacular ambiguities is afforded by a building type that is not commonly counted as vernacular but still exhibits all the characteristics. This is the form of terrace housing that flourished in Sydney, New South Wales, in the second half of the last century; though not immemorially ancestral, and given to the employment of some "Technological" materials such as cast iron, the type exhibits all the fixity of a vernacular. Its virtues are immediately apparent even from outside: the party-walls project between each house and the next to support the ends of a balcony across the façade, and the roof slopes out as far as the projecting party-walls. The façade is thus well-shaded against both overhead sun by the roof and the balcony, and against raking sun by the projecting walls, and the result is at (literal) face value, an immediately appealing form of urban housing.

However, the cultural situation which produced these elegant and efficient façades also decreed that the elaborate, or more decorated, elevation be on the front or public side of the house, toward the street. The environmental advantages that were enjoyed by the houses on the south side of the street, facing the (northern) noonday sun, were thus denied to the houses on the other side of the street, which took the blast of the noonday sun directly on their naked and unshaded backsides. This situation may seem ridiculous to rational men, but I can vouch from personal observation that the backs of these houses were clearly built without shading provision equivalent to the fronts, and I was shocked to see how such counterproductive rigidity of vernacular custom could come so close to the lives of populations who believe themselves delivered from all such ancient folly.

Architects and others, who are now recolonising these long despised terrace areas, have had to find ways of making these house-backs habitable by members of a more demanding culture than originally built them. Their solutions are diverse in a way that is strikingly different from the standardisation of the vernacular fronts. Some have opted for crude environmental power in the form of air-conditioner units; some have more or less repeated the front at the back; others have chosen to exploit the exposure by making sun-decks above to shelter indoor-outdoor spaces below, and so forth.

Such variation should be expected; these are supposed to be rational designed solutions to the problem, and the problem will vary from house to house, family to family, life-style to life-style. Conscious and professionalised architecture, as opposed to a custom-bound vernacular, should be able to reason out the unique solution to any individual problem. We should now be enough at home with our kit of parts to be effectively delivered from the tyrannies of technology, as well as those of vernaculars; all should be usable as "Modern Opportunities" in Frank Lloyd Wright's telling phrase—and he should know, for he was clearly the greatest of all opportunists surveyed in the chapters of this book.

His true inheritors are not always easy to identify; the exteriors of their works may betray nothing of the skill that has gone into their design, and even the interiors may reveal little except to those skilled in the appraisal of the environmental performance of a whole house. Architecturally, this may leave us with a curious and almost unprecedented problem—that of the building that is perfectly honest about its functioning, but offers no clues to what that functioning might be. This may only be seen as a problem in the aftermath of the International Style and Functionalism, when buildings were supposed to advertise their functions and exhibit their technical innovations, but we have to accept that we are still inhabiting that aftermath and are a long way from formulating appropriate alternative approaches.

The problem can be typified by some recent houses built in California by the Thacher and Thompson partnership. At first sight these houses are variants of revived Victorian and Edwardian styles, adhering to generally picturesque and folksy layouts, using simplified period details and planked or shingled exteriors. It would be understandable to simply dismiss these houses as current anti-modern gestures, which their designers will admit them to be. But within these pseudo-vernacular envelopes, derived from times when energy was cheap and environmental expectations were low, there lurk subtleties of thermal management which cannot be read by just looking.

"Lurk" is an improper word here; everything is visible, often on the exterior, but is it liable to come in the form of purely decorative-looking bay or oriel windows to admit solar energy and to provide vertical trunking for warmed air which may descend when cooled on the other side of the house via a stairwell or through a space contrived between a balcony and the wall to which it might otherwise be attached—a gap

**FIGURE 10–7** House in Santa Cruz, California, 1982, by Thacher and Thompson; gable and solar oriel window. [Used by permission of Thacher & Thompson Architects, Santa Cruz, California]

which first sight may interpret simply as a piece of post-modern architectural formalism? Close inspection will often reveal tiled floors as heat stores inside the bays, and slatted floors or window seats on intermediate floors to allow warmed air to circulate upwards, but many specific provisions will pass forever unnoticed.

Unlike the houses built by the official Solar movement, these do not make any display of their solarity—no sloping walls of glass facing south; no thermal storage bodies cluttering the interior spaces—yet, the architects are confident enough of their thermal performance (in an admittedly benign climate) to emulate Willis Carrier and offer an appropriately modest environmental guarantee:

> If purchasers ever use their gas-fired furnaces to heat the subject home any time within five years from the date of transfer of title . . . Thacher and Thompson will pay all gas costs related thereto.[4]

[4] Promotional brochure, Santa Cruz, California, 1982.

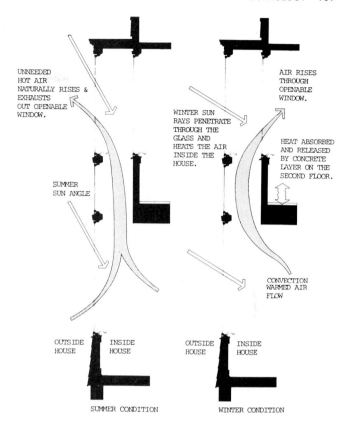

**FIGURE 10–8** House in Santa Cruz; winter and summer airflow in oriel window. [Used by permission of Thacher & Thompson Architects, Santa Cruz, California]

Deftly avoiding the climate of slogans in which most modern architecture—High-tech, Low-tech, Functionalist, Organic, Neo-vernacular, or whatever—has been created, these houses exhibit the potential for liberation and confusion that can come from the application of radical intelligence and organised knowledge to the ancient craft of building. The confusion comes directly from the liberation because what is liberated is performance from form. Or, rather, it is *expected* performance that is liberated from conventional form, as the Thacher and Thompson example shows as well as any other in this book. In some cases the liberation is almost absolute. Example: the rules of organisation, orientation and plan-form for breeze, sectional organisation for cross ventilation and cooling, that must unavoidably apply to traditional structural solutions in the hot humid climates of, say, central Africa, can become such a tyranny that the sealed, glass-walled and necessarily air-conditioned office tower begins to look an irresistible alternative, and in hot dry climates elsewhere on that continent its ability to exclude wind-borne dust from human activities that need to be kept clean makes it almost equally compelling.

Yet the present generation of experts on tropical architecture are prone to decry such solutions as mere status symbols for insecure new African regimes; for them, form and its supposed symbolisms still tend to prevail over most other considerations. They may well be succeeded by a generation of experts who will wish that our

Western civilization had been able to make an equally drastic break with its ancestral vernaculars.

For, to say explicitly what has only been implied so far, architecture, as commonly taught, practised and understood in the West, is still little more than a peasant vernacular, whatever triumphs of art it has bequeathed to the human race. It has, of course, the authority of tradition and certain hallowed forms—arch, beam, column, vault, wall, window, roof, door—whose symbolism is as well understood as their intended performance. But just as the rise of photography undermined the inherited authority of traditional modes of representation like drawing and painting, so the rise of non-structural modes of environmental management has undermined the unique force of these symbolic constructions.

Hence, the avidity with which Le Corbusier and most other Modernists purloined forms from the new mechanical technologies in search of appropriate symbols for the changed world—hence, too, the disappointments when the stolen forms proved to neither guarantee nor even *indicate* significant improvements in performance over what the older structural technologies had afforded. This, after all, was only the same older technology dressed up in borrowed finery, but not inspired by that same breath of intelligence that had created the newer technology that had delivered the new forms.

Perhaps, among the buildings discussed and illustrated in these pages, the outlines of a new and relevant language of symbolic forms may be discerned, but one must beg leave to doubt it—yet. Such a development is not in the contract, so to speak. It is probably true that only when the architecture of the well-tempered environment disposes of a language of symbolic forms as entrenched in our culture as are those of the older dispensation will it be able to hope for equal conviction and monumental authority, but that possibility seems to be excluded by the very nature of the operation which has been chronicled here. The essence of what has been done to temper the environment has been—at every single stage—the displacement of habit by experiment, and of accepted custom by informed innovation. The greatest of all environmental powers is thought, and the usefulness of thought, the very reason for applying radical intelligence to our problems, is precisely that it dissolves what architecture has been made of to date: customary forms. Our present post-Modernists who strive to restore those customary forms can do so only because environmental technology gives them the freedom to separate those forms from desired environmental performance. If this observation sounds somewhat like the comments made in the nineteenth century about those who hung irrelevant historical forms on buildings constructed out of new materials to serve new functions, then it is just that it should so sound; we see the same situation repeating itself but raised to a higher power by higher and more subtle technologies. And if this is the time when history repeats itself as farce, then it is architecture which is offering to become farcical, not the technologies that have displaced it from its ancient role. The position of these upstart technologies seems as secure as ever, in spite of the predictions of their disappearance along with fossil fuel, and the arts of making fit environments for human activities must now accept their claim to be a permanent part of the craft of architecture.

This book must no longer be filed under Technology.

选读 11
"新技术概念的一些特点"
彼得·麦克利里

在这篇文章中，麦克利里调查了技术如何产生新的建筑时空感知概念。他认为像这样一种建筑观需要一种与之相对的新的设计语言。麦克利里讨论了三种主要设计因素，这些设计因素将会影响人们看待和回应建筑形式的方法，它们包括透明性和模糊性；增强和减弱；环境适应性和对环境的使用。他要求设计师去探索"建造者及其环境之间的辩证关系是如何产生一种特殊的知识，而这种知识来自他们对技术设备、进程和理论的运用"。

READING 11

## Some Characteristics of a New Concept of Technology[1]

### Peter McCleary

In this article McCleary investigates the concepts of how technology can generate new perceptions of architectural space and time. He posits that such a view of architecture will demand a new language of design. McCleary discusses three major technological factors that will affect the ways people view and respond to architectural form: transparency and opacity; amplification and reduction; and appropriateness to and appropriation of context. He challenges the designer to explore how "the dialectical relationships between builders and their environments lead to a special kind of knowledge which comes from their use of technical equipment, processes, and theories."

*Understanding the characteristics of the new concept of technology leads to new questions and identifies new dangers. Is there an inexorable shift from transparency to opacity? Can it be reversed? Should we return to a pre-technical condition? And among the dangers that mediation presents are: a fragmentation of perception and experience; the abstract seems more real than the mundane; the persistence of the codifications of architectural languages from archaic perceptions; and the separation of the professional from the layman.*
 *The new concept of technology will generate new perceptions that might lead to new concepts of architectural space and time which will demand a new language of architecture.*

*Source:* "Some Characteristics of a New Concept of Technology" by Peter McCleary. *Journal of Architectural Education* 42, no. 1 (Fall 1988), pp. 4–9. © 1988 by the Association of Collegiate Schools of Architecture, Inc. (ACSA). Reproduced by permission of MIT Press Journals.

## INTRODUCTION

The significance of technology has returned to the discourse on the purpose and meaning of architecture. In the interim a new concept of technology has arisen, one that does not limit itself to building materials and processes, but defines technology more broadly as the understanding (skills and knowledge) of the dialectical relationship between humans and their environments (natural and built) in the production of a new superimposed built environment.

Neither the pre-modern architect as master-builder, not the Modernist coordinator of production, nor the fragmented perception of the Post-Modernist, have yielded a concept of technology useful to both designing and building. A useful, and new, concept demands a new way of thinking about the productive relationship between humans and their environments. In his "analysis of environmentality," Martin Heidegger suggests that our productive encounters with the environment are "the kind of dealing which is closest to us (and it) is not a bare perceptual cognition, but rather that kind of concern which manipulates things and puts them to use; and this had its own kind of 'knowledge.'"[2]

This implies such a knowledge derives from the activities of designing and building, that is, both reflection and action. It further recognizes that the architect, in a productive interchange with the environment experiences it through the mediation of *technics* (i.e. technical equipment) which are contextually arranged as *techniques* (i.e. technical processes); and that experience is conceptualized from the architect's reflection-in-action and then formalized as *technology* (i.e. technical theories).

If questioning is the primary tool of thinking, then new ways of thinking which lead to new concepts perhaps need to ask new questions.

This paper asks a few new "questions concerning technology" without "implying . . . the guarantee of an answer, but at least that of an informing response."[3]

The first and most general question we ask is, "what are the characteristics of the particular kind of knowledge or experience derived from the dialectic between builders and their environments and acquired through the use of technical equipment, processes, and theories?"

Just as Maurice Merleau-Ponty's blind man experiences the world at the tip of his cane, the architect-builder acquires knowledge of the environment through the mediation of equipment, processes, and theories. We will examine the levels of mediation ranging from *transparency* (where the environment is experienced through the equipment, etc.) to *opacity* (where the experience is of the equipment and not of the environment).

Examination of Martin Heidegger's claim that man lives in the space opened up by equipment shows that each mediator explores a different aspect of the environment. These aspects then stand out from the totality of the world, i.e., there is an *amplification*. At the same time there is a *reduction* in experience of the unexplored aspects. We will examine the types of mediation or "directionality" of equipment, processes and theories and the tendency to fragmentation of perception with a concomitant reduction in holistic experience.

Jose Ortega y Gasset's observation that "man finds that the world surrounds him as an intricate net woven of both facilities and difficulties" suggests that questions about the classification of the human realm and its environments might yield "informing responses" to additional characteristics of technical experience.[4]

What is "architecture without architects"? "Building" is probably the answer. Architecture that is anonymous, indigenous, vernacular, folk, and so forth, makes use of "appropriate technology." This "technology" responds to knowledge (from empirical to the natural sciences) of the physicality of the natural environment; to understanding (from craft to the applied sciences) of the built environment; and to the ethical codifications of the civil and cultural agenda of societies.

Architecture "with" architects is the product of a more autonomous act. The intentionality of the builder as architect includes aesthetic concerns along with ethical and scientific matters.

When the production limits its response to the natural and built environments, and to the civil and cultural context, the product has an *appropriateness-to* the environments.

When the architect's intentions dominate the production, the product is an *appropriation-of* aspects from the environments.

To explore this new concept of technology, we will examine the characteristics of transparency–opacity, amplification–reduction, and appropriateness–appropriation.

## THE FIRST CHARACTERISTIC: TRANSPARENCY AND OPACITY

Merleau-Ponty's blind man probes aspects of the world (in this case, the ground underfoot) at the tip of his cane. Is the ground high or low, wet or dry, hard or soft, hot or cold; is the opening wide or narrow, and so forth? It is critical to note that the characteristics the blind man seeks are those of the ground, and not of the cane; the cane, in fact, should go unnoticed. He wishes it to "withdraw," that is, to become a *transparent* technic.

If, in using a hammer, "that with which we concern ourselves primarily is the work," and it is the materiality of the "work" that we experience, then the hammer "withdraws," i.e., becomes "transparent."[5]

Perfect transparency exists only where there is no mediation, as in the "face-to-face" meeting or situation. We can experience either of two extremes: *perfect transparency* (that is, no mediation); or its opposite, the totally mediated situation in which one experiences not the world, but the machine-itself, which in turn encounters the world. In the latter case, the world is *opaque* to man, and it is not the technic that "withdraws" but rather the world itself "recedes" from man.

A vast range of experiences exist between these two extremes: from that of encountering the world *through* technics (i.e., transparency) or, as a result of a loss in transparency, to that of an experience *of* technics (i.e., opacity, where man experiences the characteristic of the technic and the technic encounters the world).

I claim that the historical development of the builders' mode of production led to a loss of transparency and a concomitant gain in opacity. A personal example: my grandfather cut grass with the short-handled sickle; my father reduced the stress in his back by using the long-handled scythe; I experienced the cutting of grass with a hand-driven lawnmower; my daughter encounters the characteristics of grass with a fuel-powered (or self-propelled) hand-guided mower; her child will use a lawn-tractor where the experience is of driving and not of cutting; my great-grandchildren will, in all likelihood use, if anything, an automaton or mechanical goat.

In making and using transparent technics (e.g., the worker's hammer and sickle), the builder designs, controls, and provides the power. The earliest machines replaced human energy with a source of power other than the human body. The introduction of servo-mechanisms (e.g., the governor, the thermostat, etc.) made human control redundant. The builder was reduced to a designer, who in turn will be replaced by self-designing automata.

We find similar losses in transparency in our move from the "primitive hut" through the log cabin to the heavy timber frame, and finally to the balloon and platform frame house. The shift occurs within the mode of production in cutting and preparing the lumber: the technics change from the maul (hammer) and wedge, to the adze, the axe, the hand-saw, the band-saw, and finally to the saw-mill.

*Technique* (or technical processes) too, has shifted towards opacity. As the skilled worker, or artisan, in the steel industry has been replaced by the chemist, metallurgist and systems engineer, the manufacture of steel products has become applied scientific. In the fabrication phase, the template shop has been replaced by computer graphics, and the chalk marking and hand cutting of steel have been replaced by robots and laser-cutting and yet fabrication remains partially empirical. The assembly of steel buildings, CPM and PERT notwithstanding, is still choreographed by the craft and experience of the contractor.

Similarly, *technology* (or technical theories) takes part in the "enframing," with the reciprocal loss in transparency. Over time, our ways of understanding have shifted from knowing through doing (i.e., *craft*) to borrowing intellectual frameworks from other disciplines or phenomena (i.e., *empiricism*) to constructing theories based on systematic, methodological observations of the material-at-hand itself (i.e., *applied science*, or engineering). Where an applied scientific theory exists, the reality of the built design must conform to that theory (albeit with a factor of safety to compensate for the lack of correlation between theory and practice). When no "true" engineering theory is available, the builder is controlled by regulations, such as standard, codes, etc., all based on the collective experience. In some cases where "good" practice was established in "pre-scientific" times, the common sense "rules-of-thumb" of craft were, and often still are, sufficient justification to build.

Our knowledge of structural steel is applied scientific, our knowledge of reinforced concrete is empirical, and our understanding of brick masonry is still based on craft. Craft knowledge of the performance characteristics of materials becomes less acceptable each day as we move inexorably towards the precision, but limited concerns, of the applied sciences. Thus, technical theories, too, have become opaque, and increasingly we can perceive the realm of building only through the filters of the theories of applied science.

Among those involved in building, the architects, many of whom favor the primacy of perception, prefer, I believe, the transparency of arts and crafts to the opacity of machine reproduction.

It would seem that many architects continue to favor: transparent passive solar heating and cooling over opaque mechanical heating, ventilation and air-conditioning systems; adze fluted stone columns to rolled and fabricated steel sections; charcoal, ink and water color drawings to computer graphics; the phenomenological investigations of Kahn, Barragan, and Scarpa to Revivalism, Modernism, Post-Modernism, and "low and high-tech"; sailing with the power of the wind and the feel of the water to the steering of a powered motor boat; the Porsche and the country road to the Mercedes and the turnpike.

Is, then, the architect's position one that says that a "good" technology is one that withdraws? Is a preference for transparency and low technology a desire for *no* technology? This preference for transparency is not unexpected, since architects derived their theories, for the most part, from buildings of the past. And since most of those buildings were built using transparent equipment, processes, and theories, the architect's perception not illogically has been, and continues to be, receptive of and encouraging to transparent modes of production.

## THE SECOND CHARACTERISTIC: AMPLIFICATION AND REDUCTION

In *Being and Time,* Heidegger notes that humans encounter the world through equipment and since they use this equipment "in-order-to do something," it has the character of "closeness," and the "closeness of the equipment has been given directionality."[6]

If the concepts of transparency and opacity explain, in part, the characteristics of "closeness," what are the characteristics of this "directionality" between equipment (technics) and the place or space that humans encounter in using equipment?

Let us take an example from visual perception. At the scale of the body, the human eye is the mediator, or technic, that receives the visual information regarding the world. Whether the world as perceived is chaos or cosmos is decided by another mediator, the schema of the brain. To encounter, visually, the micro-cosm (Gk. *kosmos* = world), the microscope (Gk. *scopos* = target, *skopeo* = to look at) is used as a technic. To encounter, visually, the macro-cosm, one uses the macro- or tele-scope.

The prefix "tele" indicates "far, or at a distance," and the scope acts in "bringing far things nearer" in the sense, here, of the visual. Similarly, in tele-communications, one encounters the telephone (sound), telegraph (drawn message), television (pictures), telephoto; additionally, there is telemechanics (mechanical movement by radio), telepathy (emotional influence), telekinesis (no material connection), and so forth.

Essentially, some aspect of the thing "at a distance" enters the space of the observer—and there is an extension, an emphasis, or *amplification* of a sense through the use of the technic.

In using the telescope, one encounters only the visual characteristics of the distant object. The activity (or object) cannot be heard (the missing technic is the telephone); it can't be touched (tele-"tact"?); smelled (tele-"fume"?), tasted (tele-"gust," or "tang"?). Phenomenologically, the greatest loss is that the "sense of place" cannot be experienced. Thus, any technic (in the above case, the telescope) in amplifying an isolated sense, or senses, brings with it the concomitant loss or *reduction* in total experience.

As with transparency and opacity, we might ask here whether techniques and technology mediate in the same manner as technics, that is, through the directionality of amplification and reduction.

In the case of the *technical processes* of production of so-called industrialized building, the means most often dominate the ends and even become the ends when the precise logic of building supersedes the imprecise concerns of dwelling. This leads to the amplification of production efficiency, tolerances, fit, modular coordination, fast-tracking and so on; and the reduction or diminution of the ends of "spaces good to be in."

Considering *technical theories:* There was a time when all theories of architecture were based on geometry. It was the time when "art" meant the craft of building and "science" meant the theory of architecture. Art as craft, and science as theory, were united through geometry which represented the science of the art, the theory of craft. Geometry could represent the dialectic between theory and practice. Also, in that time, construction was not one thing and structure another; they were two aspects of building with concern for strength, stability, and durability. The development of theories of structures and the strength and performance of materials and their separation from the act of construction is in part responsible for the separation of load-bearing and space defining elements in 19th and 20th century architecture.

Further, there was no separation between a Euclidean geometry that described the location of elements in space—and the geometry used by the carpenter in laying out the scaffolding, centering, formwork, and the stereotomy used by the mason. When the notion of measurable connected geometric spaces entered our awareness, there arose a new geometry, i.e., topology, and we shifted from an intuition of the phenomenology of spaces to a conscious exploration and articulation of those connectivity relationships.

Each particular geometry represents a conscious exploration of an aspect of the spatial relationships of elements. Technical equipment, processes, and theories all lead to an *amplification* of some aspect of, or relationship to, the world and the concomitant *reduction* of emphasis on the other aspects or relationships.

## THE THIRD CHARACTERISTIC: APPROPRIATENESS-TO AND APPROPRIATION-OF CONTEXT

Beyond the question of the relationship of technics, techniques and technology to perception, it is also "worth asking," what are the *environments* or *contexts* in which humans and their equipment are embedded; contexts, aspects of which are amplified and reduced, and which are experienced either through transparent media or which recede behind opaque media.

Since technical equipment, processes and theories mediate between individuals, and societies, and their natural worlds, the builder must respond to at least three contexts: (i) the physicality of that natural world; (ii) the civil and cultural agenda of those societies; and (iii) the intentionality of the individual builder.

In their encounters with the *physicality of the natural world,* builders find themselves surrounded by difficulties which threaten survival and with facilities available for overcoming such difficulties. When these difficulties are severe, as in extreme climates of either heat or cold, the technic seeks mainly to overcome the characteristic of the difficulty. In extreme dry heat, the builder constructs a thick wall and a courtyard (wet-heat demands a different response); in extreme cold he builds an igloo (or a fire). When the difficulty is not climatic, but concerns effort, as in spanning a great distance with a bridge or reaching a great height with a tower, the characteristic of the mediator is once more towards amplification of the special characteristic (in this case, a structural solution is the major determinant of form). If the world surrounds us with facilities, then there is need to intervene. As already noted Ortega y Gasset said, "what in reality prevails . . . man finds that the world surrounds him as an intricate net woven of both facilities and difficulties." In such a case, specific conditions do not insist on recognition, and hence there are no obvious or determinate characteristics to the technics, techniques, and technology. In such a case, what is the source of the characteristics of mediators such as a wall, a roof, or any other element of building?[7]

Consider the case of *the roof.* It is an element which mediates between several realms. Between humans, it serves as a territorial or defensive technic. Between the human and the built world there are issues of ethics and aesthetics (i.e., proportion and composition), and it must relate to both the scale of the built world and to the scale of the human (as in the external and internal domes of St. Paul's cathedral, London). Between the human realm and the natural world, there could be this "intricate net of" many difficulties and facilities.

When we consider this roof, what experience do we seek of the path of the sun, the moon, and the stars; the patterns of clouds and light; the path of the birds; the rain, hail and snow, and so forth? Since the difficulties (often, rain and snow) insist on our recognition, the response of the mediating technic to such difficulties is to "stand-out" and become the salient characteristic of the roof; while the other aspects of the world are diminished, or more often, completely dismissed.

The *civil* and *cultural agenda* of the society offers the second set of contexts. There are many known studies of the socio-economic-political-religious context, e.g., the similarities and differences in the productions of labor-intensive, capital-saving societies with those which are labor-saving, capital-intensive. Equally well-documented are the effects of the structure of a society on its modes of production and vice-versa, e.g., the discourse on historical, dialectical, and mechanical materialism. Less well studied is the dialectical relationship between the culture and its modes of production and its products.

Further examination of the roof, as mediator, shows that Northern Americans, Northern Europeans, and Japanese all have chosen to make rain the major determinant of the shape in their design. All, too, seek to control the pathway of the "run-off," in order to isolate it from the human realm. The Northern Americans and Europeans take the "run-off," in as short a distance as possible, from the three dimensional volume of the rainfall to the two dimensional plane of the roof, to the one dimensional line of gutter channels to the point of a closed-channel pipe and finally into the zero dimensional "hiddenness" of a storm-drainage system. It is as if to isolate, not only spatially but also visually, the "offending" material from their realms of senses. The Japanese traditionally have isolated the run-off spatially but amplified it visually as it cascades over the eaves of the roof. On the occasions when they have channeled rain into a linear path, it has been done with an open link chain as its guideline. And when they reduce the rain to a point it is amplified as a drop in a stone bowl. The Japanese desire a more transparent relationship to the natural elements than do Northern Americans or Europeans. Thus it would seem that cultures have a range of phenomenological relationships to the world and their degree of transparency or opacity influences their attitude to amplification and reduction.

The third context of mediation, leading to amplification of one aspect from the totality of possible experiences, derives from the *intentionality* of the designer. While the "Northerner's" encounter with rain results in a fixed set of architectural responses, there has been a more varied response to *light* as the "difficulty."

As with all aspects of our experience with the world, the encounter with light has been partitioned into the concerns of science, ethics, aesthetics and, in a previous age, metaphysics. *Science* has proposed solutions that answer to the needs of the physiology of human comfort and the physics of light. *Ethics* and societal values have struggled to free themselves from the limited concerns of human productivity and task lighting. Science and Ethics have generated a measurable and finite set of solutions to the mediation of the window, or "wind-eye." *Aesthetic* concerns have ranged from Le Corbusier's "masterly, correct and magnificent play of forms in light" to Kahn's more ordered, or perceptible, interplay between light and shadow, opening and structure, space and place, and so forth.

The *metaphysical* dimension of light would seem no longer to be part of a language of architecture. In deleting the tribune layer and attempting to integrate the triforium into the clerestorey window at Chartres, Thierry of Chartres was responding to the metaphysics of Grosseteste who said, "light is the mediator between bodiless and bodily substances" and that "the objective value of a thing is determined by the degree to which it partakes of light."[8]

The architect who followed the metaphysics of Dionysius the Pseudo-Areopagite sought no separation in science, ethics, aesthetics and metaphysics. For the architect of the 20th century there has been a fragmentation of perception, so that the transcendental has been eliminated. Of the remaining dimensions, the physical–measurable and the ethical–codifiable aspects have primacy over the aesthetic.

It is, of course, in this realm of the aesthetic that architects declare their autonomy or individual *intentionality*.

Louis Kahn's hierarchical interpenetration of the geometries of the structure, pathways of people and equipment, finishes, and so forth was composed to lock in the experience or *repose*. This "silence to light, light to silence" could be achieved only with homogeneously modulated windows.

In contrast, Le Corbusier achieved *drama* in his interiors by using articulated light scoops or heterogeneous fenestration.

Neither Kahn nor Le Corbusier limited their architectural expressions as responses to the anonymous contexts of the physical, social, and cultural. Favoring their own, i.e., autonomous, aesthetic system, they willed or appropriated a form which became a salient characteristic of their mediating roof.

In general, when the context of the physicality of the natural world and/or the socio-cultural agenda specifies what needs to be amplified and what reduced, the technics, techniques, and technology are considered *appropriate* (i.e., L. *propriare,* near; suitable, proper, belonging). In the absence of clues as to what is appropriate, the builder *appropriates* (i.e., takes possession of, L. *proprius,* own, self) a *text* from the context; that is, the builder's intentionality abstracts through amplification and reduction, and presents finally, *interpretations* of aspects of the world.

## SUMMARY AND CONCLUSIONS

The "informing response" to the question on the characteristics of knowledge derived from our experience of production says, in part, that our perception is a function of transparency and opacity, amplification and reduction, and appropriateness and appropriation. Some other characteristics to be explored in the future are: *extensions* of experience (where new aspects of the environment are presented); *transformations* of experience (where the environments are presented in a new way); *deconstruction* and reformulation of *space* and *time;* the *speed* of production and its relationship to *style;* and *spatializing* as a result of the *division of labor.*

It would be useful (for the designer who needs prescriptive rather than descriptive explanations) to address the questions to empirical cases of architectural production which are also grounded physically, socially, culturally and historically.

The characteristics, *transparency* and *opacity,* as levels of mediation, are part of the new concept of technology. As such they are responses to the original question, but our analysis makes them the source of a set of new questions. First, what could result from the seemingly inexorable shift from transparency to opacity? Newer, more opaque, technics, techniques and technologies will need new technicians. Previous shifts gave birth to the structural, mechanical, electrical, and acoustical engineers; and construction managers presently seek to legitimize their knowledge.

Second, can the loss in transparency and the concomitant gain in opacity be reversed? The revival of neo-classicism and the return to the pre-technical of vernacular are among the efforts to recapture a condition of "closeness." To achieve a true transparency and not a pastiche, the building materials, equipment, processes and theories will need to revert to a prior state. A more complex and profound belief is that the architect's task is to explore the fundamental ontology of humans which can be revealed only through the act of building. Here one seeks experience both of the world and of the technical realm, i.e., transparency and opacity at the same time, but of different realms.

Third, is it necessary to return to a pre-technical condition of transparency? Other disciplines have confronted a similar dilemma. Herbie Hancock, the jazz pianist, refutes any notion that older musical instruments are somehow more natural than their electronic counterparts. He says, "the pianist is probably farthest away from the thing that actually produces the sound . . . there's a whole series of mechanical things between the player and the sound." According to music critic John Rockwell, for Hancock and others what makes a "piano" is the mechanism by which one addresses the instrument; theoretically the hammers and strings and the resonating wooden chamber are irrelevant. Of course new compositional possibilities "open-up" where the performer is not limited by the dexterity of the body and where the logic of the mediation is "theoretically irrelevant."[9]

Can there be a space, like sound, not limited by the experience of the human body and where the logic of the building materiality and its processes of construction are theoretically irrelevant? Such space would yield a new language of architecture not influenced by building construction; an architecture where composition is everything and construction is almost nothing. Are Russian constructivism and the new deconstructivism such an architecture?

Walter Benjamin imagined a similar space when, with respect to film, he wrote, "evidently a different nature opens itself to the camera than opens to the naked eye—if only because an unconsciously penetrated space is substituted for a space consciously explored by man." If architects were to deconstruct and reconstruct their perceptions similar to "filmic" space and time, then a "different nature" would open for architecture.[10]

*Amplification* and *reduction* as further characteristics of the new concept of technology reveal some dangerous effects of mediation.

Through amplification, whether it be in response to the physicality of the natural world, the social and cultural agenda, or the intentionality of the builder, certain aspects of the world stand-out (ek-sist), that is to say that technics, techniques, and technology offer up a *drama;* and our perception converts the world into a *spectacle,* made dramatic through our mediation.

The first danger is that "isolations" of a sense are "amputations" which lead to the "blocking of perception" and a condition where humans are "fragmented by their technologies."

The second danger from forgetting the reductive effort of mediation lies not only in the "fragmentation of living" but also in the emphasis given to the amplifications. These abstracted texts, delivered through mediation, seem more precise, even perfect, and "the temptation is to take the new features . . . as 'more real' than those features which are more mundane" or worldly (that is, those features revealed by ordinary perception).[11]

A third danger is that these texts or amplifications or abstractions can be organized and even codified into a language: a language which has "silenced" the remainder of perception and a language which was codified for the perception of its time. The Classical language of architecture is such a codification. To validate its reuse is to accept its perception of the dialectic with the world, its balance of transparency and opacity, and the amplification and reduction given by *its* technical equipment, processes and theories.

A fourth danger lies in the separation of the layman from the literati. The layman experiences only the mundane; the architect, who knows the codified language, transforms reality and re-presents a text, the logic and validity of which is accessible only to those who speak or read the unembodied language.

As each culture abstracts and codifies its perception, its architects construct a language of architecture from its deconstructions. If it is a revived language, Classical or Modern, then looking at the past becomes thinking like the past and even living in the past. Such architects may be out of phase with the perception of their time.

The new concept of technology says that all technical experience is *appropriated* from contexts which are physical, social, cultural, or intentional.

Both "low-tech" and "high-tech" are "appropriate technologies"; their difference lies in the contexts from which they appropriate their products and processes. Steel was produced originally from iron ore mined from nature and today it is increasingly produced from scrap steel taken from the built environment. Similarly, while "low-technology" appropriates from natural materials and craft skills and knowledge, "high-technology" appropriates from newer human-made materials, and engineering skills and knowledge.

In general, what seems to be the case is that cultures, whether their relationship to the world is "unconsciously penetrated" phenomenological or "consciously explored" scientific, express a desire for a level of transparency and opacity which influences their choice of amplified and reduced technics, techniques and technology. Conversely, as the types of mediation become more universal, the range of cultural levels diminish and an "international" expression results. Internationalism will be modified only by the intentionality of the architect.

The concept of technology discussed in this paper is new perhaps only to the building professions. Its main schema was proposed, more than fifty years ago, by Walter Benjamin, Martin Heidegger, Ernst Mach, Michael Polanyi, and others. Benjamin began his seminal 1936 article on "The Work of Art in the Age of Mechanical Reproduction" with a quote from Paul Valery's "Aesthetics." Valery's words are worthy of repetition.

> Our fine arts were developed, their types and uses were established, in times very different from the present by men whose power of action upon things was insignificant in comparison with ours. But the amazing growth of our techniques . . . make it a certainty that profound changes are impending in the ancient craft of the Beautiful . . . For the last twenty years neither matter nor space nor time has been what it was from time immemorial. We must expect great innovations to transform the entire technique of the arts, thereby affecting artistic invention itself and perhaps even bringing about an amazing change in our very notion of art.[12]

According to the new concept of technology, perception and production are not the vanguard and the rear-guard of the dialectic between humans and their environments but they are mutually dependent experiences.

## NOTES

1. This article is derived from a paper "Metamorphosis of Perception Through Technics, Techniques, and Technology" which was written in Fall 1984 and presented to the University of Pennsylvania Faculty Mellon Seminar on Technology and Culture on April 22, 1985.
2. Martin Heidegger, *Being and Time,* trans. John Macquarrie and Edward Robinson (New York: Harper and Row, 1962), p. 95.
3. George Steiner, *Martin Heidegger* (London: Penguin Books, 1980), p. 24.
4. Jose Ortega y Gasset, "Man the Technician," in *History as a System and Other Essays toward a Philosophy of History* (New York: W. W. Norton, 1962), p. 110.
5. Heidegger, *Being and Time,* p. 99.
6. Ibid., p. 135.
7. Ortega y Gasset, "Man the Technician," p. 110.
8. Otto von Simson, *The Gothic Cathedral* (New York: Harper and Row, 1964), pp. 51–52.
9. John Rockwell, "Electronics Is Challenging Traditions of Music," *New York Times,* November 1986.
10. Walter Benjamin, "The Work of Art in the Age of Mechanical Reproduction," *Illuminations,* ed. Hannah Arendt (Great Britain: Fontana, 1982), p. 238.
11. Don Ihde, *Technics and Praxis* (Dordrecht, Holland: D. Reidel, 1979), p. 22.
12. Paul Valery, "The Conquest of Ubiquity," in *Aesthetics,* trans. Ralph Manheim (New York: Pantheon, 1964), p. 225.

## BIBLIOGRAPHY

Benjamin, Walter. "The Work of Art in the Age of Mechanical Reproduction." In *Illuminations,* ed. Hanna Arendt, pp. 219–53. Great Britain: Fontana, 1982.
Heidegger, Martin. *Being and Time.* Trans. John Macquarrie and Edward Robinson. New York: Harper and Row, 1962.
Ihde, Don. *Technics and Praxis.* Dordrecht, Holland: D. Reidel, 1979.
Ortega y Gasset, Jose. "Man the Technician." In *History as a System and Other Essays towards a Philosophy of History,* pp. 87–161. New York: W. W. Norton, 1961.
Steiner, George. *Martin Heidegger.* London: Penguin Books, 1980.
Valery, Paul. "The Conquest of Ubiquity." In *Aesthetics,* trans. Ralph Manheim. New York: Pantheon, Bollingen Series, 1964.
von Simson, Otto. *The Gothic Cathedral.* New York: Harper and Row, 1964.

选读 12

"形态抵抗结构"

马里奥·萨瓦多里

萨瓦多里把建筑结构作为建筑形式和设计过程中不可或缺的一部分进行研究。建筑师需要学习结构知识，但大量必须掌握的结构知识却让建筑师陷于抽象的数学分析和公式的苦海。萨瓦多里将建筑进程中的技术层面提升至一个新的高度，试图说明结构形式可以也应该与设计中更为主观的方面密不可分。在这篇出自《建筑物如何站起来》的文章里，萨瓦多里说明了几何形状是如何既表现了内在固有的结构准则，又可以作为强有力的设计元素来提升建筑形式的含义与意义。通过使用平板、折板、曲线形式、拱顶和相互衔接咬合的建筑部件，他提出了一种叙事技巧，这种叙事技巧以结构语言为媒介，创造出一种建筑对话。这篇文章说明了建筑师和工程师二者明显相互冲突的意图是如何融入一个统一的创造性进程之中的。

### READING 12

## Form-Resistant Structures

### Mario Salvadori

Salvadori investigates building structure as an integral part of architectural form and the design process. Much of what must be learned about structures by architects can become mired in abstract mathematical analyses and formulae. Salvadori elevates this technical dimension of the architectural process to show how structural form can and should be inseparable from the more subjective facets of design. In this article from his book *Why Buildings Stand Up,* Salvadori demonstrates how geometric shapes not only signify inherent structural principles but also can be used as powerful design elements to enhance meaning and significance in architectural form. Through the use of flat and folded plates, curvilinear forms, vaults, and interlocking building components, he provides a narrative that creates an architectural dialogue through the medium of a structural language. This article shows how the apparently conflicting intentions of the architect and the engineer can be fused into a unified creative process.

### GRIDS AND FLAT SLABS

Ever since the beginning of recorded history, and we may assume even earlier, people have gathered in large numbers for a variety of purposes be they religious, political, artistic, or competitive. The large roof, unsupported except at its boundary, arose to shelter these gatherings, evolving eventually into the huge assembly hall we know today.

*Source:* From *Why Buildings Stand Up: The Strength of Architecture* by Mario Salvadori. Copyright © 1980 by Mario Salvadori. Reprinted by permission of W. W. Norton & Company, Inc.

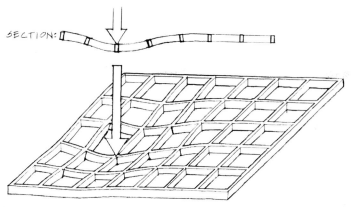

FIGURE 12–1  A rectangular grid of beams. [Saralinda Hooker and Christopher Ragus]

As we shall see, no large roof can be built by means of natural or man-made compressive materials without giving the roof a curved shape, and this is why domes were used before any other type of cover to achieve large enclosed spaces. Even wood, a material that can span relatively short horizontal distances by beam action . . ., has to be combined in conical, cylindrical or spherical shapes whenever large distances are to be spanned.

Only after the invention of inexpensive methods of steel manufacture and the recent development of reinforced concrete did large flat roofs become possible. They have obvious advantages over dome roofs: their erection is simpler, and they do not waste the upper part of the space defined by the dome which is often superfluous, unnecessarily heated or air-conditioned.

The simplest structural system for a flat rectangular roof consists of a series of parallel beams supporting some kind of roofing material. But if all four sides of the rectangle to be covered can be used to support the roof beams, it becomes more practical to set the beams in two directions, at right angles to each other, thus creating a *grid*. This *two-way system* pays only if the two dimensions of the rectangle are more or less equal. Loads tend to move to their support through the shortest possible path and if one dimension of the roof is much larger than the other, most of the load will be carried by the shorter beams, even if the beams are set in a grid pattern.

A grid is a "democratic" structural system: if a load acts on one of its beams, the beam deflects, but in so doing carries down with it all the beams of the grid around it, thus involving the carrying capacity of a number of adjoining beams. It is interesting to realize that the spreading of the load occurs in two ways: the beams parallel to the loaded beam bend together with it, but the beams at right angles to it are also compelled to twist in order to follow the deflection of the loaded beam (Figure 12-1). We thus find that in a rectangular grid loads are carried to the supports not only by beam action (bending and shear) in two directions but by an additional twisting mechanism which makes the entire system stiffer. To obtain this twisting interaction the beams of the two perpendicular systems must be rigidly connected at their intersection, something which is inherent in the monolithic nature of reinforced concrete grids and in the bolted or welded connections of steel grids. Even primitive people know how to

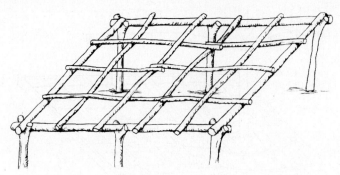

**FIGURE 12–2**   A woven grid of beams. [Saralinda Hooker and Christopher Ragus]

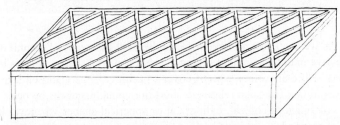

**FIGURE 12–3**   A skew grid. [Saralinda Hooker and Christopher Ragus]

obtain such twisting action by interweaving the beams of their roofs so that any displacement of one beam entails the bending and twisting displacement of all the others (Figure 12-2).

Though rectangular grids are the most commonly used, *skew grids* (Figure 12-3) have, beside aesthetic qualities, the structural and economic advantage of using equal length beams even when the dimensions of the grid are substantially different, thus distributing more evenly the carrying action between all the beams.

We have seen in Chapter 9 (of *Why Buildings Stand Up*) how grids of trusses rather than beams become necessary when spans are hundreds of feet long, and how space frames constitute some of the largest horizontal roofs erected so far, covering four or more acres without intermediate supports. We must now go one step back to discover how an extension of the grid concept has become the principle on which most of the floors and roofs of modern buildings are built.

Let us imagine that the beams of a rectangular grid are set nearer and nearer to each other and glued along their adjacent vertical sides until they constitute a continuous surface. Such a continuous surface, called a *plate* or *slab,* presents all the advantages of a grid in addition to the ease with which it can be poured on a simple horizontal scaffold when made out of concrete. Reinforced concrete horizontal slabs are the most commonly used floor and roof surfaces in buildings with both steel and concrete frames all over the world. Their smooth underside permits a number of things to hang—pipes and ducts, for instance—without having to go around beams. The setting of the slab reinforcement on flat wooden scaffolds makes the placing of the steel bars

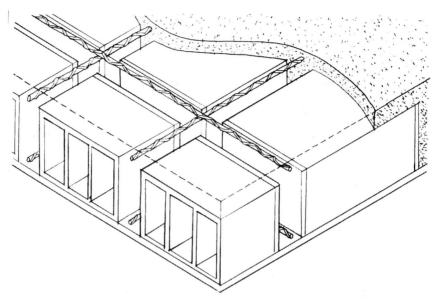

**FIGURE 12–4**  Tile–concrete floor slab. [Saralinda Hooker and Christopher Ragus]

simple and economical. In European countries concrete slabs are sometimes made lighter by incorporating hollow tiles (Figure 12-4). Through the strength of their burnt clay these tiles participate in the slab structural action, which is the same in all slabs whatever their material.

Actually slabs, besides carrying loads by bending and twisting like grids of beams, have an additional capacity which makes them even stiffer and stronger than grids. This easily understood capacity derives from the continuity of their surface. If we press on a curved sheet of material attempting to flatten it, depending on its shape, the sheet will flatten by itself or have to be stretched or sliced before it can be made flat. For example, a sheet of paper bent into a half-cylinder and then released flattens by itself [Figure 12-5(a)]. It is said to be a *developable surface* (from the idea "to unfold" contained in the verb "to develop"). But if we cut a rubber ball in half, producing a small spherical dome, the dome will not flatten by itself if we lay it on a flat surface. Neither will it become flat if we push on it. It only flattens if we cut a large number of radial cuts in it or if, assuming it is very thin, it can be *stretched* into a flat surface [Figure 12-5 (b)]. The dome (and actually all other surfaces except the cylinder) are *non-developable, unflattening surfaces.* Because they are so hard to flatten, they are also much stiffer than developable surfaces. (It will be more obvious why non-developable surfaces are better suited to build large roofs once we learn how such roofs sustain loads.)

Returning now to the behavior of a flat slab, we notice that under load it becomes "dished"—it acquires the shape of a curved surface, with an upward curvature (Figure 12-6). If it is supported only on two opposite parallel sides, it becomes a slightly curved upside-down cylinder, but if it is supported on four sides, or in any other manner, it acquires a non-developable shape. Just as the half-ball had to be stretched to be

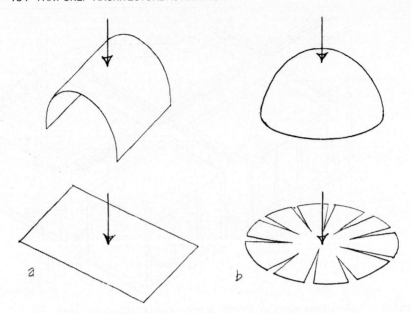

**FIGURE 12–5**   Developable (a) and non-developable (b) surfaces. [Saralinda Hooker and Christopher Ragus]

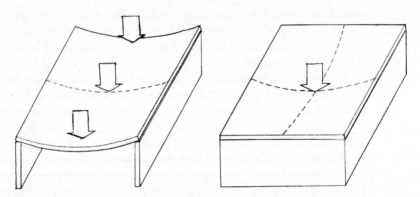

**FIGURE 12–6**   Flat slabs dished by loads. [Saralinda Hooker and Christopher Ragus]

changed from a dome to a flat surface, the plate has to be stretched to change it from a flat to a dished surface. Hence the loads on it, besides bending it and twisting it, must stretch it, and this unavoidable stretching makes the slab even stiffer. Therefore we should not be amazed to learn that plates or slabs can be made thinner than beams. While a beam spanning twenty feet must have a depth of about one-and-one-half feet, whether it is made of steel, concrete or wood, a concrete slab covering a room twenty-feet square can be made one foot deep or less.

When slabs have to span more than fifteen or twenty feet, it becomes economical to stiffen them on their underside with ribs, which can be oriented in a variety of

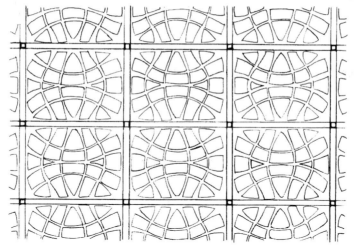

**FIGURE 12–7**  Nervi's slab with curved ribs. [Saralinda Hooker and Christopher Ragus]

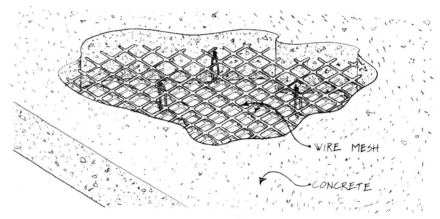

**FIGURE 12–8**  Ferrocemento mesh reinforcement. [Saralinda Hooker and Christopher Ragus]

ways. Nervi made use of Ferrocemento, a material he perfected, to build forms in which to pour slabs stiffened by *curved* ribs, which are oriented in the most logical directions to transfer the loads from the slab to the columns. These curved ribs, moreover, give great beauty to the underside of the slabs (Figure 12-7). Ferrocemento is a material consisting of a number of layers of welded mesh set at random, one on top of the other, and permeated with a concrete mortar, a mixture of sand, cement, and water (Figure 12-8). Flat or curved elements of Ferrocemento can be built only one or two inches thick, with exceptional tensile and compressive strength due to the spreading of the tensile steel mesh through the high-strength compressive mortar. First used only as a material to build complex molds in which to pour reinforced concrete elements, it later was transformed by Nervi into a structural material itself. Some of the masterpieces of Nervi owe their extraordinary beauty and efficiency to the use of Ferrocemento.

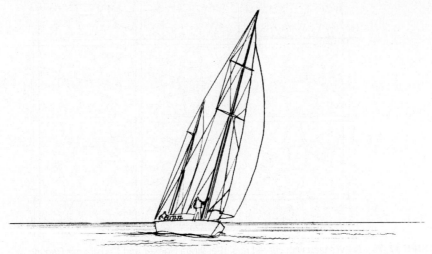

**FIGURE 12-9**  Nervi's ketch *Nennele* (of ferrocemento). [Saralinda Hooker and Christopher Ragus]

Genius often consists of an ability to take the next step, and Nervi took it by realizing that Ferrocemento would be an ideal material for building boats. His lovely ketch *Nennele* (Figure 12-9) was the first, but a large number of sailing boats have been built, mostly in Australia and the United States, with Ferrocemento hulls. They are easy to manufacture and even easier to repair in case of an accident.

## STRENGTH THROUGH FORM

The stiffness of flat slabs, like that of beams, derives from their thickness: if too thin they become too flexible to be functional. It is one of the marvels of structural behavior that stiffness and strength of sheetlike elements can be obtained not only by increasing their thickness and hence the amount of required material, but by giving them curved shapes. Some of the largest, most exciting roofs owe their resistance exclusively to their shape. This is why they are called *form-resistant structures*.

If one holds a thin sheet of paper by one of its short sides, the sheet is incapable of supporting even its own weight—the paper droops down [Figure 12-10 (a)]. But if we give the side held a slight curvature up, the same sheet of paper becomes stiffer and capable of supporting as a cantilever beam not only its weight but also the small additional weight of a pencil or pen [Figure 12-10 (b)]. We have not strengthened the paper sheet by adding material to it; we have only curved it up. This principle of strength through curvature can be applied to thin sheets of reinforced concrete and has been efficiently used to build stadium roofs that may cantilever out thirty or more feet with a thickness of only a few inches (Figure 12-11). The shape of such roofs can be shown to be non-developable and hence quite rigid, but even developable surfaces, like cylinders, show enough strength (when correctly supported) to allow their use as structural elements. To demonstrate this property, try to span the distance between two books by means of a flat sheet of paper acting as a plate. The paper will sag, fold, and slide between the book

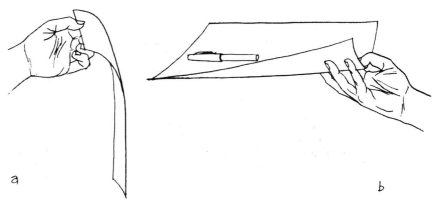

**FIGURE 12–10**  Paper sheet stiffened by curvature. [Saralinda Hooker and Christopher Ragus]

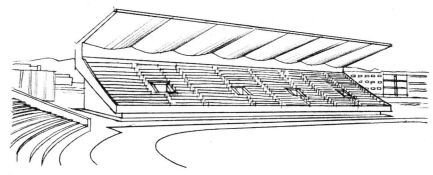

**FIGURE 12–11**  Stadium stands roof. [Saralinda Hooker and Christopher Ragus]

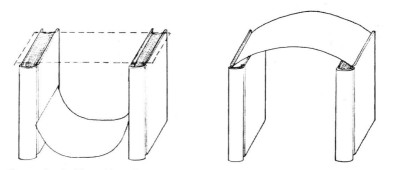

**FIGURE 12–12**  Paper sheet stiffened by cylindrical shape. [Saralinda Hooker and Christopher Ragus]

supports. If instead the sheet of paper is curved up and prevented from spreading by the book covers, it will span the distance as an arch (Figure 12-12). Again the curvature has given the thin paper its newly acquired stiffness and strength.

Nature knows well the principle of strength through curvature and uses it whenever possible to protect life with a minimum of material. The egg is a strong home for the developing chick, even though its shell weighs only a fraction of an ounce. The seashell protects the mollusk from its voracious enemy and can, in addition, sustain the pressure of deep water thanks to its curved surfaces. The same protection is given snails and turtles, tortoises and armadillos, from whom our medieval knights may have copied their curved and relatively light armor.

## CURVED SURFACES

We owe to the greatest of all mathematicians, Karl F. Gauss (1777–1855), the discovery that all the infinitely varied curved surfaces we can ever find in nature or imagine belong to only three categories, which are domelike, cylinderlike, or saddlelike.*

How do the three categories differ? Consider the dome. Imagine cutting it in half vertically with a knife. The shape of the cut is curved downwards, and if you cut the dome in half in any direction, as you do when you cut a number of ice-cream-cake wedge slices, the shape of all the cuts is still curved downward (Figure 12-13). A domelike surface has downward curvatures in all its radial directions. By the way, if instead of cutting a dome we were to cut in half a soup bowl, we would find the shape of all the cuts to be curved up, whatever their radial direction. Domes and hanging roofs, each with curvatures always in the same direction (either down or up), constitute the first of Gauss's categories. They are non-developable surfaces and have been

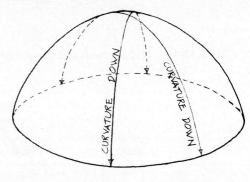

**FIGURE 12–13** Vertical cuts in dome. [Saralinda Hooker and Christopher Ragus]

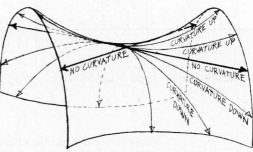

**FIGURE 12–14** Vertical cuts in saddle. [Saralinda Hooker and Christopher Ragus]

*Gauss was so great a man that he noted in a small book a number of discoveries "not worth publishing." When this booklet was found fifty years after his death, some of his "negligible" discoveries had been rediscovered and had made famous a number of his successors!

used for centuries to cover large surfaces. We will discuss their structural behavior in Chapters 13 and 15 [of *Why Buildings Stand Up*].

Let us jump to the third of Gauss's categories, the saddlelike surfaces. In a horse saddle the curve across the horse, defined by the rider's legs, is curved downward, but the curvature along the horse's spine, which prevents the rider from sliding forward or backward, is upward (Figure 12-14). Saddle surfaces are non-developable and are used as roofs because of their stiffness. The Spanish architect Felix Candela built as a saddle surface what is perhaps the thinnest concrete roof in the world. Covering the Cosmic Rays Laboratory in Mexico City, it is only half-an-inch thick.

Saddle surfaces have another property not immediately noticeable. As one rotates the saddle cuts from the direction across the horse to that along the horse, the curvature changes from down to up and, if one keeps going, it changes again from up to down. Therefore there are two directions along which the cuts are neither up nor down. They are not curved; they are straight lines (Figure 12-14). To prove this one has only to take a yardstick and place it across a saddle at its lowest point: the saddle is curved down, below the yardstick. If one then rotates the yardstick, keeping it horizontal, one finds that there is a direction along which *the yardstick lies entirely on the surface of the saddle:* in this direction the saddle has no curvature. Of course, if one rotates the yardstick in the opposite direction one locates the other no-curvature section of the saddle, which is symmetrical to the first with respect to the horse's axis. All saddle surfaces have two directions of no curvature. Cut along these directions, their boundaries are straight lines. This property makes the saddle shape an almost ideal surface with which to build roofs.

We can now go back to Gauss's second category, the cylinders. Imagine a pipe lying on the floor. If you cut vertically its top half—the half, say, with the shape of a tunnel—in any direction, you will notice that *all* of these cuts have a curvature down, *except one:* the cut along the pipe's axis is a straight line (Figure 12-15). The cylinder has no curvature in the direction of its axis. One may consider, then, the cylinder as a dividing line between the dome and the saddle. The saddles have two directions without curvatures, but as these two directions draw nearer and nearer, saddles become cylinders, with only one direction of no curvature. If this direction is now given a down curvature, the cylinder becomes a dome. If instead of considering the upper part of a cylinder, we consider its lower half—the half, say, with the shape of a gutter—we find that the vertical sections of the gutter have curvatures up in all directions except one: the direction of the axis of the gutter. Hence, gutters and tunnels belong to the same category of surfaces having only one direction of no curvature.

**FIGURE 12–15**  Vertical cuts in top half of cylinder. [Saralinda Hooker and Christopher Ragus]

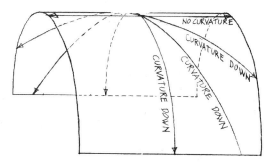

## BARREL ROOFS AND FOLDED PLATES

We have seen that cylinders are developable surfaces and, as such, are less stiff than either domes or saddles. Even so, they can be used as roofs. Actually *barrel roofs* of reinforced concrete in the shape of half-cylinders with curvatures down are commonly and inexpensively used in industrial buildings (Figure 12-16), since they can be poured on the same cylindrical formwork, which can be moved from one location to another and reused to pour a large number of barrels on the same form.

The mode of support of a barrel influences its load-carrying action. If a barrel is supported all along its two longitudinal edges [Figure 12-17 (a)] it acts as a series of arches built one next to the other and develops out-pushing thrusts, which must be absorbed by buttresses or tie-rods as in any arch. But if it is supported on its curved ends [Figure 12-17 (b)], it behaves like a beam, developing compression above the neutral axis and tension below . . . , and it does *not* develop thrust. One should not be fooled by the geometrical shape of a structure in deciding its load-carrying mechanism. Barrels should be supported on endwalls or stiff arches so as to avoid unnecessary and costly buttresses or interfering tie-rods.

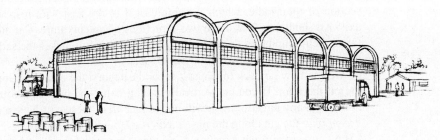

**FIGURE 12–16**   Barrel roof. [Saralinda Hooker and Christopher Ragus]

**FIGURE 12–17**   Barrel roof supports. [Saralinda Hooker and Christopher Ragus]

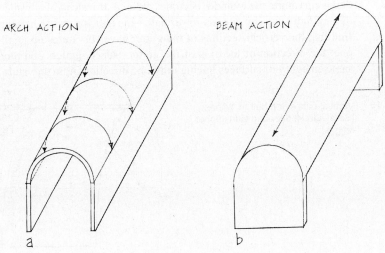

The *folded plate* roof is analogous to a series of barrels. It consists of long, narrow inclined concrete slabs, but presents a sudden *fold* or change in slope at regular intervals (Figure 12-18). Its cross-section is a zigzag line with "valleys" and "ridges." The construction of a folded plate roof requires practically no formwork, since the flat slabs can be poured on the ground and jointed at the valleys and ridges of the roof by connecting the transverse reinforcing bars of the slabs and using a good cement grout or mortar to make the slabs into a monolithic roof.

Folded plates carry loads to the supports along a twofold path. Because of the stiffness achieved by the folds, any load acting on a slab travels first up the nearest ridge or down the nearest valley, and then is carried to the end supports longitudinally by the slabs acting as beams (Figure 12-19). Folded plates must be supported at their

**FIGURE 12–18**  Folded-plate roof. [Saralinda Hooker and Christopher Ragus]

**FIGURE 12–19**  Folded-plate load paths. [Saralinda Hooker and Christopher Ragus]

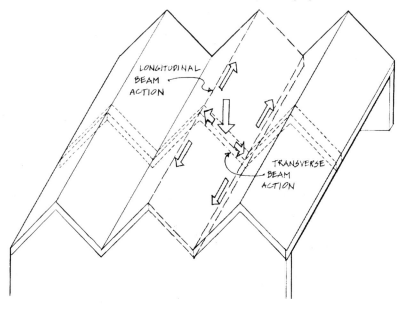

**FIGURE 12–20**  Folded-plate paper model. [Saralinda Hooker and Christopher Ragus]

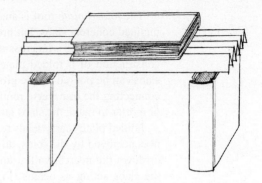

ends. Since they consist of flat surfaces and folds, they act like an accordion that can be pushed in or pulled out with little effort, and do not develop out-pushing thrusts.

It is both easy and instructive to fold a sheet of thin paper up and down, shaping it into a folded plate, and to support it between two books, possibly laying a flat sheet of paper over it (Figure 12-20). The load capacity obtained by such a flimsy piece of material through its folds is amazing: a sheet of paper weighing less than one-tenth of an ounce may carry a load of books two or three hundred times it own weight! Any reader inclined to experiment further with folded paper can take advantage of both folding and arch action by creasing a sheet of paper into a folded barrel, according to the instructions of Figure 12-21. The creased paper barrel requires buttresses to absorb its outward-acting thrusts, but its load-carrying capacity is even greater than that of a folded-plate roof and may easily reach 400 times its own weight.

## SADDLE ROOFS

Saddle surfaces, supported along their longitudinal curved edges, have a particularly elegant shape which blurs the distinction between structure and functional skin (Figure 12-22). But saddle surfaces make some of the loveliest roofs when cut and supported along those straight lines which we have seen necessarily exist on any surface with both up and down curvatures (see Figure 12-14). To visualize how a curved surface can be obtained by means of straight lines, connect by inclined straight-line segments the points of two equal circles set one above the other (Figure 12-23). The segments generate a curved surface called a *rotational hyperboloid,* used to build the enormous cooling towers of chemical plants. One of the most commonly used roof surfaces is obtained in a similar manner. Imagine a rectangle of solid struts, in which one of the corners is lifted from the plane of the other three, thus creating a frame with two horizontal and two inclined sides [Figure 12-24 (a)]. If the corresponding points of two opposite sides of this frame (one horizontal and one inclined) are connected by straight lines, for example by pulled threads, and the same is done with the other two opposite sides, the threads will describe a curved surface, although, being tensed, they are themselves straight [Figure 12-24 (b)]. This surface has a curvature

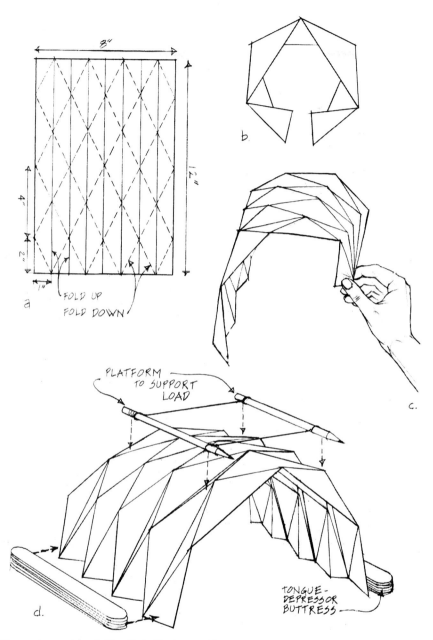

**FIGURE 12–21** Creased paper barrel. [Saralinda Hooker and Christopher Ragus]

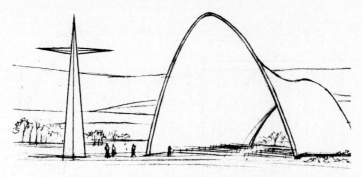

**FIGURE 12–22**   Candela's saddle-shaped chapel in Mexico. [Saralinda Hooker and Christopher Ragus]

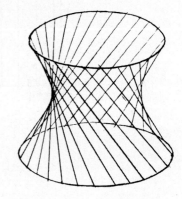

**FIGURE 12–23**   Rotational hyperboloid. [Saralinda Hooker and Christopher Ragus]

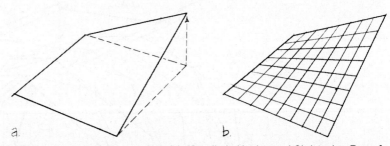

**FIGURE 12–24**   Hypar frame (a) and hypar straight-lines surface (b). [Saralinda Hooker and Christopher Ragus]

up along the line connecting the lifted corner to its diagonally opposite corner, and a curvature down in the direction of the line connecting the other two corners (Figure 12-25). It is, therefore, a saddle surface. It carries the high-sounding name of *hyperbolic paraboloid,* wisely shortened to *hypar* by our British colleagues.

One of the simplest hypar roofs is obtained by tilting the saddle and supporting it on two opposite corners. Whether the support points are on the ground or on columns, the roof looks like a butterfly ready to take off (Figure 12-26). Its structural behavior is dictated by its curvatures. Compressive arch action takes place along the sections curved downward and tensile cable action along the sections curved

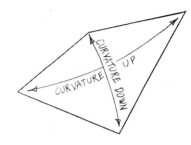

**FIGURE 12–25**  Hypar curvatures. [Saralinda Hooker and Christopher Ragus]

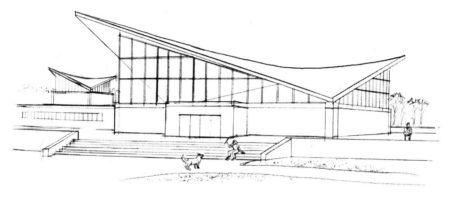

**FIGURE 12–26**  Hypar butterfly roof. [Saralinda Hooker and Christopher Ragus]

upward. The two support points must be buttressed to resist the thrusts of the arch action, while the tensile cable action at right angles to it must be absorbed by reinforcing bars, if the hypar is made out of concrete. Such is the stiffness of a hypar that its thickness need be only a few inches of concrete for spans of thirty or forty feet. The hypar has other wonderful structural properties. For example, one could fear that such a thin structure, acting in compression along its arched direction, would easily buckle, a fear quite justified were it not that the cable action at right angles to the arches pulls them up and prevents them from buckling! Finally, to make the structural engineer even more enamoured of these surfaces, under a uniform load, like its dead load or a snow load, they develop the same tension and compression everywhere. Therefore, its material, be it concrete or wood, can be used to its greatest allowable capacity all over the roof. The reader, who might not have seen too many of these magnificent roofs, may ask "Why do we not see many more of them?" The answer to this question is that there is no silver lining without a cloud and the cloud that hangs over the hypars is the cost of their formwork, as it is for all curved surfaces. More will be said later about this problem.

## COMPLEX ROOFS

The barrel and the rectangular hypar elements are the building blocks for some of the most exciting curved roofs conceived by man. Combinations of these structurally efficient components are limited only by the imagination of the architect, guided by good structural sense. It is indeed regrettable that, with a few notable exceptions, modern architecture has not used curved surfaces as glorious as some of the past and at the same time as daring as present day technology can make them. This lack of achievement is due to at least three causes. On one hand, curved surfaces are believed to be more complex to design than the flat rectangular shapes we are so used to. Usually quite the opposite is true. On the other hand, there is a gap between recent curved-structures theory and the prescriptions of the codes. A domed roof proposed for a bank in California—meant to cover a rectangular area ninety feet by sixty feet and to be only a few inches thick—was vetoed by the local building department engineer because thin curved roofs were not mentioned in the code and, hence, "did not exist." The engineer would only allow the roof to be built if two concrete arches were erected between its diagonally opposite corners "to support it." Little did he know that the thin concrete roof was so stiff that *it* would support the two heavy arches rather than be supported *by them!* Finally, one must honestly add that in the United States the ratio of labor to material costs often makes *thin shells* (as these curved roofs are usually labelled) uncompetitive with other types of construction. The situation is reversed in Europe and other parts of the world.

One of the most commonly encountered combinations of cylindrical surfaces is the *groined vault* of the Gothic cathedrals (Figure 12-27). This consists of the intersection of two cylindrical vaults at right angles to each other, supported on four boundary arches and intersecting along curved diagonal folds called *groins,* which end at the four corner columns supporting the vault. The groins have often been emphasized visually and, possibly, structurally by means of ribs but, though these ribs may be aesthetically important, they are not needed to sustain the vaults. By their curvature and folds they are self-supporting.

Among the great variety of combinations of rectangular hypar elements, two have become quite common because of their usefulness, beauty, and economy: the *hypar roof* and the *hypar umbrella.* To put together a hypar roof [Figure 12-28 (a)], consider building four hypar rectangular elements, starting as was done before with four rectangles but *lowering* (rather than lifting) one corner in each of them. The hypar roof is obtained by joining together the horizontal sides of each rectangular element so that all eight meet at the center of the area to be covered, while the lowered corners are

FIGURE 12–27  Groined vaults. [Saralinda Hooker and Christopher Ragus]

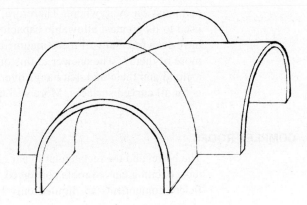

supported on four columns or on the ground at the corners of the area. The straight inclined sides of the roof act as the compressed struts of a truss, and they must be prevented from spreading outward by means of tie-rods connecting its corners, all around the covered area. The largest roof of this kind has been erected in Denver, Colorado, and measures 112 feet by 132 feet, with a three-inch thickness. It rests directly on the ground at the corners and covers a large department store.

The hypar umbrella [Figure 12-28(b)], one of the most elegant roof structures ever devised, is produced by using four rectangular hypar elements, each with a corner lowered with respect to the other three, and put together by joining the two inclined sides of each rectangular element so that all eight meet at the center of the area to be covered. The horizontal sides constitute the rectangular edge of the roof and hide the tie-rods. The shape of this hypar roof, which starts at a central point and opens up, resembles that of a rectangular umbrella and gives a visual impression of floating upward. Hypar umbrellas up to ninety-feet square have been used, for example, in the terminal building of Newark Airport (Figure 12-29).

Such is the variety of shapes which can be composed by means of hypars that the Spanish architect Felix Candela has become famous all over the world by designing and building, mostly in Mexico, roofs that use only this surface as basic element. Even though the hypar has particularly efficient structural properties when used as a horizontal roof, Candela has shown, in Mexico City, how exciting its form can be when used vertically as in the Iglesia de la Virgen Milagrosa. Nervi also has used vertical hypars as walls and roof in the monumental Cathedral of San Francisco (Figure 12-30).

## THIN SHELL DAMS

The greatest application of vertical, concrete thin shells has come not in architecture, however, but in dam construction. Although the world as a whole has so far only utilized 15 percent of the power obtainable from its natural or artificial waterfalls, the [former] USSR has exploited 18 percent of their potential, the United States 70 percent and the three European Alpine countries (France, Switzerland, and Italy) 90 percent. Dams can be built to contain water by erecting a heavy wall of earth at the end

**FIGURE 12–28** Hypar roof (a) and hypar umbrella (b). [Saralinda Hooker and Christopher Ragus]

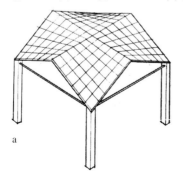
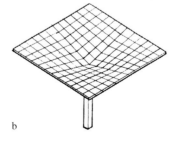

**FIGURE 12–29**  Hypar umbrellas of Newark Airport. [Saralinda Hooker and Christopher Ragus]

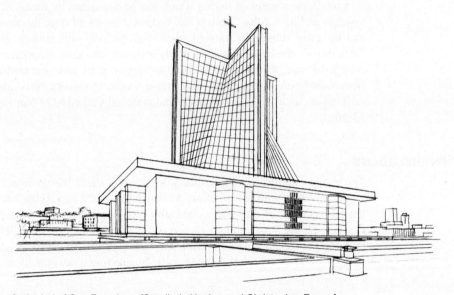

**FIGURE 12–30**  Cathedral of San Francisco. [Saralinda Hooker and Christopher Ragus]

of a valley and compacting it so as to make it watertight (Figure 12-31). These dams resist the horizontal pressure of the water behind them by means of their weight (as the weight of a building resists the horizontal pressure of the wind) and are called *gravity dams*. They are commonly used in developing countries, where labor is abundant and inexpensive and heavy earthmoving equipment rarely available. On the other hand, where valleys are deep and their sides are formed by rocky mountains, and where concrete technology is well developed, dams are often built as thin, concrete, curved surfaces, which resist the pressure of the water through their curvature (Figure 12-32). (They may be thought of as curved roofs loaded with snow, but rotated into a vertical position, so that the snow load becomes horizontal.) Some of the Alpine dams

**FIGURE 12–31** Gravity dam. [Saralinda Hooker and Christopher Ragus]

**FIGURE 12–32** Thin-shell dam. [Saralinda Hooker and Christopher Ragus]

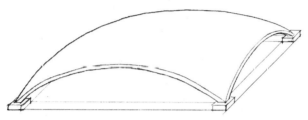

**FIGURE 12–33** Thin-shell "sail" roof. [Saralinda Hooker and Christopher Ragus]

are monumental structures reaching heights of over 1,000 feet and transmitting the thousands of tons of water pressure to the valley sides through their curvature. It may be thought ironic that such structures be called "thin," when their thickness, which increases from top to bottom, may reach ten feet. But thickness is never measured in absolute terms: what counts structurally is the *ratio* of the thickness to the radius of the curved surface, which in a dam can be as low as 1/500. It can be realized how "thin" a ten-foot-thick dam is by comparing it with the curved shell of an egg, in which the thickness to radius ratio is as much as 1/50. A dam is, relatively speaking, ten times thinner than an eggshell.

One of the questions often asked of the structural engineer is whether any of the beautiful curved surfaces encountered in nature or imagined by the fertile mind of an artist could be used to build roofs or other structures. For example, one lovely, thin-shell roof in California has been built in the shape of a square-rigger sail, blown out by the wind but then turned into a horizontal position and supported on its four corners (Figure 12-33). Although not a geometrically definable surface, it is structurally efficient. On the other hand, though an undulating surface can have a pleasant appearance, it would be quite inefficient structurally due to its tendency to fold like an accordion. We can learn a lot from nature, only if we know how to look at it with a wise and critical eye.

This chapter must end with the melancholy realization that over the last few years thin, curved shells, lovely as they may be, have not been very popular in advanced technological countries for purely economic reasons. The main obstacle to their

200  PART ONE:  ARCHITECTURE AS ARTIFACT

popularity, already mentioned, is the cost of their curved formwork. Innumerable procedures have been invented and tried to reduce the cost of the formwork or to do away with it altogether. Pneumatic forms were first used in the 1940s by Wallace Neff, who sprayed concrete on them with a spraygun. Dante Bini sets the reinforcement and pours the concrete on uninflated plastic balloons, and then lifts them by air pressure. The Bini procedure, in particular, has met with success almost all over the world in the erection of round domes of large diameter (up to 300 feet) for schools, gymnasiums, and halls. Of course, balloons are naturally efficient when round. These procedures cannot be well adapted to other thin-shell shapes.

A traditional method of construction, originating in the Catalonian region of Spain, has for centuries produced all kinds of curved thin structures without ever using complex scaffolds or formwork through the ingenious use of tiles and mortar. For example, to build a dome the Catalonians start by supporting its lowest and outermost ring of flat tiles on short, cantilevered, wooden brackets and grout to this first layer a second layer of tiles by means of a rapid-setting mortar. Once this first ring is completed and the mortar has set—in less than twelve hours—workers can erect the next ring by standing on the first and adding as many layers of tile as needed by the span of the dome, usually not more than three layers. By the same procedure, spiral staircases are erected around interior courtyards (Figure 12-34), or cylindrical barrels of groined vaults built. The Guastavino Company, whose Catalonian founder introduced this method to the United States toward the end of the nineteenth century, eventually built over 2,000 buildings in which such tile shells were used. Two of them, the dome over the crossing of the Cathedral of St. John the Divine (erected as a temporary structure while waiting for the completion of the church) and the groined vaults of the War

FIGURE 12–34    Spiral staircase of thin-shell tile construction. [Saralinda Hooker and Christopher Ragus]

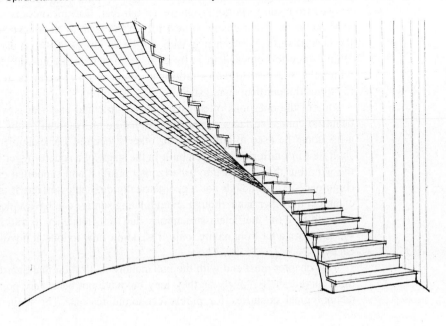

College at Fort McNair, Virginia (Figure 12-35), have by now been officially labelled United States landmarks. Unfortunately, the amount of labor required to set the tiles by hand has made even this procedure uneconomical. The last word on this method's use has not been said, however, since in the [former] USSR thin shell specialists have extended the Catalonian methodology by replacing the small tiles with large, prefabricated, curved elements of prestressed concrete. These are erected without the need for any scaffold starting at one corner of a steel or concrete structural frame (Figure 12-36).

In structures, perhaps more than in any other field of human invention, little is new under the sun, but there is always room for ingenious modifications of old ideas, as well as hope for real breakthroughs.

**FIGURE 12–35** Groin vaults of thin-shell tile at Fort McNair, Virginia. [Saralinda Hooker and Christopher Ragus]

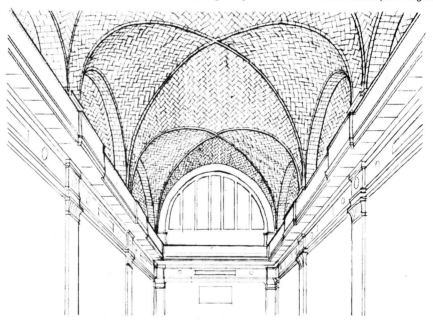

**FIGURE 12–36** Thin shell of prefabricated elements. [Saralinda Hooker and Christopher Ragus]

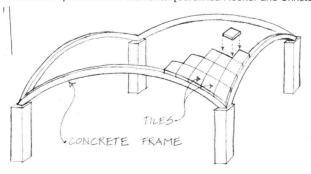

选读 13

"高层办公建筑的艺术思考"

路易斯·亨利·沙利文

沙利文将这篇选读收录进 1901 年出版的《幼儿园对话录及其他文章》中。在沙利文写作这篇文章时，他已经建起了圣路易斯的温赖特大厦、芝加哥的礼堂大厦和布法罗的保险大厦。他的这些建筑作品确立了高层建筑在美学上、结构完整性和功能方面的标准，并使美国首次在建筑类型领域有了自己的贡献。因此，这篇文章首次概述了蕴含在沙利文建筑作品中的功能主义原则。它对刚刚开始处理重视高层建筑设计问题的建筑师影响深远，也正因为这篇文章，沙利文确立了重要建筑理论家的地位。下面引自沙利文文章中的话很好地总结了他的建筑哲学："一切有机和无机的事物都要遵循的普遍法则……是形式追随功能。这就是法则。"

## READING 13

## The Tall Office Building Artistically Considered*

### Louis Henry Sullivan

Sullivan included this selection in *Kindergarten Chats and Other Writings,* published in 1901. At the time that he wrote this article, Sullivan had built the Wainwright Building in St. Louis, the Auditorium Building in Chicago, and the Guaranty Building in Buffalo. This body of work set the standard for the aesthetics, structural integrity, and functional aspects of the high-rise, and created the first American contribution to building types. Thus, this article outlined for the first time the principle of functionalism as embodied in Sullivan's work. It had a profound effect on architects who were just beginning to address the problems of high-rise design and established Sullivan as a major architectural theorist. This quote summarizes his architectural philosophy: "It is the pervading law of all things organic, and inorganic . . . that form follows function. This is the law."

The architects of this land and generation are now brought face to face with something new under the sun—namely, that evolution and integration of social conditions, that special grouping of them, that results in a demand for the erection of tall office buildings.

*Source:* "The Tall Office Building Artistically Considered" in *Kindergarten Chats and Other Writings* by Louis H. Sullivan. Dover Edition, © 1979 is an unabridged republication of the revised (1918) edition as published by Wittenborn, Schultz, Inc., New York, 1947. Reproduced by permission of Dover Publications, Inc. pp. 202–13.

*This essay was first published in *Lippincott's,* March 1896.

It is not my purpose to discuss the social conditions; I accept them as the fact, and say at once that the design of the tall office building must be recognized and confronted at the outset as a problem to be solved—a vital problem, pressing for a true solution.

Let us state the conditions in the plainest manner. Briefly, they are these: offices are necessary for the transaction of business; the invention and perfection of the high-speed elevators make vertical travel, that was once tedious and painful, now easy and comfortable; development of steel manufacture has shown the way to safe, rigid, economical constructions rising to a great height; continued growth of population in the great cities, consequent congestion of centers and rise in value of ground, stimulate an increase in number of stories; these successfully piled one upon another, react on ground values—and so on, by action and reaction, interaction and inter-reaction. Thus has come about that form of lofty construction called the "modern office building." It has come in answer to a call, for in it a new grouping of social conditions has found a habitation and a name.

Up to this point all in evidence is materialistic, an exhibition of force, of resolution, of brains in the keen sense of the word. It is the joint product of the speculator, the engineer, the builder.

Problem: How shall we impart to this sterile pile, this crude, harsh, brutal agglomeration, this stark, staring exclamation of eternal strife, the graciousness of those higher forms of sensibility and culture that rest on the lower and fiercer passions? How shall we proclaim from the dizzy height of this strange, weird, modern housetop the peaceful evangel of sentiment, of beauty, the cult of a higher life?

This is the problem; and we must seek the solution of it in a process analogous to its own evolution—indeed, a continuation of it—namely, by proceeding step by step from general to special aspects, from coarser to finer considerations.

It is my belief that it is of the very essence of every problem that it contains and suggests its own solution. This I believe to be natural law. Let us examine, then, carefully the elements, let us search out this contained suggestion, this essence of the problem.

The practical conditions are, broadly speaking, these:

Wanted—1st, a story below-ground, containing boilers, engines of various sorts, etc.—in short, the plant for power, heating, lighting, etc. 2nd, a ground floor, so called, devoted to stores, banks, or other establishments requiring large area, ample spacing, ample light, and great freedom of access. 3rd, a second story readily accessible by stairways—this space usually in large subdivisions, with corresponding liberality in structural spacing and expanse of glass and breadth of external openings. 4th, above this an indefinite number of stories of offices piled tier upon tier, one tier just like another tier, one office just like all the other offices—an office being similar to a cell in a honey-comb, merely a compartment, nothing more. 5th, and last, at the top of this pile is placed a space or story that, as related to the life and usefulness of the structure, is purely physiological in its nature—namely, the attic. In this the circulatory system completes itself and makes its grand turn, ascending and descending. The space is filled with tanks, pipes, valves, sheaves, and mechanical etcetera that supplement and complement the force-originating plant hidden below-ground in the cellar. Finally, or at the beginning rather, there must be on the ground floor a main aperture or entrance common to all the occupants or patrons of the building.

This tabulation is, in the main, characteristic of every tall office building in the country. As to the necessary arrangements for light courts, these are not germane to the problem, and as will become soon evident, I trust need not be considered here. These things, and such others as the arrangement of elevators, for example, have to do strictly with the economics of the building, and I assume them to have been fully considered and disposed of to the satisfaction of purely utilitarian and pecuniary demands. Only in rare instances does the plan or floor arrangement of the tall office building take on an aesthetic value, and this usually when the lighting court is external or becomes an internal feature of great importance.

As I am here seeking not for an individual or special solution, but for a true normal type, the attention must be confined to those conditions that, in the main, are constant in all tall office buildings, and every mere incidental and accidental variation eliminated from the consideration, as harmful to the clearness of the main inquiry.

The practical horizontal and vertical division or office unit is naturally based on a room of comfortable area and height, and the size of this standard office room as naturally predetermines the standard structural unit, and, approximately, the size of window openings. In turn, these purely arbitrary units of structure form in an equally natural way the true basis of the artistic development of the exterior. Of course the structural spacings and openings in the first or mercantile story are required to be the largest of all; those in the second or quasi-mercantile story are of a somewhat similar nature. The spacings and openings in the attic are of no importance whatsoever (the windows have no actual value), for light may be taken from the top, and no recognition of a cellular division is necessary in the structural spacing.

Hence it follows inevitably, and in the simplest possible way, that if we follow our natural instincts without thought of books, rules, precedents, or any such educational impedimenta to a spontaneous and "sensible" result, we will in the following manner design the exterior of our tall office building—to wit:

Beginning with the first story, we give this a main entrance that attracts the eye to its location, and the remainder of the story we treat in a more or less liberal, expansive, sumptuous way—a way based exactly on the practical necessities, but expressed with a sentiment of largeness and freedom. The second story we treat in a similar way, but usually with milder pretension. Above this, throughout the indefinite number of typical office tiers, we take our cue from the individual cell, which requires a window with its separating pier, its sill and lintel, and we, without more ado, make them look all alike because they are all alike. This brings us to the attic, which, having no division into office-cells, and no special requirement for lighting, gives us the power to show by means of its broad expanse of wall, and its dominating weight and character, that which is the fact—namely, that the series of office tiers has come definitely to an end.

This may perhaps seem a bald result and a heartless, pessimistic way of stating it, but even so we certainly have advanced a most characteristic stage beyond the imagined sinister building of the speculator-engineer-builder combination. For the hand of the architect is now definitely felt in the decisive position at once taken, and the suggestion of a thoroughly sound, logical, coherent expression of the conditions is becoming apparent.

When I say the hand of the architect, I do not mean necessarily the accomplished and trained architect. I mean only a man with a strong, natural liking for buildings, and a disposition to shape them in what seems to his unaffected nature a direct and simple way. He will probably tread an innocent path from his problem to its solution, and therein he will show an enviable gift of logic. If he have some gift for form in detail, some feeling for form purely and simply as form, some love for that, his result in addition to its simple straightforward naturalness and completeness in general statement, will have something of the charm of sentiment.

However, thus far the results are only partial and tentative at best; relatively true, they are but superficial. We are doubtless right in our instinct but we must seek a fuller justification, a finer sanction, for it . . .

I assume now that in the study of our problem we have passed through the various stages of inquiry, as follows: 1st, the social basis of the demand for tall office buildings; 2nd, its literal material satisfaction; 3rd, the elevation of the question from considerations of literal planning, construction, and equipment, to the plane of elementary architecture as a direct outgrowth of sound, sensible building; 4th, the question again elevated from an elementary architecture to the beginnings of true architectural expression, through the addition of a certain quality and quantity of sentiment.

But our building may have all these in a considerable degree and yet be far from that adequate solution of the problem I am attempting to define. We must now heed the imperative voice of emotion.

It demands of us, what is the chief characteristic of the tall office building? And at once we answer, it is lofty. This loftiness is to the artist-nature its thrilling aspect. It is the very open organ-tone in its appeal. It must be in turn the dominant chord in his expression of it, the true excitant of his imagination. It must be tall, every inch of it tall. The force and power of altitude must be in it, the glory and pride of exaltation must be in it. It must be every inch a proud and soaring thing, rising in sheer exultation that from bottom to top it is a unit without a single dissenting line—that it is the new, the unexpected, the eloquent peroration of most bald, most sinister, most forbidding conditions.

The man who designs in this spirit and with the sense of responsibility to the generation he lives in must be no coward, no denier, no bookworm, no dilettante. He must live of his life and for his life in the fullest, most consummate sense. He must realize at once and with the grasp of inspiration that the problem of the tall office building is one of the most stupendous, one of the most magnificent opportunities that the Lord of Nature in His beneficence has ever offered to the proud spirit of man.

That this has not been perceived—indeed, has been flatly denied—is an exhibition of human perversity that must give us pause.

One more consideration. Let us now lift this question into the region of calm, philosophic observation. Let us seek a comprehensive, a final solution: let the problem indeed dissolve.

Certain critics, and very thoughtful ones, have advanced the theory that the true prototype of the tall office building is the classical column, consisting of base, shaft and capital—the moulded base of the column typical of the lower stories of our building, the plain or fluted shaft suggesting the monotonous, uninterrupted series of office-tiers, and the capital the completing power and luxuriance of the attic.

Other theorizers, assuming a mystical symbolism as a guide, quote the many trinities in nature and art, and the beauty and conclusiveness of such trinity in unity. They aver the beauty of prime numbers, the mysticism of the number three, the beauty of all things that are in three parts—to wit, the day, subdividing into morning, noon, and night; the limbs, the thorax, and the head, constituting the body. So they say, should the building be in three parts vertically, substantially as before, but for different motives.

Others, of purely intellectual temperament, hold that such a design should be in the nature of a logical statement; it should have a beginning, a middle, and an ending, each clearly defined—therefore again a building, as above, in three parts vertically.

Others, seeking their examples and justification in the vegetable kingdom, urge that such a design shall above all things be organic. They quote the suitable flower with its bunch of leaves at the earth, its long graceful stem, carrying the gorgeous single flower. They point to the pine-tree, its massy roots, its lithe, uninterrupted trunk, its tuft of green high in the air. Thus, they say, should be the design of the tall office building: again in three parts vertically.

Others still, more susceptible to the power of a unit than to the grace of a trinity, say that such a design should be struck out at a blow, as though by a blacksmith or by mighty Jove, or should be thought-born, as was Minerva, full grown. They accept the notion of a triple division as permissible and welcome, but non-essential. With them it is a subdivision of their unit: the unit does not come from the alliance of the three; they accept it without murmur, provided the subdivision does not disturb the sense of singleness and repose.

All of these critics and theorists agree, however, positively, unequivocally, in this, that the tall office building should not, must not, be made a field for the display of architectural knowledge in the encyclopaedic sense; that too much learning in this instance is fully as dangerous, as obnoxious, as too little learning; that miscellany is abhorrent to their sense; that the sixteen-story building must not consist of sixteen separate, distinct and unrelated buildings piled one upon the other until the top of the pile is reached.

To this latter folly I would not refer were it not the fact that nine out of every ten tall office buildings are designed in precisely this way in effect, not by the ignorant, but by the educated. It would seem indeed, as though the "trained" architect, when facing this problem, were beset at every story, or at most, every third or fourth story, by the hysterical dread lest he be in "bad form"; lest he be not bedecking his building with sufficiency of quotation from this, that, or the other "correct" building in some other land and some other time; lest he be not copious enough in the display of his wares; lest he betray, in short, a lack of resource. To loosen up the touch of this cramped and fidgety hand, to allow the nerves to calm, the brain to cool, to reflect equably, to reason naturally, seems beyond him; he lives, as it were, in a waking nightmare filled with the disjecta membra of architecture. The spectacle is not inspiriting.

As to the former and serious views held by discerning and thoughtful critics, I shall, with however much of regret, dissent from them for the purpose of this demonstration, for I regard them as secondary only, non-essential, and as touching not at all upon the vital spot, upon the quick of the entire matter, upon the true, the immovable philosophy of the architectural art.

This view let me now state, for it brings to the solution of the problem a final, comprehensive formula.

All things in nature have a shape, that is to say, a form, an outward semblance, that tells us what they are, that distinguishes them from ourselves and from each other.

Unfailingly in nature these shapes express the inner life, the native quality, of the animal, tree, bird, fish, that they present to us; they are so characteristic, so recognizable, that we say, simply, it is "natural" it should be so. Yet the moment we peer beneath this surface of things, the moment we look through the tranquil reflection of ourselves and the clouds above us, down into the clear, fluent, unfathomable depth of nature, how startling is the silence of it, how amazing the flow of life, how absorbing the mystery. Unceasingly the essence of things is taking shape in the matter of things, and this unspeakable process we call birth and growth. Awhile the spirit and the matter fade away together, and it is this that we call decadence, death. These two happenings seem jointed and interdependent, blended into one like a bubble and its iridescence, and they seem borne along upon a slowly moving air. This air is wonderful past all understanding.

Yet to the steadfast eye of one standing upon the shore of things, looking chiefly and most lovingly upon that side on which the sun shines and that we feel joyously to be life, the heart is ever gladdened by the beauty, the exquisite spontaneity, with which life seeks and takes on its forms in an accord perfectly responsive to its needs. It seems ever as though the life and the form were absolutely one and inseparable, so adequate is the sense of fulfillment.

Whether it be the sweeping eagle in his flight or the open apple-blossom, the toiling work-horse, the blithe swan, the branching oak, the winding stream at its base, the drifting clouds, over all the coursing sun, form ever follows function, and this is the law. Where function does not change form does not change. The granite rocks, the ever-brooding hills, remain for ages; the lightning lives, comes into shape, and dies in a twinkling.

It is the pervading law of all things organic, and inorganic, of all things physical and metaphysical, of all things human and all things superhuman, of all true manifestations of the head, of the heart, of the soul, that the life is recognizable in its expression, that form ever follows function. This is the law.

Shall we, then, daily violate this law in our art? Are we so decadent, so imbecile, so utterly weak of eyesight, that we cannot perceive this truth so simple, so very simple? Is it indeed a truth so transparent that we see through it but do not see it? Is it really then, a very marvelous thing, or is it rather so commonplace, so everyday, so near a thing to us, that we cannot perceive that the shape, form, outward expression, design or whatever we may choose, of the tall office building should in the very nature of things follow the functions of the building, and that where the function does not change, the form is not to change?

Does this not readily, clearly, and conclusively show that the lower one or two stories will take on a special character suited to the special needs, that the tiers of typical offices, having the same unchanging function, shall continue in the same unchanging form, and that as to the attic, specific and conclusive as it is in its very nature, its function shall equally be so in force, in significance, in continuity, in conclusiveness of outward expression? From this results, naturally, spontaneously, unwittingly, a three-part division, not from any theory, symbol, or fancied logic.

And thus the design of the tall office building takes its place with all other architectural types made when architecture, as has happened once in many years, was a living art. Witness the Greek temple, the Gothic cathedral, the medieval fortress.

And thus, when native instinct and sensibility shall govern the exercise of our beloved art; when the known law, the respected law, shall be that form ever follows function; when our architects shall cease struggling and prattling handcuffed and vainglorious in the asylum of a foreign school; when it is truly felt, cheerfully accepted, that this law opens up the airy sunshine of green fields, and gives to us a freedom that the very beauty and sumptuousness of the outworking of the law itself as exhibited in nature will deter any sane, any sensitive man from changing into license, when it becomes evident that we are merely speaking a foreign language with a noticeable American accent, whereas each and every architect in the land might, under the benign influence of this law, express in the simplest, most modest, most natural way that which it is in him to say; that he might really and would surely develop his own characteristic individuality, and that the architectural art with him would certainly become a living form of speech, a natural form of utterance, giving surcease to him and adding treasures small and great to the growing art of his land; when we know and feel that Nature is our friend, not our implacable enemy—that an afternoon in the country, an hour by the sea, a full open view of one single day, through dawn, high noon, and twilight, will suggest to us so much that is rhythmical, deep, and eternal in the vast art of architecture, something so deep, so true, that all the narrow formalities, hard-and-fast rules, and strangling bonds of the schools cannot stifle it in us—then it may be proclaimed that we are on the high-road to a natural and satisfying art, an architecture that will soon become a fine art in the true, the best sense of the word, an art that will live because it will be of the people, for the people, and by the people.

## ARCHITECTURAL TECHNOLOGY—SUGGESTED READINGS

Allen, Edward. *How Buildings Work.* New York: Oxford, 1995.

Ching, Francis D. K. *Building Construction Illustrated.* New York: Van Nostrand Reinhold, 1991.

Groak, Steven. *The Idea of Building.* London: E & FN Spon, 1992.

Hopkinson, R. G., and J. B. Collins. *The Ergonomics of Lighting.* London: MacDonald, 1970.

Merritt, Frederick S., and James Ambrose. *Building Engineering and Systems Design.* New York: Van Nostrand Reinhold, 1990.

Mumford, Lewis. *Technics and Civilization.* New York: Harcourt Brace Jovanovich, 1963.

Pelletier, Louise, and Alberto Perez-Gomez. (eds.) *Architecture, Ethics, and Technology.* Montreal: McGill-Queen's University Press, 1994.

Salvadori, Mario, and Robert Heller. *Structure in Architecture.* Englewood Cliffs, NJ: Prentice Hall, 1986.

Sandaker, B. N. *The Structural Basis of Architecture.* New York: Whitney Library of Design, 1992.

Stein, Benjamin, J. S. Reynolds, and W. J. McGuinness. *Mechanical and Electrical Equipment for Buildings.* New York: J Wiley and Sons.

# PART TWO 第二部分

## THE CONTEXT OF ARCHITECTURE 建筑的环境

# 4

# THE URBAN ENVIRONMENT
# 城市环境

选读 14
"力量的场所：对洛杉矶的建议"
多洛雷斯·海登

海登在这篇文章中研究了洛杉矶少数族裔和妇女以往的政治、历史与文化。她以形成一种"场所理论"为目标，把历史研究与当下的性别、权力和设计等议题联系在一起来探讨，以期为历史保护、公共艺术和城市设计确立新的日程计划。这项研究使用经济发展这一主题来选择可供分析的场所，而分析已选取场所的方法包括口述史学法以及书面文件与视觉文献内容分析法。

## READING 14

## The Power of Place: A Proposal for Los Angeles
### Dolores Hayden

Hayden explores in this article the political, historical, and cultural past of the ethnic minorities and women in Los Angeles. Her goal is to develop a "theory of place" that connects historical research to contemporary issues of gender, power, and design in order to establish a new agenda for historic preservation, public art, and urban design. The theme of economic development is used in this research to select sites for analysis, and the methods used to analyze the places selected include oral histories and content analyses of written and visual documents.

*Source:* "The Power of Place: A Proposal for Los Angeles" by Dolores Hayden. Copyright © 1988 by The Regents of the University of California. Reprinted from *The Public Historian* 10, no. 3 (Summer 1988), pp. 5–18, by permission. Reproduced by permission of University of California Press Journals.

## THE POLITICS OF PRESERVATION—WHOSE HISTORY?

Kevin Lynch, the urban designer who converted many architects into preservationists with his influential *What Time Is This Place?* in 1972, once remarked, "Choosing a past helps us to construct a future."[1] The task of choosing a past for Los Angeles is a political act as well as a historical and cultural one. Although architectural history, currently the basis of most preservation efforts, can provide an aesthetically satisfying view of the past, it is often quite limited politically. Urban history, including economic and social history, offers complementary resources for historic preservation. Through preserving urban history, we can explain the urban design of a city and its economic growth. (This is particularly useful to residents of the newer cities of the American South and West where one does not find the traditional, legible shapes of cities built in the eighteenth or nineteenth centuries around ports or single industries, but a more complex twentieth century layering of transportation patterns and multiple industries.) In any city, as part of an urban preservation program, it is also possible to celebrate the historical experience of ethnic minorities and women, often underrepresented in architectural preservation.

No historian has yet been able to write a definitive social history of Los Angeles' multiethnic population, or a definitive economic history of the city's industries and multiethnic labor force. By the early 1980s, however, young scholars such as Richard Griswold del Castillo, Ricardo Romo and Noritaka Yagasaki were creating rich ethnic histories of Latinos and Japanese-Americans that suggested the outline which the larger urban story of Los Angeles might take.[2] At the same time, a few politically sensitive preservationists, such as Knox Mellon in California, and others nationally, were reconsidering their audiences, trying to reach beyond a small community of architectural historians and preservation-minded developers to seek broader political and financial support. Humanities councils, historical societies, public history programs, and individual scholars around the country were also seeking innovative ways to present new research in social history to the general public, especially the history of ethnic minorities and women. An exhibit on Chinese-American laundry workers in New York, a self-guided tour of women's history sites in Chicago, a Black Heritage Trail on Beacon Hill in Boston, and a survey of Black Historic sites in Kansas are a few examples of pathfinding efforts[3] that offered models for southern California.

Ethnic minorities and women are the past, present, and future majority of Los Angeles' citizens. The *pobladores* who came from Mexico to found the town in 1781 included people of Spanish, Indian, Afro-American, and *mestizo* descent. Until the transcontinental railroad arrived in 1876, Los Angeles remained what Antonio Rios-Bustamante has called a "predominantly Mexican town," even though the number of persons of Mexican descent began declining steadily relative to total population after California's statehood.[4] The original Spanish speakers were almost completely absorbed by the Anglo-Americans who arrived daily by the trainload after 1876, but at the beginning of this century, the Mexican presence increased again with large-scale migration to jobs in railroad and streetcar construction and maintenance, migratory agricultural work, the citrus industry, brick, tile and cement plants, and in general manual labor everywhere. The number of persons with Mexican heritage jumped from 90,000 in 1920 to 275,000 in 1930.[5]

Like Hispanics, Afro-Americans contributed to the founding of Los Angeles. Lawrence B. DeGraaf has noted that among the forty-four original settlers in 1781, more than half had some African ancestry. The first wave of Black migration came with the land boom of the 1880's, when, like other American-born migrants, Blacks were drawn to southern California by the possibilities of jobs in service and industry, comfortable climate and good health.[6] By 1900, 2,841 Blacks resided in Los Angeles, making it "the largest black settlement on the Pacific Coast."[7] The proportion of Blacks in the total population remained at about 2–3 percent until the post–World War II period, when it began to increase steadily to the 1980 proportion of 11 percent.[8]

Asians were also a visible minority in the city by 1900. The Chinese came to California originally to mine gold, and later to build railroads and aqueducts.[9] They settled in Los Angeles' Chinatown in large numbers between 1880 and 1900, numbering 2,111 at the turn of the century. Many ran market gardens and produce operations.[10] A labor shortage arose from federal restrictions on Chinese immigration during this period, causing an increase in Japanese immigration concentrated in the citrus industry, flower farming, agriculture, and fishing.[11] A large number of Japanese moved to Los Angeles from San Francisco following the 1906 earthquake. By 1908 over 6,000 Japanese lived in the city.[12]

The 299 Historic-Cultural Monuments currently designated in the City of Los Angeles give few hints of this diverse history. While today the urbanized County of Los Angeles numbers about eight million people, approximately one-third Hispanic, one-eighth Afro-American, one-tenth Asian American, and less than one-half Anglo-American, the landmark process has favored the history of a small minority of white, male landholders, bankers, business leaders, and their architects. When Gail Dubrow analyzed the city of Los Angeles' landmarks in 1985, she found 97.7 percent were Anglo-American and only 2.3 percent celebrated Native American and ethnic minority history. Only 4 percent were associated with any aspect of women's history, including Anglo-American women's history.[13] So, three-quarters of the current population must find its public, collective past a small fraction of the city's monuments, or live with someone else's choices about the city's history.

One reason for the neglect of ethnic and women's history is that landmark nominations everywhere in the United States frequently have been the province of passionate rather than dispassionate individuals—politicians seeking fame or favor, businessmen exploiting the commercial advantages of specific locations, and architectural critics establishing their own careers by promoting specific persons or styles. As a result few cities have chosen to celebrate the history of their citizens' most typical activities—earning a living, raising a family, carrying on local holidays, and campaigning for economic development or better municipal services.

In addition, in past decades the histories of ethnic minorities and women have been obscured by the belief that these activities are not of broad public interest and importance. In the case of ethnic minorities, some historians and preservationists have assumed that only other members of the minority group have an interest in the history. In the case of women, the stereotype that "a woman's place is in the home" has suggested that women have no significant public history. In a city with the demographic composition of Los Angeles, where ethnic minorities and women are the majority, both of these outmoded views can lead to the even more destructive generalization that Los Angeles is a city without any history. Preservation in Los Angeles has been less well funded in both the public and private sectors than in many other American cities, but it is gaining strength. New ways to make Los Angeles' diverse history visible promise to add momentum.

## LEGISLATION SUPPORTING COMMUNITY HERITAGE PRESERVATION

Using urban economic and social history to guide preservation is completely in accord with national, state and local legislation. By the Los Angeles Cultural Heritage Commission's definition, a Historic-Cultural Monument can be

> any site, building, or structure . . . in which the broad cultural, political, economic, or social history of the nation, state, or community is reflected or exemplified, [as well as] notable work of a master builder, designer, or architect.[14]

Similarly, the National Register criteria begin with

> districts, sites, buildings, structures, and objects that possess integrity of location, design, setting, materials, workmanship, feeling, and association [and] that are associated with events that have made a significant contribution to the broad patterns of our history.[15]

When these criteria were developed, the framers may have imagined that battlefields (like Concord and Gettysburg) or Presidents' homes (like Mount Vernon and Monticello) would remain the obvious selections to represent "the broad patterns of our history." In the last fifteen years, however, military and political history have been less popular than American social history and urban history. The question remains of how topics like labor history, ethnic history, and women's history can best contribute to public recognition of the community heritage.

Some significant steps toward this goal have been taken by the California Heritage Task Force. This group published a report in August 1984, calling for

> preservation of a heritage resource base for the good of the California citizenry, for the preservation of knowledge and objects as they hold value for long-term cultural coherence.[16]

The report stressed the importance of folklife, defined as "the traditional customs, art and cultural practices of a commonly united group of people."[17] At the time the Task Force report was issued, the State Department of Parks and Recreation had already spent several years developing a statewide inventory of potential ethnic landmarks. The Task Force report advocated the creation of a new State Historical Resources Commission, charged with developing a State Cultural Resource Management Plan, and the passage of Senate Bill 1252 made this possible in 1984.[18] While the structural and administrative changes outlined in the bill have now been made, the plan remains in partial draft in late 1987, and not yet implemented.[19]

## DIVERSE RESOURCES

In California, the obstacles to multicultural preservation and to the preservation of women's history are not in the realm of legislation but in the creation and implementation of workable proposals for specific places. The as-yet-unpublished survey by the State Department of Parks and Recreation, tentatively entitled "California's Ethnic Minorities Cultural Resources Survey: Afro-Americans, Chicanos/Latinos, Native Americans, Chinese-Americans, Japanese-Americans,"[20] is forthcoming in 1988. This rich, detailed survey, compiled by five research teams, is an inventory of over five-hundred potential landmarks throughout the state. But the superb research effort may raise as many questions as it answers because the resources are so diverse.

An examination of the complete survey nominations in manuscript suggests that each ethnic group's researcher has a slightly different idea of what a historic landmark should be. All Americans, they want their communities to be remembered in culturally different ways. The Japanese-American proposals for Los Angeles, for instance, stress religious and legal history: the Japanese Union Church, the Hompa Hongwanji Buddhist temple, and the Sei Fujii property where an immigrant first owned land. The Chicano/Latino proposals, in contrast, include more sites of conflict, such as police brutality on Bloody Christmas, 1951; the Garment Workers' Strike, 1933; or the destroyed community of Chavez Ravine. Some Afro-American proposals stress individuals' professional success—the office of the first Black dentist and the first Black doctor, built by a Black architect, for example. Native American resources are primarily archaeological.

At the same time that these proposals indicate new directions to challenge the uniformity of the Anglo-American history that has previously dominated local landmark selection, there are few proposals that recognize the shared experiences of different ethnic minorities at any one building or site, and there are relatively few that recognize ethnic women's experience as part of the ethnic minority experience. In addition, my windshield survey of nominations in Los Angeles county reveals many sites where buildings no longer exist and only a vacant lot remains to be marked.

Assuming suitable structures are identified, the urban physical context is still a problem for many inventoried buildings. Some structures are located in inner-city neighborhoods plagued by vandalism, abandonment, arson, and homelessness. If their preservation is to involve traditional techniques of renovation and reuse, these structures will not attract commercial developers because they will be so difficult to fund and manage. With the exception of First Street in Little Tokyo, there are few historic districts where numerous adjacent buildings could contribute to a larger whole in terms of interpretation or visual impact. The historic resources are scattered from Saugus to San Pedro, with distances to discourage even the most enthusiastic visitor. As a result, new approaches to commemorating the community heritage will be needed to ensure that more choices are available than museum use or adaptive reuse of historic structures for commercial tenants. And strategies for richer interpretations of possible landmarks, stressing the interconnections of class, race and gender, need to be developed to ensure that the largest possible audiences are addressed.

## A PROPOSAL FOR LOS ANGELES: ECONOMIC HISTORY ON A MULTIETHNIC ITINERARY

"To preserve effectively, we must know for what the past is being retained, and for whom," warned Lynch in 1972.[21] In 1982 I began work on the issue of preserving the history of ethnic minorities and women in Los Angeles by establishing a nonprofit corporation called The Power of Place, and seeking colleagues, students and donors to assist that effort. As a scholar I was interested in doing research that could connect urban history and preservation practice. As a professor in UCLA's Urban Planning Program, I wanted to connect the resources of the university to community groups that might support this effort politically.

The first step for a small, part-time, nonprofit group was to define a manageable project. The research and publication of a self-guided tour of historic places, coupled with the organization of community history workshops concerning those places, seemed to be feasible as a first step toward selecting and protecting places that both historians and citizens could agree were important to ethnic minority and women's history. For the historical tour, I established some broad criteria. First, the tour should be multicultural, explaining the diversity of Los Angeles to a general audience. Second, the tour should have a unifying theme that would display the history of many different ethnic minorities, and of women as well as men. Third, the tour should be concentrated geographically to enable a visitor to see it in one day or less. Fourth, the tour should serve as a demonstration of what other cities and towns might do.

As a broad theme that would fulfill these criteria, I selected the economic development of the city. I defined economic development in Los Angeles as a broad history of agricultural and industrial production, government and service industries, that also encompasses the reproduction of the labor force, a definition that includes both wage work and unpaid domestic labor, and captures the full range of economic activity by women as well as men.[22] The history of economic development explains the physical shape of the city over time. Economic development also explains the social composition of the labor force, as the demand for workers stimulates immigration. Whatever their social, religious, and political differences, residents in the multicultural city engage in productive labor—men, women, and frequently children, too.

Often the economic history of Los Angeles has been told in terms of consumption rather than production. Tracts of bungalows and freeways crowded with cars are said to represent the sprawling boomtowns whose residents are fascinated with automobiles and home ownership.[23] The history of consumption favors the city's outlying areas, not its core, and reinforces the stereotype of Los Angeles as "sixty suburbs in search of a city." In contrast, the history of production helps to define the historic core of the city and emphasizes the skill and energy workers have expended to feed, clothe, and house the population.

Nine major sites were selected for this itinerary.[24] Earliest are the Vignes Vineyard and Wolfskill Grove, where citrus and vines were cultivated beginning in the 1830s, first by Native Americans, then by Mexican, Chinese, and Japanese immigrants, and German and French immigrant entrepreneurs. (Figure 14–1) The produce markets run by Chinese-Americans beginning in the 1870s and the commercial flower fields established by Japanese-Americans in the 1890s came next. A Speaker's Rostrum where organizers addressed migrant Mexican and Filipino farmworkers is a related site, since these agricultural workers passed through Los Angeles on their way to other destinations. Anglo-American wildcatters explored the City Oil Field in the 1890s, along with Mexican-American laborers. Factories for garment making, furniture making, and prefabricated housing were staffed by workers from Mexico, Asia, and all over the United States and Europe, beginning about 1900 as the city became industrially more powerful. A site for service workers—the home of a well-known midwife, a former slave—and a building for civil service workers—a fire station in the Black community—complete the list (Figure 14–2).

Women's economic history is represented in many ways. Housewives' contributions are commemorated by the Pacific Readi-Cut Homes demonstration housing site where hundreds of thousands of families purchased model dwellings equipped with labor-saving devices and built-in furniture to make the housewife's day a shorter one. Women's paid work in traditional occupations is represented by midwifery, commercial flower growing, and garment manufacturing. Nontraditional occupations are represented by an oil wildcatter, produce wholesalers, and labor organizers. The work of women in creating essential social networks appears in the founding of community resources like the Afro-American Methodist Episcopal Church and the Spanish Speaking People's Congress.

**FIGURE 14–1**   View of the Wolfskill Grove, Alameda Street, Los Angeles, showing railroad train loading ranges for shipment. (Security Pacific Bank/Los Angeles Public Library) [Used by permission of Security Pacific Collection/Los Angeles Public Library]

**FIGURE 14–2**   Fire Station 30, designed by James Backus, 1913, elevation. (The Power of Place) Now a Los Angeles Cultural-Historic Landmark, nominated by The Power of Place, the building is being renovated for reuse by the Los Angeles Community Design Center. [Used by permission of Dolores Hayden]

Of the nine major sites selected, two were already Historic-Cultural Monuments that needed new, multicultural interpretation: the site of the Speaker's Rostrum at El Pueblo Historic Park, and the Embassy Theatre, where the garment workers organized a 1933 strike. Four were potential Historic-Cultural Monuments: the City Market; an 1887 house that served as an oil company headquarters; the headquarters of Pacific Readi-Cut Homes; and Fire Station 30. (The Pacific Readi-Cut Homes building would need to be moved; Fire Station 30 was approved for this designation in 1985, as a Power of Place nomination.) The last three were historic sites without surviving structures—commercial groves of vines and citrus trees, commercial flower fields, and the midwife's homestead. Thirty minor sites were also mentioned in the tour.

The decision to include the last three major sites—which are parking lots—was not obvious, but was a response to the specific conditions in downtown Los Angeles, and to the poor condition of many sites on the state's inventory. A significant proportion of downtown Los Angeles is devoted to parking lots, so it is not surprising that asphalt blankets some special historic places. To attempt to turn adversity into opportunity, I argued that these three sites offered possibilities for involving designers and artists in the overall project.[25] As an experiment, I suggested that the vacant agricultural sites be considered for new open space designs that could include historic plantings, and that the homestead be used for an experiment with public art evoking the memory of the site's history. The Community Redevelopment Agency of Los Angeles (CRA) has been initiating an extensive public art program in downtown and now requires developers of new buildings to contribute a portion of a project's cost to public art. The CRA is also developing an inventory of sites suitable for new public art and has been eager to add multiethnic historic sites to that list. New open space designs and new public art are no substitute for traditional historic preservation where sound structures exist, but these techniques do offer the possibility of augmenting the historical narrative offered on a multiethnic tour by commemorating significant missing places.

## TWO SITE-SPECIFIC EFFORTS

Publishing the walking tour map in late 1985 established some priorities for The Power of Place, and since then we have extended practical efforts to save and commemorate two sites, Fire Station 30 (a traditional preservation project) and the Biddy Mason Homestead (a new art installation). With Fire Station 30, we were able to win Historic-Cultural Monument status and persuade the city, with the backing of Councilman Gilbert Lindsay, to lease the building for $1.00 per year to The Los Angeles Community Design Center, a non-profit corporation offering architectural and planning services to many low-income groups. After a successful Power of Place workshop on the history of the building's firefighters in 1985, plans for renovation were made. Then the building was vandalized. Next, a fire destroyed parts of the second floor and the roof. Water damage followed. Funds for renovation were awarded to the Community Design Center by the state, but in late 1987 construction had not yet begun.

On the Biddy Mason site, the parking lot is about to become The Broadway-Spring Center, a ten-storey garage and retail complex, sponsored by the CRA and a private developer. The Power of Place was invited to make a proposal for public art by the

CRA. We obtained two National Endowment for the Arts grants in 1987 to plan and execute a project developed by a team, including Sheila de Bretteville, Susan King, Betye Saar, Donna Graves, and myself. The search for local matching funds has begun and is continuing. A successful community history workshop took place in 1987. Construction is tentatively scheduled for late 1988.

Perhaps the greatest impact of the Power of Place project so far has been in education. As a teaching framework, the project worked well for a class at UCLA in 1984 that included seventeen graduate students in architecture and in urban planning. Several years later, some members of that class are still working on their own projects spun off from the main research—a history of Little Manila and the Filipino community, a dissertation on women's landmarks in the United States,[26] a master's thesis on a local prefabricated housing factory and its importance to the city's residential neighborhoods in the 1920s,[27] and a walking tour of women's history in Los Angeles,[28] as well as several architectural design proposals. Community history programs for the general public, held in various settings to discuss the firefighters, the flower growers, and the midwife, Biddy Mason, have also been lively and well-attended programs.

The Power of Place walking tour, in large part a result of the research of UCLA students, and of Gail Dubrow and Carolyn Flynn, who helped me draft it, and Sheila de Bretteville, who designed it, now carries multiethnic history into other universities and high schools, where a new generation of students are thinking about the city's potential historic resources. This is still controversial material: when the Los Angeles Conservancy selected me for a 1986 Preservation Award, "for education in historic preservation," the restoration architect who was master of ceremonies introduced me as "a very provocative person" and failed to tell the assembled audience just what the award was for. But the audience found out, and most of the city's museums and bookstores now carry the publication, a small road map of downtown Los Angeles with some unusual urban history.

Reporters from *Time* recently labeled Los Angeles "The New Ellis Island," and *Time*'s cover artist showed ten foreign faces peering out of a nest of freeway ramps, a comment on current immigration without any recognition of the resonance of the past.[29] The logo of the Power of Place project is a detail of the 1849 Ord Survey of the city, showing the landscape of vineyards and groves meeting the commercial grid of business blocks, with all of the streets labeled in two languages. That multiethnic history is Los Angeles' greatest urban resource, a legacy that needs to become a physical presence in the public realm. The obstacles to change are intellectual, political, cultural and financial ones. Yet innovative preservation and public art policies can incorporate the history of ethnic minorities and women, and can help to gain majority support for these activities with votes and financial contributions.

Many new Los Angeles organizations concerned with specific ethnic minorities are now making progress with community history. Currently Little Tokyo has a new historic district and the state has funded proposals for a new Japanese-American museum in a former temple. The Chinese Historical Society of Southern California has published a historical walking tour[30] and is developing a museum for the El Pueblo complex. The recently established California Afro-American Museum mounted a historical exhibit, "Black Angelinos from 1850–1950," in mid-1988. A new museum of Latino culture and history is under discussion. With this flowering of ethnic history, a multiethnic interpretation of the city's development becomes more possible, and more urgent. The whole is more than the sum of the parts, and a new multiethnic itinerary can complement the new museums by connecting Los Angeles' diverse urban history to specific historic places.

**ACKNOWLEDGMENTS**

I would like to thank the editors and reviewers for the *Public Historian,* my UCLA research assistants Donna Graves, Drummond Buckley, Gail Dubrow, Carolyn Flynn, and Daniel Hernandez, as well as the numerous preservationists who have supplied helpful information, including Knox Mellon, Eugene Itogawa, Kathryn Gualtieri, Richard Rowe, Jay Pounds, Ruthann Lehner, and Robert Chattel.

**NOTES**

1 Kevin Lynch, *What Time Is This Place?* (Cambridge, MA: MIT Press, 1972), p. 64. Also see Dolores Hayden, "The Meaning of Place in Art and Architecture," *Design Quarterly* 122 (1983), pp. 18–20; and "The American Sense of Place and the Politics of Space," in *American Architecture: Innovation and Tradition,* ed. Robert A. M. Stern, Helen Searing, and David G. DeLong (New York: Rizzoli, 1986), pp. 184–97.

2 Perhaps the best attempt at a multiethnic approach is Carey McWilliams, *Southern California: An Island on the Land Peregrine* (Salt Lake City: Smith, 1983). For specific ethnic studies see Richard Griswold del Castillo, *The Los Angeles Barrio, 1850–1890: A Social History* (Berkeley: University of California Press, 1979); Ricardo Romo, *East Los Angeles: History of a Barrio* (Austin, TX: University of Texas Press, 1983); Noritaka Yagasaki, "Ethnic Cooperativism and Immigrant Agriculture: A Study of Japanese Floriculture and Truck Farming in California," Ph.D. Dissertation, Department of Geography, University of California, Berkeley, 1982.

3 John Kuo Wei Tchen, curator, "Eight Pound Livelihood: A History of Chinese Laundry Workers in the United States," exhibited at New York Public Library and other sites, 1982–83; Marilyn A. Domer, Jean S. Hunt, Mary Ann Johnson, Adade M. Wheeler (authors), and Babette Inglehart, ed., *Walking with Women through Chicago History: 4 Self-Guided Tours* (Chicago: self-published, 1981); *Black Heritage Trail,* Boston African American National Historic Site, National Park Service (Boston, MA: U.S. Department of the Interior, n.d.); and *Historic Preservation in Kansas: Black Historic Sites* (Topeka, KS: Kansas State Historical Society, 1977). Projects on ethnic history increased after 1980, when large federal grants were made for research in this area to many states.

4 Antonio Rios-Bustamante, "The Once and Future Majority," *California History* 60 (Spring 1981), p. 24.

5 Howard J. Nelson and William V. Clark, *Los Angeles: The Metropolitan Experience* (Cambridge, MA: Ballinger, 1976), p. 34.

6 Lawrence B. DeGraaf, "The City of Angels: Emergence of the Los Angeles Ghetto, 1890–1930," *Pacific Historical Review* 39 (August 1970), pp. 327–30.

7 Nelson and Clark, *Los Angeles,* p. 36.

8 Richard G. Lillard, "Problems and Promise in Tomorrowland," *California History* 60 (Spring 1981), p. 93.

9 Don Hata and Nadine Hata, "The Far East Meets the Far West," *California History* 60 (Spring 1981), p. 90.

10 William Mason, "The Chinese in Los Angeles," *Los Angeles Museum of Natural History Quarterly* 6 (Fall 1967), p. 16. Also see "Special Issue: The Chinese in California," *California History* 57 (Spring 1978).

11 Hata and Hata, "The Far East Meets the Far West," p. 91.
12 Icharo Mike Murase, *Little Tokyo: One Hundred Years in Pictures* (Los Angeles: Visual Communications/Asian American Studies Central, Inc., 1983), p. 8.
13 Gail Lee Dubrow, "Preserving Her Heritage: American Landmarks of Women's History," UCLA Urban Planning Program, unpublished paper (1986), pp. 23–25. One of the Afro-American monuments she counted was Fire Station Number 30, a Power of Place nomination approved in 1985.
14 *How a Property Becomes an Historic-Cultural Monument,* flyer distributed by Cultural Heritage Commission, City of Los Angeles, 1985.
15 National Park Service, U.S. Department of the Interior, *How to Complete National Register Forms* (Washington, DC: U.S. Government Printing Office, 1977), pp. 5–6.
16 California Heritage Task Force, *A Report to the Legislature and People of California, August 1984* (Sacramento, CA, 1984), p. 24. This contains a complete review of policy in preservation as well as helpful summaries of legislation and an extensive bibliography.
17 Ibid., p. 27.
18 California Senate Bill 1252, approved by the governor, September 18, 1984.
19 Personal communications with Kathryn Gualtieri, State Historic Preservation Officer, and with Donna Blitzer, assistant to Assemblyman Sam Farr, November 30, 1987
20 "California's Ethnic Minorities Cultural Resources Survey: Afro-American, Chicanos/Latinos, Native Americans, Chinese-Americans, Japanese-Americans" (Sacramento, CA: California Department of Parks and Recreation, forthcoming). Survey coordinators were Eleanor Ramsey, Jose Pitti, Lee Dixon, Nancy Wey, and Issami Waugh. The State Historic Preservation Officer was Knox Mellon, who invited me to examine the complete list of resources surveyed in manuscript in 1984.
21 Lynch, *What Time Is This Place?* p. 64.
22 Hayden, "The American Sense of Place and the Politics of Space," pp. 190–91. For "the production of space," a fuller treatment of Henri Lefebvre's ideas, see M. Gottdiener, *The Social Production of Urban Space* (Austin, TX: University of Texas Press, 1985).
23 For example, the expansion of residential areas is stressed in Reyner Banham, *Los Angeles: The Architecture of Four Ecologies* (Harmondsworth, UK: Penguin Books, 1971); David L. Clark, *Los Angeles: A City Apart* (Woodland Hills, CA: Windsor Publications, 1981); Bruce Henstell, *Los Angeles: An Illustrated History* (New York: Alfred A. Knopf, 1980).
24 Dolores Hayden, Gail Dubrow, and Carolyn Flynn, *The Power of Place: Los Angeles* (Los Angeles: The Power of Place, 1985) (also available from Faculty Publications, GSAUP, UCLA, Los Angeles, CA 90024-1467, $1.00); design by Sheila de Bretteville.
25 Dolores Hayden, "The Power of Place: The Impact of Los Angeles' Multi-Ethnic History on Public Art," *Passage,* Spring/Summer 1987, pp. 1–3.
26 The work on women's landmarks in the United States is being undertaken by Gail Dubrow, a Ph.D. student in Urban Planning at UCLA.
27 Carolyn Flynn, "A Home Every 20 Minutes: The Pacific Readi-Cut Homes Company," M.A. Thesis, Urban Planning Program, UCLA, 1985.
28 Sherry Katz, Carolyn Flynn, and Gail Dubrow, "Women's History in Los Angeles," *California Historical Courier,* February 1985, p. 11.
29 Kurt Anderson, "The New Ellis Island," *Time,* June 13, 1983, pp. 18–25. The Phrase is attributed to demographer Kevin McCarthy of The Rand Corporation.
30 Chinese Historical Society of Southern California, *Los Angeles Chinatown Walking Tour (Yesterday and Today)* (Los Angeles: Chinese Historical Society of Southern California, 1984).

选读 15
"老建筑之必要"
简·雅各布斯

雅各布斯的作品总是在赞美拥挤稠密、喧嚣多样的城市所具有的优点。她认为把她的《美国大城市的死与生》一书当作对不合理的规划和城市更新实践的回击之举丝毫不为过。这些规划和城市更新实践恶意破坏了城市生活的丰富结构。从本书选取的这篇文章里,雅各布斯赞同老旧建筑仍然具有它们的价值:"城市需要混杂老旧建筑,这样有利于培育出具备初级多样性和次级多样性特点的各种城市建筑群。特别需要指出的是,他们需要老旧建筑来酝酿形成新的初级多样性。"她同时指出老旧建筑的经济价值是无法被随意取代的:"重要的城市生活社区在年复一年中得以继承和延续是需要多样性的经济作为前提条件的。"

## READING 15

## The Need for Aged Buildings

### Jane Jacobs

Jacobs's writings have always celebrated the virtues of crowded, dense, noisy, and diverse cities. She viewed *The Death and Life of Great American Cities* as no less than an attack on the planning and urban renewal practices that anathematized the rich fabric of city life. In this selection from her book, Jacobs advocates the virtues of old structures: "Cities need a mingling of old buildings to cultivate primary-diversity mixtures, as well as secondary diversity. In particular, they need old buildings to incubate new primary diversity." She notes that the economic value of old buildings cannot be replaced at will: "This economic requisite for diversity is a requisite that vital city neighborhoods can only inherit, and then sustain over the years."

CONDITION 3: *The district must mingle buildings that vary in age and condition, including a good proportion of old ones.*

Cities need old buildings so badly it is probably impossible for vigorous streets and districts to grow without them. By old buildings I mean not museum-piece old buildings, not old buildings in an excellent and expensive state of rehabilitation—although these make fine ingredients—but also a good lot of plain, ordinary, low-value old buildings, including some rundown old buildings.

If a city area has only new buildings, the enterprises that can exist there are automatically limited to those that can support the high costs of new construction. These high costs of occupying new buildings may be levied in the form of rent, or they may be levied in the form of an owner's interest and amortization payments on the capital costs of the construction. However the costs are paid off, they have to be paid off. And for this

*Source: The Death and Life of Great American Cities* by Jane Jacobs. Copyright © 1961 by Jane Jacobs. Reprinted by permission of Random House, Inc. pp. 187–99.

reason, enterprises that support the cost of new construction must be capable of paying a relatively high overhead—high in comparison to that necessarily required by old buildings. To support such high overheads, the enterprises must be either (a) high profit or (b) well subsidized.

If you look about, you will see that only operations that are well established, high-turnover, standardized or heavily subsidized can afford, commonly, to carry the costs of new construction. Chain stores, chain restaurants and banks go into new construction. But neighborhood bars, foreign restaurants and pawn shops go into older buildings. Supermarkets and shoe stores often go into new buildings; good bookstores and antique dealers seldom do. Well-subsidized opera and art museums often go into new buildings. But the unformalized feeders of the arts—studios, galleries, stores for musical instruments and art supplies, backrooms where the low earning power of a seat and a table can absorb uneconomic discussions—these go into old buildings. Perhaps more significant, hundreds of ordinary enterprises, necessary to the safety and public life of streets and neighborhoods, and appreciated for their convenience and personal quality, can make out successfully in old buildings, but are inexorably slain by the high overhead of new construction.

As for really new ideas of any kind—no matter how ultimately profitable or otherwise successful some of them might prove to be—there is no leeway for such chancy trial, error and experimentation in the high-overhead economy of new construction. Old ideas can sometimes use new buildings. New ideas must use old buildings.

Even the enterprises that can support new construction in cities need old construction in their immediate vicinity. Otherwise they are part of a total attraction and total environment that is economically too limited—and therefore functionally too limited to be lively, interesting and convenient. Flourishing diversity anywhere in a city means the mingling of high-yield, middling-yield, low-yield and no-yield enterprises.

The only harm of aged buildings to a city district or street is the harm that eventually comes of *nothing but* old age—the harm that lies in everything being old and everything becoming worn out. But a city area in such a situation is not a failure because of being all old. It is the other way around. The area is all old because it is a failure. For some other reason or combination of reasons, all its enterprises or people are unable to support new construction. It has, perhaps, failed to hang on to its own people or enterprises that do become successful enough to support new building or rehabilitation; they leave when they become this successful. It has also failed to attract newcomers with choice; they see no opportunities or attractions here. And in some cases, such an area may be so infertile economically that enterprises which might grow into successes in other places, and build or rebuild their shelter, never make enough money in this place to do so.[1]

A successful city district becomes a kind of ever-normal granary so far as construction is concerned. Some of the old buildings, year by year, are replaced by new ones—or rehabilitated to a degree equivalent to replacement. Over the years there is,

---

[1] These are all reasons having to do with inherent, built-in handicaps. There is another reason, however, why some city districts age unremittingly, and this other reason has nothing to do, necessarily, with inherent flaws. The district may have been blacklisted, in a concerted way, by mortgage lenders, the way Boston's North End has been. This means of dooming a neighborhood to inexorable wearing out is both common and destructive. But for the moment we are dealing with the conditions that affect a city area's inherent economic ability to generate diversity and staying power.

therefore, constantly a mixture of buildings of many ages and types. This is, of course, a dynamic process, with what was once new in the mixture eventually becoming what is old in the mixture.

We are dealing here again, as we were in the case of mixed primary uses, with the economic effects of time. But in this case we are dealing with the economics of time not hour by hour through the day, but with the economics of time by decades and generations.

Time makes the high building costs of one generation the bargains of a following generation. Time pays off original capital costs, and this depreciation can be reflected in the yields required from a building. Time makes certain structures obsolete for some enterprises, and they become available to others. Time can make the space efficiencies of one generation the space luxuries of another generation. One century's building commonplace is another century's useful aberration.

The economic necessity for old buildings mixed with new is not an oddity connected with the precipitous rise in building costs since the war, and especially throughout the 1950's. To be sure, the difference between the yield most postwar building must bring and the yield that pre-Depression buildings must bring is especially sharp. In commercial space, the difference between carrying costs per square foot can be as much as 100 or 200 percent, even though the older buildings may be better built than the new, and even though the maintenance costs of all buildings, including old ones, have risen. Old buildings were a necessary ingredient of city diversity back in the 1920's and the 1890's. Old buildings will still be a necessity when today's new buildings are the old ones. This has been, still is, and will be, true no matter how erratic or how steady construction costs themselves are, because a depreciated building requires less income than one which has not yet paid off its capital costs. Steadily rising construction costs simply accentuate the need for old buildings. Possibly they also make necessary a higher *proportion* of old buildings in the total street or district mixture, because rising building costs raise the general threshold of pecuniary success required to support the costs of new construction.

A few years ago, I gave a talk at a city design conference about the social need for commercial diversity in cities. Soon my words began coming back at me from designers, planners and students in the form of a slogan (which I certainly did not invent): "We must leave room for the corner grocery store!"

At first I thought this must be a figure of speech, the part standing for the whole. But soon I began to receive in the mail plans and drawings for projects and renewal areas in which, literally, room had been left here and there at great intervals for a corner grocery store. These schemes were accompanied by letters that said, "See, we have taken to heart what you said."

This corner-grocery gimmick is a thin, patronizing conception of city diversity, possibly suited to a village of the last century, but hardly to a vital city district of today. Lone little groceries, in fact, do badly in cities as a rule. They are typically a mark of stagnant and undiverse gray area.

Nevertheless, the designers of these sweetly meant inanities were not simply being perverse. They were doing, probably, the best they could under the economic conditions set for them. A suburban-type shopping center at some place in the project, and this wan spotting of corner groceries, were the most that could be hoped for. For these were schemes contemplating either great blankets of new construction, or new construction combined with extensive, prearranged rehabilitation. Any vigorous range of diversity was precluded in advance by the consistently high overhead. (The prospects are made still poorer by insufficient primary mixtures of uses and therefore insufficient spread of customers through the day.)

Even the lone groceries, if they were ever built,[2] could hardly be the cozy enterprises envisioned by their designers. To carry their high overhead, they must either be (a) subsidized—by whom and why?—or (b) converted into routinized, high-turnover mills.

Large swatches of construction built at one time are inherently inefficient for sheltering wide ranges of cultural, population, and business diversity. They are even inefficient for sheltering much range of mere commercial diversity. This can be seen at a place like Stuyvesant Town in New York. In 1959, more than a decade after operation began, of the 32 store fronts that comprise Stuyvesant Town's commercial space, seven were either empty or were being used uneconomically (for storage, window advertising only, and the like). This represented disuse or underuse of 22 percent of the fronts. At the same time, across the bordering streets, where buildings of every age and condition are mingled, were 140 store fronts, of which 11 were empty or used uneconomically, representing a disuse or underuse of only 7 percent. Actually, the disparity is greater than this would appear, because the empty fronts in the old streets were mostly small, and in linear feet represented less than 7 percent, a condition which was not true of the project stores. The good business side of the street is the age-mingled side, even though a great share of its customers are Stuyvesant Town people, and even though they must cross wide and dangerous traffic arteries to reach it. This reality is acknowledged by the chain stores and supermarkets too, which have been building new quarters in the age-mingled setting instead of filling those empty fronts in the project.

One-age construction in city areas is sometimes protected nowadays from the threat of more efficient and responsive commercial competition. This protection—which is nothing more or less than commercial monopoly—is considered very "progressive" in planning circles. The Society Hill renewal plan for Philadelphia will, by zoning, prevent competition to its developer's shopping centers throughout a whole city district. The city's planners have also worked out a "food plan" for the area, which means offering a monopolistic restaurant concession to a single restaurant chain for the whole district. Nobody else's food allowed! The Hyde Park–Kenwood renewal district of Chicago reserves a monopoly on almost all commerce for a suburban-type shopping center to be the property of that plan's principal developer. In the huge Southwest redevelopment district of Washington, the major housing developer seems to be going so far as to eliminate competition with himself. The original plans for this scheme contemplated a central, suburban-type shopping center plus a smattering of convenience stores—our old friend, the lonely corner grocery gimmick. A shopping center economist predicted that these convenience stores might lead to diminished business for the main, suburban-type center which, itself, will have to support high overhead. To protect it, the convenience stores were dropped from the scheme. It is thus that routinized monopolistic packages of substitute city are palmed

---

[2]They are usually dropped from the plans, or indefinitely postponed, at the time when the economic realities of rents must be faced.

off as "planned shopping."

Monopoly planning can make financial successes of such inherently inefficient and stagnant one-age operations. But it cannot thereby create, in some magical fashion, an equivalent to city diversity. Nor can it substitute for the inherent efficiency, in cities, of mingled age and inherently varied overhead.

Age of buildings, in relation to usefulness or desirability, is an extremely relative thing. Nothing in a vital city district seems to be too old to be chosen for use by those who have choice—or to have its place taken, finally, by something new. And this usefulness of the old is not simply a matter of architectural distinction or charm. In the Back-of-the-Yards, Chicago, no weather-beaten, undistinguished, run-down, presumably obsolete frame house seems to be too far gone to lure out savings and to instigate borrowing—because this is a neighborhood that people are not leaving as they achieve enough success for choice. In Greenwich Village, almost no old building is scorned by middle-class families hunting a bargain in a lively district, or by rehabilitators seeking a golden egg. In successful districts, old buildings "filter up."

At the other extreme, in Miami Beach, where novelty is the sovereign remedy, hotels ten years old are considered aged and are passed up because others are newer. Newness, and its superficial gloss of well-being, is a very perishable commodity.

Many city occupants and enterprises have no need for new construction. The floor of the building in which this book is being written is occupied also by a health club with a gym, a firm of ecclesiastical decorators, an insurgent Democratic party reform club, a Liberal party political club, a music society, an accordionists' association, a retired importer who sells maté by mail, a man who sells paper and who also takes care of shipping the maté, a dental laboratory, a studio for watercolor lessons, and a maker of costume jewelry. Among the tenants who were here and gone shortly before I came in, were a man who rented out tuxedos, a union local and a Haitian dance troupe. There is no place for the likes of us in new construction. And the last thing we need is new construction.[3] What we need, and a lot of others need, is old construction in a lively district, which some among us can help make livelier.

Nor is new residential building in cities an unadulterated good. Many disadvantages accompany new residential city building; and the value placed on various advantages, or the penalties accruing from certain disadvantages, are given different weights by different people. Some people, for instance, prefer more space for the money (or equal space for less money) to a new dinette designed for midgets. Some people like walls they don't hear through. This is an advantage they can get with many old buildings but not with new apartments, whether they are public housing at $14 a room per month or luxury housing at $95 a room per month.[4] Some people would rather pay for improvements in their living conditions partly in labor and ingenuity, and by selecting which improvements are most important to them, instead of being indiscriminately improved, and all at a cost of money. In spontaneously unslumming slums, where people are staying by choice, it is easy to observe how many ordinary citizens have heard of color, lighting and furnishing devices for converting

---

[3]No, the *last* thing we need is some paternalist weighing whether we are sufficiently noncontroversial to be admitted to subsidized quarters in a Utopian dream city.

[4]"Dear, are you sure the stove is one of the 51 exciting reasons we're living in Washington Square Village?" asks the wife in a cartoon issued by protesting tenants in an expensive New York redevelopment project. "You'll have to speak up, honey," replies the husband. "Our neighbor just flushed his toilet."

deep or dismal spaces into pleasant and useful rooms, have heard of bedroom air-conditioning and of electric window fans, have learned about taking out non-bearing partitions, and have even learned about throwing two too small flats into one. Minglings of old buildings, with consequent minglings in living costs and tastes, are essential to get diversity and stability in residential populations, as well as diversity in enterprises.

Among the most admirable and enjoyable sights to be found along the sidewalks of big cities are the ingenious adaptations of old quarters to new uses. The townhouse parlor that becomes a craftsman's showroom, the stable that becomes a house, the basement that becomes an immigrants' club, the garage or brewery that becomes a theater, the beauty parlor that becomes the ground floor of a duplex, the warehouse that becomes a factory for Chinese food, the dancing school that becomes a pamphlet printer's, the cobbler's that becomes a church with lovingly painted windows—the stained glass of the poor—the butcher shop that becomes a restaurant: these are the kinds of minor changes forever occurring where city districts have vitality and are responsive to human needs.

Consider the history of the no-yield space that has recently been rehabilitated by the Arts in Louisville Association as a theater, music room, art gallery, library, bar and restaurant. It started life as a fashionable athletic club, outlived that and became a school, then the stable of a dairy company, then a riding school, then a finishing and dancing school, another athletic club, an artist's studio, a school again, a blacksmith's, a factory, a warehouse, and it is now a flourishing center of the arts. Who could anticipate or provide for such a succession of hopes and schemes? Only an unimaginative man would think he could; only an arrogant man would want to.

These eternal changes and permutations among old city buildings can be called makeshifts only in the most pedantic sense. It is rather that a form of raw material has been found in the right place. It has been put to a use that might otherwise be unborn.

What is makeshift and woebegone is to see city diversity outlawed. Outside the vast, middle-income Bronx project of Parkchester, where the standardized, routinized commerce (with its share of empty fronts) is protected from unauthorized competition or augmentation within the project, we can see such an outcast huddle, supported by Parkchester people. Beyond a corner of the project, hideously clumped on a stretch of pocked asphalt left over from a gas station, are a few of the other things the project people apparently need: quick loans, musical instruments, camera exchange, Chinese restaurant, odd-lot clothing. How many other needs remain unfilled? What is wanted becomes academic when mingled building age is replaced by the economic rigor mortis of one-age construction, with its inherent inefficiency and consequent need for forms of "protectionism."

Cities need a mingling of old buildings to cultivate primary-diversity mixtures, as well as secondary diversity. In particular, they need old buildings to incubate new primary diversity.

If the incubation is successful enough, the yield of the buildings can, and often does, rise. Grady Clay reports that this is already observable, for instance, in the Louisville sample-shoe market. "Rents were very low when the market began to attract shoppers," he says. "For a shop about twenty feet by forty feet, they were $25 to $50 a month. They have already gone up to about $75." Many a city's enterprises which become important economic assets start small and poor, and become able, eventually, to afford carrying costs of rehabilitation or new construction. But this process could not occur without that low-yield space in the right place, in which to start.

Areas where better mixtures of primary diversity must be cultivated will have to depend heavily on old buildings, especially at the beginning of deliberate attempts to catalyze diversity. If Brooklyn, New York, as an example, is ever to cultivate the quantity of diversity and degree of attraction and liveliness it needs, it must take maximum economic advantage of combinations of residence and work. Without these primary combinations, in effective and concentrated proportions, it is hard to see how Brooklyn can begin to catalyze its potential for secondary diversity.

Brooklyn cannot well compete with suburbs for capturing big and well-established manufacturers seeking a location. At least it cannot at present, certainly not by trying to beat out the suburbs at *their* game, on their terms. Brooklyn has quite different assets. If Brooklyn is to make the most of work-residence primary mixtures, it must depend mainly on incubating work enterprises, and then holding on to them as long as it can. While it has them, it must combine them with sufficiently high concentrations of residential population, and with short blocks, to make the most of their presence. The more it makes of their presence, the more firmly it is apt to hold work uses.

But to incubate those work uses, Brooklyn needs old buildings, needs them for exactly the task they fulfill there. For Brooklyn is quite an incubator. Each year, more manufacturing enterprises leave Brooklyn for other locations than move into Brooklyn from elsewhere. Yet the number of factories in Brooklyn has been constantly growing. A thesis prepared by three students at Brooklyn's Pratt Institute[5] explains this paradox well:

> The secret is that Brooklyn is an incubator of industry. Small businesses are constantly being started there. A couple of machinists, perhaps, will get tired of working for someone else and start out for themselves in the back of a garage. They'll prosper and grow; soon they will get too big for the garage and move to a rented loft; still later they buy a building. When they outgrow that, and have to build for themselves, there is a good chance they will move out to Queens, or Nassau or New Jersey. But in the meantime, twenty or fifty or a hundred more like them will have started up.
>
> Why do they move when they build for themselves? For one thing, Brooklyn offers too few attractions aside from those a new industry finds are necessities—old buildings and nearness to the wide range of other skills and supplies a small enterprise must have. For another, little or no effort has been made to plan for working needs—e.g., great sums of money are spent on highways choked with private automobiles rushing into the city and out of it; no comparable thought or money is spent on trucking expressways for manufacturers who use the city's old buildings, its docks and its railways.[6]

Brooklyn, like most of our city areas in decline, has more old buildings than it

---

[5] Stuart Cohen, Stanley Kogan and Frank Marcellino.

[6] Cost of land, conventionally assumed to be a significant deterrent today to building in the city for expanding businesses, has been steadily diminishing in ratio to construction costs, and to almost all other costs. When Time, Inc., decided to build on an expensive plot of ground near the center of Manhattan, for example, instead of on much cheaper ground near the edge, it based its decision on a host of reasons, among which was the fact that taxi fares alone for employees' business trips from the inconvenient site would come to more, per year, than the difference in land carrying costs! Stephen G. Thompson of *Architectural Forum* has made the (unpublished) observation that redevelopment subsidies frequently bring the cost of city land lower than the cost of carpet for the buildings. To justify land costs higher than carpet costs, a city has to be a *city*, not a machine or a desert.

needs. To put it another way, many of its neighborhoods have for a long time lacked gradual increments of new buildings. Yet if Brooklyn is ever to build upon its inherent assets and advantages—which is the only way successful city building can be done—many of those old buildings, well distributed, will be essential to the process. Improvement must come by supplying the conditions for generating diversity that are missing, not by wiping out old buildings in great swathes.

We can see around us, from the days preceding project building, many examples of decaying city neighborhoods built up all at once. Frequently such neighborhoods have begun life as fashionable areas; sometimes they have had instead a solid middle-class start. Every city has such physically homogeneous neighborhoods.

Usually just such neighborhoods have been handicapped in every way, so far as generating diversity is concerned. We cannot blame their poor staying power and stagnation entirely on their most obvious misfortune: being built all at once. Nevertheless, this is one of the handicaps of such neighborhoods, and unfortunately its effects can persist long after the buildings have become aged.

When such an area is new, it offers no economic possibilities to city diversity. The practical penalties of dullness, from this and other causes, stamp the neighborhood early. It becomes a place to leave. By the time the buildings have indeed aged, their only useful city attribute is low value, which by itself is not enough.

Neighborhoods built up all at once change little physically over the years as a rule. The little physical change that does occur is for the worse—gradual dilapidation, a few random, shabby new uses here and there. People look at these few, random differences and regard them as evidence, and perhaps as cause, of drastic change. Fight blight! They regret that the neighborhood has changed. Yet the fact is, physically it has changed remarkably little. People's feelings about it, rather, have changed. The neighborhood shows a strange inability to update itself, enliven itself, repair itself, or to be sought after, out of choice, by a new generation. It is dead. Actually it was dead from birth, but nobody noticed this much until the corpse began to smell.

Finally comes the decision, after exhortations to fix up and fight blight have failed, that the whole thing must be wiped out and a new cycle started. Perhaps some of the old buildings will be left if they can be "renewed" into the economic equivalent of new buildings. A new corpse is laid out. It does not smell yet, but it is just as dead, just as incapable of the constant adjustments, adaptations and permutations that make up the processes of life.

There is no reason why this dismal, foredoomed cycle need be repeated. If such an area is examined to see which of the other three conditions for generating diversity are missing, and then those missing conditions are corrected as well as they can be, some of the old buildings must go: extra streets must be added, the concentration of people must be heightened, room for new primary uses must be found, public and private. But a good mingling of the old buildings must remain, and in remaining they will have become something more than mere decay from the past or evidence of previous failure. They will have become the shelter which is necessary, and valuable to the district, for many varieties of middling-, low-and no-yield diversity. The economic value of new buildings is replaceable in cities. It is replaceable by the spending of more construction money. But the economic value of old buildings is irreplaceable at will. It is created by time. This economic requisite for diversity is a requisite that vital city neighborhoods can only inherit, and then sustain over the years.

选读 16
"回顾与展望"
刘易斯·芒福德

这篇选读是芒福德《城市发展史》一书的最后一章。芒福德回顾了从古至今的城市演进这一社会和功能现象。他向我们展示了古老仪式的过程、环境的必要性以及社会秩序如何影响了我们对当代城市生活的认识。芒福德认为城市的任务是"促使人类有意识地参与宇宙和历史进程"。芒福德的分析精彩之处在于他融合了哲学与文学的特点,并与准确无误的城市历史联系在一起。

## READING 16

## Retrospect and Prospect
### Lewis Mumford

This selection is the concluding chapter of Mumford's *The City in History*. Mumford reviews the evolution of the city as a social and functional phenomenon from antiquity to the modern era. He shows the ways that ancient processes of ritual, environmental necessity, and social order inform our contemporary visions of urban life. Mumford views the mission of the city "to further man's conscious participation in the cosmic and historic process." The brilliance of Mumford's analysis is that he combines the precision of urban history with the integrative qualities of philosophy and literature.

In taking form, the ancient city brought together many scattered organs of the common life, and within its walls promoted their interaction and fusion. The common functions that the city served were important; but the common purposes that emerged through quickened methods of communication and co-operation were even more significant. The city mediated between the cosmic order, revealed by the astronomer priests, and the unifying enterprises of kingship. The first took form within the temple and its sacred compound, the second within the citadel and the bounding city wall. By polarizing hitherto untapped human aspirations and drawing them together in a central political and religious nucleus, the city was able to cope with the immense generative abundance of neolithic culture.

By means of the order so established, large bodies of men were for the first time brought into effective co-operation. Organized in disciplined work groups, deployed by central command, the original urban populations in Mesopotamia, Egypt, and the Indus Valley controlled flood, repaired storm damage, stored water, remodeled the landscape, built up a great water network for communication and transportation, and filled the urban reservoirs with human energy available for other collective enterprises. In time, the rulers of the city created an internal fabric of order and justice that

*Source:* "Retrospect and Prospect" from *The City in History: Its Origins, Its Transformations, and Its Prospects.* Copyright © 1961 and renewed 1989 by Lewis Mumford. Reprinted by permission of Harcourt Brace & Company. pp. 568–576.

gave to the mixed populations of cities, by conscious effort, some of the moral stability and mutual aid of the village. Within the theater of the city new dramas of life were enacted.

But against these improvements we must set the darker contributions of urban civilization: war, slavery, vocational over-specialization, and in many places, a persistent orientation toward death. These institutions and activities, forming a "negative symbiosis," have accompanied the city through most of its history, and remain today in markedly brutal form, without their original religious sanctions, as the greatest threat to further human development. Both the positive and the negative aspects of the ancient city have been handed on, in some degree, to every later urban structure.

Through its concentration of physical and cultural power, the city heightened the tempo of human intercourse and translated its products into forms that could be stored and reproduced. Through its monuments, written records, and orderly habits of association, the city enlarged the scope of all human activities, extending them backwards and forwards in time. By means of its storage facilities (buildings, vaults, archives, monuments, tablets, books), the city became capable of transmitting a complex culture from generation to generation, for it marshaled together not only the physical means but the human agents needed to pass on and enlarge this heritage. That remains the greatest of the city's gifts. As compared with the complex human order of the city, our present ingenious electronic mechanisms for storing and transmitting information are crude and limited.

From the original urban integration of shrine, citadel, village, workshop, and market, all later forms of the city have, in some measure, taken their physical structure and their institutional patterns. Many parts of this fabric are still essential to effective human association, not least those that sprang originally from the shrine and the village. Without the active participation of the primary group, in family and neighborhood, it is doubtful if the elementary moral loyalties—respect for the neighbor and reverence for life—can be handed on, without savage lapses, from the old to the young.

At the other extreme, it is doubtful, too, whether those multifarious co-operations that do not lend themselves to abstraction and symbolization can continue to flourish without the city, for only a small part of the contents of life can be put on the record. Without the superposition of many different human activities, many levels of experience, within a limited urban area, where they are constantly on tap, too large a portion of life would be restricted to record-keeping. The wider the area of communication and the greater the number of participants, the more need there is for providing numerous accessible permanent centers for face-to-face intercourse and frequent meetings at every human level.

The recovery of the essential activities and values that first were incorporated in the ancient cities, above all those of Greece, is accordingly a primary condition for the further development of the city in our time. Our elaborate rituals of mechanization cannot take the place of the human dialogue, the drama, the living circle of mates and associates, the society of friends. These sustain the growth and reproduction of human culture, and without them the whole elaborate structure becomes meaningless—indeed actively hostile to the purposes of life.

Today the physical dimensions and the human scope of the city have changed; and most of the city's internal functions and structures must be recast to promote effectively the larger purposes that shall be served: the unification of man's inner and outer life, and the progressive unification of mankind itself. The city's active role in future is to bring to the highest pitch of development the variety and individuality of regions, cultures, personalities. These are complementary purposes: their alternative is the current mechanical grinding down of both the landscape and the human personality. Without the city modern man would have no effective defenses against those mechanical collectives that, even now, are ready to make all veritably human life superfluous, except to perform a few subservient functions that the machine has not yet mastered.

Ours is an age in which the increasingly automatic processes of production and urban expansion have displaced the human goals they are supposed to serve. Quantitative production has become, for our mass-minded contemporaries, the only imperative goal: they value quantification without qualification. In physical energy, in industrial productivity, in invention, in knowledge, in population the same vacuous expansions and explosions prevail. As these activities increase in volume and in tempo, they move further and further away from any humanly desirable objectives. As a result, mankind is threatened with far more formidable inundations than ancient man learned to cope with. To save himself he must turn his attention to the means of controlling, directing, organizing, and subordinating to his own biological functions and cultural purposes the insensate forces that would, by their very superabundance, undermine his life. He must curb them and even eliminate them completely when, as in the case of nuclear and bacterial weapons, they threaten his very existence.

Now it is not a river valley, but the whole planet, that must be brought under human control: not an unmanageable flood of water, but even more alarming and malign explosions of energy that might disrupt the entire ecological system on which man's own life and welfare depend. The prime need of our age is to contrive channels for excessive energies and impetuous vitalities that have departed from organic norms and limits: cultural flood control in every field calls for the erection of embankments, dams, reservoirs, to even out the flow and spread it into the final receptacles, the cities and regions, the groups, families, and personalities, who will be able to utilize this energy for their own growth and development. If we were prepared to restore the habitability of the earth and cultivate the empty spaces in the human soul, we should not be so preoccupied with sterile escapist projects for exploring inter-planetary space, or with even more rigorously dehumanized policies based on the strategy of wholesale collective extermination. It is time to come back to earth and confront life in all its organic fecundity, diversity, and creativity, instead of taking refuge in the under-dimensioned world of Post-historic Man.

Modern man, unfortunately, has still to conquer the dangerous aberrations that took institutional form in the cities of the Bronze Age and gave a destructive destination to our highest achievements. Like the rulers of the Bronze Age, we still regard power as the chief manifestation of divinity, or if not that, the main agent of human development. But "absolute power," like "absolute weapons," belongs to the same magico-religious scheme as ritual human sacrifice. Such power destroys the symbiotic co-operation of man with all other aspects of nature, and of men with other men. Living organisms can use only limited amounts of energy. "Too much" or "too little" is equally fatal to organic existence. Organisms, societies, human persons, not least, cities, are delicate devices for regulating energy and putting it to the service of life.

The chief function of the city is to convert power into form, energy into culture, dead matter into the living symbols of art, biological reproduction into social creativity. The positive functions of the city cannot be performed without creating new institutional arrangements, capable of coping with the vast energies modern man now commands: arrangements just as bold as those that originally transformed the overgrown village and its stronghold into the nucleated, highly organized city.

These necessary changes could hardly be envisaged, were it not for the fact that the negative institutions that accompanied the rise of the city have for the last four centuries been falling into decay, and seemed until recently to be ready to drop into limbo. Kingship by divine right has all but disappeared, even as a moribund idea; and the political functions that were once exercised solely by the palace and the temple, with the coercive aid of the bureaucracy and the army, were during the nineteenth century assumed by a multitude of organizations, corporations, parties, associations, and committees. So, too, the conditions laid down by Aristotle for the abolition of slave labor have now been largely met, through the harnessing of inorganic sources of energy and the invention of automatic machines and utilities. Thus slavery, forced labor, legalized expropriation, class monopoly of knowledge, have been giving way to free labor, social security, universal literacy, free education, open access to knowledge, and the beginnings of universal leisure, such as is necessary for wide participation in political duties. If vast masses of people in Asia, Africa, and South America still live under primitive conditions and depressing poverty, even the ruthless colonialism of the nineteenth century brought to these peoples the ideas that would release them. "The heart of darkness," from Livingstone on to Schweitzer, was pierced by a shaft of light.

In short, the oppressive conditions that limited the development of cities throughout history have begun to disappear. Property, caste, even vocational specialization have—through the graded income tax and the "managerial revolution"—lost most of their hereditary fixations. What Alexis de Tocqueville observed a century ago is now more true than ever: the history of the last eight hundred years is the history of the progressive equalization of classes. This change holds equally of capitalist and communist systems, in a fashion that might have shocked Karl Marx, but would not have surprised John Stuart Mill. For the latter foresaw the conditions of dynamic equilibrium under which the advances of the machine economy might at last be turned to positive human advantage. Until but yesterday, then, it seemed that the negative symbiosis that accompanied the rise of the city was doomed. The task of the emerging city was to give an ideal form to these radically superior conditions of life.

Unfortunately, the evil institutions that accompanied the rise of the ancient city have been resurrected and magnified in our own time: so the ultimate issue is in doubt. Totalitarian rulers have reappeared, sometimes elevated, like Hitler, into deities, or mummified in Pharaoh-fashion after death, for worship, like Lenin and Stalin. Their methods of coercion and terrorism surpass the vilest records of ancient rulers, and the hoary practice of exterminating whole urban populations has even been exercised by the elected leaders of democratic states, wielding powers of instantaneous destruction once reserved to the gods. Everywhere secret knowledge has put an end to effective criticism and democratic control; and the emancipation from manual labor has brought about a new kind of enslavement: abject dependence upon the machine. The monstrous gods of the ancient world have all reappeared, hugely magnified, demanding total human sacrifice. To appease their super-Moloch in the Nuclear Temples, whole nations stand ready, supinely, to throw their children into his fiery furnace.

If these demoralizing tendencies continue, the forces that are now at work will prove uncontrollable and deadly; for the powers man now commands must, unless they are detached from their ancient ties to the citadel, and devoted to human ends, lead from their present state of paranoid suspicion and hatred to a final frenzy of destruction. On the other hand, if the main negative institutions of civilization continue to crumble—that is, if the passing convulsions of totalitarianism mark in fact the death-throes of the old order—is it likely that war will escape the same fate? War was one of the "lethal genes" transmitted by the city from century to century, always doing damage but never yet widely enough to bring civilization itself to an end. That period of tolerance is now over. If civilization does not eliminate war as an open possibility, our nuclear agents will destroy civilization—and possibly exterminate mankind. The vast village populations that were once reservoirs of life will eventually perish with those of the cities.

Should the forces of life, on the other hand, rally together, we shall stand on the verge of a new urban implosion. When cities were first founded, an old Egyptian scribe tells us, the mission of the founder was to "put the gods in their shrines." The task of the coming city is not essentially different: its mission is to put the highest concerns of man at the center of all his activities: to unite the scattered fragments of the human personality, turning artificially dismembered men—bureaucrats, specialists, "experts," depersonalized agents—into complete human beings, repairing the damage that has been done by vocational separation, by social segregation, by the over-cultivation of a favored function, by tribalisms and nationalisms, by the absence of organic partnerships and ideal purposes.

Before modern man can gain control over the forces that now threaten his very existence, he must resume possession of himself. This sets the chief mission for the city of the future: that of creating a visible regional and civic structure, designed to make man at home with his deeper self and his larger world, attached to images of human nurture and love.

We must now conceive the city, accordingly, not primarily as a place of business or government, but as an essential organ for expressing and actualizing the new human personality—that of "One World Man." The old separation of man and nature, of townsman and countryman, of Greek and barbarian, of citizen and foreigner, can no longer be maintained: for communication, the entire planet is becoming a village; and as a result, the smallest neighborhood or precinct must be planned as a working model of the larger world. Now it is not the will of a single deified ruler, but the individual and corporate will of its citizens, aiming at self-knowledge, self-government, and self-actualization, that must be embodied in the city. Not industry but education will be the center of their activities; and every process and function will be evaluated and approved just to the extent that it furthers human development, whilst the city itself provides a vivid theater for the spontaneous encounters and challenges and embraces of daily life.

Apparently, the inertia of current civilization still moves toward a worldwide nuclear catastrophe; and even if that fatal event is postponed, it may be a century or more before the possibility can be written off. But happily life has one predictable attribute: it is full of surprises. At the last moment—and our generation may in fact be close to the last moment—the purposes and projects that will redeem our present aimless dynamism may gain the upper hand. When that happens, obstacles that now seem insuperable will melt away; and the vast sums of money and energy, the massive efforts of science and technics, which now go into the building of nuclear bombs, space rockets, and a hundred other cunning devices directly or indirectly attached to dehumanized and de-moralized goals, will be released for the recultivation of the earth and the rebuilding of cities: above all, for the replenishment of the human personality. If once the sterile dreams and sadistic nightmares that obsess the ruling élite are banished, there will be such a release of human vitality as will make the Renascence seem almost a stillbirth.

It would be foolish to predict when or how such a change may come about; and yet it would be even more unrealistic to dismiss it as a possibility, perhaps even an imminent possibility, despite the grip that the myth of the machine still holds on the Western World. Fortunately, the preparations for the change from a power economy to a life economy have been long in the making; and once the reorientation of basic ideas and purposes takes place, the necessary political and physical transformations may swiftly follow. Many of the same forces that are now oriented toward death will then be polarized toward life.

In discussing the apparent stabilization of the birthrate, as manifested throughout Western civilization before 1940, the writer of "The Culture of Cities" then observed: "One can easily imagine a new cult of family life, growing up in the face of some decimating catastrophe, which would necessitate a swift revision in plans for housing and city development: a generous urge toward procreation might clash in policy with the views of the prudent, bent on preserving a barely achieved equilibrium."

To many professional sociologists, captivated by the smooth curves of their population graphs, that seemed a far-fetched, indeed quite unimaginable possibility before the Second World War. But such a spontaneous reaction actually took place shortly after the war broke out, and has continued, despite various "expert" predictions to the contrary, for the last twenty years. Many people who should be vigilantly concerned over the annihilation of mankind through nuclear explosions have concealed that dire possibility from themselves by excessive anxiety over the "population explosion"—without the faintest suspicion, apparently, that the threat of de-population and that of over-population might in fact be connected.

As of today, this resurgence of reproductive activity might be partly explained as a deep instinctual answer to the premature death of scores of millions of people throughout the planet. But even more possibly, it may be the unconscious reaction to the likelihood of an annihilating outburst of nuclear genocide on a planetary scale. As such, every new baby is a blind desperate vote for survival: people who find themselves unable to register an effective political protest against extermination do so by a biological act. In countries where state aid is lacking, young parents often accept a severe privation of goods and an absence of leisure, rather than accept privation of life by forgoing children. The automatic response of every species threatened with extirpation takes the form of excessive reproduction. This is a fundamental observation of ecology.

No profit-oriented, pleasure-dominated economy can cope with such demands: no power-dominated economy can permanently suppress them. Should the same attitude spread toward the organs of education, art, and culture, man's super-biological means of reproduction, it would alter the entire human prospect: for public service would take precedence over private profit, and public funds would be available for the building and rebuilding of villages, neighborhoods, cities, and regions, on more generous lines than the aristocracies of the past were ever able to afford for themselves. Such a change would restore the discipline and the delight of the garden to every aspect of life; and it might do more to balance the birthrate, by its concern with the quality of life, than any other collective measure.

As we have seen, the city has undergone many changes during the last five thousand years; and further changes are doubtless in store. But the innovations that beckon urgently are not in the extension and perfection of physical equipment: still less in multiplying automatic electronic devices for dispersing into formless sub-urban dust the remaining organs of culture. Just the contrary: significant improvements will come only through applying art and thought to the city's central human concerns, with a fresh dedication to the cosmic and ecological processes that enfold all being. We must restore to the city the maternal, life-nurturing functions, the autonomous activities, the symbiotic associations that have long been neglected or suppressed. For the city should be an organ of love; and the best economy of cities is the care and culture of men.

The city first took form as the home of a god: a place where eternal values were represented and divine possibilities revealed. Though the symbols have changed the realities behind them remain. We know now, as never before, that the undisclosed potentialities of life reach far beyond the proud algebraics of contemporary science; and their promises for the further transformations of man are as enchanting as they are inexhaustible. Without the religious perspectives fostered by the city, it is doubtful if more than a small part of man's capacities for living and learning could have developed. Man grows in the image of his gods, and up to the measure they have set. The mixture of divinity, power, and personality that brought the ancient city into existence must be weighed out anew in terms of the ideology and the culture of our own time, and poured into fresh civic, regional, and planetary molds. In order to defeat the insensate forces that now threaten civilization from within, we must transcend the original frustrations and negations that have dogged the city throughout its history. Otherwise the sterile gods of power, unrestrained by organic limits or human goals, will remake man in their own faceless image and bring human history to an end.

The final mission of the city is to further man's conscious participation in the cosmic and the historic process. Through its own complex and enduring structure, the city vastly augments man's ability to interpret these processes and take an active, formative part in them, so that every phase of the drama it stages shall have, to the highest degree possible, the illumination of consciousness, the stamp of purpose, the color of love. That magnification of all the dimensions of life, through emotional communion, rational communication, technological mastery, and above all, dramatic representation, has been the supreme office of the city in history. And it remains the chief reason for the city's continued existence.

选读 17

"回到古希腊广场"

威廉·霍林斯沃斯·怀特

在怀特《城市》一书的最后一章，他基于多年对纽约街道和其他城市的档案分析和直接观察，提出了一个现代美国城市核心区的复兴案例。怀特最后得出的结论是古希腊广场应该作为成功设计城市中心的典范。他主要关注的问题包括城市空间如何从功能和安全两方面提升城市居所的社会健康问题以及鼓励交流、公共讨论和社会变革的问题，这些内容贯穿全书。他对街道的拐角尤为入迷。因为街角作为人们日常交易、互相道别的场所无可比拟，也是在这里，街角承担了人们各种随意随性的举止行为。

## READING 17

## Return to the Agora

### William Hollingsworth Whyte

In this final chapter of his book, *City,* Whyte presents a case for reinvigorating the core of the modern American city based on many years of researching, by archival analysis and direct observation, the streets of New York and other cities. He has concluded that the agora of ancient Greece should be used as the model in designing the successful urban center. His primary focus throughout the book is how urban spaces promote the social health of their inhabitants—from the views of both function and safety—and also encourage conversation, public discourse, and social change. He is particularly fascinated with the street corner, a place that is incomparable for bargaining, saying goodbye, and a host of other informal, unplanned human activities.

Will the center hold?

    What you see can make you doubt it. Ride the freeways and you see the consequence of a weakening center. You see a mishmash of separate centers, without focus or coherence. Taken by themselves, some of the components are well done, but it is still a mishmash that they add up to. And it is hard to see how it can do anything but worsen.

    The decentralization trend that is sending the back-office work of the center to the suburbs is strengthening. The computers have already made the move. Cities can argue that it would make much more sense to locate these functions within the city, with its services and its transit network—in low-cost Brooklyn, for example, rather than Manhattan. Few corporations are buying the argument: it's either the center or the suburbs.

    The cities of the northeastern and north-central states seem to have been hit particularly hard. A succession of demographic studies have argued that they have had it,

*Source: City: Rediscovering Its Center* by William H. Whyte. Copyright © 1989 by William H. Whyte. Used by permission of Doubleday, a division of Bantam Doubleday Dell Publishing Group, Inc. pp. 331–42.

that they are "aging" and functionally obsolete, being geared to a declining manufacturing economy and with an overpriced labor force—and that, in any event, they are in the wrong latitude. The message is clear. Go to the South and the Southwest. The cities in those regions, runs the argument, are expanding vigorously, offer lower taxes, lower-cost housing, a more tractable labor force, and a quality of life unmatched by the cold North.

But the Sunbelt cities are having their problems too. One reason taxes have been relatively low is that they have been postponed, as has been investment in infrastructures. Oil revenues are no longer taking up the slack. The spectacular population growth of some of the cities has been due in large part to the annexation of neighboring communities, but cities are running out of places to annex. They have no mass transit to speak of and are as much hostage to the car and the freeway as they are their beneficiaries. Quality of life? Migrants from the rolling green landscapes of the North have some environmental surprises in store. Those mild winters have a price.

But regional comparisons need not be invidious. The presumed decline of the aging cities of the North is belied by some significant countertrends. To a large extent, the movement to the Southwest has been a movement of maturing products and processes. Many of the Southwest's manufacturing plants will soon age too: a year there is as long as it is anywhere else. In the meantime, some presumably over-the-hill cities up North have been doing surprisingly well.

But regional differences are not as important as regional similarities. As economist John Hekman has pointed out, it is not regions that grow old, but products. They have a life cycle of their own, and they go from innovation, through development, to maturity and standardization. At each stage the resources needed by the manufacturer change, and this can prompt him to make some large geographic moves.

This is what has been happening in the moves to the Southwest. Hekman points to the computer industry as an example. "Highly sophisticated products and production processes tend to be located in the main centers of technology, like Boston, New York, Minneapolis, and, more recently, Dallas and Palo Alto." The manufacturing of peripheral items and knockoffs of systems tend to be scattered elsewhere. They do not need high technology so much as lower production costs. The Southwest has the edge on lower production costs of standardized items. New England, however, has a strong edge in its technical and research base for the development of new processes. In the birthrate of new firms, Massachusetts is second only to California.

The Middle Atlantic states are not doing so well. State and local governments in such areas often concentrate on subsidy and restriction to keep old industries from moving out. Hekman thinks that this is the worst possible mistake and that what they ought to be concentrating on is the nurturing of new firms with a strong technological base. "The key to beating the product cycle and industrial migration," he says, "is not in keeping the industries at home, but in replacing them with new industries."

The work of economist David L. Birch of MIT points to similar conclusions. Through detailed studies of job creation in a cross section of American cities, he found that most cities lose about the same percentage of their job base each year—about 8 percent on average. This is a normal consequence of entrepreneurial activity. "The culprit in declining economies," says Birch, "is not job losses but the absence of new jobs to replace the losses . . . Development strategies aimed at holding a thumb in the dike would appear to be as futile as telling the tide not to go out."

My study of the migrations from New York City provides further documentation. There was not a great deal that the city could do to stop them. To a large extent, the causes of the companies' moves were internal, the consequence of cyclical changes within the company and its industry, and were independent of the city and its pros and cons. To say that such moves are normal is not cheering to the city's defenders, but it is not all bad news. As the market-valuation comparisons demonstrated, the companies that moved out tended to be lackluster performers. The companies that stayed in the city were outstanding performers.

The city, of course, does not want to lose any companies, mediocre or not, but it most probably will lose more. Yet there are favorable aspects to the turnover. The jobs the city is losing are predominately in the older and bigger companies, and the ones that are not doing very well are the ones most apt to pack up the whole company and leave town. The jobs the city is gaining, by contrast, are predominantly in the newer and smaller firms.

In a study of San Francisco's experience between 1972 and 1984, Birch found that small firms were creating the new jobs. Larger firms with 100 employees or more were shedding jobs. Company age was a factor too: firms less than four years old produced a net gain of 30,597 jobs over the period, but those twelve years old or more had a net loss of 13,382 jobs.

New York has been enjoying similar gains. In 1986 it has a net gain of 64,000 office jobs and about the same was expected for 1987. Some of the gain has come from the expansion of local firms, most markedly in financial services. This expansion has been so buoyant as to be unsettling and has absorbed most of the space vacated by the move-outs. Firms moving into New York, including several from the Sunbelt, have brought jobs. What in time may prove the most fruitful source are the jobs created through the start-up of entirely new companies.

Conventional economic development programs concentrate on hanging on to big firms with tax abatements and other defenses. Like Hekman, Birch believes this is bad strategy. It does not meet the needs of the growth sector, and it does not do much that is effective for the nongrowth sector. Characteristically, tax abatements come on strong in the boom periods, when they are least warranted. New York has been a patsy in this respect many times. Corporate gratitude for such favors, it has learned, is a nonfactor. But the game goes on. Sponsors of big office projects press for concessions and threaten to pack up and go to Stamford if they don't get them. If the city calls their bluff, they will probably stay. If they do go, it will serve them right.

The belief that major office projects are the prime source of new jobs dies hard. Most cities have assumed that they are and have equated office building construction with the city's economic vitality. But the growth is in firms that are priced out of the tower market. They need older, somewhat beat-up quarters off to the side but not too far from the center.

"Office tower developments," Birch argues, "is not job development. It may, indeed, be something of a deterrent. The kind of firms that can afford the $35 a foot are the ones that are not producing new jobs. To make way for the tower, however, older buildings are being demolished which are affordable by the small firms which do produce new jobs."

If the number of people who can afford the high rents of the towers is diminishing, how come so many new towers are being built? It is not because of any excess of demand. Office vacancy rates across the country have been rising for some time. The impetus for construction has been financial. It has been driven by a huge supply of investment capital. What the markets give they can take away.

New York City is especially vulnerable. One of the reasons it fared well in the creation of office jobs in the mid-eighties was the prodigious growth of the financial services sector. But what markets can help create they can take away. The contraction in jobs following the 1987 crash was severe. It could get much worse.

On balance, however, the shift of jobs and people has been favorable for cities. What is most encouraging is where the growth has been taking place. The gain in jobs has been greater than the loss of them, and the gains have been mainly in newer and smaller firms—the kind where future growth is most likely to come from.

It is true that the ablest, best-performing corporations are the ones which stay in the city and that those which leave tend to be the less able. But the city cannot rejoice over this. Ill advised or no, the defecting companies take a lot of jobs with them, and the psychological hurt is perhaps the most telling.

Even the corporations that stay are moving some jobs to the suburbs. They are the more routine jobs, but the city is understandably upset to see them go. And the movement will probably continue. Headquarters remaining in the city are getting leaner; in some instances, the headquarters consist of some office suites and a handful of top executives, with the great bulk of the organization out by the freeways.

With or without a downturn, the center city may start losing more office jobs than it is gaining. But this would not necessarily be a catastrophe. Only twenty years ago, critics were jumping on the city for having too many people in it, in the center most of all. A behavioral sink, they called it, and preached the horrors of urban density. If the city now loses some density, it hardly seems fair to whack it for that, as some observers do. They would have it both ways and interpret any additional outward movement as further evidence of the city's malaise—terminal malaise.

Will this come to pass? Let us return to the Route 1 corridor and its counterparts. As we have already noted, they are supplanting the center city for white-collar work. Will they supplant it as the center as well? Some people believe so. They see these new areas as the true wave of the future—not just another kind of suburb but a new city itself. A forceful expression of this diagnosis is Robert Fishman's *Bourgeois Utopias*. "In my view," he writes, "the most important feature of postwar American development has been the almost simultaneous decentralization of housing, industry, specialized services, and office jobs; the consequent breakaway of the urban periphery from a central city it no longer needs; and the creation of a decentralized environment that nevertheless possesses all the economic and technological dynamism we associate with the city." In sum, all the advantages of the center without it—the city without tears.

How can this be? What makes this best of both worlds possible, says Fishman, is technology—specifically, "the advanced communication technology which has so completely superseded the face-to-face contact of the traditional city." It has generated urban diversity without urban concentration.

But there has been a price: scatteration. Fishman concedes that the "technoburbs," as he calls them, are a bit of a mess and will probably remain so for some time. But he believes that eventually they will be set right by regional planning and advanced traffic technology.

A premise of this optimism is that there are a good many options still open. But are they really? It is easy to overlook how preemptive early development patterns can be. In most of the new growth areas the formative decisions have already been made. There is an analogy here to the residential development of the countryside. The early subdivisions often set the character of subsequent growth long before the main body of suburbanites had arrived. In many metropolitan areas the die was cast as early as the fifties. By then, the farmers had sold off their key frontage land on the county roads; the streams had been riprapped or buried in pipes; the wooded ridges had been shaved. The names of the subdivisions had foretold the future: they were customarily named after the natural features they were about to obliterate.

But there was much hopeful talk of shaping new growth patterns. The sixties was a time of "Year 2000" plans, which were full of bold possibilities to consider: enucleated growth points, linear developments, rings of satellite communities, and the like. At countless clinics and conferences, people weighed the pros and cons of the various alternatives. It made people feel good, these exercises in ordered choice, and they went about them as solemnly as though the choices were in fact there to make.

The most celebrated was the "wedges and corridor" plan for the Washington metropolitan area. It called for channeling growth into spokelike corridors and conserving the bulk of the land in great wedges of green open space. It was a bad plan actually, the key tracts of the green wedges having been bought up by developers. But the plan persisted for some time as the region's best hope, planners preferring to go to hell with a plan than to heaven without one.

It does not follow that the new growth areas cannot be helped by regional planning. It is very much needed, late in the game though it may be. But it is hard to conceive of any drastic changes in the patterns or the lack of them. Once the freeways are built, the interchanges sited, and the malls anchored in place, future choices are constrained and not very much can be done to change matters.

There are palliatives, to be sure: jug handle turnoffs at exits of research parks, overpasses to improve access to malls, new frontage roads, additional limited-access stretches, or entirely new highways. But such measures can be extremely costly and the application of them made the more difficult by the pervading lack of centrality. There is no dominant direction. The traffic flows go every which way, and the mediation of them calls for yet more concrete.

So there are problems, the decentralizers say. But is there any real alternative? Yes, there is: the town. It was invented several millennia ago and has persisted as a remarkably consistent and useful form.

Let me cite West Chester, Pennsylvania, my hometown. It has a number of the advantages of such towns. For one thing, the settlement pattern is compact and efficient. It was laid out in the classic grid bestowed on southeastern Pennsylvania by surveyor Thomas Holme. There is a fine and complete stock of housing, ranging from single-family detached to double houses and row houses.

All of the houses are within walking distance of the center of town. That does not mean people actually walk the whole distance. They are Americans. But they could if they had to.

Downtown is intact. Back in the 1950s the town fathers turned down urban renewal—whether from torpor or foresight, no one is sure. But as a result the center was spared demolition. A few individual buildings were torn down that should not have been: such as the Turk's Head Hotel, the town's original one. But thanks to some cranks and preservationists the best old buildings stand, and by standing, have gained a new functionalism. The banks are an example.

Out by the interchanges there are drive-in banks. West Chester has *walk*-in banks. You walk right in off the street. You don't need a car to gain admittance. The architecture of the two leading banks is contextual. They have white marble Greek temple fronts. They say *bank*.

There was a clear edge between town and country, a boon that was fated to disappear. Another portent was the construction by James Rouse of the Exton Square Mall, five miles north of town. For a while it looked as if it was going to knock off West Chester's downtown. Mosteller's Department Store closed down and so did several other stores.

Rouse had said that a town such as West Chester could compete very well with malls in specialty retailing and restaurants. That is the way it has worked out. Several specialty shops have been doing well, and one, Jane Chalfant's, is outstanding. There are good places to eat. The Quaker Tea Room on East Gay Street was taken over by a French couple and reopened as La Cocotte. The Borough Hall was converted into a restaurant. So was part of the old Sharples Separator Works.

Lawyers are the principal industry of this county seat. They are located one next to another and form a stable constituency for downtown. The Chester County Courthouse itself is quite declarative. Its clock steeple and the five-story Farmers and Mechanics Building are the high points of town. You see them from far away when you are on the eastern approaches and the sight of them is oddly reassuring.

I feel very much as Russell Baker did about his Main Street. "When you stood on Main Street," he wrote, "you could tell yourself, 'This is the center, the point on which all things converge,' and feel the inexplicable but nonetheless vital comfort that results from knowing where you stand in the world and what the score is . . . On the shopping mall, people know they are standing not at the center, but somewhere vaguely off toward the edge of a center that has failed to hold."

The sense of place that a town can give is most felt by those who live within it. But it is also important for those who live beyond. A well-defined town with a tight core can give coherence to a whole countryside. It's a better place to live *in* if there is someplace to go *to*.

There is also reason to expand a town on its periphery rather than leapfrog development way beyond it. The town is an efficient model. Had West Chester been extended in the fifties and sixties about two blocks on each of its sides, some thirty years of growth could have been taken care of handily. It was not. Political realities being what they are, the expansion required several hundred square miles of land and umpteen thousand miles of utilities.

I'm not suggesting that what happened was all wrong. Some of the developments were well handled, the cluster developments especially. The county did a number of things right, such as acquiring regional parks. Thanks to an outstanding easement program, the loveliest stretches of the Brandywine were saved from development. The fact remains, however, that had more of the expansion been contiguous to the town, the end result would have been more amenable as well as more economic. So would it have been for many other towns across the country.

And so it could be in the new growth areas too. While it is too late for major shifts in direction, the application of the town principle could help tighten up a pattern that badly needs it. There has been some progress along these lines, as in the creation of combined shopping, hotel, and office centers at Forrestal Village. But the best precedent, as I noted earlier, has been Bellevue, Washington, and the transformation of an incipient office park into a city.

For those who see the suburban growth areas as the cities of the future, the key word is "technology." They have a point, but it is one that should be stood on its head. It is the genius of the center city that it is *not* high-tech. What is remarkable, indeed, is how little technology it does use. There are elevators, telephones, Xerox machines, and air-conditioning. But that is about it. The really fancy stuff is out on the periphery.

Socially, the city is a very complex place. Physically, it is comparatively simple. For the business of the center, it must have streets, buildings, and places to meet and talk. As far as essentials are concerned, it has little more than the agora of ancient Greece.

I am going to quote from R. E. Wycherley's fine study of the agora. It is not straining an analogy to see in the history of the agora some lessons for cities today. The parallels are considerable, especially so in the surprising turn the agoras finally took.

It began in a simple way: "A fairly level open space was all that was needed," writes Wycherley. "A good water supply was important and satisfactory drainage. A roughly central site was adopted, since the agora had to provide a convenient focus for city life in general and for the main streets . . . The same free space sufficed for all kinds of purposes. Here the people could assemble to be harangued; the only equipment needed was some sort of tribune for the speakers, and possibly seats for men of dignity."

As time went on, buildings were added. There would be a council house for magistrates. The stoa, or open colonnade, served as a general-purpose structure and eventually became the frame for rows of shops. The agora was a good place through which to amble: there were rows of trees for shade (plane trees mostly) and a number of convivial places at which to stop (the fountain house, for example, or the wine shops).

"No clear line," writes Wycherley, "was drawn between civic centre and market. The public buildings and shrines were in the agora; meat and fish and the rest were sold in the same agora." Booksellers had their stalls, and so did bankers, known as men-with-tables. "Marketing 'when the agora was full,' i.e., in the morning, must have been a noisy and nerve-racking business. The fishmongers had a particularly bad reputation; glared at their customers; asked exorbitant prices."

The agora was a sociable place. People would drift from spot to spot, pausing to chat under the plane trees or by one of the fountains. To some observers—Aristophanes, for one—these were vulgar, undesirable fellows. He saw no good purpose served whatsoever by "idlers," such as Socrates. Nor did Aristotle. Like some modern planners, he wanted the various functions separated out and contained. He recommended that there be two agoras—one for ordinary commerce and the other of a religious nature, free of idleness and vulgar activity.

"But the Greeks thoroughly mixed up the elements of their lives," Wycherley points out, "and for better or worse this fusion is clearly seen in the agora." It is seen physically too, for the agora was part of the street network of the city; it was not enclosed or segregated from the rest of the city but vitally linked with it.

Then, about the third century A.D., the agora began to lose its centrality. With the advent of the enclosed peristyle court, the agora was secluded from the city around it, eventually with complete enclosure. "There was a greater tendency," says Wycherley, "to plan the main agora square as a whole on this principle, to make it an enclosed building turning in on itself. City life had lost something of its old quality, and the agora had a less vital part to play, a less intimate relation with all the varied activities of the community . . . When the agora became a mere building, however grand, this meant a certain disintegration of the city."

The withdrawal and seclusion of the agora is uncomfortably similar to the direction so many cities have been taking. It does not follow that they, too, will decline, but the warning signals are worth heeding. As in Greece, many a city is moving its key public space out of the street system; it is moving the space away from the true center, putting walls around it, and making it a structure turned in on itself. For good measure, cities are adding two separators the Greeks did not have, concourses and skyways. The only encouraging note is how badly the megastructures have fared. A few more bankruptcies among such projects, and cities might conclude that they were doing something wrong.

The agora at its height would be a good guide to what is right. Its characteristics were centrality, concentration, and mixture, and these are the characteristics of the centers that work best today. Physically, there are vast differences, but in the gutty, everyday life of the street they would probably be remarkably similar. I would give anything to be able to mount a time-lapse camera atop a stoa and film the life of the agora. I would be especially interested to see if there were a considerable number of 100 percent conversations in the middle of the pedestrian flow. I would think there would be; people do not stand that way except in places where they feel comfortable. In such places, the idle gossip that so annoyed Aristotle and annoys many people today often becomes vulgar, noisy, and argumentative, but it is the true currency of the city—word of mouth.

Is word of mouth obsolete? Electronics has not dampened our love of talking. Go out to the corporate headquarters that relocated in suburbia, and you will find plenty of word of mouth there. They would go balmy if there weren't. The only trouble is that it is primarily company people talking to company people. To the disappointment of these companies, nowhere near as many people have been coming out to see them as they expected.

Increased communications and travel have not obviated face-to-face interchange; they have stimulated it. There are no reliable figures on the point, but there does seem to have been a large increase in conferences, meetings, focus groups, sensitivity workshops, and other forms of talk, the whole serviced by a league of expediters and facilitators. Outlying locations are often favored. Many meetings are convened at airport motels, which shows you how far people will go to be face-to-face.

But the city is still the prime place. It is so because of the great likelihood of *unplanned*, informal encounters or the staging of them. As I have noted, street corners are great places for this. They are also great for bargaining, with no party having an advantage over the others. I have watched scores of executive groups going through their elaborately casual postlunch good-byes, and I have to marvel at the deftness with which someone will finally get them all to the point. Elevator lobbies are good places too; so are clubs and restaurants, which provide excellent opportunities for working the room.

It is because of this centrality that the financial markets have stayed put. It had been widely forecast that they would move out en masse, financial work being among the most quantitative and computerized of functions. A lot of the back-office work has been relocated. The main business, however, is not record keeping and support services; it is people sizing up other people, and the center is the place for that. This is true of most of the major financial markets of the world. With few exceptions, they remain right where they started out.

The problems, of course, are immense. To be an optimist about the city, one must believe that it will lurch from crisis to crisis but somehow survive. Utopia is nowhere in sight and probably never will be. The city is too mixed up for that. Its strengths and its ills are inextricably bound together. The same concentration that makes the center efficient is the cause of its crowding and the destruction of its sun and light and its scale. Many of the city's problems, furthermore, are external in origin—for example, the cruel demographics of peripheral growth, which are difficult enough to forecast, let alone do anything about.

What has been taking place is a brutal simplification. The city has been losing those functions for which it is no longer competitive. Manufacturing has moved toward the periphery; the back offices are on the way. The computers are already there. But as the city has been losing functions it has been reasserting its most ancient one: a place where people come together, face-to-face.

More than ever, the center is the place for news and gossip, for the creation of ideas, for marketing them and swiping them, for hatching deals, for starting parades. This is the stuff of the public life of the city—by no means wholly admirable, often abrasive, noisy, contentious, without apparent purpose.

But this human congress is the genius of the place, its reason for being, its great marginal edge. This is the engine, the city's true export. Whatever makes this congress easier, more spontaneous, more enjoyable is not at all a frill. It is the heart of the center of the city.

Let me append a methodological note.

I have tried to be objective in this book, but I must confess a bias. In comparing notes with fellow observers, I find that I share with them a secret vice: hubris.

Observation is entrapping. It is like the scale models architects beguile you with; start lifting off the roofs and you gain a sense of power. So it is with the observation of a place: once you start making little maps of it, charting where people come and go, you begin to possess the place. You do not possess it, of course. The reality continues to exist quite independent of you or any thoughts you may project onto it. But you *feel* you possess it, and you can develop such a proprietary regard for it as to become pettily jealous if anyone else arrogates it.

A further temptation beckons. As time goes on, you become familiar with the rhythms of the various street encounters: 100 percent conversations, prolonged goodbyes, reciprocal gestures, straight man and principal. Now you can predict how they are likely to develop and, by predicting them, get the sense that you are somehow causing them as well. They are your people out there. Sheer delusion, of course, but there is nothing so satisfying as to see them all out there on the street doing what you expect they should be doing.

Three men at the corner are in a prolonged goodbye. One of them is slowly rocking back and forth on his heels. No one else is. At length, the man stops rocking back and forth. I chuckle to myself. I know that in a few moments another of the men will take up the rocking motion. Time passes. More time passes. No one budges. More time passes. At length, one of the men shifts his weight; slowly, he begins rocking back and forth on his heels. I am very pleased with myself.

**BIBLIOGRAPHY**

Bacon, Edmund. *The Design of Cities.* New York: Viking Press, 1967.
Birch, David L. *Job Creation in Cities.* Cambridge, MA: MIT Press, 1981.
Cooper-Hewitt Museum. *Cities: The Forces That Shape Them.* New York: Rizzoli, 1982.
Dubos, Rene. *So Human an Animal.* New York: Charles Scribner's Sons, 1968.
Fishman, Robert. *Bourgeois Utopias.* New York: Basic Books, 1987.
Hekman, John S. "Regions Don't Grow Old; Products Do." *New York Times,* November 5, 1979.
Jackson, J. B. *The Necessity for Ruins.* Amherst, MA: University of Massachusetts Press, 1980.
Price, Edward T. "The Central Courthouse Square in the American County Seat." *Geographical Review,* January 1968.
Redmond, Tim, and David Goldsmith. "The End of the High-Rise Jobs Myth." *Planning,* April 1986.
Sitte, Camillo. *City Planning According to Artistic Principles.* New York: Random House, 1965.
Thompson, Homer A., and R. E. Wycherley. *The Athenian Agora.* Princeton, NJ: American School of Classical Studies at Athens, 1972.
Tillich, Paul. Quoted in *The Metropolis in Modern Life,* ed. R. M. Fisher. Garden City, NY: Doubleday & Co., 1955.
Wycherley, R. E. *How the Greeks Built Cities.* Garden City, NJ: Doubleday Anchor, 1969.

**THE URBAN ENVIRONMENT—SUGGESTED READINGS**

Anderson, Stanford, ed. *On Streets.* Cambridge, MA: MIT Press, 1978.
Barnett, Jonathan. *The Elusive City.* New York: Harper and Row, 1986.
Le Corbusier. *The City of Tomorrow and Its Planning.* New York: Dover, 1987.
Cullen, Gordon. *Townscape.* New York: Reinhold, 1968.
Halprin, Lawrence. *Cities.* Cambridge, MA: MIT Press, 1972.
Hayden, Dolores. *The Power of Place: Urban Landscape as Public History.* Cambridge, MA: MIT Press, 1995.
Lynch, Kevin. *The Image of the City.* Cambridge, MA:MIT Press, 1960.
Moholy-Nagy, Sibyl. *Matrix of Man.* New York: Praeger, 1968.
Mumford, Lewis. *Sticks and Stones.* New York: Dover, 1955.
Rossi, Aldo. *The Architecture of the City.* Cambridge, MA: MIT Press, 1982.
Tunnard, Christopher. *The City of Man.* New York: Charles Scribner's Sons, 1970.

# 5

# THE NATURAL ENVIRONMENT
# 自然环境

选读 18
"美国的公共空间"
约翰·布林克霍夫·杰克逊

约翰·布林克霍夫·杰克逊在文章中描绘了如何创造一种"场所感",即通过每日的体验、自然环境与建成环境的整体性来共同进行创造。在这篇选读中,他调查了美国景观中公共场所的概念,展现了变迁之中的社会与文化标准是如何改变美国人看待和利用开放空间的方式的。他也就特定的"公共空间"概念进行了探讨。这种"公共空间"是在特别强调私有控制权的文化背景之下,存在于大量消耗自然景观的国家之中的。杰克逊认为应该把公共空间看作一种能够创造多元分化的社会场所。一个多元分化的社会有利于人们寻求心灵的慰藉,也有利于增强人们对那些有共鸣体验的情感的传递。他在文中说道:"我们在鼓励创造把娱乐、自然知识和我们的历史结合在一起的各种空间上一直做得很好。像这样的空间能够流行于世,便是它们存在的最好理由。"

## READING 18

### The American Public Space
**John Brinckerhoff Jackson**

J. B. Jackson wrote about the ways that everyday experiences and the totality of natural and built environments act together to create a "sense of place." In this selection he investigates the concept of public places in the American landscape and shows how shifting social and cultural norms have transformed the ways Americans view and use open space. He also discusses the notion of "public space" in the context of a culture that places so much emphasis on private ownership in a country that has consumed so much of the natural landscape.

*Source:* "The American Public Space," by J. B. Jackson. Reprinted with permission of *The Public Interest,* 74 (Winter 1984), pp. 52–65. Copyright © 1984 by National Affairs, Inc.

Jackson argues for the ways that public space should be viewed as places that create a venue for a diverse and fragmented society to seek solace and enhance the transmission of a common set of experiences. He states: "We do well to encourage the creation of all such spaces combining recreation with knowledge of nature and our past. Their popularity is their best reason for existing."

Those of us old enough to remember what America was like a half century ago have lived through a significant but largely unnoticed development in our landscape. By that I mean not the growth of our cities but a development that came about largely as a consequence of that growth: the great increase in the number and variety of public places all over the nation.

A public place is commonly defined as a place (or space) created and maintained by public authority, accessible to all citizens for their use and enjoyment. This tells us nothing about the different ways in which we use and enjoy them, nor about the different types of public involved, but we have only to look about us to see that they are often outside the center of town and even in the open country. Many have an educational purpose: historic zones, outdoor museums, botanical gardens. And more and more spaces are being designed to give us a brief experience of nature: hiking trails and wilderness areas and beaches, for example. When we include among the newer public spaces the parking lot, the trash disposal area, and the highway, it is evident that the public is being well provided for, not only as far as places for enjoyment are concerned, but for their use as well.

Implicit in the word "public" is the presence of other people. We know better than to resent that presence; they have as much right to be there as we have. Just the same, it is characteristic of many modern public spaces that contact between persons is likely to be brief and noncommittal. Indeed, when the public is too numerous we are made uncomfortable. We did not come here for what an earlier generation called "togetherness," we came for an individual, private experience—a sequence of emotions, perceptions, sensations, of value to ourselves. This is not to say that we are unfriendly, merely that we do not necessarily associate every public place with social intercourse. We assume, in fact, that there are special places appropriate for that. Yet when we look for them today we find they are few.

## CIVIC SPACE

Much has changed in America since the time when *every* public space was intended to be the setting for some collective, civic action. I think it can be said that beginning in the eighteenth century every public space—every piece of land controlled by the authorities—was meant to serve a public institution rather than to serve the public as an aggregate of individuals. In the newer planned towns of New England an area of public land was set aside for the support of the local church and its preacher, though not for public use. Section 16 in the townships created by the National Land Survey of 1785—the only designated public space in the township—was to support a local school. Hence its name: school section, still a feature of the western landscape. Communities more urban in character recognized the need for public spaces that the public could use; but only for the benefit of the community at large. The newly created towns of the late-eighteenth and early-nineteenth century almost invariably contained well-defined public places for the market, the drill field, the wharf, the "established" church, as well as places for a college or academy, and of course for public celebrations and public assembly. A civic function characterized them all; people were present in them to perform some public service or play some public role.

Public is a word without mystery: It derives from the Latin *populus,* and means belonging to or characteristic of the people. A public space is a people's space. But "people" as a word is less obvious. With us it simply means humanity, or a random sample of humanity, but until well into the nineteenth century it meant a specific group: sometimes the population of a nation or a town, sometimes the lowest element in that population, but always an identifiable category. Thus a common phrase in England was "the nobility, the gentry, and the public." People in this sense implied an organization and a territory; and as an organization it had an organizing or form-giving authority.

Perhaps it can be said that, as a noun, "public" implied the population, or the people, while as an adjective it referred to the authorities. Thus a public building in the eighteenth century was not a place accessible to all, for their use and enjoyment, but was the working or meeting place of the authorities.

This strictly political meaning of the word helps us interpret the kind of public place or square found in most of the new towns and cities in colonial and post-colonial America. The invaluable collections of town plans in three books by John Reps show in some detail how early planners and promoters emphasized the importance of public places.[1] Each community, each town was given a grid plan or a variation on a grid, and though many of them were little more than a cluster of square blocks, invariably there was a symmetrical array of (proposed) public buildings on a piece of public land—courthouse, market, jail, etc.

It is interesting to see how faithful colonial and frontier America was to the early Renaissance practice of according more dignity to the building than to the square in front of it. Our own perception, of course, is the opposite: In our love of open spaces we see the building as "facing" the square, the square as the focal point in the urban composition. The eighteenth-century belief was that the square, however large and imposing, derived its dignity from its association with the building, and was in fact merely the place where the inhabitants gathered to pay homage to the authorities within. Many town maps indicate proposed smaller squares, inserted into what would eventually be built-up residential areas. These were undoubtedly meant to be surrounded and dignified by public buildings, as William Penn had proposed for the four minor squares in his plan for Philadelphia. It was only in the nineteenth century that these small squares were seen and treated as parks, and this change from concentrating on the public building to concentrating on the public space would eventually produce public places bearing little or no aesthetic relationship to their urban surroundings.

Few of these neo-classical towns ever grew to resemble their paper prototypes. Many grew in an entirely unpredicted way: into monolithic compositions of identical blocks, as in Chicago, which are ideally suited to the purposes of the real estate

---

[1] John Reps, *The Making of Urban America* (Princeton: Princeton University Press, 1965); *Tidewater Towns* (Williamsburg, VA: The Colonial Williamsburg Foundation, 1972); *Cities of the American West* (Princeton: Princeton University Press, 1979).

speculator. Others did not grow at all. But their plans are no less interesting for that. They are diagrams, provincial and greatly simplified, of what Americans wanted in the way of towns: a rational, egalitarian, political ordering of spaces and structures, a sharp division between public and private, so that spaces for recreation or non-political, non-civic functions were left to the private sector to provide as best it could. The only truly public, or people's, space was the large central public square where all qualified citizens came together: a vast architectural roofless room, a stage, where all acted out their familiar assigned roles. And this, I think, is what really distinguished the traditional public space from our contemporary public space. Two centuries ago, despite the Revolution, it was still widely believed that we were *already* citizens (to the extent that we could qualify) when we appeared in public. We knew our role, our rank and place, and the structured space surrounding us merely served to confirm our status. (Much as in certain denominations we are members of the community of Christians from the moment of baptism; our subsequent participation in certain rites within the church simply *confirms* our permanent religious status.) But now we believe the contrary—that we *become* citizens by certain experiences, private as well as public. Our variety of new specialized public spaces are by way of being places where we prepare ourselves—physically, socially, and even vocationally—for the role of citizen.

## DECLINE OF THE SQUARE

There are still many among us who hold that our public spaces should perform the same civilizing, political role. Nevertheless, as the nineteenth century progressed it became evident that the public square was losing prestige. During the Revolution the centers of popular excitement had been Faneuil Hall, Independence Hall, and the New York Common. But when political oratory and political demonstrations went out of fashion, the public began to frequent the busier streets, the tree-lined promenades, and the waterfront. Other developments gave the centrifugal movement further impetus. Middle-class families felt the attraction of the suburbs or of the Independent homestead. Newcomers, many from overseas, had no sentiment for the established customs, and in newer parts of town evangelical churches competed with the established church in the center. The coming of the railroad and the factory and the mill shifted leisure time activities to the less built-up outskirts of town; and finally the public buildings themselves ceased to be the locus of real power, and gradually became office buildings for the bureaucracy. In almost every American town the traits Sam Bass Warner has identified with mid-nineteenth century Philadelphia became more and more prevalent:

> The effect of three decades of a building boom . . . was a city without squares of shops and public buildings, a city without gathering places which might have assisted in focusing the daily activities of neighborhoods. Instead of subcenters the process of building had created acres and acres of amorphous tracts—the architectural hallmark of the nineteenth and twentieth century American big city. . . . Whatever community life there was to flourish from now on would have to flourish despite the physical form of the city, not because of it.[2]

---

[2]Sam Bass Warner Jr., *Private City* (Philadelphia: University of Pennsylvania Press, 1968), p. 55.

Though the older towns and cities of the East retained their tradition of central park spaces, towns laid out in the Midwest and throughout the Great Plains were predominately grid plans of uniform blocks. Some of them, judging from Reps's *Cities of the American West,* provided for one or two "public squares" or "public grounds," yet few of these were centrally located and it is hard to find any of them associated with a public building. In many cases the larger spaces set aside for parks were soon subdivided into building lots. No doubt the reluctance to plan for public spaces in the potentially valuable downtown can be ascribed to the proprietor's eagerness to make money in downtown real estate. But by cutting straight through the average small American town and establishing its station and freightyards near main street, the railroad transformed the traditional center in ways that the railroad in Europe was never allowed to do: There the station was exiled to the outskirts. Thus the American town very early in the game developed a substitute social center (for men only) around the station and freightyards—a combination skidrow, wholesale district, and horse transportation complex that seems to have offered a variety of illicit and lower class attractions, as well as being a center for news. But this was a poor substitute for the traditional urban space where citizens could forgather and talk, a place described by the anthropologist R. Baumann "as special, isolated from others and enjoyed for its own sake, because talking there may be enjoyed for its own sake and not as part of another activity or for some special instrumental purpose." And so strong was the urge to have such places easily accessible yet detached from the workaday world that the American urban public, or a fraction of it, soon discovered a new and agreeable space on the outskirts of town—a favorite spot for relaxation and sociability. That was the cemetery.

## THE RECREATIONAL CEMETERY

The story of the development of the so-called rural cemetery in America is familiar to anyone who knows our urban history. It began in 1831 with the designing of Mt. Auburn Cemetery in Cambridge, Massachusetts, as a picturesque landscape of wooded hills, winding roads with paths, and rustic compositions of lawn and stream with pleasant views over the Charles, all in the style of the landscaped gardens fashionable in England at that time. This new type of cemetery immediately became a popular goal of excursions from the city. To quote Norman Newton's account, Mt. Auburn "soon became very popular as a quiet place in which to escape the bustle and clangor of the city—for strolling, for solitude, and even for family picnics. Following its success other cemeteries of the same type began to spring up."[3]

These rural cemeteries, usually located within easy reach of the city, attracted thousands of pleasure-seeking visitors, both before the presence of graves and tombstones, and after. A. J. Downing estimated that more than 50,000 persons visited Greenwood Cemetery in Brooklyn in the course of nine months in 1848. They came on foot and in carriages. Guidebooks in hand, they admired the monuments and the artistic planting and the views. They wandered along the lanes and paths, and rested on the expanses of lawn, sketched, ate lunch, and even practiced a little shooting. They had discovered a kind of recreation that the city had never offered.

---

[3] Norman Newton, *Design on the Land* (Cambridge, MA: Harvard University Press, 1971), p. 268.

The generally-accepted explanation for the popularity of the rural or picturesque cemetery is that it satisfied the new romantic love of nature. No doubt this had something to do with the enthusiasm, but the existence of widespread nature romanticism among working class Americans—or indeed among working class Europeans—has yet to be established. What evidence we have is largely literary, and like our contemporary environmentalism, nature romanticism seems to have been essentially a middle-class movement. In the writings of A. J. Downing, one of the most influential exponents of romanticism in architecture and landscape architecture, there are frequent suggestions that a taste for the romantic was peculiar to persons of refinement and wealth. In fact we now know enough about the fashion to recognize that a carefully-designed and well-executed picturesque landscape park called for considerable skill. It was not a "natural" space, it was (in the hands of the artist-designer) a highly structured space, a "painterly" composition whose rules and techniques were inspired by established landscape painters. It was Olmsted who best understood the canons of the picturesque and who applied them on a grandiose scale in Central Park and Prospect Park. To the average urban working-class American, relaxing in a rural cemetery, the appeal of the new landscape was quite different from that of Downing's well-to-do patrons. He was less aware of the subtleties of romantic composition, or even of the possibility of a direct contact with nature, than of the apparent *lack* of structure. The informal landscape offered the delights of spontaneous contact with other people in a setting that in no way prescribed a certain dress code or a certain code of manners. It was a public space of a novel kind: full of surprises, where emotions and pleasures were fresh and easily shared. It was not simply another artificial space; it was an *environment,* a place for new, primarily social, experiences. It represented the rejection of structure, the rejection of classical urbanism with its historical allusions, and the rejection of architectural public space.

As further evidence that nature romanticism had little or nothing to do with the acceptance of the rural cemetery, we might consider another, less familiar example of the popular American preference for unstructured public places: the camp meetings or revivals, the evangelical gatherings which were numerous in rural and frontier communities through the first half of the nineteenth century. Each of them attracted hundreds of men and women and children, black as well as white, and each lasted several nights and days. Almost always they had a forest background. Yet despite the wilderness setting and the prevailing emotionalism, it would be hard to find any trace of nature awareness in the proceedings, nor in their religious experiences. On the contrary: The forest seemed to free men and women from *any* environmental influences. Again, it was the lack of structure, the lack of behavioral design that produced the exhilaration.

## OLMSTED'S ISOLATED ART-WORKS

The park movement did not evolve out of the Great Revival, but out of the rural cemetery. It was a remarkable instance of how quickly and effectively Americans can respond to a humanitarian need—in this case the need for agreeable, healthy, and beautiful places where the urban population could enjoy itself and (of course) have contact with nature. In 1851 the New York legislature authorized the city to acquire some 840 acres for a public park. Seven years later Frederick Law Olmsted had won the design competition and work had begun on Central Park.

Ten years after that, despite the intervening war, there was not a major American city without a rural park, or the prospect of one, and many of those parks had been designed by Olmsted and his associates. Though at the beginning there were expressions of disapproval—the park would be taken over by rowdies, it was too large, it lacked the more formal qualities of the European royal parks—it was not long before parks in general were accepted as invaluable from the point of view of health, of innocent recreation, and as antidotes for the crowded and filthy city slums. Innumerable rural parks were created in smaller towns, and in the new towns of the West. It would be impossible to identify the number of college campuses, courthouses and institutional landscape designs, to say nothing of the landscaped cemeteries, that helped beautify communities throughout the country. Few of these spaces were designed by professionals. Most were the work of local amateur gardeners, and the transformation of many New England commons into neat little parks was frequently done by local women's organizations. For the most part these smaller, unpublicized parks were of no great artistic worth. They had their small lakes, their bandstands, their pretzel paths, and a monotony of elms or cottonwoods, but it must be said that they kept alive the civic tradition of public spaces at a time when the great Olmsted parks were fighting it. In those provincial parks political orators addressed the voters, band concerts were given, ethnic pageants were organized, and patriotic flowerbeds were admired. In the 1890s Frank Waugh, the landscape architect, described the typical western park as containing "race tracks, baseball grounds, camp meeting stands, carp ponds, fountains or fences."

It was this indiscriminate mixture of uses that horrified the readers of *Garden and Forest,* the organ of the Olmsted school of landscape design. For by the end of the century there had developed a very self-assured set of standards for the design of rural parks, and the most fundamental rule was that the "primary purpose of a rural park within reach of a great city is to furnish that rest and refreshment of mind and body which come from the tranquilizing influence of contact with natural scenery." This implied two restrictions: First, no building of any kind was to be erected in the park, nor (in the words of Olmsted's son John) "formal gardens, statuary, conservatories, botanical or zoological gardens, concert groves, electric fountains or the like; also popular athletic grounds, parade grounds, ball grounds for boys and facilities for boating and bathing." These installations were not themselves objectionable, and could well have been harmonized with "contact with nature." But the park was *not* nature, or a "natural environment." It was *scenery,* a whole landscape where the visitor could wander for hours over meadows and through woods and next to lakes and streams. For such was Olmsted's ideal: The park as a three-dimensional work of art. This in turn meant that the park should be visibly isolated from the surrounding city, enclosed by an impenetrable wall of greenery, so that the outside urban world would never impinge on the "rural" landscape, or on the experience of those visitors seeking rest and "refreshment of mind and body."

## OXYGEN AND VIRTUE

On aesthetic as well as on demographic grounds, there were reasons for this uncompromising isolation. But remoteness also promoted what Downing, Olmsted, and others had always considered the true role of the rural park: the physical and moral regeneration of the individual visitor. From the very beginning of the park movement there had been frequent references to the elevating influence that the rural park would have. "No one who has closely observed the conduct of the people who visit the park," Olmsted wrote, "can doubt it exercises a distinctly harmonizing and refining influence upon the most unfortunate and lawless classes of the city—an influence favorable to courtesy, self-control and temperance." To supplement this influence, Olmsted created a special park police force to control misconduct, including walking on the grass. But he also relied on the force of example to give the poor "an education to refinement and taste and the mental and moral capital of gentlemen." This was to come from observing and emulating the manners and behavior of upper-class visitors to the park. As Thomas Bender notes, "Olmsted's generation saw no difficulty in recommending that Central Park, their symbol of the democratic community, be surrounded by elegant private villas that would exert an elevating influence upon the masses who visited the park."[4]

The early vision of the rural park and its function survived intact for no more than fifty years. One of its original objectives had been the improvement of the health of city dwellers. But medical science soon proved that this was not simply a matter of fresh air and contact with nature; it was a matter of training, and the park was the appropriate place for such training. "Foul air prompts vice, " said a New York physician, "and oxygen to virtue as surely as the sunlight paints the flowers of our garden. . . . The varied opportunities of a park would educate [the slum child] and his family in the enjoyment of open-air pleasures. Deprived of these, he and his are educated into the ways of disease and vice by the character of their surroundings."

In the 1880s the well-organized playground movement, which had started by providing small playgrounds in the slums, demanded access to the park, and at much the same time a public eager for places to play various outdoor games brought pressure on the city to provide appropriate space. "It was easier to persuade the city fathers to make use of existing parks than to purchase additional land for recreation. . . . In some cases the introduction of recreational facilities was achieved with intelligence and in conformity with the original park design. In other instances the results were detrimental to the former park purpose."[5]

At the time of Olmsted's death in 1903 the park had largely ceased to be an environment in which the individual could enjoy solitary contact with nature, and had become an environment dedicated to guidance in recreation, health, citizenship, and nature knowledge. It was often crowded with cultural and recreational facilities, group activities, and increasingly populated by professional recreationists, playground supervisors, and leadership counselors. Even the definition of the park underwent drastic revision: It is now described in official documents as an open space, containing public facilities and with the appearance of a natural landscape. It is merely one element in a nation-wide ecosystem.

---

[4] Thomas Bender, *Toward an Urban Vision* (Lexington, KY: University Press of Kentucky, 1975), p. 179.
[5] George Butler, "Change in the City Parks," *Landscape* 8 (Winter 1958–59).

## THE LANDSCAPE OF DIVERSITY

In retrospect, how unpredictable and how extraordinary was the change over a period of less than a century in the American concept of public spaces! The neo-classical square, in part surrounded by public buildings, and located in the heart of the city, had been the symbol of political status, recognized by all. In the 1860s, however, the rural park began to replace it in the public perception, and the rural park not only rejected any contact with architecture and formal urbanism, but was located as far as possible from workaday activities, as if to say: Here is the true center, the place where nature establishes the laws.

But perhaps the most dramatic contrast between the two public spaces was in their respective definitions of community. The neo-classical square implied a body of people, politically and socially homogeneous, inhabiting a well-defined political territory with clear-cut class divisions. The visitors to the rural park came singly, each in pursuit of an individual contact with nature, a private experience. The ideal romantic community was the garden suburb, where reverence for the environment was the only common bond, and anything like an urban or political center was discouraged. Where can such a center be seen in Olmsted's design for Riverside, the Chicago suburb? It is a pleasant tangle of curving roads and lanes where all residences are isolated by greenery. Community did not mean homogeneity or uniformity, it meant diversity, and nothing was more gratifying to Olmsted than to see a diverse public in Central Park. Again and again he described "the persons brought closely together, poor and rich, young and old, Jew and Gentile." But this was a diversity of *individuals*. When the various recreational activities invaded the park in the 1880s, what ensued was a diversity of *groups:* age groups, ethnic groups, sports groups, neighborhood groups. That was quite another kind of diversity, and the spaces occupied by these groups, if only temporarily, constituted so many public places in the strictest sense: places where like-minded people came together to share an identity.

The emergence of these hitherto non-existent groups was probably the greatest contribution of the rural park. Long after the old-fashioned solitary pursuit of the contact with nature had vanished from the scene, these miniature societies with special identities continued to flourish and to acquire increasing public recognition over the years. Eventually they expanded beyond the park, and I think it was the automobile that encouraged this dispersal.

Though it is common practice to blame the automobile for having destroyed many territorial communities—particularly rural ones—the car has made it possible for us to come together over great distances and in a shorter length of time than ever before and this in turn has made possible the creation of many new and different public spaces throughout the landscape. Without the automobile countless recreational areas, monuments, national parks, to say nothing of remoter sections of our cities, would never have become part of those experiences which Americans always pursue. Quite aside from what they teach us, they serve as way-stations, as it were, in our American, essentially Protestant, pilgrimage of self-perfection through endless education.

We do well to encourage the creation of all such spaces combining recreation with knowledge of nature and of our past. Their popularity is their best reason for existing. Yet many of us are aware that another, no less important public space, the one where we seek out and enjoy the company and stimulation of others, has been much neglected. We have outgrown the classical monumental square with its political overtones. We now seldom congregate as citizens, and when we do, it is more to protest than to celebrate our collective identity. We have learned from experience that such oversize public spaces—and I would certainly include the Olmsted style park with its oversize natural landscapes—eventually can be subverted by the authorities and used to indoctrinate us with some establishment philosophy: the Wonders of Nature, the Wonders of Art, the Wonders of Physical Fitness, or the supreme Wonders of the Commissars looking down at us from their podium. Small, more intimate, less structured spaces are what we now prefer. "This loss of the natural impulse to monumentality," John Summerson observed, "should not be a matter of regret. It is a perfectly natural reflection of the change which is taking place in the whole character of western culture. All those things which suggested and supported monumentality are in dissolution. The corporate or social importance of religion was one of them. The sense of the dominance of a class—of the exclusive possession of certain privileges by certain groups—was another."[6]

To these things supporting monumentality in the past must be added the concept of a monolithic Public, the concept of a homogeneous People. For the public now is a composition of constantly shifting, overlapping groups—ethnic groups, social groups, age groups, special interest groups—and each of them, at one time or another, needs its own space, distinct from the surrounding urban fabric, where its own special social forms, its own special language and set of relations, can flourish, a space which confers a brief visibility on the group. We have too few of these spaces today, too few resembling the Prado in the North End of Boston, or the ad hoc open air social spaces which often evolve in urban ethnic or racial neighborhoods and do so much to maintain a sense of local identity and custom. With great taste and infinite goodwill, landscape architects have designed many mini-parks to relieve the monotony of our towns and cities. But the Olmsted tradition persists, and what we all too often have are overelaborate spaces with the inevitable display of vegetation, the inevitable ingenious fountain, and the inevitable emphasis on individual isolation. Yet contact with other people, not contact with nature, is what most of us are really after.

## A RETURN TO THE STREET

Despite our current admiration for the formal square as a feature of the urban scene, despite our attempts to introduce it in our residential areas and shopping centers and in urban renewal projects, the time is approaching, I suspect, when we will turn our attention elsewhere. There are in fact many signs that the street, or a given fragment of the street, will be the true public space of the future.

If in fact this is the case, we will be reverting, unconsciously of course, to a medieval urban concept which long preceded the Renaissance concept of the public square. In the Middle Ages it was the street—tortuous, dirty, crowded—and not the public space identified with the church or castle or market, that was the center of economic and social life. The street was the place of work, the place of buying and selling, the place of meeting and negotiating, and the scene of the important religious and civic ceremonies and processions.

[6]John N. Summerson, *Heavenly Mansions* (New York: Norton Library, 1963).

But its most significant trait was its blending of domestic and public life, its interplay of two distinct kinds of space. The narrow, overcrowded buildings bordering it spilled over into the street and transformed it into a place of workshops, kitchens, and merchandising, into a place of leisure and sociability, and confrontation of every kind. It was this confusion of functions, the confusion of two different realms of law and custom, that made the medieval street a kind of city within a city, and the scene of innovations in policing, maintenance, and social reform. Until the eighteenth century the street was actually something far more extensive than the traveled space between the houses. It was the matrix of a community, always alive to threats of intrusion, jealous of privileges and customs, and conscious of its own unique character.

For many economic, social, and aesthetic reasons, we are now beginning to think of the street and its relation to its inhabitants in a way that recalls the medieval concept. Robert Gutman wrote:

> The revival of interest in the urban street has been accompanied by a wholly new emphasis in the view of the street's primary social function. Put simply, what sets the contemporary idea apart from previous definitions is the conviction that the street should be designed and managed for the benefit of its residents. . . . These impulses to make the street work, to make it into a community, some of which are specific to the situation of the city today, have gained strength because . . . the residents of urban streets until recently regarded themselves as a relatively homogeneous population. This important point is often overlooked. We are concerned about the street community in large part because for the very first time in the history of cities the simple virtues and joys of urban life have been diminished for all social groups; and we connect this reduction in our level of satisfaction and safety with the breakdown of the community.[7]

Not every street can be defined as an essential spatial element in a community. The majority will continue to be public utilities. But insofar as certain streets will be seen as public places, as being closely related to their immediate built environment, they will be playing the social role we have long associated with the traditional public square: the place where we exhibit our permanent identity as members of the community. The learning experience, the experience of contact with new sensations, new people, new ideas belongs elsewhere. The street as the public space of a community, modest in size, simple in structure, will serve a strictly traditional purpose. It will be where, in the words of Paul Weiss, we recognize and abide by "a mosaic of accepted customs, conventions, habitual ways of evaluating, responding, and acting . . . men must, to be perfected, become social beings. They must act to make the structure of the group an integral part of themselves and a desirable link with others."

选读 19
"场所的浪费"
凯文·林奇

凯文·林奇是研究环境质量和城市形态的先驱，他在这篇选读中谈到了浪费这

---

[7]Robert Gutman, "The Street Generation," in *On Streets,* ed. Stanford Anderson (Cambridge, MA: MIT Press, 1978).

一人类社会的普遍过程。他在文中写道"巨大的浪费反复发生在人类的居所之中",并记录下了不同形式的浪费。他谈到建筑破坏者的作用和不断发展的现代拆除产业、建材回收、废品买卖业务以及破坏"非个人的"公共或机构财产的诱人之处。林奇对遭到废弃的内城居所也十分关心,资本的流动和我们不断迁移的历史风俗加剧了其被弃置的过程。林奇最后得出结论,浪费是"暗流涌动中引领我们沿途而上的一种特征,变化无常是事物的常态。就短期而言,浪费是物体的浪费;而就长期来看,浪费则是场所的浪费。二者各有其自身的特点。"

## READING 19

## The Waste of Place

### Kevin Lynch

Kevin Lynch, a pioneer in studies of the quality of the environment and urban form, in this selection writes about waste as a pervasive process in human society. He writes that "great wastings are echoed in the human settlement" and notes the many different forms of waste. He notes the role of the building wrecker and the growth of the modern demolition industry, the recycling of building materials and the salvage business, and the attraction of vandals to "impersonal" public or institutional property. Lynch is also concerned about the abandonment of inner-city housing, a process that is reinforced by mobility of capital and our historic custom of moving on. He concludes that wasting is "a feature of the underlying flux that carries us along, the everlasting impermanence of things. There is a short-term wasting of objects, and a long-term wasting of place. Each has its own characteristics."

## THE NATURAL CYCLE

Sewage, smog, garbage, scrap, litter and trash make up the daily waste flow of the city. There are more protracted wastings in nature. Supernovas explode, and the shell of debris racing out into space sweeps up the dust and gas from which new stars condense, to reignite the atomic furnaces. The sun wastes its substance, and the mountains wear away. They are thrown up with a vomiting of magma, gas, and ashes, which destroy living communities and then convert to the fertile volcanic soil that supports new life. Carbon is extracted from the air by plants, and is locked by their death in beds of coal, or pools of oil. The calcium in the shells of sea creatures drifts to the depths, there to be sealed off in deep limestone layers. The smooth circling of the ecological system is only an aspect of more pervasive change—more protracted, more violent, more wasteful of matter and energy, more an irretrievable flight than a placid turning.

*Source:* "The Waste of Place," in *Wasting Away* by Kevin Lynch. Copyright © 1990 by Catherine, David, Laura, and Peter Lynch. Reproduced by permission of Catherine, David, Laura, and Peter Lynch. Sierra Club Books, pp. 81–117.

The earth is intensively energetic, compared with other satellites of the solar system. Its surface is broken into separate plates that grind against each other. The ocean crusts are pushed down into the mantle, and boil up again at the rift valleys. This dynamism torments us with earthquakes, volcanoes, tsunamis, and hurricanes, but also gives us rich concentrated resources, formed and still forming. Life itself needs more than carbon, hydrogen, nitrogen, and oxygen for survival. At least 20 other elements are essential, in definite concentrations. Old, stable areas of the continental shields, subject to tropical weathering, are agriculturally unproductive, since they lack some of these important elements in the soil. Active volcanic regions, whose new rocks expose a wide spectrum of trace metals, are far more fertile.

Man is now a significant agent in the transfer of material in this dynamic system and may soon be the dominant one. If the rate of garbage production in North America were to be equated throughout the world, then that mass transfer would somewhat exceed the rate of volcanic upwelling that has built the mountains of the Pacific rim. If the per capita rate of the use of new minerals that characterizes the modern industrial world should be adopted by even 15 percent of the world's population by the year 2000, then it will amount to 20 billion tons per year, which is a mass comparable to such global processes as mountain-building, erosion, ocean crust formation (estimated at 30 billion tons per year), or the recycling of all the earth's biomass (estimated at 60 billion tons per year). Combustion adds 50 billion tons of carbon to the atmosphere each year, which is 1 percent of the total already present. We now discharge such metals as iron, copper, zinc, and lead into the oceans at rates that exceed natural processes by an order of magnitude.

## HUMAN WASTE OF PLACE

These great wastings are echoed in the human settlement. Buildings are abandoned, moved, or demolished; whole areas are cleared and rebuilt. Materials weather and age, are broken up and reused. Vandalism and arson render sound structures useless. Inner-city regions may be deserted—at first slowly, then with increasing speed. Lands fall vacant or derelict. Abhorred, unwanted uses are shunted out to marginal areas. Entire cities may decline or gradually be abandoned.

Kyoto was once a capital of 400,000 people, and it contains 700,000 today. Yet it shrank to a village in between. A wooden city, it went through repeated fires and savage civil wars. Buildings were thrown up by forced labor, abandoned, ruined, burnt, moved, or given away in pieces. Palaces were occupied for only a few years, or even for a few days in the year. Emperors and nobility moved about between houses and temples as their palaces were destroyed. The waste of habitat was on a grand scale. Here, as in many other ancient societies, buildings and even settlements were wasted deliberately, as a symbol of royal prestige and purity, just as emperors were served more than they could eat, and possessed more clothes than they could ever wear. New cities and palaces were at one time built at every accession, and abandoned at each royal death.

## WRECKERS AND SCAVENGERS

Superficially, the building wrecker is like the saprophyte of the natural system, which reduces dead organisms to their simpler elements to speed the recycling of matter. But the likeness is only superficial. The saprophytes break an organization down into simpler compounds, in order to make use of the material and the energy released. Wreckers also break up old patterns, but they make little use of the energy so released. The salvaged material is only incidental to their work, and much of it is more intimately mixed when they are done. They are recyclers only secondarily, and certainly they are not remanufacturers. They are paid to clear a space, and not to prey on the dead.

It is the vandal stripper who more closely resembles the natural saprophyte. In ancient cities, old monuments and buildings were routinely mined for their stones, beams, or roofing material. A Roman imperial rescript to the Count of the East, in A.D. 397, instructs him to use the material from demolished pagan temples to maintain the public bridges, highways, aqueducts, and wells. The ruined aqueducts of Rome were closed in to make squatters' dwellings. A heap of old rubbish from the Great Fire, which encumbered eight and a half acres of central London for half a century, was shipped as fill to create the new Russian city of St. Petersburg. One advantage of living in any ruined city is its concentrated wealth of material, as well as the half-built spaces it affords.

Modern demolition is an organized trade, intended to create a site. Decorative materials are salvaged first, as well as metals, doors, windows, plumbing, good lumber, hardware, pipe, clean brick, and wire. Then the structure is reduced to rubble, and the rubble is carted away. Once it would have been burned on the site to reduce its volume, converting building waste into air pollution. Now on-site burning is prohibited, and the bulky rubble must, at substantial cost, be carried off in huge trucks. The dumps for which it is destined lie at increasing distances, and so illegal dumping by small demolition contractors has become profitable. The demolition sites themselves must then be fenced and guarded against hit-and-run depositors. If the rubble is disposed on site, as happened in the ancient cities, then ground levels gradually rise, producing those elevated "tells" that mark the locations of most former cities in the Middle East. In medieval Winchester, the accumulation of rubbish in the course of 150 years caused one street to rise five feet. A more modern example is Berlin, which has built its famous "Mt. Junk" out of the rubble of its wartime ruins.[1] Similarly, the abandoned apartment structures of Breezy Point, Long Island, have been converted into lookout mounds in the new national park.[2]

One soils scientist in the Netherlands asserts that construction in his small country must stop altogether by the year 2000, unless builders are able to check the flow of material from the sources of aggregate in the upland east, through the new buildings, to eventual demolition and deposit in landfills in the lowland west. Forty square kilometers of eastern excavation are gouged out each year, and one-and-one-quarter square kilometers of new landfill are piled up in the urban west. The existing elevation differential in the nation is slowly being reduced. To check this flow, building rubble must be recycled; structures that can be dismantled and reused must become the norm. Architects must begin to think about holes in the ground, about flows of material, and about the topographic inversion of their country.

Masonry rubble is commonly reused as landfill. Broken concrete can also be used as coarse aggregate in new concrete—at some reduction in strength. But if the old concrete is contaminated with gypsum or other substances, it is of very little use. The combustibles, once so easily burned off, now mix into the whole mass of rubble and complicate its use as a stable, compactible fill. Due to the labor costs of sorting today, and to the new synthetic building materials, little salvage is economically feasible other than the special decorative items. But recycling plants can make 60 percent of all demolition wastes reusable, if they can locate at least 200,000 tons of it per year, and can be within 20 kilometers of the supply, their market, and a landfill site. Thus they are economical in cities of over one million people.

Demolition uses special techniques and machines: wrecking balls, pusher arms, explosives, bursters, thermic lances. The most delicate work is still done by the skilled "topmen," who are standing on or next to what they are taking down, and thus in constant risk of falls, collapses, fumes, dust, nails, and bad footing. The rate of injury is very high: wrecking firms may pay one-third of their payrolls to cover workmen's insurance. Demolition is temporary and irregular work, but also dangerous and highly skilled. It is not unionized. In England, it tends to be carried down a family line. The topmen migrate from job to job, personally known to one another and to the contractors. In its camaraderie, its stigma and danger, its pride in special skill, its migrant nature and its personal links, it resembles the medieval wandering crafts of builder and smith.

Demolition is usually an afterthought, a minor event between site acquisition and new construction. Yet it steadily becomes more difficult, due to the greater restrictions placed on it and to the swelling use of materials and forms that are either intrinsically difficult to break up—such as reinforced concrete, or tall buildings—or are so new that there has been no prior experience with their demolition. Demolition contractors plead that building designers consider the eventual break-up of their fabric, and file specifications for its dissolution as well as its creation. Since these pleas are not heeded, wreckers face unpleasant surprises. Unpleasantness may also arise from the previous use of a building. One old London house, of a type quite easy to tear down, proved to have been at one time a factory for painting luminous clock faces. It contained 50 million curies of radiation, and the demolition costs soared to thousands of pounds. There were unexpected problems, such as the risk to the public of any small object stolen from the site while the building was coming down.

## SALVAGE

Some communities have acted to increase the recycling of building material. Baltimore, for example, operates a public salvage depot that stores decorative material collected from city-owned buildings being demolished. Any city residents engaged in restoring their own homes can buy these items at cost. New York has created a similar cash-and-carry operation that saves pieces from 150 buildings per year, or about 10 percent of those torn down each year. This public depot focuses on the more prosaic external parts, leaving the high-value internal items to exploitation by private dealers.

Specialist firms hold large stocks of fine interior elements. One firm in Portland, Oregon, acquired much of its stock from old New York houses and transported it across the country for reuse in the West. Stanford White, the fashionable 19th-century New York architect, picked up the mantelpieces for his expensive houses in the wreckers' yards of the East River. The Anonymous Art Recovery Society (the so-called "gingerbread snatchers") has been collecting building ornaments from wreckers and dumps in New York for 20 years, and has a permanent showing in the Brooklyn Museum sculpture garden. As the market for these items has developed,

**Figure 19–1** Demolition is a highly specialized and dangerous activity requiring special techniques and equipment. [© Kevin Lynch, used by permission of Catherine, David, Laura, and Peter Lynch]

demolition supervisors, who used to give the stuff away, are now alert to sell it.

Other firms save and stock more mundane items. There are over 200 wrecking and salvage firms in New England alone. A typical yard and warehouse of this kind has extensive sheds full of old plumbing, brick, doors, windows, and miscellaneous hardware. Much of this is the residue of the 1950s, which were boom years for rebuilding, and yet also a time when it was still profitable to wreck and sort by hand. With the speed and mechanization of contemporary demolition, it is now much less advantageous to salvage. The typical firm is reduced to wrecking small houses, or to stripping easily salvageable parts as an initial step in larger jobs. Meanwhile, their stock of older pieces continues to rise in value.

But if one is willing to undertake the hand labor, and risk the danger, then old buildings have much useful material, particularly lumber and brick, in addition to the usual special salvage items, which are commonly taken off only in the first stages of building removal. Some young builders become "demolition addicts," finding a pleasure in tearing down, and an equal pleasure in rescuing and reusing secondhand material. After the experience of using old wood, they say the new wood seems raw, and without character. Gathering their material first, they design a building that takes advantage of what they have. Thus the new building acquires a certain patina, a particular character and sense of history. They cruise the city, spotting reusable material in alleys, trash boxes, construction and remodeling sites, dumps, beaches, and disaster areas.

## MARKS OF TIME

Great Britain's Building Research Station has issued several technical bulletins on the weathering and deterioration of building material.[3] In perusing the numerous illustrations in these bulletins, it is interesting to see how weathering increases the expressiveness of old surfaces. "Counter-shading," for example, is a process in which dirt drifts down and is deposited on upward-facing surfaces. This gives an unexpected effect of light projected from below, highlighting detail in a dramatic way. Hidden structure is exposed by stains. Rains streak the surfaces. Projections, equipment, or adjacent buildings cast permanent shadows, as in the phenomenon of "shuttermarks." Cracks and discolorations appear, and differences in tone and grain are exaggerated. The orientation of a wall to wind, rain, and sun is expressed in differential weathering. The research station sees these phenomena as defects, and yet they confer a richer, more particular character. The well-modeled surfaces of older buildings seem to take these traces more happily than those newer skins whose esthetic it is to be smooth and clean. On the latter, a streak is a disfigurement. Some metals, like copper, oxidize attractively; others, like aluminum, are dulled by time. Brickwork mellows with age, unless it should effloresce. Timbers darken or silver, and become eloquent of grain. But concrete cracks and discolors in meaningless forms. Might materials also be chosen for their qualities in old age, and surfaces be detailed so that the marks of time make them more expressive and diverse?[4]

## VANDALISM

Vandalism is more powerful than weathering. Like demolition, it creates waste deliberately. Vandalism first meant the willful and ignorant destruction of beautiful or venerable things by invading barbarians. Now it means the willful destruction of any property. It is widespread, but not meaningless, and may arise in varied circumstances.

At times, vandalism is the by-product of an illegal livelihood gained by the stripping and resale of valuable parts. Occasionally, it may be the unintended by-product of mere play, of exuberant action. More often, it is deliberate destruction, aimed at a person or institution that has injured the vandal. Or it is part of some large struggle, and so it is "sabotage." Most often—and this is the motive that both fascinates and frustrates the nonvandal—it is an intentional act not directed toward any definite end: the expression of a generalized hostility, or of a sheer pleasure in destruction. Therefore, it is called "mindless." But it is not mindless. It is quite mindful, and because of that, quite difficult to prevent.

Especially for those young males whose future is restricted and meaningless—but also for other similarly placed groups in society—vandalism is just the ticket, the ideal form of rule breaking. In a world that seems indifferent to their existence, vandalism is expressive of their feelings, and also instrumental, since by its means the world is forced to respond to them. There are risks, which add a spice of danger, but the risks are not great. The familiar and alien environment is restructured and played with according to the vandal's own rules. Mostly absent in traditional, controlled societies, or in those that are in hopeful transition, vandalism is present throughout the relatively affluent world, both capitalist and socialist. Some vandalism is condoned ("students will have their fling"), or it is hidden within institutions, where it is expected and provided for. Elsewhere it is a common scandal.

Vandalism is more likely to be inflicted on public or institutional property, where the owner is an impersonal "they." Like litter magnets, places are more likely to be vandalized when they already exhibit signs of dilapidation, low supervision, or uncertain ownership, and when repairs are not made quickly. Vandals are attracted to things that smash well, like glass, or are easy or valuable to strip, or are intended to frustrate their action, such as a fence or bar. Vandalism is usually the work of 10- to 12-year-olds at play, or of disaffected older adolescents. However, there is no "vandal type." Juvenile vandalism has one of the lowest reconviction rates of any offense. In one study, it was the only unfavorable child symptom that was not predictive of any later personality disturbance or psychiatric disorder[5] (which I am pleased to hear, having smashed some street lights in my own time).

An experiment by Philip Zimbardo vivifies these acts.[6] He left an automobile at a curb on a street across from the Bronx campus of New York University. It had no license plates, and its hood was raised. He then observed events by means of a hidden camera. Within ten minutes the first strippers had attacked the car, and within 24 hours they had removed battery, radiator, air cleaner, radio antenna, wipers, chrome strips, hubcaps, tires, jumper cables, gas can, and car wax. The strippers were well-dressed adult whites, the saprophytes of the city streets. In the next 9 hours, young people commenced the random destruction of the car, first smashing the windows. In three days, the car was a battered hulk, no longer worthy of attack, but in use as a receptacle for other waste. Up to that point, almost one-third of those who passed the car had attacked it in some way.

In contrast, a similar car was abandoned in an affluent area of Palo Alto, California. In over a week, there was no incident. One passer-by closed the hood during a rain. At that point, the experimenter forced the issue. He directed his students to strike the car with a sledge hammer. They began sheepishly, then were carried away by delight. Observers joined in. Once battered, the car was prey to vandals in the same progression as before. Signals release the vandal activity, and they must be stronger where social control is stronger. First the car is stripped, then smashed in easy ways, and then with difficulty. Finally, it is ignored, except as a target for other waste.

Vandalism can be dealt with in different ways. It may be accepted as inevitable, an expected deterioration, a reminder of the need for regular repair. Repair must be prompt, or the place is broadcast as not being under control. The glass in a deserted building is very quickly broken out. The alternative is to harden or to police the environment, to prevent destructive acts. The costs of policing or hardening may easily be higher than the costs of repair. Moreover, the vandals, being human and ingenious, will find ways to circumvent the hardening. They may even be stimulated by it, and will show how they can bend steel and shatter concrete. A counterstrategy, therefore, is to tenderize a place, making it fragile and soft in hope of deflecting the feelings of malice. This is risky, but at times it works, especially when maintenance is impeccable, and particularly if local people had a hand in the making of the place, and thus a stake in its protection.

Retribution is still another response: jail terms, fines, or commitment to labor teams sentenced to clean and repair the damaged environment. In the latter case, some of the costs of vandalism are shifted to the vandals. Yet, except in the case of adults engaged in stripping the environment for its valuables, retribution may do little to deter. Vandalism for noncommercial ends is largely committed by groups of young people on unpremeditated occasions, when the threat of future punishment does not have great weight. "Public education" efforts have very little effect. The vandals themselves are not moved by such appeals, which do little more than heighten the general public awareness of what is going on.

The only remaining strategy is to move against the causes of the phenomenon. But since the phenomenon is as plural as its causes, strategy must vary according to circumstance. Where vandalism is the stripping of parts, then one must deal with the market for parts. Where it is sheer play, then there must be better scope for the activity of children. Where it is vindictive, one goes to the sources of conflict, to see if mediation is possible. But where it is a product of generalized frustration, as it so often is, the only direct solution is to deal with those features of school, family, or economy that create that alienation—not an easy solution. Here one may be tempted to agree with Bakunin that "the urge to destroy is also a creative urge"—to see vandalism as a healthy reaction in some ways, a rebellion of youth against a society to which their parents have become resigned.[7] There is no need to romanticize this, to fall in love with violence and graffiti, in order to look for the roots of vandalism and to think of ways of turning that force and skill to creative ends. There are times when the vandalized place may itself be part of the solution, if the vandals can be given a responsible and challenging role in making and protecting it.

**ABANDONMENT**

Inner-city housing is being abandoned at an increasing rate. Almost 5 percent of all U.S. inner-city dwelling units are now boarded up. In New York City alone, 59,000 buildings containing 700,000 apartments are in tax arrears, and presumably on their way to abandonment. Deserted buildings are not new in history, but in the past they retained some value, and were held for a favorable turn of the market. Now vandalism, demolition, and arson are destroying this potentially useful stock, and destroying whole communities in the process. The city of St. Louis is an extreme case: over 8 percent of its housing was vacant in the 1960s (compared with less than 2 percent in the 1950s), and 17 percent of its entire supply was demolished in that decade. The steady long-term improvement in the housing of low-income families in this country may have been reversed some time in the 1960s, despite the continued rise of real incomes. There has been a surge of deterioration and abandonment in the existing inventory to which these families are confined. In consequence, inner-city, low-income tenants living in substandard quarters pay roughly the same rents as outer-city, moderate-income tenants living in standard housing.

Sporadic abandonment occurs in relatively sound inner areas, as operating costs rise due to vacancies, arrears, insurance, utility costs, and misuse or mismanagement. Returns on capital fall below savings bank interest, or even down to negative cash flows. Vandalism or accident may initiate the process. Fearful, confused owners refuse to invest in rehabilitation, even where it could be profitable. Absentee but amateur landlords are particularly prone to this irrational course. It happens more rarely under experienced, large-scale management, or where structures are owner-occupied.

When sporadic abandonment accumulates, it may become contagious; expectations shift, professional vandals begin to operate, and landlords turn to arson. When their properties fall to zero value, landlords have them torched in order to collect the insurance, before giving them up. Lives are lost, and personal possessions are destroyed. Insurance rates rise, and buildings which by minor repairs could have been made useful again are rendered unsalvageable. Wasting accelerates. Whole inner areas may be cleared, as if by a forest fire, or urban redevelopment gone wild.

Teenagers break into closed buildings, to use them as clubhouses. They set fires and initiate water damage. Local junk dealers facilitate wasting by fencing stripped hardware, metals, and fixtures. Troubled families, unwanted elsewhere, are dumped into these unfavorable areas, and responsible tenants move away if they can. A basically sound housing stock, capable of modest repair, is reduced to lines of gutted shells, as if it had been subjected to aerial bombardment. The last available response is to accelerate the process and clear the land—perhaps to leave it empty, perhaps to build on it once more. In 1979, New York City, having some 10,000 buildings foreclosed and beyond salvage, asked the U.S. Army for troops and military technicians to help demolish these structures—a one-time effort to sweep away the backlog. But the army was dubious about diverting its men for the task, and the building unions resisted.

There are public programs that encourage the reuse of these abandoned structures. One is the popular "homesteading" procedure, in which tax-abandoned houses are sold by the city for nominal sums. With a subsidized loan, the new owner then rehabilitates the house, which thus is returned to use and to the tax rolls. The prospective owner-occupier must repair and move into the house within a given time in order to perfect his title, which is the analogy with the historic "homesteading" of the public lands.

This has been a reasonably successful, if small-scale, means of recycling sound single-family units in desirable close-in areas. At times, it has initiated a more general upgrading of a neighborhood, as the well-to-do begin to move back in. Originally intended to ease the housing crisis of the poor via the use of their own "sweat equity," it has more often been used by young middle-class families, who have the physical energy—and also the capital, leisure, and skill—to work on their own house.

Vacant buildings may also be rehabilitated directly by the city, for resale or for low-income rental. Baltimore recycles some 350 units per year by this practice, and no longer demolishes any vacant house unless it is structurally unsafe. Whole abandoned blocks are boarded up, fenced in, and so stockpiled until they are in demand. It is not clear whether such recycled units are cheaper to produce than new units, despite the obvious social advantages. The costs are obscure, since it is difficult to disentangle the effect of the multiple public subsidies involved.

In Portland, Oregon, old but sound houses scheduled for demolition are now regularly purchased and moved to new locations on inner-city vacant lots, where they re-

**FIGURE 19–2** House moving, once a common activity, is reappearing in some cities to provide affordable inner city housing. [© Rajeev Bhatia, used by permission of Rajeev Bhatia]

sume their useful life. Moving houses was quite common, in a day when their value was high relative to a normal income. The old houses of Martha's Vineyard were shifted from farm to farm by ox team as new families were formed, and the barracks of Camp Meigs in Boston—the training camp for the first black regiment in the Civil War—were later sold for a token sum and moved to scattered lots in Hyde Park as houses. Several of these were still in use in the 1920s.

Apartment houses, once abandoned for taxes, have been more difficult to put back into service without a major public investment. Private capital is difficult to enlist. If put up for sale, they are picked up by speculators at low prices in public auctions. They are milked briefly for current income without further investment, and then abandoned once more, now in a far more degraded condition. Although buildings can be moved or repaired, and to some extent their material can be recycled, it is difficult to remanufacture them, since even today they are not usually the product of an assembly line, and their components are not easily separated. Might it be possible, however, to apply that idea to the mass-produced trailer, or "mobile home," which so quickly loses its value, and which begins to be abandoned as obsolete in increasing numbers? Could mobile homes be designed with such reprocessing in mind?

The disposition of publicly owned real estate unwanted for public purpose has always been controversial in this capitalist nation. Should the city or federal agency return the property as quickly as possible to private use, thus realizing what it can, renewing the tax base, and relieving itself of an embarrassing administrative burden? Or should it do so only with deliberation, making sure that there is in fact no public use in prospect, and controlling the private use to achieve some desired community outcome? Or might such properties be left in public ownership, to be used for some subsidized purpose, or even for rental income? As public land falls out of use, and especially as tax delinquencies multiply, these real estate leavings pile up. In some central areas, outside of the central business district itself, governments are being driven towards socialization of the land. By our ideology this is a fearful outcome, and yet it may prove the most workable alternative.

## URBAN DECLINE

In recent times in the United States, there has been a marked decline in certain entire urban areas, a process backed by our high mobility of capital and young labor, and our historic custom of moving on. Europe, on the other hand, has shown a less marked decline of its old cities, due to repeated public interventions and to the national barriers against free population movement.[8] Mobility means freedom, and efficiency in the use of resources—at least in the short term. The 1980 President's Commission for a National Agenda proposed that national policy should encourage this mobility, rather than seeking to check it.[9] The poor should be given incentives to move to where the jobs are, to go from Rustbelt to Sunbelt. Older cities should be allowed to shrink. Present subsidies to declining places, in their opinion, only trap the poor, since they tempt them to stay and survive, when they might move and prosper. Moving on and abandoning things is the American Way, the expression of our free spirit.

In 1975, Edgar Rust studied the declining metropolitan areas of the U.S.[10] He found an increase in metropolitan areas with less than 1 percent population growth in a decade: from 5 in the 1940s, to 10 in the 1950s, and to 26 in the 1960s. Between 1970 and 1972, 27 actually declined, and the trend has since become more marked. The shift from the north to the southwest is common talk, but in fact these losing areas can be found in any section of the nation.

Typically, a city in decline is one that boomed in the past, dominated by a single economic activity in which it specialized. When that activity faded, or found a more advantageous locale, the city failed to shift to new enterprise. At times, decline was reinforced by some major disaster—fire, flood, or earthquake; or by the loss of some transport connection—canal, port, or railway. The original boom may have been founded on commerce, on servicing westward settlement, on resource extraction, heavy industry, military procurement, consumer products and services, or even, as at present, on the attraction of some preferred climate or landscape.

The larger administrative centers, with their multiple economic bases and concentrations of headquarters offices, remain stable in this flux. Headquarter locations monopolize capital and skill, and can afford to hold on to them in hard times, or to take the gamble of a shift to some new activity. They sit at the major nodes of transportation systems, so that they are not easily isolated. Administrative functions tend to persist. Moreover, a tradition of civic pride may have encouraged businesses to invest in the public environment in flush times, leaving a heritage of amenity that continues to hold their skilled people in times of adversity.

Rust finds such migration influenced more by pull than push. High wages elsewhere attract the skilled and mobile young. The old and the poor stay behind. If ownership is centered elsewhere, it is relatively easy to disinvest in real estate by withholding maintenance and taxes. Such capital can be written off in a few years, and the plant closed without loss—without loss to the entrepreneur, that is. Public services must continue to serve a diminished clientele in the old place while they are duplicated in the new. Social ties are disrupted as the young disperse and forsake the old. The remaining labor force is committed to work. Populations and public capital cannot match the free flow of private capital.

The mismatch is sharpened when the first economic boom in the new city, marked by an influx of the young, is followed by a second population boom, as the young newcomers raise their children. If this second boom is not met by continued economic growth and the creation of new enterprises, then these children leave when their time comes, and the area goes into a sharp decline, ages rise, incomes level off, and risk capital goes elsewhere.

This is followed by a protracted secondary decline, marked by underemployment, falling incomes, increasing nonlocal control of activity, risk-avoiding management, restricted access to new ideas and markets, a shortage of trained professionals and managers, and a diversion of public expenditures to the shoring up of specialized but obsolete activity or to the attraction of transient, low-wage firms. Large corporations locate their low-skill branch plants in such places: the very plants that are most sensitive to future economic swings. Young people are educated at local cost, and then depart. Resignation replaces the earlier attempts to respond to the challenge of the first decline. There is a long stagnation, a resistance to change, until people, things, and institutions have eroded sufficiently to permit a new turn, or at least to establish an equilibrium at some lower level.

Declining areas have their own values: low housing costs, less crowding, and a relatively placid, stress-free world. Church, family and ethnic ties are strong, even if the mature children are gone. But the environment is likely to be of low quality (with a few splendid survivals), and expectations and self-esteem are depressed.

Public policies that treat of decline as a local disease, or come too late, or encourage growth in other places, can be ineffective or damaging. Typically, significant efforts are rarely made to address decline at its roots: to create flexibility and diversity at an early stage; to invest in the public amenity that will stabilize a place; to compensate for the social costs of mobility; to put the control of enterprise in local hands; to capitalize on the hidden benefits of stability, stagnation, and decline.

If the government is serious about responding to the mismatch of people and capital, then it must propose far more radical actions: the transshipment of entire communities, and not just the mobile young; the invention of transferable infrastructure and institutions; and a humane closing down of abandoned settlements. Such a policy could be an interesting speculation. It might even be rational, if expensive and

**FIGURE 19–3**   The Bingham copper mine near Salt Lake City is the largest surface copper mine in North America. It covers, 1,050 acres and contains about 175 miles of railroad track. After the copper has been mined what might be done with the excavation? [*The Bettmann Archive,* used by permission of Corbiss: Bettmann]

politically distasteful. At least, it would make evident the hidden costs of uneven growth.

## DERELICTION

Derelict land is even more extensive than derelict building. Derelict land is often defined as land so damaged by development that it is incapable of beneficial use without further treatment. Note that this definition excludes land abandoned because of changes in the market, like an empty millyard; land that is simply unpleasant or dangerous; and land naturally unusable or made so by natural cause. If it pays, it isn't derelict. If it doesn't pay, due to some human devilment, and once did pay, then it is derelict.

Surface and subsurface mining creates much land of this kind. It may destroy topsoil and vegetation, leave pits and holes, cause subsidence and flooding, and pollute the ground with brines, spoil heaps, acid wastes, and slimes. Modern mechanical mining may accelerate this dereliction—causing deeper subsidence, or more extensive gashing—since it works to greater depths at greater speeds, and it discards larger percentages of waste piling in taller, looser heaps. Some manufacturing processes, such as steel works, smelters, power plants, gas works, or industrial chemicals, also pollute the land with their deposits of ashes, slags, metals, chemical wastes, radioactive material, and other toxic substances, as well as by their massive foundations and utilities. At the same time, there have been technical advances in land reclamation, using heavy earth-moving machinery to reshape the waste heaps or recover the strip mines, and new methods for reestablishing vegetation by drainage, rebuilding the soil, and progressive planting of selected species. Reclaimed sites need 20 years to mature to a stable state. Meanwhile, they still *appear* derelict, and so may continue to attract illegal dumping, and the destruction of their cover.

Some wastes can find an economic use, such as ashes for fill or for building blocks, or slag for road metal or for fertilizer. On occasion, therefore, it may become profitable to rework the old waste deposits, as recovery techniques improve or raw material prices rise. Or the disturbed land form itself may be made useful: old pits and quarries may be converted into recreational water bodies, for example. Many other wastes are persistent, and difficult to rework or detoxify. Radioactive materials are notorious. Chemical slimes must be held indefinitely in expansive ponds, always liable to accidental release. The soft residue of the Le Blanc process for making soda and potash has remained a noxious presence for 60 years. The extremely deep pits made by china clay extraction are difficult to refill. Old waste heaps may collapse, as happened in the Aberfan disaster. The contour mines and waste reservoirs of Appalachia are unstable: they slip and block water courses, or release sudden floods, as occurred at Buffalo Creek in West Virginia. (The buffalo having been wasted long before.)

## ABANDONED TRANSPORT

Abandoned transport also results in derelict lands, but these will more easily find new uses. Railroad closures have left substantial mileages of unused lines which turn out to be useful secondary recreational routes. The old city terminals and yards, being extensive and close-in, are valuable for urban development. The old canals, especially in Great Britain, have proved ideal for pleasure boating.[11] Some remote bulk seaports are truly derelict, but the old urban waterfronts from which modern container shipping has largely receded are now prime development sites. The wartime airfields of Great Britain, predicted to become permanent scars, are now largely returned to other uses. Runways are mined for aggregate, broken up, and returned to fields, or used as farm roads and hardstands. Roads themselves rarely become derelict. They continue to be used for access, although some of their structures, such as our recent elevated highways, may in good time have to be torn down. Any connected spatial network, once acquired, continues to be useful for many modes of flow: trains, pipes, cars, cycles, horses, wires, walkers, canals, whatever. When Los Angeles abandoned its extensive street railway system, it also dismembered the rights-of-way. The city has had ample occasion to regret the miscalculation.

## THE GREAT WASTERS

Once their lords have been banished, the domains of the great space wasters—the kings, armies, and extractive industries—can become the parks and gardens of their humble successors. The arrogant dead enrich the ground. In American cities, the former military reservations, along with old railroad yards and dumps, are a primary source of renewable land. The city parks of London are there because the kings reserved hunting grounds next to their palaces, once a very inefficient use of city land.

War and the sudden exploitation of resources are champion wasters. Entire regions are ransacked and emptied; vast works are built for brief use. In his history of the Mediterranean, Braudel tells of the constant building, demolition, and rebuilding of fortifications.[12] Warfare evolved rapidly, and each new fort was obsolete when completed. To build city walls entailed a gigantic civic effort. Then they became an awkward encumbrance, then a mine of material, and finally, once they had laboriously been leveled, a valuable open space or circuit road. Our own war material still litters the islands of the Pacific. The barracks of the two world wars, and of our Civil War, were, for their time, a principal source of secondhand lumber and used buildings. A substantial surplus of temporary buildings from World War II still waits to be demolished or reused. The MX missile scheme as first planned would have wasted 4,000 to 8,000 square miles of ground, or about 1 percent of the area of the eight Great Basin states. By past experience, it would soon be obsolete, or if useful, would then make *us* obsolete. What would it or our remains be useful for?

The Industrial Revolution began in Great Britain, and industrial waste has accumulated there to greater depth than in any other nation. It was only in the 19th century, however, that the industrial scale enlarged sufficiently, and abandonment became frequent enough, to create substantial derelict areas. As the century wore on, the wasting rate went up, greater damage was accomplished by more powerful machinery, and land subsidence began to occur. After 1920, old mines and plants closed on a significant scale, and open-cast mines became common. After World War II, closures accelerated, and derelict land appeared as a national problem. Some 2,000 hectares of derelict land are still being created every year, a rate that is 300 hectares greater than the rate of reclamation. Only some 0.3 percent of the national territory is derelict, but this incidence is more severe in some places: 1.8 percent of the area of Cornwall, for example, or up to 13 percent in some local areas.

In most U.S. cities having a population of more than 100,000, 20 to 25 percent of the land is vacant. Of this vacant land, 20 to 25 percent once again is unbuildable, due to size or shape, or to slope, instability, or flooding. Some of the remainder is in confused ownership, but most of it is held for speculation, or as an institutional or corporate reserve. If this "waste" land could be utilized, it would relieve the development pressures at the city fringe. These half-deserted inner cities are visual symbols of despair. They color our view of city life. On the other hand, it is easy to find sites for new public facilities, and there is room for the activities of children. The tightly planned and rapidly built new suburb might be an example of too little waste. It constricts the individual, and fixes the pattern of activity for a generation. Thinking of the resources that its construction consumes, and the emptying of the older stock that it induces, large-scale new building may simply be another manifestation of the empty center. In a well-known article on "scatteration," Jack Lessinger points out the values of scattered growth, with its plentiful holes for later infill.[13]

## REGULATION OF DERELICT LAND

Great Britain and West Germany, scarred so early by industrial development, now carefully regulate the production of derelict land. They require that strip mines recover and revegetate as they go, and that deep mines replace their wastes

CHAPTER 5: THE NATURAL ENVIRONMENT 273

(a)

(b)

**FIGURE 19–4** It is difficult to kill a city, not only because of its concentration of physical structures, but especially because of the memories, desires, and skills of its residents. Like most parts of Warsaw, its old town square was severely damaged in World War II (a) but by 1965, 20 years later, it had been largely reconstructed, duplicating street patterns and facades as closely as possible (b). [*Polish Interpress Agency,* used by permission of Polish Interpress Agency]

underground. In the United States, on the other hand, there is very little control of such activities. Surface mining produces an estimated 500,000 acres of abandoned land each year. Unfortunately, it is doubtful if the reclamation of this land for agriculture or forestry is economical, unless the cost is borne as an original cost of production. It is usually cheaper for the mining enterprise, if left to its own accounting, to pay damages for subsidence or simply to buy the land and throw it away once mined, than it is to take the action necessary to restore it to its former usefulness. In a present-oriented, strictly economic evaluation, it will be cheaper to increase agricultural or forest production simply by upgrading the undisturbed land that remains. Reclaiming derelict land for dense housing, urban recreation, or industrial use, on the other hand, is more likely to pay off.

Despite the new practices, the complete reclamation of all the derelict land in Great Britain would require, if accomplished over the next 20 years, a public expenditure of about one-half of all the public sums spent on housing and environmental services—a staggering bill. One is cynically tempted to let the underdeveloped countries do our mining for us, preserving the amenities at home while they cope with our degradation.

Such analyses take no account of the long-term consequences of the loss of arable land, nor of the satisfactions to be derived from a handsome landscape. Reclamation is as often motivated by community pride as by economic calculation. In some cases, the waste heaps themselves are historic landmarks, to be reshaped and reclaimed but not to be removed. Indeed, some historians fear that wholesale land reclamation risks obliterating our industrial past. Much of the 19th century industrial dereliction has with time become romantically attractive.

## WASTES OF SUCCESSIVE OCCUPATIONS

Landscapes shift from one function to another, are abandoned and reoccupied, take on new forms, revert, and occasionally are changed irretrievably. The wastes of successive occupations accumulate and become part of the nature of the land. The European occupation of New England is a capsule example. From subsistence farms on newly cleared forest land, farmers turned to cash crops of wheat. In 1824 the wheat weevil arrived, and in 1825 the Erie Canal brought midwestern grain to the cities. A population exodus began, and the wheatfields were converted to sheep pasture. Removal of the wool tariff in 1840 caused a shift from sheep to dairy cattle, and emigration was hastened by the loss of men in the Civil War. People moved down from the upland farms and towns to the water power, the railroad lines, and the meadows of the river flats. Later, the cows no longer grazed for themselves, but were fed in the barnyard, and so the upland pastures grew back to cherry, maple, blackberry, and alder: "puckerbush" in the local tongue.[14]

The farmers thought it was sinful to let those fields revert to the original forest, and yet the process continues, leaving stone fences and cellar holes in the woods, and a skeleton of narrow forest roads. "I sold the cows when the barn floor collapsed, loaned out one horse when its mate died, sold the sheep when the fences got too weak, and now I rent the tillage land and have a town job," recounted one Vermont farmer. "Somewhere in every New England mind there is an abandoned farm," wrote poet David McCord.

Today, the landscape is being reconverted to recreation. The houses are going uphill and out of town again. The roads are reused, and the stone fences emerge once more. Land is cleared to gain a view instead of to raise a crop. Ski runs are cut down the steep slopes, and new commercial activities appear along the river-bottom highways. In time, future successions will again remake these highways, these new second homes, these narrow ribbons of rye grass cascading down the slopes. The landscape changes, accumulating historic waste.

## THE MAORI

The Maori made garden soils over extensive areas of New Zealand by laborious digging, the addition of sand, weeding, and burning. Lands partially in grass were by fire converted to continuous prairies, unleashing severe erosion and silting the river mouths. Once these economic resources were established, they turned to war, built massive fortified settlements, and abandoned much of their garden land. Whole regions were depopulated and went back to waste. Many settlements were sacked. The flightless moa was slaughtered and driven to extinction, leaving bone deposits as dense as 800 skeletons to the acre. The Maori mined these sites for tools, and then the Europeans carted the bones to mills to make fertilizer. These ruins, boneyards, soils, grasslands, siltings, erosions, new and vanished species—along with the usual massive changes brought on by the European settlers—are all part of the productive landscape of New Zealand today.

## NEGEV

Whole regions have been abandoned, and then reoccupied after lying empty for long periods. The arid Negev in Palestine has filled and emptied at least five times in human history. The Chalcolithic, Bronze Age, Abrahamitic, Judean, Nabataean, and current settlements have each been separated by a long vacant period. This is a difficult land; its use depends on careful water management. Its abandonment and reuse did not follow fluctuations in the climate, but changes in the social order. Raids and warfare emptied it, and external power refilled it, since any repopulation was dependent on capital, concerted effort, and security. The Nabataean water-capturing devices were especially elaborate: cisterns, terraces, dams, tunnels, and rock mounds on which dew could form. Much of this is still operable today, the wastes of an older civilization in a wasted land, a land being reclaimed once more.

Not all changes are gradual and reversible. The open roasting of copper sulfide ore in a single plant in southeastern Tennessee, stopped by legal action in 1910, destroyed the forest cover over an area of 25,000 acres by its emission of sulfur dioxide. Seven thousand acres of this land are still desert, and are eroding in the heavy rains; the remainder has been converted to permanent grassland. The Salton Sea is another such human artifact, carried out on nature's ample scale. An illegal canal, built to Mexicali to prevent flooding in the delta of the Colorado, through miscalculation overflowed into a part of the Imperial Valley lying below sea level. The result was a permanent salt lake 30 miles long that drowned out towns and railroad lines. Now a recreation industry is based on that undrainable flood.

In other cases, we have silted up rivers and lakes, or driven species to extinction, just as the aboriginal New Zealanders did. The changes that men accomplish seem most thorough and irreversible when they speed some natural evolutionary process taking place at a more majestic pace: the erosion of continents, the extinction of species, the eutrophication of lakes, the changing of river courses. We can see a minor example in New York's Central Park. The Pond, a creation of Olmsted's plan, is now silting up, due to accelerated runoff from the open park lands and periodic disturbances of construction. This is a natural evolution for any pond, but with our help it goes faster. Now a debate arises: should it be restored to Olmsted's artificial design as part of a "clean lakes" program, or be conserved as it is as a man-made "wetland"? It has a different smell, and its water is no longer clear, but it harbors new plants and new bird species. It is pollution, decay, and new life.

But this is small stuff. Lately, we aspire to acts more nearly divine: the contamination of the seas or of the global atmosphere, or the deposit of radioactivity for the ages.

## THE PERSISTENCE OF CITIES

Abandoned cities—as distinct from derelict lands and ghost villages or small towns—are not as numerous as one might think, despite the vivid role their awesome ruins play in our imagination. Babylon, Nineveh, Chan Chan, Troy: the names of lost cities are magical and nostalgic. But if we analyze the listings by Chandler and Fox of those ancient cities founded between 1360 B.C. and A.D. 620, we find that out of a total of 69, 31 survive today.[15] A 45 percent survival rate, across a time span that now averages over 2,500 years, is hardly a sign of evanescence. (If their listing is incomplete, the actual survival rate may be somewhat lower.) If we look at their more complete listing of cities built in the last 1,000 years, and count only the mature, major settlements—that is, all those that reached a size of at least 40,000 before 1900, or, in continents other than Asia (where larger cities appeared earlier), of at least 20,000 before 1600—we have a list of some 905 places. Of these, only 30 do not exist today, and 21 of the 30 were in Africa and the Americas, where the failure rate was near 10 percent. Moreover, only 20 of the remaining 875 survivors have dwindled to populations of below 5,000. Urban settlements seem to have sticking power, despite (could it be because of?) the concentrated wastes they generate.

Isolated disasters have not often caused a permanent abandonment, unless they were natural shifts that destroyed the economic base (such as extensive soil erosion, siltation, or change in sea level, which destroy a harbor, or long-continued drought), or unless they were purposeful devastations, executed with malice and power. Above all else, settlements need to provide safe transport, and security from human predators. The principal killers have been war, disorder, and the shift of trade. Final abandonment comes only after a long series of disasters that exhaust the will and capital of the survivors. Knossos, the great city of Crete, was devastated by an earthquake in 1700 B.C., and rebuilt as large as ever on the ruins. After the terrible volcanic explosion of Thera in 1500 B.C., Knossos was rebuilt again, but on a somewhat lesser scale. Taken by the Mycenaeans in 1450, it was levelled once more by cataclysm and fire in 1400. When rebuilt this time, it was a backwater settlement, and yet the written records do not cease. About 1200, it was destroyed once more by the Dorians, and finally abandoned. Salamis, another Cretan city, whose necropolis is twice the area of the city itself, endured three great earthquakes, siltation, a Jewish revolt, and Arab pillage, before its citizens abandoned it, transporting the old stones to be rebuilt at Famagusta.

Antioch, in Syria, founded in 307 B.C., was one of the great cities of the Hellenistic and Roman empires. Its collapse into a small provincial town came only after a 100-year period that included: a great fire in A.D. 525, followed by recurrent outbreaks for six months; the earthquake of 526, in which 250,000 persons died, almost all the buildings fell down, ruins and corpses were ransacked by thieves, trade ceased, and citizens emigrated; aftershocks, and another major quake in 528, in which all surviving buildings and walls fell and 5,000 were killed; capture and sack by the Persians in 540, after severe street fighting, burning of the city and suburbs, and deportation of the inhabitants to Persia; bubonic plague in 542, and recurring outbreaks thereafter; earthquakes in 551 and 557, when the walls came down again; cattle plague in 557; bubonic plague again in 560 and 561; earthquakes in 557 and 588, with 60,000 killed in the latter; loss of all the olive trees, a vital permanent crop, in the drought of 599; devastation of the crops by weevils in 600; capture again by the Persians in 611, and evacuation by them in 628; capture by the Arabs in 638, after the collapse of Roman power in Syria. At that point, Great Antioch was at last reduced to a minor settlement.

Or consider Baghdad: sacked by the Mongols in 1258; suffered the great plague of 1348; taken by Tamerlane in 1393, and retaken in 1401 with a great massacre; reported to be in ruins in 1437, but taken by the Safavids in 1508, then by the Ottomans in 1534, and again in 1638. It underwent floods, epidemics, mutinies, looting. The city survived, and is a great capital today.

Systematic destruction of a city in war has perhaps been a more common cause of final abandonment than natural disaster. Even then, a city is hard to kill, in part because of its strategic geographic location, its concentrated, persisting stock of physical capital, and even more because of the memories, motives, and skills of its inhabitants. The destruction of Carthage was an unusual success (although the site is now recommended for a new town), but the attempt to obliterate Poland's capital after the Warsaw Uprising was an instructive failure. The German army was ordered to destroy the city forever; no usable fragment was to remain. First, those who had survived the Nazi atrocities were evacuated. Section by section, the city was fired, to reduce its mass, and then blown up by demolition teams. A large, sophisticated military force organized a tremendous effort, over a period of several weeks. All the buildings fell, but a surprising amount of structure survived, both underground and at ground level. The very rubble of the explosions protected the remains, and blocked the movement of the fire and the demolition teams. Even more intractable was the consuming desire of the Polish people to restore their remembered city. So Warsaw reappeared.[16]

Atlanta was taken by Sherman's army in 1864; after a forced evacuation, it was burnt to the ground in two weeks of November. It had a population of 17,000 when the siege began, zero at the end of 1864, and 20,000 again by 1866, of whom 5,000 were widows. By 1869 it had reached 22,000, and was on its way to becoming a major industrial and rail center of the South.

If there is liquid capital, and society is organized to use it, rebuilding can be rapid, especially if the disaster is local and recovery has external support. The waste is quickly converted to fill or to new buildings. Like a homeless swarm of bees, there is surplus energy, and a strong motivation to remake the hive. Much depends on attitudes and on coordination: good communications, coherent values, and hope for the future. An economic boom is frequently set off by reconstruction, and in the end the physical plant is better than it was, and the economy at a higher level. Meanwhile, the stress reveals itself in personal relations, in bursts of nostalgia, in a resort to magic rituals, and scapegoating.

## DISASTER AND SOCIAL CHANGE

The explosion of a munitions ship in the harbor of Halifax, Nova Scotia, in 1917, killed 9,963 people, injured 9,000, and destroyed a two-and-one-half square mile area, causing $35 million in damage (1917 dollars). Previously, Halifax had been a static provincial town, but the effort to recover from this disaster set off a chain reaction. A new port was built, the retail section improved, the hospital enlarged, a new health center and central park created, a new street railway built, and telephone connections were laid to the rest of Canada and to the United States. Other changes included an influx of workmen, a union of churches, a new housing commission, and important initiatives in city planning, zoning, medical clinics, and sanitary inspection. Women tram conductors appeared for the first time. There were also new neaurasthenias and frequent rumors. As with any social upheaval, people were on edge, without knowing why.

The structure of society is rarely revolutionized by these wasteful events, although particular social groups may gain or lose and social shifts already in progress can be accelerated. After the Black Death in Europe, which caused a loss of one-third to one-half of the population in some areas, there were pogroms, outbreaks of hedonism, and a dancing mania. Some impetus was added to the liberation from feudal society, in places where it was already under way. Prices and wages rose; land rents fell. Church and manor were for a time disorganized. There were revolts, and they were suppressed. In the end, the social shifts were additions to changes in process.

## INEQUITIES OF WASTING

Not infrequently, however, inequities increase after a wasting, since disaster bears heavily on the poor, who have the narrowest margin of energy and capital with which to rebuild. New classes on the way up, on the other hand, may find special opportunities in the chaos. Observers who came to Managua, in Nicaragua, just after the earthquake of 1972 found that the worst burden of the recovery process fell on the poorest people, unable to rebuild, who suffered severe stress in very crowded living spaces. They were reduced to casual or part-time employment by the loss of the small center-city workshops. They were forced to move to the periphery of the city, from which the journey to work might be two hours each way. (After the Great Fire of London, in 1666, when the old City was rebuilt for the well-to-do, the poor were also driven to the outskirts.) The displaced Managuans had to learn new patterns of transport and activity since the old ones had been completely disrupted. They had lost contact with kin and friends, whose support they urgently needed. They were separated from the markets for cheap food. There were too few hours in the day to cover the commute to work and the effort to reestablish social contacts. The city, meanwhile, spread out and became more segregated, peppered with vacant lots and heaps of rubble. The old center was cordoned off, and a new middle-class shopping district arose farther out.

To go back three-quarters of a century, the great San Francisco earthquake and fire of 1906 had similar effects. Within five years, the city was repopulated and regarded as rebuilt. But there had been an exodus from the old center. The city had spread out, and was more segregated by class. By 1915, the new housing on the periphery alone occupied twice the area of the old housing destroyed. The upper classes were quickly reestablished, but the poor were on the move for years. The journey to work shortened for the former, and lengthened for the latter. Low-income jobs were lost due to the destruction of loft buildings, and purchasing power fell as prices rose. The city boomed, and so did the divorce rate.

## OTHER PLACES OF WASTE

Abandonment, dereliction, and destruction are not the only breeders of waste ground. There are uses not welcome in any settled community, but essential to the larger region. These include accommodation for people on the fringe of society in one way or another: halfway houses, mental hospitals, or low-income housing projects. There are also facilities that have some direct nuisance effect: highways, airports, truck and bus terminals, distribution centers, quarries, power plants, and heavy industry. There are some that need cheap quarters, or cannot or will not pay for public services: marginal industries, squatter housing, storage yards, and tax-exempt institutions. And there are the abhorred wasting facilities themselves: dumps, incinerators, sewage plants, and outfalls. Communities always applaud their location somewhere else in the region. We avoid them and yet depend on them.

In no contemporary new town in America, where every inch is planned, is there any provision for dump or burial ground. Indeed, there were very few cemeteries in the initial layouts of our 19th-century cities, although the churchyard had been a standard component of the colonial town. Now we keep death at a distance, and the thought that cemeteries are part of our disposal system is quite disturbing. We rarely enter them alive, except for the ceremony of interment. The old tradition of visiting the family graves on special days of the year is fading away. Yet graveyards were once the parks of the city, places of quiet escape and social recreation. In a few cases, they remain so today. The vast cemeteries surrounding Cairo were used on holidays by everyone. Now they are squatter settlements. Our own park-building movement began with such landscaped burial grounds as the Mt. Auburn Cemetery in Cambridge, Massachusetts, and the Spring Grove Cemetery in Cincinnati. Today, a cemetery is the nighttime haunt of adolescents, who are also in many ways at the margin of society. Moreover, these graveyards are a refuge for wild fauna and uncultivated plants.

## URBAN WILDERNESS

Wilderness will develop in almost any untended land. The site of an old railroad station in the heart of West Berlin, once the largest passenger station in Europe, is now a rich landscape of ruined walls, tracks, and bridges, overgrown by thickets and wildflower meadows. The site, bombed out in World War II, contains examples of one-third of all the flora of the region, including rare and endangered species and some indigenous forms, but particularly the exotic urban ornamental run wild. Half of this will be conserved as a lightly managed city wilderness. Fitter describes a similar vegetative seizure of the bombed-out sites of London in his wonderful *London's Natural History,* and also discusses the influence of refuse disposal on the plant and animal life of the city.[17] Every fall, polar bears invade Churchill, Manitoba, to feed on its garbage. The economic base of this small town rests partly on the scientists and tourists who come to see the bears.

Urban wilds, cemeteries, and city dumps move farther and farther out, as settlement spreads. The quest for disposal room grows more pressing, and is more fiercely resisted by the outlying suburban towns. The acceptable location of the margin becomes a regional issue. The swelling inflow of goods and energy into the dense urban area makes it increasingly difficult to return the waste to any productive cycle, or even to put it down where it will not offend. It was this very concentration of resources in cities, of course, that first provoked the great wastings of war. Now, as the region continues to expand, wastelands reappear at the city center, in the form of vacant lots, boarded housing, junked cars, and exhausted slums. The rural poverty and rural waste heaps of the past are being encompassed as underused land and marginal groups within the city itself. Remote or central, these wastelands are also the places where discarded ways of life survive, and where new things begin.

Within any city littered yards are used for low-cost storage and low-value activity, and fragmented, masterless spaces are used for disposal. Grady Clay has named them "sinks." Linwood Avenue, in inner Somerville, Massachusetts, is typical of such marginal areas. Isolated behind the elevated McGrath highway, it is accessible only by a single indirect entrance. Its low, repatched, concrete block buildings, spotted with signs, are closed in on themselves. These are warehouses, service industries, and repair depots. They stand within ragged dirt and asphalt yards, full of discarded objects. The broad streets, surfaced in cracked and oily paving, have no regular edges, but are

FIGURE 19–5  Urban "sinks," however unattractive, have their own values and delights. They are relatively free of social control and provide habitats where outdated things can survive and new ones may gain a foothold. [© Michael Southworth, used by permission of Michael Southworth]

sporadically lined with broken chain-link fences. Trucks and cars are double- and triple-parked, or nose into the yards. The workers are male, in rough and dirty clothes. An ugly, polluted, yet tolerant place, where the workers seem at ease (a remnant left by a carelessly planned highway), it is a refuge for infant and relict enterprises. However unseemly, these urban remnants are also freer places, where one is momentarily relieved of the pressures of status, power, explicit purpose, and strict control. These shabby careless backsides, these rear yards, outhouses, and urban ratholes, have their own delights.

There is another, older example in the Boston region. From its early days, South Boston has been a dead-end appendage of the mother city—a vermiform appendix—into which the Irish were excreted. City lands on the north slope of that peninsula were used in succession for a dump, an almshouse, a lunatic asylum, and a jail. That the city persisted in loading its wastes on their territory was a constant source of "Southie's" anger. So as the Irish gained political power, they converted that city land into Independence Square, and pushed the unwanted uses farther down the hill. Now Southie has elected mayors of its own: the vermiform appendix has burst, in an electoral uprising of the local neighborhoods.

Just out to sea, the harbor of Boston and its multitude of islands, once used for safe pasture, became a dumping ground. The Nonantum Indians died there on an island concentration camp, and there sat the prison camps of the successive wars. Refuse was burned on one island for decades, and sewage still flows out into the harbor waters, laying down a thick deposit of bottom sludge. Today, the harbor islands are being converted to recreation, as the city slowly turns round to face the sea once more. But Long Island, central to the group, still supports its ruined fort, and a dilapidated hospital for alcoholics and the chronically ill poor.

Marginal islands are always fair game. Randall's and Ward's Islands in New York's East River were repositories for city refuse throughout the city history. They were the site of garbage dumps, potter's fields, and almshouses in the 18th and 19th centuries. By 1934, their occupants included a sewage disposal plant, a City Hospital for the Feeble Minded and Tubercular, a "house of refuge" of the Society for the Reformation of Juvenile Delinquents, the Manhattan State Hospital for the Insane, a military hospital, and the piers of an unbuilt bridge.

The elevation in Rome called the Testaccio was produced by the heaping up of broken containers that accumulated behind the port area of ancient Rome. This low, empty hill, for centuries known as "the field of the Roman people," lay just outside the ancient city wall. In the 19th century, it also lay beyond the barrier created by the excavations of the archeological zone, and in the opposite direction from the city's growth. Here, in the 1880s, there was created a district for workers' housing and for the "*arti clamorose*": a brickyard, warehouses, gasworks, a central market, and various industrial and storage yards. It was Rome's first example of deliberate activity zoning, a kind of apartheid. Twenty years later, those tenements were desperately overcrowded; the streets were unpaved; there were no schools, clinics, or baths. The unbuilt lots were used for dumping trash and spoil. Infant mortality exceeded 50 percent. This was the area chosen in 1913 for building some of the first public, low-income housing in the city. Just as, back in Boston again, the swampy Columbia Point with its sewage outfall and its open trash heaps, lying on the edge of the city below South Boston, was chosen as the proper location for a large public housing project. It is a disaster area now being painfully recovered.[18]

Wastes are traditionally dumped at the edges of settlement—in areas where the powerless live, where land claims are weak, and where controls are soft. We find this phenomenon of the margin at many scales. In the house, things of small value are put in the cellar, the attic, or the garage. In a well-kept suburb, the compost heap, the brush pile, and the trash can are located at the lot line. When searching for the public dump or for nuisance industry in any New England town, look first along its boundaries with adjacent towns. An 18th-century manual on the founding of towns in Connecticut provides space for a ring of waste at a distance from the center of the settlement.[19] Just at the edge of Las Vegas, where the desert begins, there is a notable belt of old tins and glass. Older cities were surrounded by rings of trash; the defensive moat beyond the wall was an ideal dumping ground. (Unfortunately, dumping in the moat not only rendered the water noxious, but made an easy path for a future enemy.) Nineteenth-century travellers give vivid descriptions of the mounds of rubbish on the fringes of Alexandria and Cairo. In our Southwest, the trash of the pueblos flows down the precipitous sides of the ancient mesas, a spectacular showing, against an ancient backdrop, of the goods of our industrial society. At the scale of nations, outlaws and unwanted people live in the border mountains, swamps, and islands.

The bayou country of the Mississippi delta is another example of the margin. These marshes are fringe areas both socially and geographically. It is a fluctuating habitat on the lower end of a great river system, always dependent on the state and management of that lordly water, at some times subject to too much water or silt and at others to too little; subject to the making and unmaking of delta land, to the intrusion of salt water or fresh, and to the catastrophic effect of violent storms from the Gulf. It has been the refuge of the Acadians twice dislodged, of slaves, of ruined French aristocrats and poor Chinese. Its economy whips about at the tail of outside markets; it jumps and stumbles with the demand for crayfish, oysters, shrimp, furs, oil, or agricultural land. The habitat is frequently overturned by the clearance, drainage, and abandonment of cropland, or by the careless introduction of nutria, muskrat, or water hyacinth. Its Cajun and Sabine people, anxious and uncertain, live a free and exploitative life. They plunder the marsh, as they are plundered by the outside world. They rapidly use up, and visibly discard, manufactured goods of all kinds. Abandoned equipment surrounds their shacks: generators, boats, refrigerators, stoves. Their livelihoods—principally trapping for fur, or the gathering of shellfish—create vast quantities of organic waste. Usually, this work is only seasonal or intermittent, a peripheral to some work in a nearby town. The bayou is a landscape of waste, the ass-end of a great river valley, a marginal place for marginal people. It has its own beauty, its own free spirit, and its people are fiercely attached to it.

## A TANGLED MIX

Wasting is a pervasive (if valiantly ignored) process in human society, just as it is elsewhere in the living system. It is a feature of the underlying flux that carries us along, the everlasting impermanence of things. There is a short-term wasting of objects, and a long-term wasting of place. Each has its own characteristics. The rate fluctuates, and the flow is cyclical or directed, depending on circumstances. It threatens our health, our comfort, and our feelings. It interferes with the efficiency of our enterprises. Still, it has its own values. If we seek to preserve things, it is a ceaseless threat. If we look for continuity and not permanence, on the other hand, then wasting might be turned to account. Rarely has any accumulation of waste caused the abandonment of a settled place, unless it has served to hasten some natural evolution. Only occasionally has the environment been pushed to some truly irreversible dead end. Wasting has not usually caused fundamental social change, but it accelerates changes already under way, and shifts the distribution of burdens. It seems to us a tangled mix of good and evil, and mostly the latter. Hidden behind the polite facade of living its presence preoccupies us: it is an affair of the mind. Might there be pleasures in it, and practical opportunities? Could we be at ease with it?

**NOTES**

1 And Frankfurt has its "Monte Scherbelino" (literally, mountain of broken glass), made of war debris.
2 Donald G. McNeil Jr., "Deserted Buildings Turn into Hills at Breezy Point," *New York Times,* January 27, 1979, p. 1.
3 "Design and Appearance—1 and 2," *Building Research Station Digest,* no. 45.
4 The early design guidelines for Sea Ranch specified ungalvanized nails so that the rusting nails would ensure streaking patterns on the wood siding.
5 Colin Ward, ed., *Vandalism.*
6 Philip G. Zimbardo, "A Field Experiment in Auto Shaping."
7 Mikhail Bakunin, *Oeuvres,* vol. 1, p. 288.
8 It should be noted that decentralization of the American city has been heavily subsidized by the federal government, especially by means of highway construction and backing of home mortgages.
9 *Report of the President's Commission for a National Agenda for the Eighties* (Washington, DC: U.S. Government Printing Office, 1980).
10 Edgar Rust, "Development without Growth: Lessons Derived from the U.S. Metropolitan Experience."
11 The old canals of Lowell, Massachusetts, built to power the textile mills, have now been recycled along with the mills and machinery to become the framework for an educational and historical park.
12 Fernand Braudel, *The Mediterranean and the Mediterranean World in the Age of Philip II.*
13 Jack Lessinger, "The Case for Scatteration—Some Reflections on the National Capital Region *Plan for the Year 2000.*"
14 Sheafe Satterthwaite, "Puckerbrush, Cellar Holes, Rubble: Observations on Abandonment in Vermont."
15 Tertius Chandler and Gerald Fox, *3000 Years of Urban Growth.*
16 By the end of World War II, about 90 percent of the historic buildings, three-quarters of all residential buildings, and one-third of the streets of Warsaw had been destroyed. Both old and new towns of Warsaw were reconstructed building by building and almost brick by brick. In the old town, the reconstruction program included not only all the churches, but also the burghers' houses, dating from the 15th to 17th centuries. The interiors were rebuilt according to the extant architectural plans and facades were restored on the basis of old photographs and drawings. The maze of old town streets and squares and public foci such as the Fukier wine-shop were carefully reconstructed. The Royal Road, lined by historic residences, churches, and monuments, representing diverse architectural and sculptural styles from the 16th to 20th centuries, was rebuilt

meticulously, along with the two royal residences, which have a very important place in people's memories. Juliusz W. Gomulicki, *Warsaw* (Warsaw: Arkady, 1967).

17  Richard Sidney Richmond Fitter, *London's Natural History*.
18  As a member of the planning team, Kevin Lynch did much to recycle Columbia Point into a livable place (renamed Harbor Point) in the early 1980s. See "A Community Revitalization Plan for Columbia Point," a planning report prepared by Carr/Lynch.
19  There are no references to this manual in Lynch's notes, and discussions with the Connecticut Historical Society and the Hartford Athenaeum have failed to locate the source. However, an anonymous pre-1638 paper, "Essay on the Ordering of Towns" (found among the papers of John Winthrop, governor of the Massachusetts Bay Colony 1630–1649) describes the ideal Puritan town plan. It consisted of 6 concentric circles set within a six-mile square. At the center was the meeting house, which was surrounded by successive rings of houses, common fields, livestock farms, and estates. Outside the fifth ring were the "swamps and rubbish waste grounds" which were to be owned but not occupied by the town. [See John R. Stilgoe, "The Puritan Townscape: Ideal and Reality," *Landscape* 20, no. 3 (Spring 1976), pp. 3–7.]

选读 20

"价值观"

伊恩·伦诺克斯·麦克哈格

麦克哈格在《设计结合自然》这本书中调查了生态响应和环境敏感性问题。现在人们认为他的作品对当代"绿色建筑"和可再生建筑过程的思想发展具有里程碑式的意义。在这篇选读中，麦克哈格分析了致使物质环境恶化的西方哲学先例。他在分析中直言不讳地写道："这就是我们继承的许多破烂的古老观点，这些观点大多数出自于无知，使人产生恐惧和敌对情绪，肯定会产生破坏，无助于创造。"

## READING 20

# On Values

**Ian Lennox McHarg**

McHarg investigates in *Design with Nature* the issues of ecological responsiveness and environmental sensitivity. His work is now recognized as a significant milestone in the development of our contemporary concerns for "green architecture" and renewable construction processes. In this selection McHarg analyzes the precedents in Western philosophy that have led to the degradation of the physical environment. He does not mince words in his analysis: "Such is our inheritance. A ragbag of ancient views, most of them

*Source:* "On Values" from *Design with Nature* by Ian L. McHarg. Copyright © 1971 by Ian L. McHarg. Reproduced by permission of Ian L. McHarg. Doubleday, pp. 66–77.

CHAPTER 5: THE NATURAL ENVIRONMENT 285

**FIGURE 20–1**   Taos Mountain and Pueblo. [New Mexico State Tourist Bureau]

breeding fear and hostility, based on ignorance, certain to destroy, incapable of creation."

Who can imagine that virgin continent of America accumulating in age and wealth, inordinately stable, rich beyond the dreams of avarice in everything that man could desire. Moreover, it was a new world, some ten thousand years ago, that land which

FIGURE 20–2  [California Anti-Litter League]

we can never see again. Yet, the men who followed the trails of the huge herbivores during some interglacial period from Siberia across the land bridge to America, probably discerned little change; the new environment was much like the old. It could not have been until they made deep penetrations to the south that profound changes were discernible. Yet the passage of time must have been great and the memories of distant ice sheets would have been only tribal sagas, intermeshed with fancies and mythology.

The men who first entered this new world were another type of predator, in some ways little different from the sabertooth. Like any other predator, they were limited by the numbers of their prey—but man introduced a new and powerful tool which proved that he was no ordinary hunter.

The emergence of the seed in Jurassic times precipitated the explosion of the flower. The embryonic plant encapsulated in the seed was infinitely more enduring and mobile than the early naked seeds or the older spore. It was this new flowering plant with its fleshed embryo that colonized the world and made available a food source unknown in previous eras. Of all the repercussions of the angiosperm, none was more dramatic than the clothing of the prairies with grasses. It was these which sustained the enormous populations of large, fleet-footed, far-ranging creatures, in turn the prey of the predators, chief among which—one day—would be the human hunters. The grasses covered the prairies; the herbivores were sustained by them and proliferated. Onto this scene entered the new predator with a tool more powerful than required, beyond his power to control and of enormous consequence—a criticism as topical for atomic man.

This new and devastating tool was fire. It was no novelty in the prairie—lightning-caused fire was common and indeed the prairie climax was a response to it. But the induced fire of the hunter was more frequent than the natural occurrence. The prairies were burned to drive the bison and deer, the mammoth and mastodon into closed valleys or over precipices. This was a time of climatic adversity, threatening to the creatures as well as to men. It is thought that it was the combination of human hunters and a hostile climate that resulted in the extinction of this first great human inheritance in North America, the prairie herbivores. Firelike the grasses spread, firelike the herds of grazing animals swept to exploit the prairies—and it was the fire of the aboriginal hunter that hastened or accomplished their extinction. This was the first major impact of man on the continent during the aboriginal occupation.

In the subsequent millennia there were, as far as is known, no comparable depredations. If one can infer from the ways of the North American Indians, there evolved a most harmonious balance of man and nature. The gatherer and hunter learned to adapt his cropping to the capacity of the crop and prey. In this evolution there must have developed an understanding of creatures and their habits. Hunting must respond to the breeding seasons, be protective of pregnant females, cull the surplus males. This is a major step in human evolution. The first ancestral tree shrews were puny creatures among the gigantic, predatory dinosaurs. The fire-wielding human hunter was no longer puny—he had equaled the depredations of *Tyrannosaurus rex*—but the hunter who adapted his hunting practices to the habits and capacity of his prey was truly thinking man and this was the first testament to brains as a device to manage the biosphere. This is then no longer simply man, the speaking animal, the maker of stone tools or man, the agent of fire—it is man, the thinking hunter. Yet we must take care not to make too large a claim. Many other creatures, whose brains are less vaunted, have also been able to regulate their populations to available prey.

Considerations of "primitive" men have been obscured by the wide divergence of views that range between the idealized "noble savage" on one hand and the conception of aboriginal peoples being "missing links" on the other. It seems clear that simplicity neither ensures nor denies nobility, that the mental equipment of the primitive man is indistinguishable from that of his most sophisticated brethren. The supremacy of the latter, in his own terms, lies in the inheritance of the tools, information and powers from his predecessors. While there are a few exceptions, "wild nature" seldom provides an ideal environment and those men who live in primitive societies are susceptible to disease, suffer from a short lifespan and are vulnerable to extremes of heat and cold, drought, starvation and exposure. They often suffer from fears and superstitions, but they have frequently acquired an astonishing empirical knowledge of their environment, its creatures and their processes. This is absorbed into religion or superstitions. Indeed it might well be said that their success, their adaption, is precisely this understanding. Societies that sustained themselves for these many millennia are testimony to this understanding; it is indeed the best evidence in support of brain, the presumptive manager of the biosphere.

Paganism is an unnecessarily pejorative term; pantheism is a better word. Who knows God so well that he can reject other likenesses? With Voltaire one asks to see his credentials. Animism, which permeates pantheism, involves the theory of the existence of immaterial principle, inseparable from matter, to which all life and action are attributable. In the pantheist view the entire phenomenal world contains godlike attributes: the relations of man to this world are sacramental. It is believed that the actions of man in nature can affect his own fate, that these actions are consequential, immediate and relevant to life. There is, in this relationship, no non-nature category—nor is there either romanticism or sentimentality.

The Iroquois view is typical of Indian pantheism. The Iroquois cosmography begins with a perfect sky world from which falls the earth mother, arrested by the birds, landing upon the back of a turtle, the earth. Her grandchildren are twins, one good and the other evil. All that is delightful and satisfactory derives from the first: twin streams that flow in both directions, fat corn, abundant game, soft stones and balmy climate. The evil twin is the source of bats and snakes, whirlpools and waterfalls, blighted corn, ice, age, disease and death. The opposition of these two forces is the arena of life; they can be affected by man's acts in the world of actuality. Consequently all acts—birth and growth, procreating, eating and evacuating, hunting and gathering, making voyages and journeys—are sacramental.

In a hunting society the attitude to the prey is of vital significance. Among the Iroquois the bear was highly esteemed. It provided not only an excellent hide and meat, but also oil that was used for cooking and could be stored. When the hunted bear was confronted, the kill was preceded by a long monologue in which the needs of the hunter were fully explained and assurances were given that the killing was motivated by need, not the wish to dishonor. Now if you would wish to develop an attitude to prey that would ensure stability in a hunting society, then such views are the guarantee. Like the crystal of potassium permanganate in a beaker of water, diffusing into equilibrium, in steady state, the hunter who believes that all matter and actions are sacramental and consequential will bring deference and understanding to his relations with the environment. He will achieve a steady state with this environment—he will live in harmony with nature and survive because of it.

It is deep in history that we abandoned such a view. The conception of man—exclusively divine, given dominion over all life and non-life, enjoined to subdue the earth—contained in the creation story of Genesis represents the total antithesis of the pantheist view. While the Greeks conceived not only of man Gods, but nature Gods as well, this survived only marginally into the humanism of the Renaissance and pantheism has been lost to the western tradition; in Europe it persists only with the Lapps. Yet, as leading theologians retreat in consternation from the literality of Genesis—Buber and Heschel, Tillich and Weigel and even more Teilhard de Chardin, offended by its arrogant transcendence—the more quietly deferential view of the pantheists seems to present a better beginning, at least a working hypothesis. If divinity there is, then all is divine. If so, then the acts of man in nature are sacramental.

In Central and South America the aboriginal societies developed great cultures—Maya, Aztec, Tolmec, Toltec. In the north, there were no such products. Here, very simple hunting and gathering societies with a primitive agriculture evolved upon the land, thinking predators who managed to sustain equilibrium in the system for many thousands of years. They developed a great acuity to nature and its processes and institutionalized this in a variety of pantheist cosmologies. These may well be unacceptable to modern western man, but they were effective as a view of man-nature for these societies and their technology.

Generally the members of these aboriginal societies could promise their children the inheritance of a physical environment at least as good as had been inherited—a claim few of us can make today. They were, in the history of America, the first occupants and they could claim to have managed their resources well. Life and knowledge have become more complicated in the intervening centuries, but, whatever excuses we offer, it is clear that we cannot equal this claim.

It is quite impossible to recreate the awesome sense of discovery experienced by Columbus and Cortés, Cabot and Cartier, Frobisher and Drake. It is difficult to sense the wonder of the following thousands who too encountered lands and prospects as yet untouched, unseen by western man, they who came in search of refuge, land, gold, silver, furs or freedom and found, whether they knew it or not, the last great cornucopia of the world's bounty. Who again can experience with Balboa a new continent and a new ocean?

But still much remains untouched where men have only seen or left some footprints on the ground. Those great preservationists—inaccessibility and poverty—have ensured that there is still an image of an earlier time when the men of the west came to make this continent their own. Consider Mount McKinley and the Athabaska Glacier, the North Atlantic beating on the rocky coast of Maine, the Kilauea Volcano in Hawaii, the glory of Yosemite and the Tetons, the park landscape of Texas and Oklahoma, the extensive painted deserts of Arizona and New Mexico, the palmetto, the mangrove swamps, the sculpture of Zion and Bryce, the geological fantasy of the Grand Canyon, the sandbars of Hatteras, the Appalachian Plateau, heartland of the great eastern forest, the gigantic redwoods fringing the Pacific, the fogs that gather about them, Crater Lake, Nantucket, the Columbia, the Sangre di Cristo Mountains, the rain forest of the northwest, the rich beauty of the Adirondacks, the continent-draining Mississippi and its delta.

Sad losses there have been, but the grizzly is still with us as are the bison, elk, moose, caribou and antelope, the wild goat, mountain lion, cougar, lynx, bobcat, coyote, the bald eagle and osprey, the great heron, the whales breeding off Baja California, seals and sea lions, sharks, porpoise and dolphin, sailfish and tuna.

In Bandolier are the habitations of early Americans, at Mesa Verde the works of their successors, while at Zuni, Taos and Acoma they live today. But of the jewels of this great inheritance, that which most of all justifies the title cornucopia, there remains only a whimpering trace. The fringe of the railroad and the uncultivated hedgerow are the descendants of those grasses which built the prairie sod, deep and fertile—a geological deposit of a richness exceeding all the dreams of gold and silver, coal and iron. Of the prairies there is hardly any trace and little more of the great beasts that once dominated them.

**FIGURE 20–3**   Villa d'Este.

**FIGURE 20–4**  Parterre de Broderie.

**FIGURE 20–5**  Versailles. [Historic Urban Plans, Ithaca]

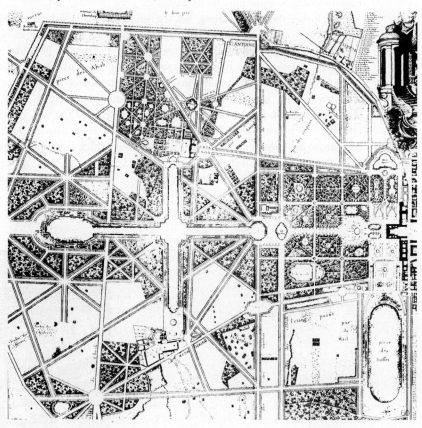

When Columbus, Ponce de León, Cortés, Cabrillo, and Coronado arrived in America, they brought with them the Iberian tradition. Cabot, Frobisher, Drake, Hudson and Baffin and their men transmitted the mores of England while Cartier, Marquette and Joliette were the vanguard for the culture of France. While these and their fellow nationals who followed were united in the zeal for exploration and conquest, there were important distinctions in the attitudes each brought to bear upon this primeval continent.

If one looks through that narrow aperture of history, at the attitudes to the land which these cultures brought, it is apparent that there are four distinct divisions and that each has national associations. The first explorations in the 16th century were reflections of the great release of Renaissance humanism. This originated in Italy and it was here that is to be found the humanist expression of man and nature.

This assumption of power by man, rejecting the cosmography of the Middle Ages, is seen in a procession of projects. The first of these are the villas and gardens of Florence, after which the epicenter of expression moved to Rome and Tivoli. Bramante, Ligorio, Raphael, Palladio and Vignola created the symbolic expression of humanism upon the land, to be seen in the Villa Medici, Poggio a Cajano, the Villa d'Este and the Villa Lante, the Villa Madama and the Boboli Gardens and, in the final phase, the Villa Aldobrandini and Mondragone. In these the authority of man was made visible by the imposition of a simple Euclidean geometry upon the landscape, and this is seen to increase within the period. Man imposes his simple, entertaining illusion of order, accomplished with great art, upon an unknowing and uncaring nature. The garden is offered as proof of man's superiority.

The second stage occurred a century later—at the time of the earliest colonial settlements—but the locus of power and expression had moved to France. Here the same anthropomorphic simplicity was applied at larger scale upon a flat and docile landscape. So at Vaux-le-Vicomte and Versailles one sees the French baroque expression through the works of André Le Nôtre, the zenith of Euclid upon the land. Louis XIV lay transected by the twin axes at Versailles, king by divine right, the ordered gardens below testimony to the divinity of man and his supremacy over a base and subject nature. Or so it seemed.

In the western tradition, with the single exception of the English 18th century and its extensions, landscape architecture has been identified with garden making—be it Alhambra, the Abbey of St. Gall, d'Este or Versailles. In this tradition, decorative and tractable plants are arranged in a simple geometry as a comprehensible metaphysical symbol of a submissive and orderly world, created by man.

Here the ornamental qualities of plants are paramount—no ecological concepts of community or association becloud the objective. Plants are analogous to domestic pets, dogs, cats, ponies, canaries and goldfish, tolerant to man and dependent upon him; lawn grasses, hedges, flowering shrubs and trees, tractable and benign, are thus man's companions, sharing his domestication.

This is the walled garden, separated from nature: a symbol of beneficence, island of delight, tranquillity and introspection. It is quite consistent that the final symbol of this garden is the flower.

Not only is this a selected nature, decorative and tame, but the order of its array is, unlike the complexity of nature, reduced to a simple and comprehensible geometry.

**FIGURE 20–6** Vaux-le-Vicomte. [Used by permission of Ian McHarg]

This is then a selected nature, simply ordered to create a symbolic reassurance of a benign and orderly world—an island within the world and separate from it. Yet the knowledge persists that nature reveals a different form and aspect beyond the wall. Loren Eiseley has said that "the unknown within the self is linked to the wild." The garden symbolizes domesticated nature, the wild is beyond. It is indeed only the man who believes himself apart from nature who needs such a garden. For the pantheist nature itself best serves this role.

Each century saw a migration of power and in the 18th century it moved to England in the third phase, where there arose that unlikely efflorescence which is the beginning of a modern view. Believing that some unity of man-nature was possible and could not only be created but idealized, a handful of landscape architects took the dreams of writers and poets, images of painters of the period and the hints of a quite different order from the Orientalist Sir William Temple and, through the successive hands of William Kent, Humphrey Repton, Lancelot ("Capability") Brown, Uvedale Price, Payne Knight, and William Shenstone made over that raddled landscape of England into the fair image we can see today. Never has any society accomplished such a beneficent transformation of an entire landscape. It is the greatest creation of perception and art of the western world and it is a lesson still largely unlearned.

In the 18th century in England landscape architects "leaped the fence, and saw that all nature was a garden."* Beyond the fence "Men saw a new creation before their eyes." The leap did not occur until a new view of nature dispelled the old and a new aesthetic was developed consonant with the enlarged arena.

Starting with a denuded landscape, a backward agriculture and a medieval pattern of attenuated land holdings, this landscape tradition rehabilitated an entire countryside, allowing that fair image to persist to today. It is a testimony to the prescience of Kent, Brown, Repton and their followers that, lacking a science of ecology, they used native plant materials to create communities that so well reflected natural processes that their creations have endured and are self-perpetuating.

The functional objective was a productive, working landscape. Hilltops and hillsides were planted to forest, great meadows occupied the valley bottoms in which lakes were constructed and streams meandered. The product of this new landscape was the extensive meadow supporting cattle, horses and sheep. The forests provided valuable timber (the lack of which Evelyn had earlier deplored) and supported game, while freestanding copses in the meadows provided shade and shelter for grazing animals.

The planting reflected the necessities of shipbuilding. The preferred trees—oak and beech—were climax species but they were planted *de novo*. On sites where these were inappropriate—northern slopes, thin soils, elevations—pine and birch were planted. Watercourses were graced with willows, alders, and osiers, while the meadows supported grasses and meadow flowers.

The objective, however, was more complex than function alone. Paintings of the Roman Campagna by Claude Lorraine, Poussin and Salvator Rosa, a eulogy of nature, which obsessed poets and writers, had developed the concept of an ideal nature. Yet, it clearly had not existed in the medieval landscape. It had to be created. The ruling principle was that "nature is the gardener's best designer"—an empirical ecology. Ornamental horticulture, which had obtained within garden walls, was disdained and a precursory ecology replaced it. The meadow was the single artifice—the remaining components were natural expressions, their dramatic and experiential qualities exploited, it is true, but deriving in the first place from that observed in nature.

Nature itself produced the aesthetic; the simple geometry—not simplicity but simple-mindedness—of the Renaissance was banished. "Nature abhors a straight line." The discovery of an established aesthetic in the Orient based upon the occult balance of asymmetry confirmed this view. In the 18th century landscape began the revolution that banished the classical image and the imposition of its geometry as a symbol of man-nature.

This tradition is important in many respects. It founded applied ecology as the basis for function and aesthetics in the landscape. Indeed before the manifesto of modern architecture had been propounded—"form follows function"—it had been superseded by the 18th-century concept, in which form and process were indivisible aspects of a single phenomenon. It is important because of the scale of operation. One recalls that Capability Brown, when asked to undertake a project in Ireland, retorted,

---

*Horace Walpole. *Anecdotes of Painting in England with Some Account of the Principal Artists*, collected and digested by George Vertue (London: Henry G. Bohn, 1849). Vol. III, p. 801.

**FIGURE 20–7**  Central Park, New York. [J. Clarence Davis Collection, Museum of the City of New York]

"I have not finished England yet." Another reason for its importance lies in the fact that it was a creation. Here the landscape architect, like the empiricist doctor, found a land in ill health and brought it to good heart and to beauty. Man the artist, understanding nature's laws and forms, accelerated the process of regeneration so well indeed that who today can discern the artifice from the untouched? Nature completed man's works.

It is hard to find fault with this tradition: but one must observe that while the principles of ecology and its aesthetic are general, the realization of this movement was particular. It reflects an agricultural economy, principally based upon cattle, horses and sheep. It never confronted the city, which, in the 18th century, remains the Renaissance prototype. Only in the urban square, in parks, circles, and natural plantings is the 18th-century city distinguishable from its antecedents.

The rejection of nature as crude, vile—the lapsed paradise—and the recognition of the land as the milieu of life, which could be made rich and fair, is the great *volte face* of the western world. It did undoubtedly have some strange advocates; it encased the illusion of the noble savage and many other views, indeed it succumbed to an excess of romanticism—hired hermits standing picturesquely beside grottos and broken Greek urns—but it was a precursory ecology, its practitioners were more perceptive

**FIGURE 20–8**   Myrtle Court, Alhambra. [Used by permission of Ian McHarg]

and capable than its theorist advocates. And it has endured.

Yet this entirely novel view, the best of all for those who would open a great natural treasure house, did not enter the American consciousness until the mid-19th century, when the gothic preoccupations attending its final phase were advocated by Andrew Jackson Downing. It was not until the end of the century that the English landscape tradition found a worthy advocate in Frederick Law Olmsted, but it was too late to affect the American ethos in any profound way; the west had been opened and the great depredations were not to be halted. Yet it was from this source that the National Park System, the parkway, the college campus and the humane suburb were all derived.

But only in the smallest part was the American style affected by the great 18th-century experiment. The dominant intention was to conquer nature and the resulting form is either the evidence of despoliation itself or, if it is symbolized, in the simple-mindedness of a Euclidean geometry. The 18th-century landscape tradition exists in those reserves in which great natural beauty persists and in the small but precious oases that redeem the city.

Older than all of these—and, in a certain sense, a living tradition—is yet another quite different view: that derived from Islam and absorbed into the Spanish and the Hispano-American tradition. From the 9th to the 12th centuries the Moors civilized North Africa and the Iberian peninsula, offering testimonies of a culture undreamed of by their laggard European neighbors. They survived the uncultivated

Crusaders, but in Spain succumbed to that great iconoclast Charles V, who relentlessly destroyed Muslim art and architecture to replace it with the parochial crudities of the Spanish Renaissance.

The attitude of Islam to nature derived from exactly the same source as the barbarism of Charles V. Both came from Genesis. The Moors emphasized the second chapter, with the injunction to dress the garden and keep it—man the steward—and developed the belief that man could make a garden of nature; paradise could be created by wise men and realized by artists. Moreover the paradise garden was an ingredient of urban form.

It would be charitable to suggest that this most benign, unchristian, Asian view permeated Spanish thought, but it does persist as a particularly felicitous adaptation to hot arid climates and provides a most direct and beautiful expression of which its Islamic prototypes, Alhambra and the Generalife, are the most brilliant testimonies. But the great Islamic tradition is all but dead, its present image composed of the decadent urban forms of the Ecole des Beaux Arts, with the new intrusions of an inappropriate International Style architecture.

The final phase includes the 19th and 20th centuries. In largest part it represents the age-old attitude of conquest, but now powered by larger and ever larger tools. Its great contribution is to the increasing concern for social justice—but as to the land, nothing has changed. We see the descendant of the small, cowering primitive animal, rather poorly endowed, omniverously eating carrion, roots, birds' eggs, and the occasional kill, who has built a great cultural antagonism to a beneficent nature. The instincts that had sustained his ape ancestors and the empiric knowledge of his later human ancestors were lost and his brain was still inadequate to allow him to eat from the cornucopia: his hostility increased. Today he can savor the benison of the land, but his hostility remains like a vestigial tail or appendix.

Our injunction is not ambiguous: man is exclusively divine, given dominion, enjoined to subdue the earth. Until Aaron David Gordon proposed, as a purpose of Zionism, that Jews return to the land to rediscover God, Judaism showed no contrary views. The medieval Christian Church introduced otherworldliness, which only exacerbated the consequence of the injunctions of the old law. Life on earth was seen as a probation for the life hereafter. The earth and nature were carnal, they constituted temptations of the devil. It was a lapsed earth, fallen from Eden—nature shared man's original sin: indeed it represented his temptation and the reason for his fall from grace. There were contrary views: Duns Scotus and Erigena sought to show nature to be a manifestation of God, while Francis of Assisi sought to love nature rather than to conquer it. But the view was not well received and on his death Francis received his reward—his Order was given to one of the most venal men in Christendom.

Within the Protestant movement there are two distinct variations. The Lutherans emphasized the here and now, the immanence of God, which required perception rather than action. In contrast, the Calvinists were determined to accomplish God's work on earth, to redeem nature through the works of godly man. Calvin believed that his role was to conquer carnal, bestial nature and make it subject to man, the servant of God.

In this perusal there are two clear paradoxes. The same Semitic people, living in the same arid and hostile environment, deriving their religious views from the same source in Genesis, developed two quite distinct views of man-nature. The first, represented in Islam, emphasized that man could make paradise on earth, make the desert bloom, that he was the creator and the steward. The Jews and the later Christians emphasized conquest.

The 18th century in England saw an astonishing efflorescence in which developed the view of all nature as a garden: man could make the earth at once rich and beautiful. Within a century this new view transformed the medieval face of an impoverished England, with the most backward agriculture in Europe, into its leader. Yet, this same England, with a mainly Anglican population—more akin in its views to Luther than to Calvin—was the cradle of the industrial revolution and became the leader in the conquest and despoliation of nature.

There remains that aberrant theme, the pagan view never completely suppressed, evident in classical Greece, widespread in Rome, vestigial in the Middle Ages—where its celebrations, incorporated into Christian festivals, retained some of their older connotations—and the naturalist theme in the 18th century. The neolithic memory persists perhaps most strongly today in that movement which is called Conservation. It seems clear that, whatever religion its adherents espouse, their devotion to nature and its cherishing and nurture derives little from either Judaism or Christianity.

The attitudes of Jews in the wilderness or in simple settlements had little immediate effect upon nature. The same attitudes in the medieval Christian Church were of as little consequence. Medieval cities huddled behind walls while nature surrounded them like a mighty ocean. Inside the city walls their paeans could rise in the high vaulted sublimity of Gothic architecture, but nature was unaffected. In the Renaissance the views of humanism produced many beautiful gardens based on a most inadequate view of man and nature, which (if not taken seriously as metaphysical symbols) can only delight. But in the French Renaissance, where the same theme was spoken with a louder voice, one begins to fear for the consequence of this great illusion. Its bearers are about to discover the ends of the earth and bring their conqueror's creed to other peoples and to all the waiting lands.

The 18th century produced the new view, the emancipated man: but while this transformation affected a nation, it did little or nothing to modify the attitudes of all of the *conquistadores* who spread to rape the earth. Indeed it was not sufficient to temper the next generation of Englishmen, who so eagerly espoused the industrial revolution. If Stowe and Woodstock, Rousham and Leasowes are the symbols of the 18th century, then the dark satanic mills, the Manchesters and Bradfords, are the symbols of its successor.

Such is our inheritance. A ragbag of ancient views, most of them breeding fear and hostility, based on ignorance, certain to destroy, incapable of creation. Show me the prototypical anthropocentric, anthropomorphizing man and you will see the destroyer, atomic demolition expert, clear feller of forests, careless miner, he who fouls the air and the water, destroys whole species of wildlife: the gratified driver of bulldozers, the uglifier.

The early colonists who came to this continent were truly pre-Copernican, their ignorance cannot be our excuse. Their rapacity was understandable if deplorable. Their whole inheritance had seemed a war against nature; they were determined to conquer this enemy. They were unaware that it had been the selfsame depredations, accomplished by the same ignorance, that had depleted their historic homelands. Yet this was their heritage and their view—nature bestial, savage, rude, the arena of the carnal, the temptation of the flesh, the antithesis of the aspiration to godliness. We might well ask whence came this astonishing illusion, this most destructive of all views, a testimony to a profound inferiority complex, reflected in aggression. The aboriginals whom they confronted bore no such resentment. They had other views of human destiny and fulfillment.

We have looked at the attitudes that our ancestors of many races and creeds brought with them to this waiting continent. We can see today the consequences of these views—they are written on the land, our institutions, and the cities. Much that can be seen is remarkable and the greatest of testimonies to this people. It is the arena of the only successful social revolution. Consider the disillusion of the justifiable Russian revolt, the tragedy of China embarking on a Russian adventure half a century later with nothing learned. The French Revolution was inconclusive and class conflict persists. The great glory of Madison and Hamilton, Jefferson and Washington was that they engineered the first successful social revolution. It is incomplete on several counts, but it remains the great example for the world to see.

Parallel to this great accomplishment runs a countertheme. During this same period when the streams of colonists and refugees exercised their industry and inventiveness, when the fruits of this labor were increasingly disbursed, there occurred the most wanton, prodigal despoliation of resources that the world has ever seen. More, the products of these efforts, made visible in cities and towns, increasingly preempted the exclusive title of the greatest uglification and vulgarity in world history. Much smaller nations—Switzerland and Sweden, Norway and the Netherlands—could offer to the world's view vastly superior evidence as land custodians and builders of cities.

The ransacking of the world's last great cornucopia has as its visible consequence the largest, most inhumane and ugliest cities ever made by man. This is the greatest indictment of the American experiment. Poverty can exercise a great constraint on vulgarity—and wealth is its fuel: but this alone cannot explain the American failure. It is clear that a profound ignorance, disdain, and carelessness prevails. It is because of these that we are unable to create a handsome visage for the land of the free, the humane and life-enhancing forms for the cities and homes of the brave.

## THE NATURAL ENVIRONMENT—SUGGESTED READINGS

Francis, Mark, and Randolph T. Hester Jr. eds. *The Meaning of Gardens*. Cambridge, MA: MIT Press, 1990.
Hunt, John Dixon, and Peter Willis, eds. *The Genius of the Place*. Cambridge, MA: MIT Press, 1988.
Jackson, John B. *Discovering the Vernacular Landscape*. New Haven, CT: Yale University Press, 1984.
Jellicoe, Geoffrey, and Susan Jellicoe. *The Landscape of Man*. London: Thames and Hudson, 1995.
Jensen, Jens. *Siftings*. Baltimore: Johns Hopkins Press, 1990.
Lynch, Kevin, and Gary Hack. *Site Planning*. Cambridge, MA: MIT Press, 1984.
Moore, Charles; William J. Mitchell; and William Turnbull. *The Poetics of Gardens*. Cambridge, MA: MIT Press, 1988.
Stea, David, and Mete Turan. *Placemaking*. Aldershot: Avebury, 1993.
Tuan, Yi-Fu. *Topophilia*. Englewood Cliffs, NJ: Prentice Hall, 1974.

# 6

# THE HUMAN ENVIRONMENT
# 人文环境

选读 21
"作为自我象征的房子"
克莱尔·库珀·马库斯

克莱尔·库珀·马库斯应用卡尔·荣格（Carl Jung, 1875—1961, 瑞士心理学家）的心理学理论——集体无意识、原型和符号——来探知"房子"的含义。她在文中发问："但是与其他空间相比，为什么我们在这个特殊的盒子里更能发现本我？它看起来就好像是伴随我们同行的个人空间气泡，几乎是自我的有形延伸，这个气泡不断膨胀，让我们相信我们选定好的房子其实就是我们自己。"库珀·马库斯敦促设计师研究调查他们自己的个人偏好并开始"理解现在的自我形象是如何在无意识间被设计具体化的"。作者在总体上想传递给读者的观点是可能在无视科学方法的领域能够找到更为深刻的真理。

## READING 21

## The House as Symbol of the Self
**Clare Cooper Marcus**

Clare Cooper Marcus applies the psychological theories of Carl Jung—the collective unconscious, the archetype, and the symbol—to ascertain the meaning of "house." She asks: "But why in this particular box should we be ourselves more than in any other? It seems as though the personal space bubble which we carry with us and which is an almost tangible extension of our self expands to embrace the house we have designated as ours." Cooper Marcus urges designers to investigate his or her own biases and to begin "to understand how present self-images are being unconsciously concretized in design." The author's overall

*Source:* "The House as Symbol of the Self," by Clare Cooper Marcus, University of California, Berkeley, reprinted from Jon T. Lang, ed., *Designing for Human Behavior* (New York: Dowden, Hutchinson & Ross, Inc., 1974), pp. 130–46. Reproduced by permission of Clare Cooper Marcus.

message is that the deeper truths may be found in areas that defy the scientific method.

## INTRODUCTION

My work of the last few years comprised sociological surveys of people's responses to the designs of their houses and communication of the resultant guidelines to architects. But I have experienced a nagging doubt that I was merely scratching the surface of the true meaning of "the house." There seemed to be something far deeper and more subliminal that I was not admitting, or that my surveys and investigations were not revealing. The exciting personal discovery of the work of the psychologist Carl Jung has opened a door into another level of my own consciousness which has prompted me to consider the house from a wholly different viewpoint. This paper is a tentative initial exploration into the subject.

The reader must expect no startling, all-embracing conclusion; there is none. This is a speculative think piece and is deliberately left open-ended in the hope that it will motivate the reader, and the author, to think further and more deeply in this area.

## JUNG'S CONCEPTS OF THE COLLECTIVE UNCONSCIOUS, THE ARCHETYPE, AND THE SYMBOL

Three of the most significant contributions of Carl Jung to the understanding of the human psyche are the concepts of the *collective* unconscious, the archetype, and the symbol. Sigmund Freud postulated an *individual* unconscious in which are deposited the suppressed and repressed memories of infancy and childhood. Theoretically, the psyche keeps these memories in storage until they are reawakened into consciousness by the medium of the dream, or its waking equivalent, free association.

Initially embracing Freud's theories, Jung became increasingly dissatisfied as his studies of persistent motifs in his patients' dreams and fantasies, and in primitive mythology and folk tales, revealed what seemed to be *universal* patterns which could not be accounted for solely by the theory of an *individual* unconscious. He began to postulate the theory of an individual unconscious plus a universal or collective unconscious linking man to his primitive past, and in which are deposited certain basic and timeless nodes of psychic energy, which he termed *archetypes*.

Jolande Jacobi has termed the archetype "a profound riddle surpassing our rational comprehension."[1] It precedes all conscious experience and therefore cannot be fully explained through conscious thought processes. Perhaps one of the simplest analogies is that offered by Jacobi of a kind of "psychic mesh" with nodal points within the unconscious, a structure which somehow has shaped and organized the myriad contents of the psyche into potential images, emotions, ideas, and patterns of behavior. The archetype can only provide a potential or possibility of representation in the conscious mind, for as soon as we encounter it through dreams, fantasies, or rational thought, the archetype becomes clothed in images of the concrete world and is no longer an archetype: it is an *archetypal image* or *symbol*. As Jacobi has written:

> Man's need to understand the world and his experience in it symbolically as well as realistically may be noted early in the lives of many children. The symbolic imaginative view of the world is just as organic a part of the child's life as the view transmitted by the sense

organs. It represents a natural and spontaneous striving which adds to man's biological bond a parallel and equivalent psychic bond, thus enriching life by another dimension—and it is eminently this dimension that makes man what he is. It is the root of all creative activity...[2]

If we can think of the archetype as a node of psychic energy within the unconscious, then the symbol is the medium by which it becomes manifest in the here and now of space and time. Thus a symbol, although it has objective visible reality, always has behind it a hidden, profound, and only partly intelligible meaning which represents its roots in the archetype.

Although impossible for most of us to define or describe, we are all aware of the existence of something we call "self": the inner heart of our being, our soul, our uniqueness—however we want to describe it. It is in the nature of man that he constantly seeks a rational explanation of the inexplicable, and so he struggles with the questions: What is self? Why here? Why now? In trying to comprehend this most basic of archetypes—self—to give it concrete substance, man grasps at physical forms or symbols which are close and meaningful to him, and which are visible and definable. The first and most consciously selected form to represent self is the body, for it appears to be both the outward manifestation, and the encloser, of self. On a less conscious level, I believe, man also frequently selects the house, that basic protector of his internal environment (beyond skin and clothing) to represent or symbolize what is tantalizingly unrepresentable.

The French philosopher Gaston Bachelard has suggested that just as the house and the nonhouse are the basic divisions of geographic space, so the self and the nonself represent the basic divisions of psychic space.[3] The house both encloses space (the house interior) and excludes space (everything outside it). Thus it has two very important and different components: its interior and its façade. The house therefore nicely reflects how man sees himself, with both an intimate interior, or self as viewed from within and revealed only to those intimates who are invited inside, and a public exterior (the *persona* or *mask,* in Jungian terms) or the self that we choose to display to others.[4]

Most of us have had the experience of moving from one house to another, and of finding the new abode initially strange, unwelcoming, perhaps even hostile. But with time, we get used to the new house and its quirks, and it seems almost as though it gets used to us; we can relax when we return to it, put our feet up, become ourselves. But why in this particular box should we be ourselves more than in any other? It seems as though the personal space bubble which we carry with us and which is an almost tangible extension of our self expands to embrace the house we have designated as ours. As we become accustomed to, and lay claim to, this little niche in the world, we project something of ourselves onto its physical fabric. The furniture we install, the way we arrange it, the pictures we hang, the plants we buy and tend, all are expressions of our image of ourselves, all are messages about ourselves that we want to convey back to ourselves, and to the few intimates that we invite into this, our house. Thus, the house might be viewed as both an avowal of the self—that is, the psychic messages are moving from self to the objective symbol of self—and as a revelation of the nature of self; that is, the messages are moving from objective symbol back to the self. It is almost as if the house–self continuum could be thought of as both the negative and positive of a film, simultaneously.

Figure 21–1 [Photo credit: Mitchell Payne. Used by permission of Clare Cooper Marcus]

## THE HOUSE AS SYMBOL-OF-SELF: EXAMPLES FROM CONTEMPORARY ARCHITECTURE

Man was a symbol-making animal long before he was a toolmaker: he reached high degrees of specialization in song, dance, ritual, religion, and myth before he did in the material aspects of culture. Describing the rich symbolism of the man-made environment in part of Africa, Amos Rapoport notes:

> Among the Dogon and Bambara of Mali every object and social event has a symbolic as well as a utilitarian function. Houses, household objects, and chairs have all this symbolic quality, and the Dogon civilization, otherwise relatively poor, has several thousand symbolic elements. The farm plots and the whole landscape of the Dogon reflect this cosmic order. The villages are built in pairs to represent heaven and earth, and fields are cleared in spirals because the world has been created spirally. The villages are laid out in the way the parts of the body lie with respect to each other, while the house of the Dogon, or paramount chief, is a model of the universe at a smaller scale.[5]

Rapoport concludes significantly that "man's achievements have been due more to his need to utilize his internal resources than to his needs for control of the physical environment or more food."[6]

It would seem that there is an inverse relationship between technological advances and the cultivation of symbol and ritual. For so-called civilized man, the conscious recognition of the symbolism of what we do, how we live, and the houses we live in, has been all but lost. But if we start to delve beneath the surface, the symbolism is still there.

In a recent study of how contemporary California suburbanites chose their homes, Berkeley sociologist Carl Werthman concluded that many people bought houses to

**Figure 21–2** [Photo credit: Mitchell Payne. Used by permission of Clare Cooper Marcus]

bolster their image of self—both as an individual and as a person in a certain status position in society.[7] In one large suburban development near San Francisco, for example, he noted that extroverted, self-made businessmen tended to choose somewhat ostentatious, mock-colonial display homes, such as in Figure 21–1, while people in the helping professions, whose goals revolved around personal satisfaction rather than financial success, tended to opt for the quieter, inward-looking architect-designed styles conforming to current standards of "good design," such as that in Figure 21–2.

In the contemporary English-speaking world, a premium is put on originality, on having a house that is unique and somewhat different from the others on the street, for the inhabitants who identify with these houses are themselves struggling to maintain some sense of personal uniqueness in an increasingly conformist world. On the other hand, one's house must not be too way-out, for that would label the inhabitant as a nonconformist, and that, for many Americans, is a label to be avoided.

The house as symbol-of-self is deeply engrained in the American ethos (albeit unconsciously for many), and this may partly explain the inability of society to come to grips with the housing problem—a problem which is quite within its technological and financial capabilities to solve and which it persistently delegates to a low level in the hierarchy of budgetary values. America is the home of the self-made man, and if the house is seen (even unconsciously) as the symbol of self, then it is small wonder that there is a resistance to subsidized housing or to the State's providing houses *for* people. The frontier image of the man clearing the land and building a cabin for himself and his family is not far behind us. To a culture inbred with this image, the house–self identity is particularly strong. In some barely conscious way, society has decided to penalize those who, through no fault of their own cannot build, buy, or rent their own housing. They are not self-made men.

Numbers of studies in England, Australia, and the United States have indicated that when asked to describe their ideal house, people of all incomes and backgrounds will tend to describe a free-standing, square, detached, single-family house and yard. For example, in a recent survey of 748 men and women in thirty-two metropolitan areas in the U.S. 85 percent said they preferred living in a single-family house rather than in an apartment.[8] It is difficult to say whether the attachment to this form is the form itself, or the fact that it subsumes territorial rights over a small portion of the earth, or the fact that apartments can rarely be owned. But we do know that almost universally the image of the high-rise building for family living is rejected. An apartment is rarely seen as home, for a house can only be seen as a free-standing house-on-the-ground.

One could argue that people have been conditioned to want this through advertising, model homes salesmanship, and the image of the good life portrayed on television. To a certain extent this must be true, but these media are in turn only reflecting what seems to be a universal need for a house form in which the self and family unit can be seen as separate, unique, private, and protected.

The high-rise apartment building is rejected by most Americans as a family home because, I would suggest, it gives one no territory on the ground, violates the archaic image of what a house is, and is perceived unconsciously as a threat to one's self-image as a separate and unique personality. The house form in which people are being asked to live is not a symbol-of-self, but the symbol of a stereotyped, anonymous filing-cabinet collection of selves, which people fear they are becoming. Even though we may make apartments larger, with many of the appurtenances of a house, as well as opportunities for modification and ownership, it may still be a long time before the majority of lower- and middle-income American families will accept this as a valid image of a permanent home.[9] It is too great a threat to their self-image. It is possible that the vandalism inflicted on high-rise housing projects is, in part, an angry reaction of the inhabitants to this blatant violation of self-image.

The mobile hippie house-on-wheels is another instance of a new housing form greatly threatening people's image of what a house—or by implication, its inhabitants—*should* be. The van converted to mobile home and the wooden gable-roofed house built in the back of a truck are becoming common sights in a university community such as Berkeley and drop-out staging grounds, such as San Francisco. It is tempting to speculate that this house form has been adopted by hippies, not only because of its cheapness as living accommodation, but also because its mobility and form are reflections of where the inhabitants are in psychic terms—concerned with self and with making manifest their own uniqueness, convinced of the need for inward exploration and for freedom to move and swing with whatever happens. Hippies view themselves as different from the average person, and so they have chosen to live in self-generated house forms—converted trucks, tree-houses, geodesic domes, Indian teepees—which reflect and bolster that uniqueness.

It was perhaps to be expected that eventually the establishment would react. In February 1970, the city of Berkeley passed an ordinance making it illegal to live in a converted truck or van; the residents of these new houses mobilized and formed the Rolling Homes Association, but it was too late to prevent the ordinance from being passed.[10] When others too openly display the appurtenance (clothes, hair-styles, houses) of a new self-image, it is perceived as a threat to the values and images of the majority community. The image of the self as a house-on-wheels was too much for the establishment to accept.

Even the edge-of-town mobile home park occupied by the young retireds and the transient lower middle class is somehow looked down upon by the average American home owner as violating the true image of home and neighborhood. A person who lives in a house that moves must somehow be as unstable as the structure he inhabits. Very much the same view is held by house owners in Marin County, California, about the houseboat dwellers in Sausalito. They are "different," "Bohemian," "nonconformists," and their extraordinary choice of dwelling reflects these values.

The contrasting views which people of different socioeconomic classes in the U.S. have of their houses reflects again the house as a symbol-of-self in a self–world relationship. The greater are people's feelings of living in a dangerous and hostile world with constant threats to the self, the greater is the likelihood that they will regard their house as a shell, a fortress into which to retreat. The sociologist Lee Rainwater has shown that this image of the self, and of the house, is true for low-income blacks (particularly women) in the ghettoes and housing projects of this country.[11] With increasing economic and psychic stability (and in some cases, these are linked), a person may no longer regard his house as a fortress-to-be-defended, but as an attractive, individual expression of self-and-family with picture windows so that neighbors can admire the inside. Thus, for many in the middle-income bracket, the house is an expression of self, rather than a defender of self. The self-and-environment are seen in a state of mutual regard, instead of a state of combat.

The fact that the decoration of the house interior often symbolizes the inhabitants' feelings about self is one that has long been recognized. It has even been suggested that the rise in popularity of the profession of interior decorating is in some way related to people's inability to make these decisions for themselves since they're not sure what their self really is. The phenomenon of people, particularly women, rearranging the furniture in their house at times of psychic turmoil or changes-in-self, is a further suggestion that the house is very intimately entwined with the psyche.

The pregnant woman—in a very special psychological and physiological state of change—is especially likely to identify with the house, both in dreams and in reality:

> Sudden compulsive urges to do thorough house cleaning seem common among pregnant women. They are, on the one level, practical attempts to prepare for the coming baby; but when the house is already amply clean and delivery is impending, there may be a second, more significant level. The woman may be acting out her unconscious identification of the house with her own body. She may feel that if she cleans out the house and puts everything in order, she is in some way doing something about that other living space, the "house" of her unborn child. For her, it is an object rather than a word, which has taken on secret meanings.[12]

An interesting contemporary development is the interior decoration of the urban commune. In a number of examples in the Berkeley–Oakland area visited by the author, it was very noticeable that the bedrooms, the only private spaces of residents, were decorated in an attractive and highly personal way symbolic of the self whose space it was, as shown in Figure 21–3. The living rooms, the communal territory of six or eight or more different personalities, however, were only sparsely decorated, as exemplified by the one in Figure 21–4, since, presumably, the problem of getting

306 PART TWO: THE CONTEXT OF ARCHITECTURE

**Figure 21–3** [Photo credit: Mitchell Payne. Used by permission of Clare Cooper Marcus]

**Figure 21–4** [Photo credit: Mitchell Payne. Used by permission of Clare Cooper Marcus]

agreement on taste from a number of disparate and highly individual selves was too great to overcome. Interestingly, the more normal family house may display an opposite arrangement, with bedrooms functionally but uninterestingly decorated, and the living room, where guests and relatives are entertained, containing the best furniture, family mementos, art purchases, photos, and so on, and representing the collective family self. The only exception to this pattern may be the teenager's room—highly personalized as a reflection of his struggle to become an individual with a personality separate from his parents.

In a recently published study of living rooms, Edward Laumann and James House have found that the presence or absence of certain objects are good if not perfect clues to status and attitudes. It is the living room rather than any other room in the house which provides these clues because

> The living room is the area where "performances" for guests are most often given, and hence the "setting" of it must be appropriate to the performance. Thus we expect that more than any other part of the home, the living room reflects the individual's conscious and unconscious attempts to express a social identity.[13]

For example, they looked at a random sample of 41 homes from among 186 respondents (all of which were one-and-two-family home dwellers in Detroit) who had annual incomes over $15,000 and presumably had enough money to decorate any way they wanted. They found that those with a traditional decor—French or Early American furniture, wall mirrors, small potted plants and/or artificial flowers, paintings of people or still lifes, clocks—tended to be the white Anglo-Saxon establishment, occupying occupations and status positions similar to their fathers. Those with a more modern decor, characterized by modern furniture, wood walls, abstract paintings, solid carpets, and abstract designed curtains, tended to be upwardly mobile, non Anglo-Saxon Catholics whose families had migrated to the United States from southern and eastern Europe after 1900.

> The *nouveaux riches* have a strong need to validate their new found status, yet they are not acceptable socially by the traditional upper classes. Since their associations do not clearly validate their position, they turn to conspicuous consumption.... The *nouveaux riches,* then, spurn the style of the traditional upper class in favor of the newer fashions. This serves a double purpose: to establish their tastefulness and hence status, while symbolically showing their disdain for the "snobby" traditionals.[14]

The findings of this study of decorative styles of living rooms seem to tie in well with the result of Werthman's study of choices of house styles, for in both cases there appears to be a strong correlation between the style selected and the self-image of the consumer. The house façade and the interior design seem often to be selected so that they reflect how a person views himself both as an individual psyche, and in relation to society and the outside world, and how he wishes to present his self to family and friends.

These are just a few examples of how the house-as-self linkage becomes manifest in individual and societal behavior and attitudes; no doubt the reader can add many more instances from his personal experience. The thesis is not a new one: but it seems that the Jungian notions of the collective unconscious, the archetype and the symbol, may offer a useful conceptual structure to tie these examples together. Since the house–self symbolism seems to arise again and again, in many disparate settings, and since there appears to be little conscious sharing of this phenomenon, it seems reasonable to suggest that it is through the medium of the collective unconscious that people are in touch with an archaic and basically similar archetype (the self) and with a symbol for that archetype that has changed little through space and time (the house). Perhaps we can comprehend the essence of the house–self analogy more easily by looking at evidence from literature, poetry, and dreams—forms of expression that may get closer to true unconscious meanings than sociological surveys or similar empirical investigations.

## THE HOUSE-AS-SELF AS MANIFESTED IN LITERATURE, POETRY, AND DREAMS

One doesn't have to look farther than the very words that are sometimes used to describe houses—austere, welcoming, friendly—to see that we have somehow invested the house with human qualities. In a book describing his experiences while cleaning and repairing a country cottage to live in, Walter Murray wrote:

> So I left the cottage, swept if not yet garnished, and as I looked back at it that quiet evening with the sunset all aglow behind it, it seemed that somehow it was changed. The windows were clean, and the soul of a house looked out of its eyes; sweet cottages peep, old houses blink and welcome. Now Copsford, which had at first defied, gazed after me at least as an acquaintance, and months later was even friendly. But I never knew a smile to wrinkle the hard corners of its eyes.[15]

Although one might perhaps sneer at its cute anthropomorphizing of the environment, it is passages such as this which reveal what may be profound and barely recognized connections with, and projections onto, that environment.

In her introspective autobiography, written in the form of a diary, Anais Nin saw quite clearly both the security and sustenance that can ensue from living in a house that reflects one's own self-image, and the phenomenon of projecting onto the home one's inner fears and anxieties:

> When I look at the large green iron gate from my window it takes on the air of a prison gate. An unjust feeling, since I know I can leave the place whenever I want to, and since I know that human beings place upon an object, or a person, this responsibility of being the obstacle when the obstacle lies always within one's self.
>
> In spite of this knowledge, I often stand at the window staring at the large closed iron gate, as if hoping to obtain from this contemplation a reflection of my inner obstacles to a full, open life. . . . But the little gate, with its overhanging ivy like disordered hair over a running child's forehead, has a sleepy and sly air, an air of being always half open.
>
> I chose the house for many reasons.
>
> Because it seemed to have sprouted out of the earth like a tree, so deeply grooved it was within the old garden. It had no cellar and the rooms rested right on the ground. Below the rug, I felt, was the earth. I could take root here, feel at one with house and garden, take nourishment from them like the plants.[16]

In a short passage from a popular newsmagazine description of the German writer Günter Grass, the image of his style of writing, his way of working, his clothes, and the house he lives in—all reflect the inner character, the self, of this man:

> Grass is a fanatic for moderation. He is a moderate the way other men are extremists.

He is a man almost crazy for sanity.

Balance is Grass's game. He is in love with the firm, the tangible. He has a peasant's instinct for the solid ground, an artisan's feeling for materials. His West Berlin home—described by one visitor as "a god-awful Wilhelmian house"—is solid as a fort. The furniture is reassuringly thick-legged. The floors are bare. There are no curtains. In lean, wrinkled, absolutely undistinguished clothes—open necked shirts are the rule—Grass walks from room to room with workmanlike purpose. He looks like a visiting plumber who has a job to do and knows quite well that he can do it.[17]

The notion of house as symbol of mother or the womb is one fairly common in literature, and indeed has been the inspiration of a number of organic architects who have tried to re-create this safe, enclosed, encircling feeling in their designs. In the following fictional account, we see how the house takes on a symbolic maternal function in response to the fear of the man within and the storm outside:

The house was fighting gallantly. At first it gave voice to its complaints; the most awful gusts were attacking it from every side at once, with evident hatred and such howls of rage that, at times, I trembled with fear. But it stood firm. . . . The already human being in whom I had sought shelter for my body yielded nothing to the storm. The house clung to me, like a she wolf, and at times I could smell her odor penetrating maternally to my very heart. That night she was really my mother. She was all I had to keep and sustain me. We were alone.[18]

Here, in the unusual circumstances of a storm, one can see how this human, protective symbol of the house might well be conceived. But what of ordinary circumstances? How does the house-as-self symbol first begin to take root? Undoubtedly, one must look for the roots in infancy. At first, the mother is its whole environment. Gradually, as the range of senses expands, the baby begins to perceive the people and physical environment around it. The house becomes its world, its very cosmos. From being a shadowy shell glimpsed out of half-closed eyes, the house becomes familiar, recognizable, a place of security and love.

The child's world then becomes divided into the house, that microspace within the greater world that he knows through personal discovery, and everything that lies beyond it, which is unknown and perhaps frightening. In a sense, the child's experience reflects the assessment of known space as made by preliterate societies. As Mircea Eliade has written:

One of the outstanding characteristics of traditional societies is the opposition that they assume between their inhabited territory and the unknown and indeterminate space that surrounds it. The former is world (more precisely, our world), the cosmos; everything outside it is no longer a cosmos but a sort of "other world," foreign, chaotic space, peopled by ghosts, demons, foreigners. . . .[19]

As the child matures, he ventures into the house's outer space, the yard, the garden, then gradually into the neighborhood, the city, the region, the world. As space becomes known and experienced, it becomes a part of his world. But all the time, the house is home, the place of first conscious thoughts, of security and roots. It is no longer an inert box; it has been experienced, has become a symbol for self, family, mother, security. As Bachelard has written, "geometry is transcended."

As we become more ourselves—more self-actualized, in Maslow's terms—it seems that the house-as-symbol becomes even less tied to its geometry. A writer quoted by Bachelard describes his house thus:

> My house is diaphanous but it is not of glass. It is more of the nature of vapor. Its walls contract and expand as I desire. At times, I draw them close about me like protective armor.... But at others, I let the walls of my house blossom out in their own space, which is infinitely extensible.[20]

The symbol has become flexible, expandable according to psychic needs. For most people, the house is not actually changeable, except by such measures as opening and closing drapes and rearranging furniture to suit our moods. For one French poet, these alternate needs of expansion and contraction, extroversion and introspection, openness and withdrawal were made physical realities in the design of his dream home—a Breton fisherman's cottage around which he constructed a magnificent manor house.

> In the body of the winged manor, which dominates both town and sea, man and the universe, he retained a cottage chrysalis in order to be able to hide alone in complete repose.... The two extreme realities of cottage and manor... take into account our need for retreat and expansion, for simplicity and magnificence.[21]

Perhaps the suburban home buyers' yen for both an opulent façade with picture-window view and colonial porch and for a private secluded den is a modern manifestation of this need.

A recent news story suggests, in somewhat startling fashion, what may be strong evidence for the significance of house or home to the psyche:

> When both his parachutes failed in a recent jump from a plane 3,300 feet above the Coolidge, Ariz. airport, sky diver Bob Hall, 19, plummeted earthward and hit the ground at an estimated 60 m.p.h. Miraculously, he survived. A few days later, recovering from nothing more serious than a smashed nose and loosened teeth, he told reporters what the plunge had been like: "I screamed. I knew I was dead and that my life was ended. All my past life flashed before my eyes, it really did. I saw my mother's face, *all the homes I've lived in* [italics added], the military academy I attended, the faces of friends, everything."[22]

Surely, the fact that images of "all the homes I've lived in" flashed through the mind of a man approaching almost certain death, must indicate a significance of that element of the physical environment far beyond its concrete reality.

If we start to consider the messages from the unconscious made manifest through dreams, we have even more striking evidence of the house-as-self symbol. Carl Jung in his autobiography describes quite vividly a dream of himself as house, and his explorations within it.

> I was in a house I did not know, which had two storeys. It was "my house." I found myself in the upper storey, where there was a kind of salon furnished with fine old pieces in rococo style. On the walls hung a number of precious old paintings. I wondered that this should be my house, and thought, "Not bad." But then it occurred to me that I did not know what the lower floor looked like. Descending the stairs, I reached the ground floor. There everything was much older, and I realized that this part of the house must date from about the fifteenth or sixteenth century. The furnishings were medieval; the floors were of red brick. Everywhere it was rather dark. I went from one room to another thinking, "Now I really must explore the whole house." I came upon a heavy door and opened it. Beyond it, I discovered a stone stairway that led down into the cellar. Descending again, I found myself in a beautifully vaulted room which looked exceedingly ancient. Examining the walls, I discovered layers of brick among the ordinary stone blocks, and chips of brick in the mortar. As soon as I saw this I knew

that the walls dated from Roman times. My interest by now was intense. I looked more closely at the floor. It was on stone slabs, and in one of these I discovered a ring. When I pulled it, the stone slab lifted, and again I saw a stairway of narrow stone steps leading down into the depths. These, too, I descended, and entered a low cave cut into the rock. Thick dust lay on the floor, and in the dust were scattered bones and broken pottery, like remains of a primitive culture. I discovered two human skulls, obviously very old and half disintegrated. Then I awoke.[23]

Jung's own interpretation of the dream was as follows:

> It was plain to me that the house represented a kind of image of the psyche—that is to say, of my then state of consciousness, with hitherto unconscious additions. Consciousness was represented by the salon. It had an inhabited atmosphere, in spite of its antiquated style.
>
> The ground floor stood for the first level of the unconscious. The deeper I went, the more alien and the darker the scene became. In the cave, I discovered remains of a primitive culture, that is the world of the primitive man within myself—a world which can scarcely be reached or illuminated by consciousness. The primitive psyche of man borders on the life of the animal soul, just as the caves of prehistoric times were usually inhabited by animals before man laid claim to them.[24]

Jung describes here the house with many levels seen as the symbol-of-self with its many levels of consciousness; the descent downward into lesser known realms of the unconscious is represented by the ground floor, cellar, and vault beneath it. A final descent leads to a cave cut into bedrock, a part of the house rooted in the very earth itself. This seems very clearly to be a symbol of the collective unconscious, part of the self-house and yet, too, part of the universal bedrock of humanity.

Jung, unlike Freud, also saw the dream as a possible prognosticator of the future; the unconscious not only holds individual and collective memories but also the seeds of future action. At one period of his life Jung was searching for some historical basis or precedent for the ideas he was developing about the unconscious. He didn't know where to start the search. At this point he started having a series of dreams which all dealt with the same theme:

> Beside my house stood another, that is to say, another wing or annex, which was strange to me. Each time I would wonder in my dream why I did not know this house, although it had apparently always been there. Finally came a dream in which I reached the other wing. I discovered there a wonderful library, dating largely from the sixteenth and seventeenth centuries. Large, fat folio volumes bound in pigskin stood along the walls. Among them were a number of books embellished with copper engravings of a strange character, and illustrations containing curious symbols such as I had never seen before. At the time I did not know to what they referred; only much later did I recognize them as alchemical symbols. In the dream I was conscious only of the fascination exerted by them and by the entire library. It was a collection of incunabula and sixteenth century prints.
>
> The unknown wing of the house was a part of my personality, an aspect of myself; it represented something that belonged to me but of which I was not yet conscious. It, and especially the library, referred to alchemy of which I was ignorant, but which I was soon to study. Some fifteen years later I had assembled a library very like the one in the dream.[25]

Thus here in another dream Jung sees an unexplored wing of the house as an unknown part of himself and a symbol of an area of study with which he would become very absorbed in the future, and which permitted him to expand his concepts of the transformation of the self.

From many house dreams I have collected, two will suffice here to further emphasize the point. In the first one, the dreamer had, in reality, just lost a close friend in an auto accident. She reports the dream thus:

> I was being led through a ruined house by a tall, calm man, dressed all in white. The house was alone in a field, its walls of rubble, the layout and doorways no longer visible. The man guide led me slowly through the house pointing out how it used to be, where rooms were connected, where doorways lead to the outside world.
>
> My interpretation of this dream is that, the tall man is a part of me, maybe my masculine, strong, calm side, and he is pointing out that despite the fact that my self-life-house appear to be in ruins right now, due to my shock and grief at A's death, there is part of me that calmly and clearly will know how to find my way through the chaos. It was a very comforting dream at a time of great stress.

In another dream, the dreamer was in reality under much pressure from students and colleagues in his academic job. He described his dream thus:

> There was a house, a large English stately home, open to the public to look at and traipse through. But on this day, it was temporarily closed, and visitors were disappointedly reading the notices and turning away. I was in the basement of the house, sorting through some oil paintings, to see if there was anything there of value.

With the aid of a therapist, skilled in the interpretation of dreams, he saw the following message within the dream:

> I need to "close up shop," take a vacation from all the pressures and human input I'm experiencing right now, and have time to sort through some ideas in my unconscious (the basement of the house) to see if any are of value in guiding my future direction.

Returning to Jung's autobiography, he describes how, later in his life, he made manifest in stone the symbol which had at times stood for self in his dreams. He describes how he yearned to put his knowledge of the contents of the unconscious into solid form, rather than just describe them in words. In the building of his house—the tower at Bollingen on Lake Zurick—he was to make "a confession of faith in stone":

> At first I did not plan a proper house, but merely a kind of primitive one-storey dwelling. It was to be a round structure with a hearth in the center and bunks along the walls. I more or less had in mind an African hut where the fire, ringed with stone, burns in the middle, and the whole life of the family revolves around this centre. Primitive huts concretise an idea of wholeness, a familial wholeness in which all sorts of domestic animals likewise participate. But I altered the plan even during the first stages of building, for I felt it was too primitive. I realized it would have to be a regular two-storey house, not a mere hut crouched on the ground. So in 1923 the first round house was built, and when it was over I saw that it had become a suitable dwelling tower.
>
> The feeling of repose and renewal that I had in this tower was intense from the start. It represented for me the maternal hearth.[26]

Feeling that something more needed to be said, four years later Jung added another building with a tower-like annex. Again, after an interval of four years, he felt the need to add more and built onto the tower a retiring room for meditation and seclusion where no one else could enter; it became his retreat for spiritual concentration. After another interval of four years he felt the need for another area, open to nature and the sky, and so added a courtyard and an adjoining loggia. The resultant quanternity pleased him, no doubt because his own studies in mythology and symbolism had provided much evidence of the completeness and wholeness represented by the figure four. Finally, after his wife's death, he felt an inner obligation to "become what I myself am," and recognized that the small central section of the house

> which crouched so low and hidden was myself! I could no longer hide myself behind the "maternal" and "spiritual" towers. So in the same year, I added an upper storey to this section, which represents myself or my ego-personality. Earlier, I would not have been able to do this; I would have regarded it as presumptuous self-emphasis. Now it signified an extension of consciousness achieved in old age. With that the building was complete.[27]

Jung had thus built his house over time as a representation in stone of his own evolving and maturing psyche; it was the place, he said, where "I am in the midst of my true life, I am most deeply myself." He describes how:

> From the beginning I felt the Tower as in some way a place of maturation—a maternal womb or a maternal figure in which I could become what I was, what I am and will be. It gave me a feeling as if I were being reborn in stone. It is thus a concretisation of the individuation process. . . . During the building work of course, I never considered these matters. . . . Only afterwards did I see how all the parts fitted together and that a meaningful form had resulted: a symbol of psychic wholeness.[28]

In examining at some length Jung's own reflections on the house as dream-symbol and the building of his own house as a manifestation of the self, we are not just examining one man's inner life; hopefully, there is something here of the inner symbolism of all men. Jung, perhaps more than any other thinker or writer of this century, has fearlessly examined his own unconscious and delved into a great range of disciplines which together aided him in his quest to build a theory of the unconscious and the self.

We must return again to Jung's concept of the collective unconscious. It should be possible if his notion of an unconscious stretching through space and time beyond the individual is correct to find comparable indications of the house–self linkage in places and times far removed from contemporary Western civilization. If there is indeed an archetype self, then perhaps in other places and times, the house has become one (though not necessarily the only) symbol for that indefinable archetype in the physical world. For, as Jung has confirmed with ample evidence, the older and more archaic the archetype, the more persistent and unchanging the symbol.

## MAKING SPACE SACRED

In the opening chapter of his book *The Sacred and the Profane: The Nature of Religion* entitled "Sacred Space and Making the World Sacred,"[29] the noted historian of religion, Mircea Eliade describes how for many preliterate societies, space was not homogenous; inhabited parts were seen as sacred while all other space around was a formless, foreign expanse. In settling a new territory, man was faced with both a horizontal expanse of unknown land, and a complete lack of vertical connections to other cosmic levels, such as the heavens and the underworld. In defining and consecrating a spot as sacred, be it shrine, a temple, a ceremonial house, man gave himself a fixed point, a point of reference from which to structure the world about him. In doing so, he consciously emulated the gods who, many believed, created the world by starting at a fixed point—for example, an egg, or the navel of a slain monster—then moving out to the surrounding territory. As Hebrew tradition retells it: "The Most Holy One created the world like an embryo. As the embryo grows from the navel, so God began to create the world by the navel and from there it spread out in all directions."[30] Through finding a sacred space, generally with the aid of signs or the revelations of animals, man began to transform the shapeless, homogeneous chaos of space into his world.

Once located, the sacred space had to be consecrated, and this very often took the form of a construction which had at its center a pillar, pole, or tree. This was seen as a symbol for the cosmic axis and the means by which communication was made possible from one cosmic level to another. Whether seen as a ladder, as in Jacob's dream, or as a sacred pillar, as worshipped by the Celts and Germans before their conversion to Christianity, the vertical upright was an almost universal symbol for passage to the worlds of the gods above and below the earth.

Having created a sacred place in the homogeneity of space, man erected a symbol for the cosmic axis and thus centered this place at the Center of the World. But, Eliade maintains, there could be many Centers of the World, and indeed the Achilpa people of the Arunta tribe of Australian aborigines always carried the sacred pole with them so as not to be far from the Center or its link with other worlds. The religious man of fixed settlements, although he knew that his country and village and temple all constituted the navel of the universe,

> also wanted his own house to be at the Center and to be an "imago mundi"... (He) could only live in a space opening upward, where the break in plane was symbolically assured and hence communication with the "other world," the transcendental world, was ritually possible. Of course the sanctuary—the Center par excellence—was there, close to him ... but he felt the need to live at the Center *always*....[31]

Thus it was that the house, like the temple and the city, became a symbol of the universe with man, like God, at its center and in charge of its creation. The house, like the temple or shrine, was sanctified by ritual.

Just as the entrance to the temple was, and still is, regarded as the dividing line between the sacred and the profane worlds and is suitably embellished to ward off evil spirits which might attempt to enter the inner sanctum, so the threshold of the house is regarded as one of the most important dividing lines between inner private space and the other public world. Even if few living in the Western World would admit today to a belief in household spirits, there are still parts of the world where there are strong beliefs about how the house should be entered (right foot first among country dwellers in Finland, Syria, Egypt, and Yorkshire), and the custom of carrying the bride over the threshold is widespread throughout the world and has been recorded since ancient Roman times. Among contemporary city dwellers, the sanctity of the threshold is still revered by such behavior as removing one's hat and wiping one's shoes before entering the dwelling, or in Arab houses, by removing one's shoes. In China, the orientation of the door toward the south, and in Madagascar toward the west, are examples of the importance of a felicitous orientation of the door to the cosmos.[32] Among orthodox Jews, the Commandments are attached to the doorpost of the house, for they have been ordered: "Thou shalt write them on the posts by thy house and on thy gates" (Deuteronomy VI: 9). In northern England working class districts, the daily routine of polishing the front door knob and whitening the doorstep is a further contemporary example of special, almost ritualistic, attention paid to the threshold.

The location of the threshold varies in different cultures,[33] and it may well be that this location vis-à-vis the outside world is symbolic of how the people as individuals relate to the rest of society. In the American house, the front yard is generally unfenced and part of the streetscape, and may be viewed as semipublic territory; the real threshold to the house is the front door itself. This may reflect an American interpersonal trait of openness to strangers and of (initial at least) friendliness to people they hardly know. In England, however, the fenced front garden with a gate puts the initial threshold at some distance from the house itself, and is probably symbolic of the greater English reserve at inviting strangers into their houses and at opening up to people before they know them very well. The compound of a Moslem house puts the threshold even more forcibly and deliberately at some distance from the house, and reflects the extreme privacy required by individuals, particularly women, from strangers and neighbors.

Traditionally one of the principal tasks of the woman of the house was to keep the hearth fire perpetually burning. Lord Raglan in his study of the origins of the house[34] suggests that the hearth was originally conceived as a microcosm of the sun. Cooking took place outside, or in a separate building, and the sacred hearth was seen as a parallel to the sacred flame in the temple, not something to be cooked on, but a symbol of the sun which must never be allowed to go out for fear the sun itself would go out.[35]

It is probable that fire existed before man built his first dwellings. Pierre Deffontaines has suggested that the house originated as a shelter for this sacred fire that must not be allowed to go out.[36] Among the ancient Greeks the sacred fire was first enclosed in a special precinct, which later was surrounded by the living quarters of the family. The dwelling thus came into existence to protect the fire, and the Greeks maintain it was the sacred hearth that inspired man to build the house. In the houses of northern China, the kang, a large central hearth of brick and earth, is thought of and referred to as "the mother of the dwelling." Deffontaines reports that until recently in houses in rural Sardinia, the hearth fire was kept perpetually alight and only extinguished when someone died, for the period of mourning. The belief that the house had its traditional beginning in the protection of fire is still maintained in Madagascar, where fire must be the first item brought into a newly completed dwelling.[37]

The hearth was, until very recently, still the focus of family life in England, where wives left behind by their soldier husbands in World War I were enjoined to "keep the home fires burning." Although central heating is becoming more and more common in England, and antipollution laws prevent the burning of coal in open fires in most parts of the country, many families have replaced the perpetual hearth with an electric heater displaying artificial smouldering "logs." It is not easy, after many centuries of veneration of the hearth, to replace it overnight with concealed hot air vents and to feel that something of the home has not been lost. An interesting parallel reported in the *San Francisco Chronicle* in May 1971 told of the demolition of a soup kitchen in the Mission District where the only item to be saved for incorporation in a new old men's hospital was the much loved symbolic hearth.

The ritual of keeping the hearth alight because it represents the sun can be termed a cosmic ritual. Such rituals are based upon the belief that one can affect the macrocosm by acting upon a microcosm. There are many indications that temples of various faiths have been built as symbols of the universe, with the dome or high vaulted roof as symbolic of the heavens, and the floor symbolic of earth below. Raglan reports "in the rituals of the Pawnees the earth lodge is made typical of man's abode on earth; the floor is the plain, the wall the distant horizon, the dome the arching sky, the central opening, the zenith, dwelling place of Tirawa, the invisible power which gives life to all created beings."[38]

Since one of the most widespread primitive beliefs about the creation of the world was that it originated from an egg, so many of the first cosmic manifestations in temples and houses were round or spherical in shape. Lord Raglan has suggested that an original belief in the world as circular began to be replaced by a belief in the world as square, and starting in Mesopotamia and Egypt, and spreading later to China, India, Rome, North America, and Africa, the temple and the house as cosmic manifestations began to be built on a square or rectangular plan, instead of a circular one.[39] People as far apart as the Eskimos, Egyptians, Maoris, and tribes of the North Cameroons believed that the sky or heavens were held by four corner posts which had to be protected from decay or damage, and whose guardian deities had to be placated by ritual. The weathercock on the roof, which is believed in parts of England to crow to wind spirits in the four quarters and ward them off, is one of the few contemporary western manifestations of the ancient cosmic significance of the square and the four cardinal points.

In most parts of the world, the rectangular house predominates today, but the circular shape has often been retained in the form of the dome for religious or important secular buildings (for example, city hall, the state capitol, the opera house), recalling much earlier times when the circle had specific cosmic significance.

To summarize Raglan's thesis, he suggests that house forms were derived from the forms of temples (the houses of the gods), and symbolize man's early beliefs concerning the form and shape of the universe. Drawing conclusions from his studies of myth and folklore, rather than buildings, Eliade comes to similar conclusions.

> By assuming the responsibility of creating the world that he has chosen to inhabit, he not only cosmicizes chaos but also sanctifies his little cosmos by making it like the world of the gods. . . . That is why settling somewhere—building a village or merely a house—represents a serious decision, for the very existence of man is involved; he must, in short, create his own world and assume the responsibility of maintaining and renewing it. Habitations are not lightly changed, for it is not easy to abandon one's world. The house is not an object, a "machine to live in"; it is a universe that man constructs for himself by imitating the paradigmatic creation of the gods, the cosmogony.[40]

## THE SELF-HOUSE/SELF-UNIVERSE ANALOGY

It seems that consciously or unconsciously, then, many men in many parts of the world have built their cities, temples, and houses as images of the universe. My contention is that somewhere, through the collective unconscious, man is still in touch with this symbolism. Our house is seen, however unconsciously, as the center of *our* universe and symbolic of *the* universe. But how does this connect with my earlier arguments regarding the house-as-symbol-of-self? Primitive man sees his dwelling as symbolic of the universe, with himself, like God, at its center. Modern man apparently sees his dwelling as symbolic of the self, but has lost touch with this archaic connection between house–self–universe.

The phenomenon of dreaming or imagining the self as a house—that package outside our own skin which encloses us and in which we feel most secure—is perhaps the first glimmering of the unconscious that the "I" and the "non-I" are indeed one and the same. As Alan Watts has so eloquently written in *The Book: On the Taboo Against Knowing Who You Are*,[41] the notion that each individual ego is separate (in space) and finite (in time) and is something different from the universe around him is one of the grand hoaxes of Western thought. Although virtually impossible for most of us nonmystics to grasp in more than a superficial way, this knowledge of our indivisibility from the environment is buried deep within the collective unconscious and becomes manifest symbolically (often without our recognizing it) in fantasies, flashes of intuition, dreams, poems, painting, and literature.

The so-called mentally ill may in fact be more closely in touch with these lost connections between self and environment than any of us realize. After a long career working with schizophrenics, Harold Searles noted:

> It seems to me that, in our culture, a conscious ignoring of the psychological importance of the nonhuman environment exists simultaneously with a (largely unconscious) *over-dependence* upon that environment. I believe that the actual importance of that environment to the individual is so great that he dare not recognize it. Unconsciously it is felt, I believe, to be not only an intensely important conglomeration of things outside the self, but also a large and integral part of the self. . . .[42]

> The concreteness of the child's thinking suggests for him, as for the member of the so-called primitive culture and for the schizophrenic adult, the wealth of nonhuman objects about him are constituents of his psychological being in a more intimate sense than they are for the adult in our culture, the adult whose ego is, as Hartman and Werner emphasize, relatively clearly differentiated from the surrounding world, and whose development of the capacity for abstract thinking helps free him . . . from his original oneness with the nonhuman world.[43]

Perhaps it is the so-called normal adult who, having been socialized to regard self and environment as separate and totally different, is most out of touch with the essential reality of oneness with the environment, which small children, schizophrenics, preliterate people, and adherents of certain Eastern religions understand completely. There are certain religions, for example Buddhism, that regard the apparent separation of the individual and the universe as a delusion. My contention is that in thinking, dreaming, or fantasying about self and house as somehow being inextricably intertwined, as being at some level one and the same thing, man may be taking the first step on the path towards what Zen adherents would term enlightenment. He is ridding himself of the delusion of the separation of man from his environment.

## CONCLUSION

If there is some validity to the notion of house-as-self, it goes part of the way to explain why for most people their house is so sacred and why they so strongly resist a change in the basic form which they and their fathers and their fathers' fathers have lived in since the dawn of time. Jung recognized that the more archaic and universal the archetype made manifest in the symbol, the more universal and unchanging the symbol itself. Since self must be an archetype as universal and almost as archaic as man himself, this may explain the universality of its symbolic form, the house, and the extreme resistance of most people to any change in its basic form.

For most people the self is a fragile and vulnerable entity; we wish therefore to envelop ourselves in a symbol-for-self which is familiar, solid, inviolate, unchanging. Small wonder, then, that in Anglo-Saxon law it is permissible, if necessary, to kill anyone who breaks and enters your house. A violation of the self (house) is perhaps one of man's most deep-seated and universal fears. Similarly, the thought of living in a round house or a houseboat or a mobile home is, to most people, as threatening as is the suggestion that they might change their basic self-concept. A conventional house and a rigidly static concept of self are mutually supporting. Perhaps with the coming of age of Reich's Consciousness III generation, and the social movements (civil rights, women's liberation, human potential movement, etc.) which are causing many to question the inviolate nature of old self-concepts, we can expect an increased openness to new housing forms and living arrangements, the beginnings of which are already apparent in the proliferation of communes and drop-out communities.

This long statement on house-as-symbol-of-the-self brings me back to my original problem: how to advise architects on the design of houses for clients who are often poor, whom they will never know, let alone delve into their psychic lives or concepts of self. I have no pat answer, but if there is some validity to the concept of house-as-self, we must learn ways—through group encounters, resident-meetings, participant observation, interviews—of empathizing with the users' concepts of self, and we must devise means of complementing and enhancing that image through dwelling design. If in new housing forms we violate this image, we may have produced an objective reality which pleases the politicians and designers, but at the same time produced a symbolic reality which leaves the residents bewildered and resentful.

Certainly, one area that every architect involved with house design can and should investigate is his or her own biases based on images of self. Bachelard, in his very thought provoking study *The Poetics of Space,* suggests somewhat fancifully, that along with psychoanalysis, every patient should be assisted in making a topoanalysis, or an analysis of the spaces and places which have been settings for his past emotional development. I would go further and say this exercise should be required of every designer. He or she should begin to understand how present self-images are being unconsciously concretized in design, and how scenes of earlier development (particularly childhood between the ages of about 5 and 12) are often unconsciously reproduced in designs in an effort, presumably, to recall that earlier often happier phase of life.

In the past few years, as a teacher in the College of Environmental Design at Berkeley, I have had students draw, in as much detail as they can remember, their childhood environments. After an interval of a few weeks, they have then drawn what for each of them would be an ideal environment. The similarities are often striking, as also are the similarities they begin to observe between these two drawings, and what they produce in the design studio. The purpose of the exercise is not to say that there is anything wrong with such influences from the past, but just to point out that they are there, and it may well be to his advantage as a designer to recognize the biases they may introduce into his work.

In the field of man's relationship with his environment, the type of approach which might be termed intuitive speculation seems to have been lost in a world devoted to the supposedly more scientific approach of objective analysis. As Alan Watts has speculated, this emphasis on the so-called objective may indeed be a sickness of Western man, for it enables him to retain his belief in the separateness of the ego from all that surrounds it. Although certain objective facts have been presented in this paper, it is hoped by the author that its overall message is clear: allow yourself to be open to the consideration of relationships other than those that can be proved or disproved by scientific method, for it may well be in these that a deeper truth lies. Perhaps no one has stated it more eloquently than Watts, and it is with a quotation from his *Nature, Man and Woman* that I will end this paper:

> The laws and hypotheses of science are not so much discoveries as instruments, like knives and hammers, for bending nature to one's will. So there is a type of personality which approaches the world with an entire armory of sharp and hard instruments, by means of which it slices and sorts the universe into precise and sterile categories which will not interfere with one's peace of mind.
>
> There is a place in life for a sharp knife, but there is a still more important place for other kinds of contact with the world. Man is not to be an intellectual porcupine, meeting his environment with a surface of spikes. Man meets the world outside with a soft skin, with a delicate eyeball and eardrum and finds communion with it through a warm melting, vaguely defined, and caressing touch whereby the world is not set at a distance like an enemy to be shot, but embraced to become one flesh, like a beloved wife. . . . Hence the importance of opinion, of instruments of the mind, which are vague, misty, and melting rather than clear-cut. They provide possibilities of communication, of actual contact and relationships with nature more intimate than anything to be found by preserving at all costs the "distance of objectivity." As Chinese and Japanese painters have so well understood there are landscapes which are best viewed through half-closed eyes, mountains which are most alluring when partially veiled in mist, and waters which are most profound when the horizon is lost, and they are merged with the sky.[44]

## NOTES AND REFERENCES

1 Jolande Jacobi, *Complex, Archetype, Symbol in the Psychology of C. G. Jung* (New York: Pantheon Books, 1957).
2 Ibid., p. 47.
3 Gaston Bachelard, *The Poetics of Space* (Boston: Beacon Press, 1969).
4 For the purposes of this paper, we will accept the Jungian view of "self," which he saw as both the core of the unconscious *and* the totality of the conscious and the unconscious. To illustrate with a diagram:

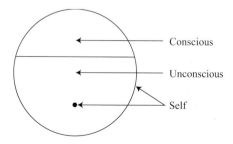

5 Amos Rapoport, *House Form and Culture* (Englewood Cliffs, NJ: Prentice Hall, 1969), p. 50.
6 Ibid., p. 43.
7 Carl Werthman, "The Social Meaning of the Physical Environment," Ph.D. dissertation in sociology, University of California, Berkeley, 1968.
8 William Michelson, "Most People Don't Want What Architects Want," *Transaction* 5 (July–August 1968), pp. 37–43.
9 The urban rich accept apartments because they generally have a house somewhere else; the elderly seem to adapt well to apartments because they offer privacy with the possibility of many nearby neighbors, minimum upkeep problems, security, communal facilities, etc.; and for mobile young singles or childless couples the limited spatial and temporal commitment of an apartment is generally the ideal living environment.
10 A similar ordinance was passed in San Francisco in March 1971.
11 Lee Rainwater, "Fear and House-as-Haven in the Lower Class," *Journal of the American Institute of Planners* 32 (January 1966), pp. 23–31; and *Behind Ghetto Walls* (Chicago: Aldine-Atherton, 1970).
12 Arthur Colman and Libby Colman, *Pregnancy: The Psychological Experience* (New York: Herder and Herder, 1971).
13 Edward Laumann and James House, "Living Room Styles and Social Attributes: Patterning of Material Artifacts in an Urban Community," in *The Logic of Social Hierarchies,* ed. Laumann, Siegel, and Hodges (Chicago: Markham, 1972), pp. 189–203.
14 Ibid.
15 Walter J. C. Murray, *Copsford* (London: Allen and Unwin, 1950), p. 34.
16 Anais Nin, *The Diary of Anais Nin, 1931–34* (New York: Harcourt, 1966).
17 "The Dentist's Chair as an Allegory of Life," *Time,* April 13, 1970, p. 70.
18 Henri Bosco, *Malicroix,* as quoted in *The Poetics of Space,* G. Bachelard (New York: The Orion Press, 1964), p. 45.
19 Mircea Eliade, *The Sacred and the Profane: The Nature of Religion* (New York: Harcourt, 1959).
20 George Spyridaki, *Mort Lucide,* as quoted in Bachelard, *The Poetics of Space,* p. 51.
21 Bachelard, *The Poetics of Space,* p. 65.
22 "The Pleasure of Dying." *Time,* December 4, 1972, pp. 44–45.
23 Carl Jung, *Memories, Dreams and Reflections* (London: Collins, The Fontana Library Series, 1969), pp. 182–83.
24 Ibid., p. 184.
25 Ibid., p. 228.
26 Ibid., p. 250.
27 Ibid., p. 252.
28 Ibid., p. 253.
29 Eliade, *The Sacred and the Profane.*
30 Ibid., p. 4.
31 Ibid., p. 43.
32 Pierre Deffontaines, "The Place of Believing," extracted from "Géographie et religions," in *Landscape* 2 (Spring 1953), p. 26.
33 Rapoport, *House Form and Culture,* p. 80.
34 Lord Raglan, *The Temple and the House* (London: Routledge & Kegan Paul, 1964).
35 In most parts of the world, cooking was one of a number of activities (others included childbirth and death) which could not take place within the house.
36 Deffontaines, "The Place of Believing," p. 26.
37 Ibid.
38 Raglan, *The Temple and the House,* p. 138.

39 Ibid., p. 158.
40 Eliade, *The Sacred and the Profane,* pp. 56–57.
41 Alan Watts, *The Book: On the Taboo Against Knowing Who You Are* (New York: Macmillan, Collier Books, 1966). p. 43.
42 Harold F. Searles. *The Nonhuman Environment in Normal Development and in Schizophenia* (New York: International Universities Press, 1960), p. 395.
43 Ibid., p. 42.
44 Alan W. Watts, *Nature, Man and Woman* (New York: Random House, Vintage Books, 1970), pp. 80–81.

选读 22

"空间人类学：一种组织模式"

爱德华·特威切尔·霍尔

在这一章里，霍尔就个人感知层面（个人、小群体和城市居民）和人类文化层面体验物质环境的不同方式做出了一种行为学解释。通过兼用科学研究和个人体验的方法，霍尔建立了一套人们如何看待和响应他们周边环境的理论，并且给出了建筑空间所塑造的社会进程的基本定义。他用个人居所来举例说明不同文化如何来命名和使用房子里各不相同的"房间"。霍尔是最早向建筑师指明空间视觉图像和场所心理形象不同之处的理论家之一。明显的空间视觉图像是指人看到了什么，而场所心理形象则是指人如何看，是每个人终其一生经过自身不断吸收加工的产物。

## READING 22

## The Anthropology of Space: An Organizing Model
### Edward Twitchell Hall

In this chapter Hall presents a behavioral explanation of the ways physical environments are experienced at the level of the individual senses, by a single person, by small clusters of people, by the inhabitants of cities, and at the level of human culture. By using both scientific studies and personal experience, Hall establishes a rationale for the ways people view and respond to their surroundings, and he gives fundamental definitions for the social processes that are shaped by architectural space. He uses the individual dwelling, for example, to illustrate how different cultures name and use distinct "rooms" within the house. Hall was one of the first theorists to instruct architects in the differences between overt visual patterns of space (*what* one sees) and the mental images of place that each person assimilates throughout a lifetime (*how* one sees).

*Source: The Hidden Dimension* by Edward T. Hall. Copyright © 1966, 1982 by Edward T. Hall. Used by permission of Doubleday, a division of Bantam Doubleday Dell Publishing Group, Inc. pp. 101–12.

Territoriality, spacing, and population control were discussed earlier in this book. *Infra-culture* is the term I have applied to behavior on lower organizational levels that underlie culture. It is part of the proxemic classification system and implies a specific set of levels of relationships with other parts of the system. As the reader will remember, the term *proxemics* is used to define the interrelated observations and theories of man's use of space.

Chapters IV, V, and VI [of *The Hidden Dimension*] were devoted to the senses, the physiological base shared by all human beings, to which culture gives structure and meaning. It is this *pre*cultural sensory base to which the scientist must inevitably refer in comparing the proxemic patterns of Culture A with those of Culture B. Thus, we have already considered two proxemic manifestations. One, the *infra*cultural, is behavioral and is rooted in man's biological past. The second, *pre*cultural, is physiological and very much in the present. The third, the *micro*cultural level, is the one on which most proxemic observations are made. Proxemics as a manifestation of microculture has three aspects: fixed-feature, semifixed feature, and informal.

Although proper translation from level to level is ordinarily quite complex, it should be attempted by the scientist from time to time if only for the sake of perspective. Without comprehensive systems of thought which tie levels together, man develops a kind of schizoid detachment and isolation that can be very dangerous. If, for example, civilized man continues to ignore the data obtained on the infracultural level about the consequences of crowding, he runs the risk of developing the equivalent of the behavioral sink, if indeed he has not already done so. The experience of James Island deer chillingly recalls the Black Death which killed off two-thirds of Europe's population in the mid-fourteenth century. Though this great human die-off was due directly to *Bacillis pestis,* the effect was undoubtedly exacerbated by lowered resistance from the stressfully crowded life in medieval towns and cities.

The methodological difficulty in translating from level to level stems from the *essential indeterminacy of culture,* which I discussed in *The Silent Language.* Cultural indeterminacy is a function of the many different levels on which cultural events occur and the fact that it is virtually impossible for an observer to examine simultaneously with equal degrees of precision something occurring on two or more widely separated analytic or behavioral levels. The reader can test this for himself by simply concentrating on the phonetic details of speech (the way sounds actually are made) and at the same time trying to talk eloquently. I do not mean simply to enunciate clearly but to think about where you place your tongue, how you hold your lips, whether your vocal chords are vibrating or not, and how you are breathing with each syllable. The indeterminacy referred to here requires additional comment. All organisms are highly dependent on redundancy; that is, information received from one system is backed up by other systems in case of failure. Man himself is also programmed by culture in a massively redundant way. If he weren't, he could not talk or interact at all; it would take too long. Whenever people talk, they supply only part of the message. The rest is filled in by the listener. Much of what is *not* said is taken for granted. However, cultures vary in what is left unsaid. To an American, it is superfluous to have to indicate to a shoeshine boy the color of the paste to be used. But in Japan, Americans who do not indicate this may send out brown shoes only to have them returned black! The function of the conceptual model and the classification system, therefore, is to make explicit the taken-for-granted parts of communications and to indicate relationships of the parts to each other.

What I learned from my research on the infracultural level was also very helpful in the creation of models for work on the cultural level of proxemics. Contrary to popular belief, territorial behavior for any given stage of life (such as courting or rearing the young) is quite fixed and rigid. The boundaries of the territories remain reasonably constant, as do the locations for specific activities within the territory, such as sleeping, eating, and nesting. The territory is in every sense of the word an extension of the organism, which is marked by visual, vocal, and olfactory signs. Man has created material extensions of territoriality as well as visible and invisible territorial markers. Therefore, because territoriality is relatively fixed, I have termed this type of space on the proxemic level *fixed-feature space*. The next section will be devoted to fixed-feature space, followed by discussions of semifixed feature and informal space.

## FIXED-FEATURE SPACE

Fixed-feature space is one of the basic ways of organizing the activities of individuals and groups. It includes material manifestations as well as the hidden, internalized designs that govern behavior as man moves about on this earth. Buildings are one expression of fixed-feature patterns, but buildings are also grouped together in characteristic ways as well as being divided internally according to culturally determined designs. The layout of villages, towns, cities, and the intervening countryside is not haphazard but follows a plan which changes with time and culture.

Even the inside of the Western house is organized spatially. Not only are there special rooms for special functions—food preparation, eating, entertaining and socializing, rest, recuperation, and procreation—but for sanitation as well. *If,* as sometimes happens, either the artifacts or the activities associated with one space are transferred to another space, this fact is immediately apparent. People who "live in a mess" or a "constant state of confusion" are those who fail to classify activities and artifacts according to a uniform, consistent, or predictable spatial plan. At the opposite end of the scale is the assembly line, a precise organization of objects in *time* and *space*.

Actually the present internal layout of the house, which Americans and Europeans take for granted, is quite recent. As Philippe Ariès points out in *Centuries of Childhood,* rooms had no fixed functions in European houses until the eighteenth century. Members of the family had no privacy as we know it today. There were no spaces that were sacred or specialized. Strangers came and went at will, while beds and tables were set up and taken down according to the moods and appetites of the occupants. Children dressed and were treated as small adults. It is no wonder that the concept of childhood and its associated concept, the nuclear family, had to await the specialization of rooms according to function and the separation of rooms from each other. In the eighteenth century, the house altered its form. In French, *chambre* was distinguished from *salle*. In English, the function of a room was indicated by its name—bedroom, living room, dining room. Rooms were arranged to open into a corridor or hall, like houses into a street. No longer did the occupants pass through one room into another. Relieved of the Grand Central Station atmosphere and protected by new spaces, the family pattern began to stabilize and was expressed further in the form of the house.

Goffman's *Presentation of Self in Everyday Life* is a detailed, sensitive record of observations on the relationship of the façade that people present to the world and the self they hide behind it. The use of the term façade is in itself revealing. It signifies recognition of levels to be penetrated and hints at the functions performed by architectural features which provide screens behind which to retire from time to time. The strain of keeping up a façade can be great. Architecture can and does take over this burden for people. It can also provide a refuge where the individual can "let his hair down" and be himself.

The fact that so few businessmen have offices in their homes cannot be solely explained on the basis of convention and top management's uneasiness when executives are not visibly present. I have observed that many men have two or more distinct personalities, one for business and one for the home. The separation of office and home in these instances helps to keep the two often incompatible personalities from conflicting and may even serve to stabilize an idealized version of each which conforms to the projected image of both architecture and setting.

The relationship of fixed-feature space to personality as well as to culture is nowhere more apparent than in the kitchen. When micropatterns interfere as they do in the kitchen, it is more than just annoying to the women I interviewed. My wife, who has struggled for years with kitchens of all types, comments on male design in this way: "If any of the men who designed this kitchen had ever worked in it, they wouldn't have done it this way." The lack of congruence between the design elements, female stature and body build (women are not usually tall enough to reach things), and the activities to be performed, while not obvious at first, is often beyond belief. The size, the shape, the arrangement, and the placing in the house all communicate to the women of the house how much or how little the architect and designer knew about fixed-feature details.

Man's feeling about being properly oriented in space runs deep. Such knowledge is ultimately linked to survival and sanity. To be disoriented in space is to be psychotic. The difference between acting with reflex speed and having to stop to think in an emergency may mean the difference between life and death—a rule which applies equally to the driver negotiating freeway traffic and the rodent dodging predators. Lewis Mumford observes that the uniform grid pattern of our cities "makes strangers as much at home as the oldest inhabitants." Americans who have become dependent on this pattern are often frustrated by anything different. It is difficult for them to feel at home in European capitals that don't conform to this simple plan. Those who travel and live abroad frequently get lost. An interesting feature of these complaints reveals the relationship of the layout to the person. Almost without exception, the newcomer uses words and tones associated with a personal affront, as though the town held something against him. It is no wonder that people brought up on either the French radiating star or the Roman grid have difficulty in a place like Japan where the entire fixed-feature pattern is basically and radically different. In fact, if one were to set out to design two systems in contrasts, it is hard to see how one could do better. The European systems stress the lines, which they name; the Japanese treat the intersecting points technically and forget about the lines. In Japan, the intersections but not the streets are named. Houses instead of being related in space are related in time and numbered in the order in which they are built. The Japanese pattern emphasizes hierarchies that grow around centers; the American plan finds its ultimate development in the sameness of suburbia, because one number along a line is the same as any other.

In a Japanese neighborhood, the first house built is a constant reminder to the residents of house #20 that #1 was there first.

Some aspects of fixed-feature space are not visible until one observes human behavior. For example, although the separate dining room is fast vanishing from American houses, the line separating the dining area from the rest of the living room is quite real. The invisible boundary which separates one yard from another in suburbia is also a fixed-feature of American culture or at least some of its subcultures.

Architects traditionally are preoccupied with the visual patterns of structures—what one sees. They are almost totally unaware of the fact that people carry around with them internalizations of fixed-feature space learned early in life. It isn't only the Arab who feels depressed unless he has enough space but many Americans as well. As one of my subjects said: "I can put up with almost anything as long as I have large rooms and high ceilings. You see, I was raised in an old house in Brooklyn and I have never been able to accustom myself to anything different." Fortunately, there are a few architects who take the time to discover the internalized fixed-feature needs of their clients. However, the *individual* client is not my primary concern. The problem facing us today in designing and rebuilding our cities is understanding the needs of large numbers of people. We are building huge apartment houses and mammoth office buildings with no understanding of the needs of the occupants.

The important point about fixed-feature space is that it is the mold into which a great deal of behavior is cast. It was this feature of space that the late Sir Winston Churchill referred to when he said: "We shape our buildings and they shape us." During the debate on restoring the House of Commons after the war, Churchill feared that departure from the intimate spatial pattern of the House, where opponents face each other across a narrow aisle, would seriously alter the patterns of government. He may not have been the first to put his finger on the influence of fixed-feature space, but its effects have never been so succinctly stated.

One of the many basic differences between cultures is that they extend different anatomical and behavioral features of the human organism. Whenever there is cross-cultural borrowing, the borrowed items have to be adapted. Otherwise, the new and the old do not match, and in some instances, the two patterns are completely contradictory. For example, Japan has had problems integrating the automobile into a culture in which the lines between points (highways) receive less attention than the points. Hence, Tokyo is famous for producing some of the world's most impressive traffic jams. The automobile is also poorly adapted to India, where cities are physically crowded and the society has elaborate hierarchical features. Unless Indian engineers can design roads that will separate slow pedestrians from fast-moving vehicles, the class-conscious drivers' lack of consideration for the poor will continue to breed disaster. Even Le Corbusier's great buildings at Chandigarh, capital of Punjab, had to be modified by the residents to make them habitable. The Indians walled up Corbusier's balconies, converting them into kitchens! Similarly, Arabs coming to the United States find that their own internalized fixed-feature patterns do not fit American housing. Arabs feel oppressed by it—the ceilings are too low, the rooms too small, privacy from the outside inadequate, and views non-existent.

It should not be thought, however, that incongruity between internalized and externalized patterns occurs only between cultures. As our own technology explodes, air conditioning, fluorescent lighting, and soundproofing make it possible to design houses and offices without regard to traditional patterns of windows and doors. The new inventions sometimes result in great barnlike rooms where the "territory" of scores of employees in a "bull pen" is ambiguous.

## SEMIFIXED-FEATURE SPACE

Several years ago, a talented and perceptive physician named Humphry Osmond was asked to direct a large health and research center in Saskatchewan. His hospital was one of the first in which the relationship between semifixed-feature space and behavior was clearly demonstrated. Osmond had noticed that some spaces, like railway waiting rooms, tend to keep people apart. These he called sociofugal spaces. Others, such as the booths in the old-fashioned drugstore or the tables at a French sidewalk café, tend to bring people together. These he called sociopetal. The hospital of which he was in charge was replete with sociofugal spaces and had very few which might be called sociopetal. Furthermore, the custodial staff and nurses tended to prefer the former to the latter because they were easier to maintain. Chairs in the halls, which would be found in little circles after visiting hours, would soon be lined up neatly in military fashion, in rows along the walls.

One situation which attracted Osmond's attention was the newly built "model" female geriatrics ward. Everything was new and shiny, neat and clean. There was enough space, and the colors were cheerful. The only trouble was that the longer the patients stayed in the ward, the less they seemed to talk to each other. Gradually, they were becoming like the furniture, permanently and silently glued to the walls at regular intervals between the beds. In addition, they all seemed depressed.

Sensing that the space was more sociofugal than sociopetal, Osmond put a perceptive young psychologist, Robert Sommer, to work to find out as much as he could about the relationship of furniture to conversations. Looking for a natural setting which offered a number of different situations in which people could be observed in conversations, Sommer selected the hospital cafeteria, where 36 by 72-inch tables accommodated six people. As the figure below indicates, these tables provided six different distances and orientations of the bodies in relation to each other.

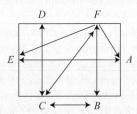

- F–A  Across the corner
- C–B  Side by side
- C–D  Across the table
- E–A  From one end to the other
- E–F  Diagonally the length of the table
- C–F  Diagonally across the table

Fifty observational sessions in which conversations were counted at controlled intervals revealed that: F–A (cross corner) conversations were twice as frequent as the C–B (side by side) type, which in turn were three times as frequent as those at C–D (across the table). No conversations were observed by Sommer for the other positions. In other words, corner situations with people at right angles to each other produced six times as many conversations as face-to-face situations across the 36-inch span of the table, and twice as many as the side-by-side arrangement.

The results of these observations suggested a solution to the problem of gradual disengagement and withdrawal of the old people. But before anything could be done, a number of preparations had to be made. As everyone knows, people have deep personal feelings about space and furniture arrangements. Neither the staff nor the patients would put up with outsiders "messing around" with their furniture. Osmond, as director, could order anything he wanted done, but he knew the staff would quietly sabotage any arbitrary moves. So the first step was to involve them in a series of "experiments." Both Osmond and Sommer had noted that the ward patients were more often in the *B–C* and *C–D* relationships (side by side and across) than they were in the cafeteria, and they sat at much greater distances. In addition, there was no place to put anything, no place for personal belongings. The only territorial features associated with the patients were the bed and the chair. As a consequence, magazines ended up on the floor and were quickly swept up by staff members. Enough small tables so that every patient had a place would provide additional territoriality and an opportunity to keep magazines, books, and writing materials. If the tables were square, they would also help to structure relationships between patients so that there was a maximum opportunity to converse.

Once the staff had been cajoled into participating in the experiments, the small tables were moved in and the chairs arranged around them. At first, the patients resisted. They had become accustomed to the placement of "their" chairs in particular spots, and they did not take easily to being moved around by others. By now, the staff was involved to the point of keeping the new arrangement reasonably intact until it was established as an alternative rather than an annoying feature to be selectively inattended. When this point had been reached, a repeat count of conversations was made. The number of conversations had doubled, while reading had tripled, possibly because there was now a place to keep reading material. Similar restructuring of the dayroom met with the same resistances and the same ultimate increase in verbal interaction.

At this point, three things must be said. Conclusions drawn from observations made in the hospital situation just described are not universally applicable. That is, across-the-corner-at-right-angles is conducive *only* to: (a) conversations of certain types between (b) persons in certain relationships and (c) in very restricted cultural settings. Second, what is sociofugal in one culture may be sociopetal in another. Third, sociofugal space is not necessarily bad, nor is sociopetal space universally good. What *is* desirable is flexibility and congruence between design and function so that there are a variety of spaces, and people can be involved or not, as the occasion and mood demand. The main point of the Canadian experiment for us is its demonstration that the structuring of semifixed features can have a profound effect on behavior and that this effect is measurable. This will come as no surprise to housewives who are constantly trying to balance the relationship of fixed-feature enclosures to arrangement of their semifixed furniture. Many have had the experience of getting a room nicely arranged, only to find that conversation was impossible if the chairs were left nicely arranged.

It should be noted that what is fixed-feature space in one culture may be semifixed in another, and vice versa. In Japan, for example, the walls are movable, opening and closing as the day's activities change. In the United States, people move from room to room or from one part of a room to another for each different activity, such as eating, sleeping, working, or socializing with relatives. In Japan, it is quite common for the person to remain in one spot while the activities change. The Chinese provide us with further opportunities to observe the diversity of human treatment of space, for they assign to the fixed-feature category certain items which Americans treat as semifixed. Apparently, a guest in a Chinese home *does not move his chair* except at the host's suggestion. To do so would be like going into someone else's home and moving a screen or even a partition. In this sense, the semifixed nature of furniture in American homes is merely a matter of degree and situation. Light chairs are more mobile than sofas or heavy tables. I have noted, however, that some Americans hesitate to adjust furniture in another person's house or office. Of the forty students in one of my classes, half manifested such hesitation.

Many American women know it is hard to find things in someone else's kitchen. Conversely, it can be exasperating to have kitchenware put away by well-meaning helpers who don't know where things "belong." How and where belongings are arranged and stored is a function of microcultural patterns, representative not only of large cultural groups but of the minute variations on cultures that make each individual unique. Just as variations in the quality and use of the voice make it possible to distinguish one person's voice from another, handling of materials also has a characteristic pattern that is unique.

## INFORMAL SPACE

We turn now to the category of spatial experience, which is perhaps most significant for the individual because it includes the distances maintained in encounters with others. These distances are for the most part outside awareness. I have called this category *informal space* because it is unstated, not because it lacks form or has no importance. Indeed, as the next chapter will show, informal spatial patterns have distinct bounds, and such deep, if unvoiced, significance that they form an essential part of the culture. To misunderstand this significance may invite disaster.

## BIBLIOGRAPHY AND REFERENCES

Ariès, Philippe. *Centuries of Childhood.,*New York: Alfred A. Knopf, 1962.

Goffman, Erving. *The Presentation of Self in Everyday Life.* Garden City, NY: Doubleday & Company, Inc., 1959.

Hall, Edward T. *The Silent Language.* Garden City, NY: Doubleday & Company, Inc., 1959.

Mumford, Lewis. *The City in History.* New York: Harcourt, Brace, 1961.

Osmond, Humphrey. "The Relationship Between Architect and Psychiatrist." In *Psychiatric Architecture,* ed. C. Goshen. Washington, DC: American Psychiatric Association, 1959.

———."The Historical and Sociological Development of Mental Hospitals." In *Psychiatric Architecture,* ed. C. Goshen. Washington, DC: American Psychiatric Association, 1959.

———."Function as the Basis of Psychiatric Ward Design." *Mental Hospitals* (Architectural Supplement), April 1957, pp. 23–29.

Sommer, Robert. "The Distance for Comfortable Conversation: A Further Study." *Sociometry* 25 (1962).
———. "Leadership and Group Geography." *Sociometry* 24 (1961).
———. "Studies in Personal Space." *Sociometry* 22 (1959).
Sommer, Robert, and H. Ross. "Social Interaction on a Geriatric Ward." *International Journal of Social Psychology* 4 (1958), pp. 128–33.
Sommer, Robert, and G. Whitney. "Design for Friendship." *Canadian Architect*, 1961.

选读 23
"论建筑的文化反应"
阿摩斯·拉普卜特

在这篇文章中，拉普卜特以他先前在环境行为学领域的研究成果为依托，着力探讨了建成环境和人类文化之间的关系。他在文中阐明了文化是如何对经过设计的环境产生影响的概念，并从人类在不同情况下"建筑响应"的角度考虑了这一关系。他认为传统建筑方法并不能有效解决建筑使用者的问题，这主要是因为建筑师的价值观与个人喜好并不能与人们的具体需要、渴求以及社会的总体需求、希望产生共鸣。

READING 23

## On the Cultural Responsiveness of Architecture
### Amos Rapoport

Rapoport in this article approaches the relationship between built form and human culture from the perspective of his previous work in the field of environment-behavior studies. He clarifies the concept of how culture influences the designed environment, and considers this relationship in terms of the "responsiveness of architecture" to the human condition. He argues that traditional architectural approaches to the problems of building users have not been effective, primarily because the values and preferences of architects have become unresponsive to the needs and aspirations of human beings in particular and society in general.

## INTRODUCTION

Starting with *House Form and Culture* (1969), I have had much to say about the relation between culture and the built environment. Most recently I have synthesized

*Source:* "On the Cultural Responsiveness of Architecture" by Amos Rapoport. *Journal of Architectural Education* 41, no. 1 (Fall 1987), pp. 10–15. Copyright © 1987 by the Association of Collegiate Schools of Architecture, Inc. (ACSA). Reproduced by permission of MIT Press Journals.

some of these matters.[1] Underlying much of this work has been the notion of *culture-specific design*. This means design which responds to and supports the specific cultural characteristics of various user groups. In this recent work, I have discussed the problem of how the concept of "culture" is to be used and have also tried to clarify concepts such as "environment" and "design."[2] These discussions will be briefly summarized below.

The question of how the environment responds to and fits culture has not, however, received comparable attention. Part of the problem may be that it is difficult to describe or assess fit or responsiveness to culture. But that is not an uncommon problem in anthropology and related fields. Frequently one can only do comparative work: in the present case, for example, I judge that given environments fit culture more or less well than others. In addition, even for such comparative rankings it is necessary to know what might be possible indicators of fit or congruence of environments with culture. This I take to be the meaning of the term "cultural responsiveness of architecture," which was the title of the Northeast Regional ACSA meetings in Montreal in October 1984 where I was asked to discuss this topic. That meeting raised an important and neglected question to which this revised paper is a very preliminary answer.

I have often discussed environment-behavior studies in terms of what I call the three basic questions:

**1** How do people shape their environment—what characteristics of people as individuals or groups are relevant to the shaping of particular environments?

**2** How and to what extent does the physical environment affect which people, in what ways and in which contexts?

**3** What are the mechanisms which link people and environments in this two-way interaction?

One can then suggest that "culture" is an aspect of question 1, the nature of "environment" is an aspect of question 2, while "responsiveness" has to do with question 3—the mechanisms that link culture and environment. Mechanisms are, of course, extremely important since they make concepts and theories more convincing. They do this by showing how the abstract systems of constructs called theories can link to empirical reality.

Clearly, before one can discuss congruence of culture and environment those terms need to be discussed. While I have written extensively about both, and readers could refer to published work, it seemed useful to summarize at least the principal points here.

## CULTURE

I will not try to argue for the importance of culture in understanding how people and environments interact. Rather I will try briefly to clarify the meaning of this concept. Two principal points will be made.

**1** There is an extremely large and complex literature on this topic, with many apparently contradictory definitions. Yet, if they are seen as complementary they give an idea of what culture is. Also, too "tight" a definition is not wanted, since it is not yet clear which aspects of culture are the most important regarding built environments and in which circumstances.

I have argued elsewhere[3] that the many different definitions fall into one of three general views of what culture is. I quote:

> One defines it as a way of life typical of a group, the second as a system of symbols, meanings, and cognitive schemata transmitted through symbolic codes, the third as a set of adaptive strategies for survival related to ecology, and resources. Increasingly, these three views are seen not as being in conflict but rather as complementary. Thus, designed environments of particular cultures are settings for the kind of people which a particular group sees as normative, and the particular life-style which is significant and typical, distinguishing the group from others. In creating such settings and lifestyles, an order is expressed, a set of cognitive schemata, symbols, and some vision of an ideal are given form—however imperfectly; finally, both the lifestyle and symbolic system may be part of the group's adaptive strategies within their ecological setting.

For our purposes, then culture may be said to be about a group of people who have a set of values and beliefs which embody ideals, and are transmitted to members of the group through enculturation. These lead to a world view—the characteristic way of looking at the world and, in the case of design, of shaping the world. The world is shaped by applying rules which lead to systematic and consistent choices (to be discussed below) whether in creating a life-style, a building style, a landscape, or a settlement.

One can also ask what cultures *do,* as opposed to what they *are.*[4] One view implicit in the above is that cultures or their constituents may be seen as *properties of populations,* i.e. the distinctive means by which such populations maintain their identity and relate to their environment.[5] Another view is of culture as *a control mechanism* for everything humans do.[6] A third is to look at culture as *that structure which gives meaning to particulars.*[7] By considering the views of culture discussed here, and asking how they relate to built form, some insights might be gained.

**2**  In addition, however and whatever culture is or does, one can question the utility of that concept in trying to understand built form and how it is used. It can be suggested that "culture" is both too *global* and *too abstract* to be useful. It is often helpful, as I have argued since the mid-70s, to clarify excessively abstract and broad concepts by "dismantling" them and then studying the components and their various interrelationships with each other and with other variables. One cannot "design for culture" but only specific parts of environments for specific components of culture. In response to "culture" being too global, I have proposed a sequence from culture, through world view, values, images/schemata, lifestyle to activity systems, suggesting that the last two offer a useful starting point. Note that activities comprise the activities themselves, how they are carried out, how grouped into systems and their meaning. The latter three aspects are increasingly variable with culture and account for the extraordinary variety of environments created for many fewer activities.[8] In response to the concept being too abstract, it is possible to make it more operational by discovering the social manifestations of culture which are potentially observable, such as family and kinship groups, family structures, institutions, social relationships, status and other roles, rituals, food habits and many others. These can then be studied and can be related to the built environment, influencing the latter and being influenced by it. Starting with these more specific, more concrete expressions of culture offers another way of relating it to built form.[9]

I have also distinguished between the *core* of a culture and its periphery, particularly in conditions of rapid cultural change.[10] What I mean by that is that certain elements (peripheral) are given up not only willingly but eagerly for new ones, but that others (core) are retained until the bitter end. What these are need to be discovered.

In any case, these and other successive definitions, clarifications, reconceptualizations and "dismantlings" of the concept of culture enable us, at least in principle, both to understand and analyze cultural aspects of environment-behavior relationships and to "design for culture." Clearly, much more work is needed.

## ENVIRONMENT AND DESIGN

I begin with a point I have made frequently: that it may be more useful to use a term such as environmental design rather than architecture, since one cannot consider buildings in isolation. People live in systems of settings ranging from smaller than rooms through buildings, open spaces, neighborhoods, cities, regions and so forth.[11] One then needs to ask how "environment" and "design" may best be conceptualized.

"Environment" initially is another rather global term which needs dismantling before it becomes useful. This, again, can be done in various ways. For example, environment can be conceptualized as a system of settings within which human life takes place. These systems of settings, the components which make them up and the linkage and separations in space and time, their meaning etc. are highly variable culturally and need to be *discovered* rather than assumed.[12]

Secondly, the environment can be conceptualized in a variety of ways, given that it consists not only of things, but also of people. At the very least, then, it is a set of relations between people and people, people and things and things and things. More specifically various versions of this have been proposed which I will not review here.[13] One that I find useful which combines *design* and *environment* is that the design of the environment is the organization of four things: space, time, meaning and communication.[14] It then becomes possible to consider, discuss and investigate the implications of each for a given culture. This does, of course, raise the question of what "design" means.

"Design" needs to be seen as any purposeful change of the above four components and their physical expression. Thus almost anything humans do to alter the face of the earth and anything they build is design—not just what professionals do. In fact, as I have long argued, preliterate, vernacular, popular and other such environments are particularly illuminating regarding cultural responsiveness; they also comprise most of what has ever been built so that they must be considered if any valid generalizations about environment-behavior relations are to be made.[15]

As in all this discussion, *specifics* become crucial, whether about culture or the environment. For example, space organization (and form) are not the same as shape and, in most cases, more important.[16] In some cases, however, shape may be more important, for example in terms of identity, religion, cosmology, etc. Thus, once again, what is important needs to be *discovered* not assumed.

## CULTURAL RESPONSIVENESS

The current importance of this topic, as of culture-environment relations generally, is due to certain changes in the context of design. Traditionally, built form has responded effectively to culture—for example, in preliterate and vernacular contexts. Such environments communicated effectively and fully to users whereas currently there is a concern that environments do not respond and do not communicate effectively. One can still observe effective communication in certain traditional settings, spontaneous ("squatter") settlements and, to a degree, in popular environments (although partly distorted and weakened by controls, regulations, etc.). In the case of professionally designed environments this seems to work less well: frequently they do not communicate to users.[17]

The problem of cultural responsiveness is complicated by pluralism, the presence of multiple groups and subcultures. Most traditional environments were for homogeneous groups and of much smaller scale. In such cases the congruence of culture and built form was much simpler and easier to achieve. To consider merely communication, in the present situation cues may not be noticed, if noticed they may not be understood; if both noticed and understood some users may refuse to conform. A consequence is that in complex, pluralistic, large-scale situations the problem is much more difficult.[18]

What is new, other than this pluralism and large scale, is that firstly many environments which traditionally were vernacular, popular or "spontaneous" are now designed by *professionals* who typically have not dealt with them and really do not know how to tackle them. Second, in the past designers and clients shared culture, they were typically members of the same group. Moreover clients and users were identical. Currently, none of these conditions apply: designers are very different indeed to users in lifestyle, values, etc.—i.e., culture. While I have only touched on a single set of points, I would argue that it is circumstances such as these which lead to the perceived problems, the consequent interest in culture/environment relationships and hence the importance of culturally responsive environments.

To clarify the concept of the cultural responsiveness of built environments, consider a series of questions about it:

**a. Why Culturally Responsive?** That environments should be culturally responsive is, of course, a normative statement. Yet one can ask whether, and why, this quality is necessary, since it is not a self-evident objective. The Modern Movement, for one, rejected this view *implicitly* by emphasizing "universality" and ignoring context and cultural specificity. Others have questioned this position *explicitly*[19] saying, in effect, "let them adapt." Of course, "cultural responsiveness" is still a minority position in architecture.

Before normative statements can be made one needs understanding of the problem: explanatory theory needs to precede normative statements which should be based on such explanatory theory. One must understand what a thing is supposed to do, and why, before one proceeds with implementing it and before one can judge whether it does it well. The validity of objectives must be evaluated before one evaluates whether objectives have been achieved.[20] Clearly we cannot go into this here, so let me just say that there is a good deal of inescapable evidence for an intimate relationship between culture and built form and accept it as a valid goal for the purpose of this discussion. (In fact a lot of my work is specifically in support of this position, hence I certainly support it.)

While this is mainly a research question, both in general terms and in any given case, it is also partly an ethical and a value question, in two senses. First, there is an implicit (or explicit) commitment to pluralism and the importance and value of helping a variety of cultures and groups to survive.[21] Secondly, there is an imperative to set objectives on the basis of the best available knowledge, research and *user's* needs and wants.[22]

The possible importance of the environment in this connection is also an open and researchable question. At the theoretical, explanatory level it has to do with issues of criticality, the question of environmental determinism, culture core, the potential role of environments in the preservation of specific components of culture and many other issues such as the relative importance of built form vis-à-vis other aspects of culture. All these I have discussed at length elsewhere and they cannot be elaborated here. It is, in fact, becoming possible to specify the conditions under which culturally responsive environments may become important to the very survival of cultures and those circumstances where they are less important. One can also begin to specify which features or elements of the environment can be expected to be important in any given case and why environment can be expected to be important in any given case.[23] This is essential since even those who believe in this position do not always succeed because the problem, requirements and means have not been fully understood. One must also remember that while "culture" is a very broad concept, the built environment is a small subset of it and architecture an even smaller one. They are *not* equals. Thus many other subsystems of culture are equally, and frequently even more, important. Moreover environmental design is embedded in culture rather than being a separate and equal element.[24]

**b. What Is Responsiveness?** The term "responsive" may not be the best one to use since, in fact, only living things respond, except metaphorically. Thus it is difficult to ask: to what does environment respond? What responds? What does "respond" mean? How does one know an environment is responsive?

It may be useful, then, to reformulate this objective. This can be done in several ways which, again, are not contradictory but merely different and possibly complementary. Each one contributes to a clarification of what it is one is trying to achieve.

One way of looking at the problem is to think of environments as being *congruent with culture* or, rather, as we have seen, in terms of certain parts of environments (which?) being congruent with certain parts of culture (which?). One still needs to know what "congruence" is and how one evaluates it, which, as mentioned above, is not an easy task.

Another way is to think of design as being *culture-specific,* i.e. that built forms will have certain qualities which will differ for different cultures, as they relate to certain parts of the culture, particularly its core.

One could reformulate the objective to state that responsive environments are those which can be manipulated or changed as culture changes, i.e. open-ended, flexible and adaptive environments. That is certainly one meaning of "responsive." On the other hand, continuity, stability and guidance are also necessary. One can then begin to find out which parts of the environment need to be open-ended (this will be culturally variable); who specifically needs to do the manipulating; when in the process it needs to be done, etc. The responsiveness of open-endedness (and hence participation) is not enough—it is necessary but not sufficient.²⁵

A responsive environment can be understood as one which can be manipulated in another sense—as an active component of culture, used as an element of that culture rather than as a passive container of culture, as it were.

Finally, responsiveness can be reformulated in a way which I have recently found rather useful—as a *supportive environment*. It is then possible to specify the conditions, of high criticality, under which high supportiveness is essential²⁶ and then one can ask:

What is being supported?
How is it being supported?
By what is it being supported?

This is, I think, an approach which can be used in design, for example in the case of developing countries.²⁷ I mention developing countries neither because of their intrinsic importance nor because they are necessarily of major interest to readers of *JAE (Journal of Architectural Education)*. They provide a useful starting point for getting into the issues we have been discussing. This is because it is a useful technique, used in anthropology, psychology and science generally, to look first at extreme situations and, having understood the principles, to move on to more subtle ones. Many of the things I have been discussing are seen most clearly, not to say starkly, and best studied, in such contexts. There criticality is high, hence the effects of environment on culture and vice-versa are strong. Rapid culture change enables one to identify core vs. peripheral elements, those elements changing rapidly vs. those remaining relatively constant. The critical cultural and physical elements can be identified; failures are also more catastrophic. Finally, traditional vernacular and spontaneous environments abound and provide valuable lessons.²⁸

**c. Culturally Responsive to Whom?** This issue has already been raised. From the previous discussion it follows that when one speaks of "culturally responsive architecture" two immediate interpretations of the underlying problem suggest themselves:

i. The fact that designers, clients and users effectively constitute different cultures.
ii. That users themselves are variable in most present day situations, i.e. there are present numerous groups of users with different cultures.

Both, of course, are aspects of a single question which some of my work has addressed, on and off, for over a decade: *Whose culture?* In order to answer it one needs to define those groups which are relevant for design. The definition and identification of relevant groups and those of their characteristics which are significant regarding the environment is a major challenge and an underresearched question. The "groups" whose "culture" is relevant and to whom the environment is supposed to be responsive (or supportive) need to be discovered. Clearly this means designing for (and with) specific *users* rather than for oneself or one's peers.

**d. How Are Environments Responsive?** Using the idea of a supportive environment one can rephrase this in terms of how particular environmental elements support certain cultural mechanisms which link people and environments (part of the third of the three basic questions of environment-behavior studies). For example instrumental aspects of activities; meaning, status and identity; institutions (family, social units, religion, etc.), tradition and continuity, temporal orientations, etc.

By identifying which of these form part of the culture core, and what specific form they take, and which elements of the physical environment support them, and how, one can begin to achieve such goals, and to suggest the consequences of failing to do so.

Also, *maximizing choice* is critical. The major effect of environment on behavior is through habitat selection: people leave unsatisfactory environments for those they regard as better. Blocked habitat selection is a major factor in unresponsiveness.[29] There is strong evidence that environments which are *chosen* are inherently supportive, and hence responsive, as opposed to identical environments which are *imposed*. The fact of having been chosen may be as important as what is chosen. But if one studies choice itself, one finds it also to be highly variable: *cultural specificity continues to operate.* One chooses particular profiles of environmental quality—but environmental quality is culturally defined.[30]

## CONCLUSION

There are very many other important issues which have not even been mentioned. This is clearly a vast topic which cannot be developed further here. Those interested will have to turn to the literature. But some of the implications of what I have been saying need to be traced.

Architects are very different to other people. There is evidence that architects' values, lifestyles and preferences are different to most users'. There are also many different groups of users. The result is that a large variety of different types of environments need to be designed which are, to designers, counterintuitive and unlikely a priori. Their very unfamiliarity and incongruence with what designers like may make them appear disordered or chaotic. They are not—their order is merely one designers do not like, do not understand or that is incongruous for them.[31]

Let me make two deliberately provocative statements which follow from the notion of designing for various users' culture. The first is that the ultimate expression of a culturally responsive environment may be a designed environment which we do not like or even hate!! The second is the assertion that some of the best new environments today are to be found among the spontaneous settlements in developing countries; some of the worst are designed by architects. A profoundly important and disturbing question follows: how can people who may often be illiterate and who possess minimal resources and power often design so much better than professional designers?

Cultural responsiveness can be seen in terms of the individual designer or design; it can also be part of the general theories, objectives, values, reward systems, education, etc., of the profession.[32] It is what we could term the public professional aspects which are critical. There can be responsive design, open-endedness, good design and so forth on the part of enlightened individuals, but these will remain isolated instances without corresponding changes at the public, institutional level.

The failure of the theoretical approach here presented, and the knowledge already available, to influence design can be attributed to several things. One is that culture-specific design, which is a logical consequence of cultural responsiveness, is politically difficult to advocate and to implement.[33] Another is that while it is necessary to know what is already known, use research and know the literature, that is not sufficient. "Facts" alone are not influential. Whether they are used depends on how one sees things, i.e. defines the domain and the problem.[34] This argument has recently received empirical confirmation in a study of landscape architecture students. Although they themselves had done research on the cultural characteristics of particular user groups none of that research influenced their design work supposedly based on it. Design followed traditional arbitrary, formal and "aesthetic" criteria.[35] Thus, for design to become more culturally responsive, changes are essential in what architecture does, what design is seen to be and what theory is understood to be. The emphasis shifts to problem understanding, clarification and definition before problem solving. There will need to be a concern with *what* is to be done and *why* (based on the best available theory and knowledge). Explicit objectives ("what") will need to be set, their validity judged and justified ("why"?). Are the objectives valid? How do we know? How can we find out? Having accepted the validity of these objectives one *then* turns to *how* one achieves them (which currently is the major concern). Here also much more explicitness is necessary: what means are suitable? Why? How do we know? Then we ask: have we achieved our objectives? How can we find out? All these are clearly related to what I have called the *public* aspects of design, the framework within which individual designers work and designs occur.

The conclusion is that for designed environments to become culturally responsive *one needs to change the professional culture*. This also follows from one of the interpretations of culture discussed above, which defines it as that structure that gives meaning to particulars.[36] In that sense I found it encouraging that the 1984 Northeast Regional ACSA meetings addressed the topic of "The Cultural Responsiveness of Architecture." Could it possibly mean that our professional culture may at long last be beginning to change?

**NOTES**

1 Amos Rapoport, "Culture and Built Form: A Reconsideration," keynote address at Conference on Built Form and Culture Research, University of Kansas, Lawrence, Kansas, November 1986, proceedings in press.
2 Amos Rapoport, *Human Aspects of Urban Form* (Oxford, UK: Pergamon, 1977); and Amos Rapoport, *The Meaning of the Built Environment: A Nonverbal Communication Approach* (Beverly Hills, CA: Sage, 1982).
3 Amos Rapoport, "Cross-Cultural Aspects of Man-Environment Studies," in *Human Behavior and Environment,* Vol. 4: *Environment and Culture,* ed. I. Altman, A. Rapoport, and J. F. Wohlwill (New York: Plenum, 1980), pp. 7–46; and Rapoport, "Culture and Built Form."
4 Ibid.
5 Roy A. Rappoport, *Ecology, Meaning and Religion* (Richmond, CA: North Atlantic Books, 1979); and Amos Rapoport, "Identity and Environment: A Cross-Cultural Perspective," in *Housing and Identity: Cross Cultural Perspectives,* ed. J. S. Duncan (London: Croom-Helm, 1981), pp. 6–35.
6 See Clifford Geertz, *Local Knowledge* (New York: Basic Books, 1983).
7 Jerome Bruner, *In Search of Mind* (New York: Harper and Row, 1983), p. 195.
8 Amos Rapoport, "Socio-Cultural Aspects of Man-Environment Studies," in *The Mutual Interaction of People and Their Built Environment (A Cross-Cultural Perspective),*

ed. Amos Rapoport (The Hague: Monton, 1976), pp. 3–35; Rapoport, *Human Aspects of Urban Form,* Fig. 1.9, p. 20; and Rapoport, *The Meaning of the Built Environment.*

9 Amos Rapoport, "Systems of Activities and Systems of Settings," in *Domestic Architecture and Use of Space—An Interdisciplinary Perspective,* ed. S. Kent (in press).

10 Amos Rapoport, "Development, Culture Change and Supportive Design," *Habitat International* 7, no. 5/6, pp. 249–68.

11 See Rapoport, *Human Aspects of Urban Form;* Amos Rapoport, "Urban Design and Human Systems—On Ways of Relating Buildings to Urban Fabric," in *Human and Energy Factors in Urban Planning: A Systems Approach,* ed. P. Laconte, J. Gibson, and A. Rapoport, NATO Advanced Study Institute, series D. Beh. and Soc. Sci. No. 12 (The Hague: Nijhoff, 1982), pp. 161–84; and Rapoport, "Systems of Activities and Systems of Settings."

12 Rapoport, *Human Aspects of Urban Form;* Rapoport, "Cross-Cultural Aspects of Man-Environment Studies;" Rapoport, "Urban Design and Human Systems;" and Rapoport, "Systems of Activities and Systems of Settings."

13 Rapoport, *Human Aspects of Urban Form,* p. 8.

14 Ibid.; Rapoport, "Cross-Cultural Aspects of Man-Environment Studies;" and Rapoport, *The Meaning of the Built Environment.*

15 See Amos Rapoport, "Environment and People," in *Australia as Human Setting,* ed. A. Rapoport (Sydney: Angus and Robertson, 1972), pp. 3–21; Rapoport, "Cross-Cultural Aspects of Man-Environment Studies;" and Amos Rapoport, "Culture and the Urban Order," in *The City in Cultural Context,* ed. J. Agnew, J. Mercer, and D. Sopher (Boston: Allen and Unwin, 1984), pp. 50–75.

16 Rapoport, *Human Aspects of Urban Form,* pp. 9–11.

17 Rapoport, *The Meaning of the Built Environment,* fig. II, p. 83.

18 Ibid. and Rapoport, *Human Aspects of Urban Form.*

19 J. Sonnenfeld, "Imposing Environmental Meaning: A Commentary," in *Environmental Knowing,* ed. G. T. Moore and R. G. Golledge (Stroudsburg, PA: Dowden, Hutchinson and Ross, 1976), pp. 254–57.

20 Amos Rapoport, "Debating Architectural Alternatives," *RIBA Transactions,* no. 3 (1983), pp. 105–9.

21 Rapoport, "Cross-Cultural Aspects of Man-Environment Studies."

22 Rapoport, "Development, Culture Change and Supportive Design;" and Rapoport, "Debating Architectural Alternatives."

23 Amos Rapoport, "The Effect of Environment on Behavior," in *Environment and Population: Problems of Adaption* (New York: Praeger, 1983), pp. 200–201.

24 Rapoport, "Culture and Built Form."

25 Amos Rapoport, "Culture, Site-Layout and Housing," *Architectural Association Quarterly* 12, no. 1 (1980), pp. 4–7; and Rapoport, "Development, Culture Change and Supportive Design."

26 Rapoport, "The Effect of Environment on Behavior."

27 Rapoport, "Development, Culture Change and Supportive Design."

28 Ibid. and Amos Rapoport, "Environmental Quality, Metropolitan Areas and Traditional Settlements," *Habitat International* 7, no. 3/4 (1983), pp. 37–63.

29 Rapoport, "The Effect of Environment on Behavior."

30 Rapoport, *Human Aspects of Urban Form;* Rapoport, "Cross-Cultural Aspects of Man-Environment Studies;" and Amos Rapoport, "Thinking About Home Environments: A Conceptual Framework," in *Home Environments,* Vol. 8: *Human Behavior and Environment,* ed. I. Altman and C. Werner (New York: Plenum, 1985), pp. 255–86.

31 Rapoport, "Culture and Urban Order;" and Rapoport, "Culture and Built Form."

32 Rapoport, "Debating Architectural Alternatives."

33 Rapoport, "Culture and Built Form."
34 Rapoport, "Environment and People;" and Rapoport, "Debating Architectural Alternatives."
35 S. M. Low, *Professional Culture: The Boundary Between Theory and Practice in Design,* NIE Education Resource Center (ERIC), No. ED 219890 (Washington, DC: U.S. Department of Health, Education, and Welfare, 1982).
36 Bruner, *In Search of Mind,* p. 185.

选读 24

"空间—时间"

罗伯特·萨默

建筑师用来处理物质空间的传统工具通常与视觉和造型艺术语言相关。萨默在相关研究和著作中阐述了建筑师在设计过程中应该怎样依靠社会科学的方法和理论。"空间—时间"一章出自萨默的《设计意识》一书，萨默在这一章中说明了为什么对设计师来说时间维度和三维世界同样重要。他认为只有理解了什么是时间，设计师才能欣赏到建成形式的动态本质。人们体验建筑，并不是通过静态的媒介，而是在不停变化之中的一个个瞬间的集合里获得体验。各种文化变体和建筑意图可以缓和人对环境动态的感知程度，而它们是如何做到这一点的正是他文章中的一个重要主题。萨默运用社会科学家的语言和观点来使设计过程得到优化。

## READING 24

## Space-Time

**Robert Sommer**

The traditional tools that the architect employs to manipulate physical space are usually associated with the language of the visual and plastic arts. Sommer has researched and written about the ways that the architect should also rely on the methods and theories of the social sciences during the design process. In "Space-Time," a chapter in his book *Design Awareness,* Sommer shows how the dimension of time should be as important to the designer as the three-dimensional world. He argues that only through an understanding of time can the designer appreciate the dynamic nature of built form. People experience architecture not through static media but within a constantly changing temporal continuum. A primary theme of his narrative is the ways that a person's perception of the dynamics of the environment are tempered by cultural variations and architectural intent. He employs the language and perspective of the social scientist to enhance the process of design.

*Source:* "Space-Time" in *Design Awareness* by Robert Sommer. Copyright © 1972 by Rinehart Press. Reproduced by permission of Robert Sommer, University of California, Davis. Rinehart Press, pp. 66–81.

*The size of a park is directly related to the manner in which you use it. If you are in a canoe traveling at three miles an hour, the lake on which you are paddling is ten times as long and ten times as broad as it is to the man in the speed boat going thirty. . . . Every road that replaces a footpath, every outboard motor that replaces a canoe paddle, shrinks the area of the park.*

Paul Brooks

Architects, city planners, and other form makers are acknowledged to be competent in making space allocations, but there is no single group of professionals responsible for designing time-worlds. Historians record the rise and fall of civilizations over time; yet time measured by the notches on the stick or the calendar markings on the wall remains a backdrop to experience. The horologist is a gadgeteer who has little interest in the thing being measured.[1] In Western thought, time is as important a resource as space. Americans are always short of time; a San Francisco radio station announces the exact time 932 times a week.

Discussions of time are limited by our lack of a suitable vocabulary. Our words are static and structured, whereas time requires movement and fluidity. Time is built into every sentence, and every object has time attributes.[2] It is so pervasive a dimension that it is difficult to objectify in spatial terms and difficult to consider apart from oneself. Time has been identified with space,[3] with movement through space,[4] and the unreality of both space and time have been asserted.[5] In another sense, it is not time itself that is of concern to us, but the exchange between the person and his environment. Time is what we do over time—our sequential acts. Time is only a resource because it limits our transactions. Adaptation to a color is a response over time, not because of time.

Time can be fixed at a particular point by means of photography, sound recording, and the daily newspaper. A moment is captured; with imagination, it can be recreated at other places and at other times. One can also slow down the flow of time and extend the present by use of drugs, hypnosis, and auto-suggestion. The present is neither a point nor a plane but it is a pontoon bridge resting upon the river of time. The present cannot interrupt the forward flow of time but it can provide another plane of experience. Time is seen as a medium, a fluid, a great void. It is like lava, hardening as it cools, with the red-hot part continuing to stretch forward. Time is ubiquitous, insidious, and pervasive in human experience. It infiltrates the language and is the backdrop for all action. All buildings are seen and experienced in a time frame. It is tempting to continue in this vein, but it is *not* the purpose of this chapter to discuss the nature of time. Philosophers and physicists have territorial claims on this topic; I gladly leave it to them. I will devote myself to the relevance of time to environmental experience and action. The movement to maintain and improve environmental quality needs to operate

---

[1] W. Zelinsky, "Of Time and the Geographer," *Landscape,* Winter 1965–66, pp. 21–22.
[2] W. V. Quine, *Word and Object* (New York: John Wiley & Sons, Inc., 1960).
[3] R. Taylor, "Spatial and Temporal Analogies and the Concept of Identity," *Problems of Space and Time,* ed. J. C. C. Smart (New York: The Macmillan Company, 1964).
[4] W. D. Ross, ed., *The Student's Oxford Aristotle* (New York: Oxford University Press, Inc., 1942), p. 219.
[5] F. H. Bradley, *Appearance and Reality* (Oxford: Clarendon Press, 1930).

on several different time scales simultaneously—from an immediate assault on pollution problems and unchecked population growth to a long-range reconstruction of society in a better relation with environment.[6]

In visual perception, a two-dimensional image is projected onto the retina; we infer the third dimension of depth. It is not often recognized that a fourth dimension—time—is also built into our percepts. I disagree with J. B. Priestley who, if I read him correctly, denies perceptual characteristics to temporal dimensions.[7] He doubts that past, present, or future can be regarded as qualities of objects or events. I believe that we perceive something with the knowledge that it is there *now*, and that this is different from our memories of something we have seen in the past or our ideas of what we may see in the future. The nowness of perception does not contradict the idea that I see something as "an old building" or as a design that is ahead of its time. If objects and events are perceived in a time frame, as I believe they are, it seems correct to speak of time as a perceptual quality—not simply duration, but temporal experience in the broadest sense as an ingredient of immediate perception. I looked for an English term that might make this time-dimension more explicit but was handicapped by the fact that in the English language *time* is expressed by verbs rather than by nouns. In German, compound verb-nouns are more acceptable. The term *Verlaufwahrnelunung* ("perception of the passage of time") comes close to describing how we see process as well as structure.

The politician acknowledges that there are still problems in the ghetto but declares, "Look how far you people have come in the last 20 years." The conservationist sees Lake Tahoe as it was when John Muir gazed upon it, as it is presently being developed by real estate promoters, and as it is likely to look in 50 years if development is not controlled. The hologram is a laser photograph which shows an object as seen from all 360 degrees. It would be useful to have time holograms that would show objects or events from their origin to their decay.

The Western notion that every act can be fixed in time with a beginning and an end helps explain the views that architecture is great hollow sculpture and that the city is a great stage on which the human drama is played. Without denying the aesthetic or educational values of good design, we must state clearly that architecture, city planning, industrial design, and all the other design professions differ from the fine arts in important respects. Unlike the sculptor, the architect or planner does not attempt to create an object that will endure unchanged through time. A painting, photograph, or a ceramic vase represent [sic] final solutions. One cannot improve on a Van Gogh painting by adding to it or by changing the colors. The owner preserves it under conditions that will minimize change, and he will restore the painting if it is damaged. Individual buildings can be preserved in deference to former occupants or to their designers; an entire city like Port Townsend, Washington, can be an architectural museum. To convert a building or a city into a monument is to forge a link with the past. In contrast to monuments or museum pieces whose time milieu remains fixed, a design solution for a particular client's problem is an experiment. Insofar as it succeeds, it can be

---

[6]Kenneth J. Hare, cited in John S. Steinhart and Stacie Cherniack, *The Universities and Environmental Quality*, a report of the Office of Science and Technology, Executive Office of the President, September 1969, p. 6.

[7]J. B. Priestley, *Man and Time*, (Garden City, NY: Doubleday & Company, Inc., 1964).

replicated and improved; but, since the client's needs, resources, and the available technology are always in flux, there is no final solution to a design problem. The time perspective of the designer must be diverted from rigidly-bound intervals and fixed timeless products to fluid and dynamic processes with myriad influences upon them.

> The Darwinian revolution . . . is based on the concept that life is a process unfolding over vast spans of time. To formulate laws based on a static concept of time, such as "balance of nature," is to misconstrue the essence of the process. Natural selection—the force that shapes man and other existing forms of life—involves a dynamic change in which malleable organisms interact with a shifting environment. There are no stable elements in the system. Nowhere does the process sustain its existing forms. There is no balance. The scales are always tipping under new weights.[8]

One of the major contributions of James Hutton[9] to modern geology was his view of world as time. Writing that "waterdrops have worn the stones of Troy and swallowed up cities," he was able to perceive the power that rain might have over a thousand-year span. The fourth dimension of time altered his way of looking at water, wind, life, and death. A world without the experience of time is impoverished and lacks an essential quality of human experience. Through hypnosis one can experience a world without depth—a flat, drab, and lifeless plane. A world without time lacks an equally important dimension of human experience.

Some aspects of the ecological perspective have mystical connotations to laymen—time-as-flux is one of these. The meaning of this conception becomes clear when it is applied to a practical problem. First of all, any environmental change is likely to benefit certain individuals and institutions, to worsen the condition of others, and to force some to change, die, or move away. One of my students is currently studying the new pedestrian mall in Sacramento, California. Less than a year after the mall opened, some businesses evidently were helped by the absence of cars; others were seriously hurt. The mall was attracting its own clientele—pedestrian traffic, mothers and children enjoying the play areas, and older retired people on the benches watching the action. Although it is of practical interest to ask about the effects of the mall on existing stores, in the long run, this is an unimportant question. Over a period of years, the mall will attract its own clientele and its own stores suited to pedestrian activities.

Time is a major constituent of architectural experience. People's reactions to a building are influenced by the past, the present, and the future as well as by its physical dimensions, color, material, and style. A building can be programmed temporally as well as spatially to link the present with the past or the present with the future. A college library can emphasize the continuity of learning and the cumulative nature of scholarly research, as well as express the traditions of the academic community. The visitor to Oxford University is struck by the thought that 400 years earlier scholars walked down the same halls. One can design a newspaper office to enhance the here and now, the immediacy of the ephemeral present, disappearing as soon as it is portrayed in banner headlines. A scientific laboratory, bereft of familiar materials and forms, can challenge its occupants to imagine the world of tomorrow.

---

[8]Loren Eiseley, cited in Paul Fleischman, "Conservation: The Biological Fallacy," *Landscape,* 18, no. 2 (1969), pp. 23–26.
[9]Cited in Loren Eiseley, *The Firmament of Time* (New York: Atheneum Publishers, 1966), p. 25.

Prison designers have been admonished to pay particular attention to future trends because it is difficult to obtain funds to replace outmoded institutions. A community may outgrow its hospitals and schools and look forward to building new ones, but in the case of prisons and jails, the tendency has been to expand or to overcrowd existing buildings rather than to build new ones.[10]

The noted social psychologist Kurt Lewin[11] wrote, "The life-space of an individual includes his future, his present, and his past. Actions, emotions, and certainly his morale at any instant depend upon his total time perspective." Environmental experience is affected by such attributes as *duration*—the length of time one spends in a building or a city; *tempo*—how quickly people move or events pass (a lake looks very different at 70 miles an hour than it does while walking alongside it); *sequence*—certain routes will provide contrasts and surprises and make a building seem exciting and alive; *chronicity*—several brief visits will produce a different environmental experience than one long visit; and *familiarity*—the visitor and the old-time resident may share space but their experiences are different. Familiarity means orientation in time as well as in space—knowing how long it takes to get from one place to another, knowing that it takes 20 minutes to get to the airport, knowing whether it is a rush hour or a slack period. Studies of environmental perception cannot be based solely on casual visitors, although these "tourists" represent a valuable source of data to determine environmental experience before habituation has set in. Some of the most revealing comments about buildings may come from those people who are unable to adjust to them and depart, whereas some of the least informative may come from those who remain by "tuning out" or adapting to unsatisfying surroundings.

## SOCIO-CULTURAL TIME

Other cultures conceive of time in ways unfamiliar to most Westerners; these differences are often related to the structure of space. Indian psychiatrist Shashi K. Pande describes the Westerner's view of time as a "unique opportunity, to be utilized and be filled to the utmost with engagements, events, and endeavors in order to capture the richest share of life."[12] He describes the visit of a Malayan prince who saw London for the first time when he attended the coronation of Edward VII in 1902. The prince remarked to his escort that he understood for the first time why Europeans valued time so highly: "In England each day is so packed with living that if a man misses so much as a quarter of an hour, never again will he catch up with the minutes which have escaped him. With us life saunters: here it gallops as if it were pursued by devils." In contrast to this is the profligate attitude of Western man toward space—he spreads himself around as if the earth were infinite. One wonders how the recent view of earth as a finite ball will affect the Western view of time. Perhaps the Westerner may loosen his grip on time—releasing experience from its temporal bonds—and enjoy the trip to the moon as well as the landing and splashdown.

---

[10] R. D. Barnes, "Modern Prison Planning," in *Contemporary Corrections*, ed. Paul W. Tappen (New York: McGraw-Hill Book Company, 1951), p. 272.

[11] Kurt Lewin, "Time Perspective and Morale," in *Resolving Social Conflicts* (New York: Harper & Row, Publishers, 1948).

[12] Shashi K. Pande, "From Hurried Habitability to Heightened Habitability," *Proceedings of the First National Symposium on Habitability*, Los Angeles, May 11–14, 1970.

The democratic ideal rejected the theory of historical causation and believed that the poor could rise out of slums, colonies could free themselves, and minorities could become integrated into the mainstream of society. Hard work would be rewarded and nothing was inevitable except death and taxes. History was a dry chronicle of past events, the names of victorious kings and queens, dates of wars and treaties, and the sites of battles and discoveries. There was no common thread running through the past nor any indication what form the future might take. It was a succession of events and individuals without a causal force. The American antipathy to long-term planning can thus be understood. Looking too far back denies our freedom of action; planning too far ahead will hinder the freedom of our grandchildren to determine their own present. As a rule of thumb, a man is supposed to plan for his children but not for his grandchildren—that is his children's job.

Psychiatrist Humphrey Osmond describes the Vietnam conflict as a war between two opposing forces with different time-space perspectives.[13] General Giap counts off the conflict in decades. He opposes the mighty American spaceman who, though he roams widely, is always short of time. The Americans are straining at the bit to "get it over with," and they complain that the war is "endless." One cannot foresee a quick solution to a war in which one opponent is committed to winning space while the other is trying to gain time. With their evolutionary view of events, Marxist theorists believe it is unrealistic to fix the status quo in a treaty or agreement since historical forces move by their own internal dynamics. It would be like passing a law that no acorn could become an oak or that snow could not become water. The only way to enforce such laws would be to destroy all acorns or to prevent them from coming in contact with the soil; or, in the case of snow, to keep it forever at a freezing temperature. In these cross-cultural contacts, the effects of differences in time perspectives are most clearly felt.[14]

Perception can be defined as the interpretation of a stimulus, or as giving a stimulus a meaning; time concepts are intrinsic to this process. Consciously or unconsciously, all scenes are perceived in a time frame. There is an interaction between external cues and internal ones, and a relativity between one's own shifting time-coordinates and those of the outside world. All this has particular relevance for those who design or manage environmental settings. Blueprints, drawings, and photographs, even if they are three-D holograms, capture time at an instant and portray a static reality. Just as it takes a special act of consciousness to translate two-dimensional drawings into the mental image of a three-dimensional building, it takes similar mental effort, using somewhat different concepts, to add a fourth dimension of time to a three-dimensional building.

The sociologist Sorokin[15] maintains that each culture and academic discipline has its own conception of time—geological time with vast spans incomprehensible to most laymen, biological time based on rhythms and internal clocks, historical time based on events and cycles—and these differ in important ways from physico-mathematical time. Sorokin's particular concern is *socio-cultural time* which uses social

---

[13] Comments contained in unpublished and highly stimulating memoranda and letters, 1968–71.
[14] E. T. Hall, *The Hidden Dimension* (Garden City, NY: Doubleday & Company, Inc., 1966).
[15] P. A. Sorokin, *Socio-Cultural Causality, Space, Time* (New York: Russell and Russell, Publishers, 1964).

events as points of reference and has these characteristics: (a) It does not flow evenly—one year in an urban society is packed with more changes than 50 years of existence in an isolated primitive tribe; (b) The appearance of markers and events is irregular—there are critical moments as well as dead time; (c) It is not infinitely divisible—one can rent a hotel room by the day or by the week but rarely for an hour or a minute; (d) It is determined by social conditions, and reflects the rhythms and pulsations of life in a given group. Within the same geographic area, different occupational and cultural groups will have different socio-cultural times. Every year Harvard University issues a calendar, which not only differs in important ways from the Harvard calendar of 50 years ago, but is vastly different from the calendar of workers in a local factory.

Socio-cultural time cannot be replaced by purely quantitative time without devitalizing it and making orientation in time virtually impossible. If all socio-cultural time conventions were removed—the beginning of the year, the week, the month, Christmas, special dates and holidays—we would be lost in time. There would be an unlimited number of units of mathematical time at our disposal—billions of seconds to use in measuring duration—but we would not know where to start and where to end. Mathematical time is continuous and flows evenly, there are no physical dates or points of reference, the units are all identical and colorless, ignoring the qualitative aspects of time experience. By contrast, socio-cultural time contains markers and divisions which are sinful, holy, happy, times of harvest, times for sowing, for labor, and for rest, which in our experience are inseparable from time since they are tied in with the rhythms and pulsations of individual and group experience.

The same reasoning applies to *socio-cultural* space as distinct from physical-mathematical space. Notation systems for analyzing spatial experience are intrinsically bound with concepts of flow, duration, and periodicity.[16] Using these notations, a highway designer should be able to plan the experience of a driver or of a pedestrian just as a composer writes out a symphony as it will sound—not to the people who play it—but to the audience who hear it. Philip Thiel believes that skills similar to those of the composer or film editor will be required—a sense of rhythm, tempo, meter, and a feeling for the interrelationship of sequence.

Environmental experience does not flow evenly in equal time units—rather it begins and ends with certain events—entering a building, opening a door to one's apartment, leaving the apartment, leaving the building. The time worlds of different occupational groups in a single building are not likely to coincide. Secretaries and salesmen will have different time-coordinates—coffee breaks, lunch periods, and vacation times; different penalties for arriving late or leaving early, and different conceptions of how long each will remain at the same desk or at the same job. The physical structure of a building creates the distinction between inside and outside time. A secretary's day passes slowly when business is slack, but she anticipates an active social life in the evening. Her boss may be under extreme pressure to increase sales; there are not enough minutes in the day for him. Occasionally, to preserve his sanity, he dreams of

---

[16]See, for example, D. Appleyard, K. Lynch, and J. R. Myer, *The View from the Road* (Cambridge, MA: MIT Press, 1964); or Philip Thiel, "Notes on the Description, Scaling, Notation, and Scoring of Some Perceptual and Cognitive Attributes of the Physical Environment," in *Environmental Psychology,* ed. H. Proshansky, W. H. Ittelson, and Leanne G. Rivlin (New York: Holt, Rinehart and Winston, Inc., 1970).

the leisurely pace of his vacation when there is "nothing to do." The purpose of certain space dividers may be to keep people with different time-coordinates out of one another's way. Research scientists who operate without deadlines should not be mixed with technical people who must produce answers to fit the rhythms of the production line. Hospitals typically segregate the long-term or chronic patients from newly-admitted patients; and jails do the same for inmates.

Sometimes the objective of a building is to take a person out of one time and enclose him in another. An archeological museum moves a person back centuries, whereas a science and technology museum projects him forward. A theater must be prepared to create and change time worlds rapidly. Resorts and retirement communities convey a freedom from outside time definitions. Good design means protecting the visitor against too abrupt a change in time worlds. The museum lobby may prepare the visitor for the shock of moving from the outside traffic into the Pleistocene era; the stodgy formality of a bank lobby slows the customer down to a dignified and unhurried tempo.

There is a difference between a preserved historical building and the faked gimmickry of Disneyland. There is no comparison between the experience of standing in awe in a redwood forest and realizing that these trees were in America before Columbus landed and of seeing a concrete replica of such a forest in Florida or in Los Angeles. One of the most important differences between the Disneyland building and the historical building is the time experience of the visitor—his ability to sense the continuity between the people who had actually lived in the building and himself, and his appreciation of what it meant to hew the timbers and fashion the nails by hand, and to bring the furniture around the Horn to San Francisco and transport it by land to the Pacific Northwest.

There has never been a more eloquent advocate of historical preservation than John Ruskin:

> Do not let us talk then of restoration. The thing is a Lie from beginning to end. You may make a model of a building as you may of a corpse, and your model may have the shell of the old walls within it as your cast might have the skeleton. . . . But the old building is destroyed, and that more totally and mercilessly than if it had sunk into a heap of dust, or melted into a mass of clay. . . . We have no right whatever to touch buildings of the past. They are not ours. They belong partly to those who built them, and partly to all generations of mankind who are to follow us.[17]

Geographer Wilbur Zelinsky[18] questions why two intervals between the ticks of a clock should be any more similar than two rectangular patches of land on a map. In the year 1800, Liverpool, Lima, and Tunis were at different levels of development, and each was stuck in its own time rut. Traveling from California to Kansas, one goes over one thousand miles east, and also goes back ten years in manners, morals, and symbols of popular culture.[19] Traveling across a small American city, there is a peeling back of layers of house styles and types from the new suburbs to the central city, and back again on

---

[17] John Ruskin, *The Seven Lamps of Architecture* (London: G. Allen 1904), pp. 185–186.
[18] Zelinsky, "Of Time and the Geographer."
[19] The nineteenth-century German poet Heinrich Heine is alleged to have said, "If the world comes to an end, I shall go to Holland. There everything happens 50 years later."

the other side. Moving to another apartment is a time trip as one sorts through acquisitions and treasures, some of which have not been seen since the last move. There are high school yearbooks, old photograph albums, bundles of letters, tax receipts, unused wedding presents, and clothing which has been saved in the hope that the styles will come back. Decisions have to be made about what is to be saved intact, what is to be altered, and what is to be given away. Just as each man is his own reservoir of time, holding back immense quantities of experience, his possessions are layers from different periods of his life.

A room, a park, or a highway tells a visitor "Stay awhile" or "Get past me as quickly as possible." Duration messages apply not only to circulation areas but also to destinations. The poignant sign outside some motels and restaurants, announcing "This is the place," has a deep emotional significance. The intended message is that the traveler has driven all the way through *nowhere* and has finally arrived *somewhere*. He has reached the island in the sea, the oasis in the desert, the refuge in the wilderness. Too many buildings and public spaces lack a sense of place. A man drives on a freeway, which looks like any freeway, in a car, which looks like any car, and even the gas stations and motels are indistinguishable one from another. Airports are probably the ultimate of placelessness. The traveler parks his car, enters a vast funnel, walks through one tunnel into another, does not meet anyone, goes through a small doorway, sits down in a long tube, and perhaps does not even see the outside of the plane in which he flies. There is a slick sameness to airport shops—the Hertz and Avis girls in their neat uniforms, the insurance counters, the restaurants with identical menus, the newsstands, the souvenir counters with the sign, "What did you bring me?"[20]

According to J. B. Priestley, the air traveler exchanges time for space, and loses place in the bargain:

> Let us say that I travel by car, at 60 miles an hour, across the five miles of the Little Puddlefield District. I see a church, two farms, four bungalows, and an inn, successively within five minutes; I have a Time relation with this region. On the next occasion I fly over it in a jet plane on a clear day, look down and see all at once the church, the two farms, the four bungalows, the inn, and what was in Time is now in Space.[21]

The placeless environment means that more people will be spending more of their time in transit—never getting anywhere, never arriving.[22] The placeless environment resists all efforts at personalization; it does not change in response to user inputs, a man can leave his litter but not his mark. This homogenized setting tends to blur personal time relationships. The nowhere quality of the airport, the sense of being out of

---

[20] Robert Sommer, "The Lonely Airport Crowd," *Air Travel,* April 1969, pp. 16–18.
[21] Priestley, *Man and Time,* p. 105.
[22] A good description of this attitude is found in the song "Goodbye and Hello" by Tim Buckley (© 1968 by THIRD STORY MUSIC, INC. All rights reserved):

> The velocity addicts explode on the highways
>    Ignoring the journey and moving so fast
> Their nerves fall apart and they gasp but can't breathe
>    They run from cops of the skeleton past
> Petrified by tradition in a nightmare they stagger
>    Into nowhere at all and then look up aghast
>      And I wave goodbye to speed
>      And smile hello to a rose

place, also means that the person is out of time. A well-designed building will give a person a sense that he is somewhere, not only to improve his orientation, but also to increase his feelings of personal relatedness to his surroundings, to enhance and legitimize the quality of his immediate experiences, and, to put it in plain language, to give him something to talk about afterward. Sequence and tempo are time-related concepts that must be included in the building program. An event will be affected by what preceded it and by what comes after it. The endless tunnels and funnels of an airport could be justified if they culminated in a grand panorama of openness and sky. One would feel exhilarated upon reaching the end of the last tunnel and finally stepping into open space. Instead one despondently finishes his journey strapped into a seat in a long narrow tube.

## HYPNOSIS AND SPACE-TIME

Some of the most relevant data on the way time concepts are built into ordinary perceptions are found in research using hypnosis. The sense of time can be changed through specific instructions by the hypnotist to the subject. By means of hypnosis, a person can go forward and backward in time; the passage of time can be speeded up or slowed down; and individual items in a person's past can be temporarily obliterated or extended. Hypnotic regressions have been used diagnostically, to learn about repressed or forgotten memories, and therapeutically, to enable the individual to work out his difficulties in the time period in which they occurred. Aaronson[23] has undertaken a fascinating series of studies, in which he hypnotically removed or expanded the past, present, and future, and changed the experience of depth, size, and distance. He found that removing depth produced withdrawal and irritability and also the sense of being hemmed in and separate from other people. Disturbances of gait were also evident. Increasing the experience of depth produced the feeling that objects and scenes were very beautiful; all subjects were exuberantly happy and compared the session to a psychedelic experience.

Reducing the size of objects produced feelings of anxiety and isolation. Each person sought methods of getting himself "back into scale" with his surroundings. One person solved this by reducing the scale of his own body through suggestion, and another, by imagining himself as a child. The subjects used similar scale-fitting devices when the world was increased in size. One reacted to the immense world with fear, but subsequently he fell asleep and dreamed he was standing in the nave of a large cathedral where he grew in size until he filled it completely. He awoke in tune with the world, as large as everything else, and at peace in a world in great magnificence.

Decreasing the subject's sense of distance produced profound pathological reactions in the first two subjects, and the investigator hesitated to use it any further. One man became afraid he would be attacked and hid all sharp and pointed objects from the experimenter. He accused people of stealing his air. The second subject felt that the walls were closing in on him and retired to his bed for the remainder of the ses-

---

[23]B. S. Aaronson, "Distance, Depth, and Schizophrenia," *American Journal of Clinical Hypnosis* 9 (1967), pp. 203–7; "Lilliput and Brobdignag—Self and World," *American Journal of Clinical Hypnosis* 10 (1968), pp. 160–66; "Hypnotic Alterations of Space and Time," *International Journal of Parapsychology* 10 (1968), pp. 5–36.

sion. He could be roused to activity only with great difficulty. Increasing the distance between the person and the world produced reactions of isolation and withdrawal, also quite distressing to the subjects. It is clear that the person's location in the world has profound implications for his self-concept—too little distance between himself and others is encroaching, too much is isolating.

It is an intriguing hypothesis that many of the behaviors observed in schizophrenia and other pathological states result from perceptual disturbances. Kinzel reported that prison inmates involved in frequent fights and aggressive incidents had relatively large bubbles of personal space and therefore needed more free space around them.[24] The autobiographies of mental patients show how supposedly bizarre behaviors make good sense if one understands how a patient perceives the world.[25] If people can't be trusted it may be prudent to withdraw and say nothing; if the world seems unreal, and all efforts to make contact result in failure, it makes sense to retreat and wait for things to improve. It is possible that the extreme sensory distortions found in some cases of schizophrenia also occur in varying degrees among the normal population and account for individual differences in spatio-temporal worlds. An imposing entrance in a city hall may impress the middle-class business man, but it may frighten the welfare client.

Aaronson found that removing the subject's past through hypnosis produced confusion and disorientation. Expanding the past produced happy reminiscences and absorption in past events. Removing the present resulted in withdrawal and even immobility to a pathological degree. Expanding the present produced involvement and sensory enhancement. Objects and sounds became clear, luminous, and fascinating. Removing the future reduced anxiety since there was nothing to worry about and increased the person's interest in the immediate perceptual experience. These findings raise interesting questions about the most effective time frame to be used in solving a design problem. It could be argued that if we want designers to be absorbed in the world around them, it might be desirable to expand their awareness of the present and reduce thoughts about the future. This approach would be self-defeating since a planned structure is not a here-and-now enterprise but rather a future possibility to be used by people who either do not exist or who may change by the time the building is completed. The time frame of the designer must span buildings of the past and present, and how they work in practice, as well as be aware of the dialectical growth forces in society and the forms that these are likely to take.

Aaronson[26] has also studied the semantic connections between space and time. Depth was most clearly associated with time, width to a lesser extent, and height least of all. In three-dimensional space, past is located to the *back,* the *bottom,* and the *left,* future to the *front, top,* and *right*. Guilford[27] did a similar study and asked his students to make drawings to represent the past, present, and future. Most showed lines going

---

[24] A. F. Kinzel, "Violent Persons More Afraid of Attack," *Hospital Tribune,* September 23, 1968.
[25] Robert Sommer and Humphrey Osmond, "Autobiographies of Former Mental Patients," *Journal of Mental Science* 106 (1960), pp. 648–62.
[26] B. S. Aaronson, "Behavior and the Place Names of Time," *American Journal of Clinical Hypnosis* 9 (1966), pp. 1–17.
[27] J. P. Guilford, "Spatial Symbols in the Apprehension of Time," *American Journal of Psychology* 37 (1928), p. 420.

from left to right, with the past below and the future above, and sometimes the present was pictured as the crest of a wave.

My colleague Charles Tart,[28] who is an accomplished hypnotist, undertook a small investigation of the connection between personal space and tempo. When people were told that their personal space bubbles had doubled in size, they walked very slowly around the room to avoid colliding with other people. When their space bubbles were reduced in size, people walked faster, since there was more free space in the room and therefore less risk of colliding with others. Expanding personal space to three times its normal size produced a euphoric reaction and a feeling of being protected. Expanding it to the size of the whole room generally dissolved the effect, since personal space was no longer sensed. Retracting the boundaries of personal space to the individual's body had a dysphoric effect—people felt constricted, rigid, and unprotected.

In this chapter I have attempted to share what I have learned from my concern with time as an aspect of environmental experience. Some of the more salient points are:

1 Time is an inseparable ingredient of environmental experience. In the same way that we perceive color, texture, and shape, we perceive objects and situations in a time frame.

2 A building can be programmed temporally as well as spatially. A good designer aims at including certain time experiences in his building and at excluding others.

3 Many programs involving long-range planning have foundered upon the Western view of time as a short-run commodity. Frequently the time-coordinates dictated by existing political processes do not coincide with long-range planning needs.

4 Design education has overemphasized both the past and the future—the historical buildings of antiquity and the utopias of the future—at the expense of the present. Serious evaluation of existing buildings will provide valuable guidelines for designing new ones.

## THE HUMAN ENVIRONMENT—SUGGESTED READINGS

Altman, Irwin, and Martin Chemers. *Culture and Environment.* Monterey, CA: Brooks/Cole, 1980.
Arnheim, Rudolf. *New Essays on the Psychology of Art.* Berkeley: University of California, 1986.
Bechtel, Robert; Robert Marans; and William Michelson. *Methods in Environmental and Behavioral Research.* New York: Van Nostrand Reinhold, 1987.
Cooper, Clare. *Easter Hill Village: Some Social Implications of Design.* New York: Free Press, 1975.
Gutman, Robert, ed. *People and Buildings.* New York: Basic Books, 1972.
Newman, Oscar. *Defensible Space.* New York: Macmillan, 1972.
Quantrill, Malcolm. *The Environmental Memory.* New York: Schocken Books, 1987.
Rapoport, Amos. *Human Aspects of Urban Form.* New York: Pergamon, 1977.
Sommer, Robert. *Social Design.* Englewood Cliffs, NJ: Prentice-Hall, 1983.
Zeisel, John. *Inquiry by Design.* New York: Cambridge University, 1984.

---

[28]Charles T. Tart, "The Hypnotic Investigation of Personal Space," unpublished manuscript, 1970.

# PART THREE 第三部分

## THE PROCESS OF ARCHITECTURE 建筑过程

# 7

## THE DESIGN PROCESS
## 设计过程

选读 25
"良好的吻合度"
克里斯托弗·亚历山大

20 世纪 60 年代初出版的《形式综合论》一书确立了克里斯托弗·亚历山大在全面和连贯的建筑设计理论研究上的代表地位。此书出版后，他又对书中的一些概念加以完善，并且不再认同把固定的"设计方法"这一简化概念作为指导实践的工具。但是有一个概念贯穿全书，含义也未曾改变，这个概念就是人的需求和建成形式之间的"吻合度"。"吻合度"这一概念之后在他支持和提高人类生活现状的物质实体的"模式"研究中也有涉及。在"良好的吻合度"一章中，物质环境能够创造真正意义上的人的使用需求与建筑形式的平衡，而亚历山大通过研究发现了像这样的物质环境所应具备的特征。而与发现的这些特征相伴相生，作者在本章中也勾勒出了模式理论的理论基础。他还强调了设计分析方法"示意图"这一概念。在他之后的《建筑模式语言》一书中，"示意图"的概念思想得到了进一步的强调与深化。

### READING 25

### Goodness of Fit

#### Christopher Alexander

The publication of *Notes on the Synthesis of Form* in the early 1960s established Christopher Alexander as a major voice in the search for a coherent and comprehensive design theory in architecture. He has since refined the concepts found in the book, and he has moved away from the simplistic notion of fixed "design methods" as prescriptive practice devices. The one concept that has

*Source:* "Goodness of Fit" from *Notes on the Synthesis of Form* by Christopher Alexander. Cambridge, MA: Harvard University Press. Copyright © 1964 by the President and Fellows of Harvard College. Reproduced by permission of Harvard University Press. pp. 15–27.

remained constant throughout his work has been the notion of "fit" between human needs and built forms. This idea has been referred to in his later work as the search for "patterns" of physical objects that support and enhance the human condition. In the chapter "Goodness of Fit," Alexander outlines the basis for such a theory of patterns as he searches out the characteristics of physical settings that create a true balance between human use and architectural form. He also emphasizes the concept of the "Diagram" as a method of design analysis, an idea that will take on increased importance in his later work, *A Pattern Language*.

The ultimate object of design is form.

The reason that iron filings placed in a magnetic field exhibit a pattern—or have form, as we say—is that the field they are in is not homogeneous. If the world were totally regular and homogeneous, there would be no forces, and no forms. Everything would be amorphous. But an irregular world tries to compensate for its own irregularities by fitting itself to them, and thereby takes on form.[1] D'Arcy Thompson has even called form the "diagram of forces" for the irregularities.[2] More usually we speak of these irregularities as the functional origins of the form.

The following argument is based on the assumption that physical clarity cannot be achieved in a form until there is first some programmatic clarity in the designer's mind and actions; and that for this to be possible, in turn, the designer must first trace his design problem to its earliest functional origins and be able to find some sort of pattern in them.[3] I shall try to outline a general way of stating design problems which draws attention to these functional origins, and makes their pattern reasonably easy to see.

It is based on the idea that every design problem begins with an effort to achieve fitness between two entities: the form in question and its context.[4] The form is the solution to the problem; the context defines the problem. In other words, when we speak of design, the real object of discussion is not the form alone, but the ensemble comprising the form and its context. Good fit is a desired property of this ensemble which relates to some particular division of the ensemble into form and context.[5]

There is a wide variety of ensembles which we can talk about like this. The biological ensemble made up of a natural organism and its physical environment is the most familiar: in this case we are used to describing the fit between the two as well-adaptedness.[6] But the same kind of objective aptness is to be found in many other situations. The ensemble consisting of a suit and tie is a familiar case in point; one tie goes well with a certain suit, another goes less well.[7] Again, the ensemble may be a game of chess, where at a certain stage of the game some moves are more appropriate than others because they fit the context of the previous moves more aptly.[8] The ensemble may be a musical composition—musical phrases have to fit their contexts too: think of the perfect rightness when Mozart puts just *this* phrase at a certain point in a sonata.[9] If the ensemble is a truckdriver plus a traffic sign, the graphic design of the sign must fit the demands made on it by the driver's eye. An object like a kettle has to fit the context of its use, and the technical context of its production cycle.[10] In the pursuit of urbanism, the ensemble which confronts us is the city and its habits. Here the human background which defines the need for new buildings, and the physical environment provided by the available sites, make a context for the form of the city's growth. In an extreme case of this kind, we may even speak of a culture itself as an ensemble in which the various fashions and artifacts which develop are slowly fitted to the rest.[11]

The rightness of the form depends, in each one of these cases, on the degree to which it fits the rest of the ensemble.[12]

We must also recognize that no one division of the ensemble into form and context is unique. Fitness across any one such division is just one instance of the ensemble's internal coherence. Many other divisions of the ensemble will be equally significant. Indeed, in the great majority of actual cases, it is necessary for the designer to consider several different divisions of an ensemble, superimposed, at the same time.

Let us consider an ensemble consisting of the kettle plus everything about the world outside the kettle which is relevant to the use and manufacture of household utensils. Here again there seems to be a clear boundary between the teakettle and the rest of the ensemble, if we want one, because the kettle itself is a clearly defined kind of object. But I can easily make changes in the boundary. If I say that the kettle is the wrong way to heat domestic drinking water anyway, I can quickly be involved in the redesign of the entire house, and thereby push the context back to those things outside the house which influence the house's form. Alternatively I may claim that it is not the kettle which needs to be redesigned, but the method of heating kettles. In this case the kettle becomes part of the context, while the stove perhaps is form.

There are two sides to this tendency designers have to change the definition of the problem. On the one hand, the impractical idealism of designers who want to redesign entire cities and whole processes of manufacture when they are asked to design simple objects is often only an attempt to loosen difficult constraints by stretching the form–context boundary.

On the other hand, this way in which the good designer keeps an eye on the possible changes at every point of the ensemble is part of his job. He is bound, if he knows what he is doing, to be sensitive to the fit at several boundaries within the ensemble at once. Indeed, this ability to deal with several layers of form–context boundaries in concert is an important part of what we often refer to as the designer's sense of organization. The internal coherence of an ensemble depends on a whole net of such adaptations. In a perfectly coherent ensemble we should expect the two halves of every possible division of the ensemble to fit one another.

It is true, then, that since we are ultimately interested in the ensemble as a whole, there is no good reason to divide it up just once. We ought always really to design with a number of nested, overlapped form–context boundaries in mind. Indeed, the form itself relies on its own inner organization and on the internal fitness between the pieces it is made of to control its fit as a whole to the context outside.

However, since we cannot hope to understand this highly interlaced and complex phenomenon until we understand how to achieve fit at a single arbitrarily chosen boundary, we must agree for the present to deal only with the simplest problem. Let us decide that, for the duration of any one discussion, we shall maintain the same single division of a given ensemble into form and context, even though we acknowledge that the division is probably chosen arbitrarily. And let us remember, as a corollary, that for the present we shall be giving no deep thought to the internal organization of the form as such, but only to the simplest premise and aspect of that organization: namely, that fitness which is the residue of adaptation across the single form–context boundary we choose to examine.[13]

The form is a part of the world over which we have control, and which we decide to shape while leaving the rest of the world as it is. The context is that part of the world which puts demands on this form; anything in the world that makes demands of the form is context. Fitness is a relation of mutual acceptability between these two. In a problem of design we want to satisfy the mutual demands which the two make on one another. We want to put the context and the form into effortless contact or frictionless coexistence.

We now come to the task of characterizing the fit between form and context. Let us consider a simple specific case.

It is common practice in engineering, if we wish to make a metal face perfectly smooth and level, to fit it against the surface of a standard steel block, which is level within finer limits than those we are aiming at, by inking the surface of this standard block and rubbing our metal face against the inked surface. If our metal face is not quite level, ink marks appear on it at those points which are higher than the rest. We grind away these high spots, and try to fit it against the block again. The face is level when it fits the block perfectly, so that there are no high spots which stand out any more.

This ensemble of two metal faces is so simple that we shall not be distracted by the possibility of multiple form–context boundaries within it. There is only one such boundary worth discussion at a macroscopic level, that between the standard face (the context), and the face which we are trying to smooth (the form). Moreover, since the context is fixed, and only the form variable, the task of smoothing a metal face serves well as a paradigm design problem. In this case we may distinguish good fit from bad experimentally, by inking the standard block, putting the metal face against it, and checking the marking that gets transferred. If we wish to judge the form without actually putting it in contact with its context, in this case we may also do so. If we define levelness in mathematical terms, as a limitation on the variance which is permitted over the surface, we can test the form itself, without testing it against the context. We can do this because the criterion for levelness is, simultaneously, a description of the required form, and also a description of the context.

Consider a second, slightly more complex example. Suppose we are to invent an arrangement of iron filings which is stable when placed in a certain position in a given magnetic field. Clearly we may treat this as a design problem. The iron filings constitute a form, the magnetic field a context. Again we may easily judge the fit of a form by placing it in the magnetic field, and watching to see whether any of the filings move under its influence. If they do not, the form fits well. And again, if we wish to judge the fit of the form without recourse to this experiment, we may describe the lines of force of the magnetic field in mathematical terms, and calculate the fit or lack of fit. As before, the opportunity to evaluate the form when it is away from its context depends on the fact that we can give a precise mathematical description of the context (in this case the equations of the magnetic field).

In general, unfortunately, we cannot give an adequate description of the context we are dealing with. The fields of the contexts we encounter in the real world cannot be described in the unitary fashion we have found for levelness and magnetic fields. There is as yet no theory of ensembles capable of expressing a unitary description of the varied phenomena we encounter in the urban context of a dwelling, for example, or in a sonata, or a production cycle.

Yet we certainly need a way of evaluating the fit of a form which does not rely on the experiment of actually trying the form out in the real world context. Trial-and-error design is an admirable method. But it is just real world trial and error which we are trying to replace by a symbolic method, because real trial and error is too expensive and too slow.

The experiment of putting a prototype form in the context itself is the real criterion of fit. A complete unitary description of the demands made by the context is the only fully adequate nonexperimental criterion. The first is too expensive, the second is impossible: so what shall we do?

Let us observe, first of all, that we should not really expect to be able to give a unitary description of the context for complex cases: if we could do so, there would be no problems of design. The context and the form are complementary. This is what lies behind D'Arcy Thompson's remark that the form is a diagram of forces.[14] Once we have the diagram of forces in the literal sense (that is, the field description of the context), this will in essence also describe the form as a complementary diagram of forces. Once we have described the levelness of the metal block, or the lines of force of the magnetic field, there is no conceptual difficulty, only a technical one, in getting the form to fit them, because the unitary description of the context is in both cases also a description of the required form.

In such cases there is no design problem. *What does make design a problem in real world cases is that we are trying to make a diagram for forces whose field we do not understand.*[15] Understanding the field of the context and inventing a form to fit it are really two aspects of the same process. It is because the context is obscure that we cannot give a direct, fully coherent criterion for the fit we are trying to achieve; and it is also its obscurity which makes the task of shaping a well-fitting form at all problematic. What do we do about this difficulty in everyday cases? Good fit means something, after all—even in cases where we cannot give a completely satisfactory field-like criterion for it. How is it, cognitively, that we experience the sensation of fit?

If we go back to the procedure of leveling metal faces against a standard block, and think about the way in which good fit and bad fit present themselves to us, we find a rather curious feature. Oddly enough, the procedure suggests no direct practical way of identifying good fit. We recognize bad fit whenever we see a high spot marked by ink. But in practice we see good fit only from a negative point of view, as the limiting case where there are no high spots.

Our own lives, where the distinction between good and bad fit is a normal part of everyday social behavior, show the same feature. If a man wears eighteenth-century dress today, or wears his hair down to his shoulders, or builds Gothic mansions, we very likely call his behavior odd; it does not fit our time. These are abnormalities. Yet it is such departures from the norm which stand out in our minds, rather than the norm itself. Their wrongness is somehow more immediate than the rightness of less peculiar behavior, and therefore more compelling. Thus even in everyday life the concept of good fit, though positive in meaning, seems very largely to feed on negative instances; it is the aspects of our lives which are obsolete, incongruous, or out of tune that catch our attention.

The same happens in house design. We should find it almost impossible to characterize a house which fits its context. Yet it is the easiest thing in the world to name the specific kinds of misfit which prevent good fit. A kitchen which is hard to clean, no place to park my car, the child playing where it can be run down by someone else's car, rainwater coming in, overcrowding and lack of privacy, the eye-level grill which spits hot fat right into my eye, the gold plastic doorknob which deceives my expectations, and the front door I cannot find, are all misfits between the house and the lives and habits it is meant to fit. These misfits are the forces which must shape it, and there is no mistaking them. Because they are expressed in negative form they are specific, and tangible enough to talk about.

The same thing happens in perception. Suppose we are given a button to match, from among a box of assorted buttons. How do we proceed? We examine the buttons in the box, one at a time; but we do not look directly for a button which fits the first. What we do, actually, is to scan the buttons, rejecting each one in which we notice some discrepancy (this one is larger, this one darker, this one has too many holes, and so on), until we come to one where we can see no differences. Then we say that we have found a matching one. Notice that here again it is much easier to explain the misfit of a wrong button than to justify the congruity of one which fits.

When we speak of bad fit we refer to a single identifiable property of an ensemble, which is immediate in experience, and describable. Wherever an instance of misfit occurs in an ensemble, we are able to point specifically at what fails and to describe it. It seems as though in practice the concept of good fit, describing only the absence of such failures and hence leaving us nothing concrete to refer to in explanation, can only be explained indirectly; it is, in practice, as it were, the disjunction of all possible misfits.[16]

With this in mind, I should like to recommend that we should always expect to see the process of achieving good fit between two entities as a negative process of neutralizing the incongruities, or irritants, or forces, which cause misfit.[17]

It will be objected that to call good fit the absence of certain negative qualities is no more illuminating than to say that it is the presence of certain positive qualities.[18] However, though the two are equivalent from a logical point of view, from a phenomenological and practical point of view they are very different.[19] In practice, it will never be as natural to speak of good fit as the simultaneous satisfaction of a number of requirements, as it will be to call it the simultaneous nonoccurrence of the same number of corresponding misfits.

Let us suppose that we did try to write down a list of all possible relations between a form and its context which were required by good fit. (Such a list would in fact be just the list of requirements which designers often do try to write down.) In theory, we could then use each requirement on the list as an independent criterion, and accept a form as well fitting only if it satisfied all these criteria simultaneously.

However, thought of in this way, such a list of requirements is potentially endless, and still really needs a "field" description to tie it together. Think, for instance, of trying to specify all the properties a button had to have in order to match another. Apart from the kinds of thing we have already mentioned, size, color, number of holes, and so on, we should also have to specify its specific gravity, its electrostatic charge, its viscosity, its rigidity, the fact that it should be round, that it should not be made of paper, etc., etc. In other words, we should not only have to specify the qualities which distinguish it from all other buttons, but we should also have to specify all the characteristics which actually made it a button at all.

Unfortunately, the list of distinguishable characteristics we can write down for the button is infinite. It remains infinite for all practical purposes until we discover a field description of the button. Without the field description of the button, there is no way of reducing the list of required attributes to finite terms. We are therefore forced to economize when we try to specify the nature of a matching button, because we can only grasp a finite list (and rather a short one at that). Naturally, we choose to specify those characteristics which are most likely to cause trouble in the business of matching, and which are therefore most useful in our effort to distinguish among the objects we are likely to come across in our search for buttons. But to do this, we must rely on the fact that a great many objects will not even come up for consideration. There are, after all, conceivable objects which are buttons in every respect except that they carry an electric charge of one thousand coulombs, say. Yet in practice it would be utterly superfluous, as well as rather unwieldy, to specify the electrostatic charge a well-matched button needed to have. No button we are likely to find carries such a charge, so we ignore the possibility. The only reason we are able to match one thing with another at all is that we rely on a good deal of unexpressed information contained in the statement of the task, and take a great deal for granted.[20]

In the case of a design problem which is truly problematical, we encounter the same situation. We do not have a field description of the context, and therefore have no intrinsic way of reducing the potentially infinite set of requirements to finite terms. Yet for practical reasons we do need some way of picking a finite set from the infinite set of possible ones. In the case of requirements, no sensible way of picking this finite set presents itself. From a purely descriptive standpoint we have no way of knowing which of the infinitely many relations between form and context to include, and which ones to leave out.

But if we think of the requirements from a negative point of view, as potential misfits, there is a simple way of picking a finite set. This is because it is through misfit that the problem originally brings itself to our attention. We take just those relations between form and context which obtrude most strongly, which demand attention most clearly, which seem most likely to go wrong. We cannot do better than this.[21] If there were some intrinsic way of reducing the list of requirements to a few, this would mean in essence that we were in possession of a field description of the context: if this were so, the problem of creating fit would become trivial, and no longer a problem of design. We cannot have a unitary or field description of a context and still have a design problem worth attention.

In the case of a real design problem, even our conviction that there is such a thing as fit to be achieved is curiously flimsy and insubstantial. We are searching for some kind of harmony between two intangibles: a form which we have not yet designed, and a context which we cannot properly describe. The only reason we have for thinking that there must be some kind of fit to be achieved between them is that we can detect incongruities, or negative instances of it. The incongruities in an ensemble are the primary data of experience. If we agree to treat fit as the absence of misfits, and to use a list of those potential misfits which are most likely to occur as our criterion for fit, our theory will at least have the same nature as our intuitive conviction that there is a problem to be solved.

The results of this chapter, expressed in formal terms, are these. If we divide an ensemble into form and context, the fit between them may be regarded as an orderly condition of the ensemble, subject to disturbance in various ways, each one a potential misfit. Examples are the misfits between a house and its users, mentioned on page 357. We may summarize the state of each potential misfit by means of a binary variable. If the misfit occurs, we say the variable takes the value 1. If the misfit does not occur, we say the variable takes the value 0. Each binary variable stands for one possible kind of misfit between form and context.[22] The value this variable takes, 0 or 1, describes a state of affairs that is not either in the form alone or in the context alone, but a relation between the two. The state of this relation, fit or misfit, describes one aspect of the whole ensemble. It is a condition of harmony and good fit in the ensemble that none of the possible misfits should actually occur. We represent this fact by demanding that all the variables take the value 0.

The task of design is not to create form which meets certain conditions, but to create such an order in the ensemble that all the variables take the value 0. The form is simply that part of the ensemble over which we have control. It is only through the form that we can create order in the ensemble.

## NOTES

1 The source of form actually lies in the fact that the world tries to compensate for its irregularities as economically as possible. This principle, sometimes called the principle of least action, has been noted in various fields: notably by Le Chatelier, who observed that chemical systems tend to react to external forces in such a way as to neutralize the forces; also in mechanics as Newton's law, as Lenz's law in electricity, again as Volterra's theory of populations. See Adolph Mayer, *Geschichte des Prinzips der kleinsten Action* (Leipzig, 1877).
2 D'Arcy Wentworth Thompson, *On Growth and Form,* 2nd ed. (Cambridge, 1959), p. 16.
3 This old idea is at least as old as Plato: see, e.g., *Gorgias* 474–75.
4 The symmetry of this situation (i.e., the fact that adaptation is a mutual phenomenon referring to the context's adaptation to the form as much as to the form's adaptation to its context) is very important. See L. J. Henderson, *The Fitness of the Environment* (New York, 1913), p. v: "Darwinian fitness is compounded of a mutual relationship between the organism and the environment." Also E. H. Starling's remark, "Organism and environment form a whole, and must be viewed as such." For a beautifully concise description of the concept "form," see Albert M. Dalcq, "Form and Modern Embryology," in *Aspects of Form,* ed. Lancelot Whyte (London, 1951), pp. 91–116, and other articles in the same symposium.
5 At later points in the text where I use the word "system," this always refers to the whole ensemble. However, some care is required here, since many writers refer to that part of the ensemble which is held constant as the environment, and call only the part under adjustment the "system." For these writers my form, not my ensemble, would be the system.
6 In essence this is a very old idea. It was the first clearly formulated by Darwin in *The Origin of Species,* and has since been highly developed by such writers as W. B. Cannon, *The Wisdom of the Body* (London, 1932), and W. Ross Ashby, *Design for a Brain,* 2nd ed. (New York, 1960).

7 Wolfgang Köhler, *The Place of Value in a World of Facts* (New York, 1938), p. 96.
8 A. D. de Groot, "Über das Denken des Schachspielers," *Rivista di psicologia,* 50 (October–December 1956), pp. 90–91; Ludwig Wittgenstein, *Philosophical Investigations* (Oxford, 1953), p. 15.
9 See Max Wertheimer, "Zu dem Problem der Unterscheidung von Einzelinhalt und Teil," *Zeitschrift für Psychologie* 129 (1933), p. 356, and "On Truth," *Social Research* 1 (May 1934), p. 144.
10 K. Lönberg Holm and C. Theodore Larsen, *Development Index* (Ann Arbor, 1953), p. Ib.
11 Again, this idea is not a new one. It was certainly present in Frank Lloyd Wright's use of the phrase "organic architecture," for example, though on his tongue the phrase contained so many other intentions that it is hard to understand it clearly. For a good discussion see Peter Collins, "Biological Analogy," *Architectural Review* 126 (December 1959), pp. 303–6.
12 This observation appears with beautiful clarity in Ozenfant's *Foundations of Modern Art* (New York, 1952), pp. 340–41. Also Kurt Koffka, *Principles of Gestalt Psychology* (London, 1935), pp. 638–44.
13 The idea that the residual patterns of adaptive processes are intrinsically well organized is expressed by W. Ross Ashby in *Design for a Brain,* p. 233, and by Norbert Wiener in *The Human Use of Human Beings* (New York, 1954), p. 37.
14 See note 2.
15 The concept of an image, comparable to the ideal field statement of a problem, is discussed at great length in G. A. Miller, Eugene Galanter, and Karl H. Pribram, *Plans and the Structure of Behavior* (New York, 1960). The "image" is presented there as something present in every problem solver's mind, and used by him as a criterion for the problem's solution and hence as the chief guide in problem planning and solving. It seems worth making a brief comment. In the majority of interesting cases I do not believe that such an image exists psychologically, so that the testing paradigm described by Miller et al. in *Plans* is therefore an incorrect description of complex problem-solving behavior. In interesting cases the solution of the problem cannot be tested against an image, because the search for the image or criterion for success is actually going on at the same time as the search for a solution.

Miller does make a brief comment acknowledging this possibility on pp. 171–72. He also agreed to this point in personal discussions at Harvard in 1961.
16 It is not hard to see why, if this is so, the concept of good fit is relatively hard to grasp. It has been shown by a number of investigators, for example, Jerome Bruner et al., *A Study of Thinking* (New York, 1958), that people are very unwilling and slow to accept disjunctive concepts. To be told what something is not is of very little use if you are trying to find out what it is. See pp. 156–81. See also C. L. Hovland and W. Weiss, "Transmission of Information Concerning Concepts through Positive and Negative Instances," *Journal of Experimental Psychology* 45 (1953), pp. 175–82.
17 The near identity of "force" on the one hand, and the "requiredness" generated by the context on the other, is discussed fully in Köhler, *The Place of Value in a World of Facts,* p. 345, and throughout pp. 329–60. There is, to my mind, a striking similarity between the difficulty of dealing with good fit directly, in spite of its primary importance, and the difficulty of the concept zero. Zero and the concept of emptiness, too, are comparatively late inventions (clearly because they too leave one nothing to hold onto in explaining them). Even now we find it hard to conceive of emptiness as such: we only manage to think of it as the absence of something positive. Yet in many metaphysical systems, notably those of the East, emptiness and absence are regarded as more fundamental and ultimately more substantial than presence.

This is also connected with the fact, now acknowledged by most biologists, that symmetry, being the natural condition of an unstressed situation, does not require explanation, but that on the contrary it is asymmetry which needs to be explained. See D'Arcy Thompson, *On Growth and Form,* p. 357; Wilhelm Ludwig, *Recht-links-problem im Tierreich und beim Menschen* (Berlin, 1932); Hermann Weyl, *Symmetry* (Princeton, 1952), pp. 25–26; Ernst Mach, "Über die physikalische Bedeutung der Gesetze der Symmetrie," *Lotos* 21 (1871); pp. 139–47.

18  The logical equivalence of these two views is expressed by De Morgan's law, which says essentially that if *A, B, C,* etc., are propositions, then [(Not *A*) and (Not *B*) and (Not *C*) . . .] is always the same as Not [(*A* or *B* or *C* or . . .)].

19  For the idea that departures from closure force themselves on the attention more strikingly than closure itself, and are actually the primary data of a certain kind of evaluative experience, and for a number of specific examples (not only ethical), see Max Wertheimer, "Some Problems in Ethics," *Social Research* 2 (August 1935), p. 352ff. In particular, what I have been describing as misfits are described there as *Leerstellen* or emptinesses. The feeling that something is missing, and the need to fill whatever is incomplete (*Lückenfüllung*), are discussed in some detail.

20  Any psychological theory which treats perception or cognition as information processing is bound to come to the same kind of conclusion. For a typical discussion of such information-reducing processes, see Bruner et al., *A Study of Thinking,* p. 166.

21  It is perhaps instructive to note that both the concept of organic health in medicine and the concept of psychological normality in psychiatry are subject to the same kind of difficulties as my conception of a well-fitting form or coherent ensemble. In their respective professions they are considered to be well defined. Yet the only definitions that can be given are of a negative kind. See, for instance, Sir Geoffrey Vickers, "The Concept of Stress in Relation to the Disorganization of Human Behavior," in *Stress and Psychiatric Disorder,* ed. J. M. Tanner (Oxford, 1960).

22  In case it seems doubtful whether all the relevant properties of an ensemble can be expressed as variables, let us be quite clear about the fact that these variables are not necessarily capable of continuous variation. Indeed, it is quite obvious that most of the issues which occur in a design problem cannot be treated numerically, as this would require. A binary variable is simply a formal shorthand way of classifying situations; it is an indicator which distinguishes between forms that work and those that do not, in a given context.

## 选读 26
### "建筑设计中的先验知识和启发性推理"
彼得·罗

在这篇选读中，罗从人们希望建筑师解决的各类问题的角度来探索设计过程的本质。如果不能正确认识想要达到的目的、解决问题的合理方法，那么只能将建筑问题定义为"恶劣"二字；在设计过程中无所谓明确的起点，也无所谓明确的终点。他在文中写道："解决这类问题需要建筑师具有初始的洞察力，还要练习制定临时性条款，勤于推理，能够做出看上去比较可行的策略。换言之，建筑师要大量运用启发性推理方法。"罗认为建筑师在设计过程前期萌生的最初概念和想法将会对最终建筑设计方案产生深远影响，建筑师也应该把设计过程本身看作各种解决问题方法的搭配运用。

**READING 26**

# A Priori Knowledge and Heuristic Reasoning in Architectural Design

## Peter Rowe

Rowe explores in this selection the nature of the design process in terms of the kinds of problems architects are expected to solve. Architectural problems are defined as "wicked" in that both the ends and means of solution are ill-defined; there are no clean starting or stopping points in the design process. He states: "Tackling a problem of this type requires some initial insight, the exercise of some provisional set of rules, inferences, or plausible strategy, in other words, the use of heuristic reasoning." Rowe argues that initial ideas and concepts generated by the architect early in the design process will have profound effects on the final building solution, and the process itself should be seen as a co-mingling of problem-solving methods.

A distinction can be made in the world of problems between those that are well defined and those that are ill defined.[1] In solving the former kind the "ends" are known and one has to find the "means." In the latter kind, that includes most architectural design problems, both the "ends" and the "means" are unknown at first and one has to define the problem. Architectural design problems can also be referred to as being "wicked problems" in that they have no definitive formulation, no explicit "stopping rule," always more than one plausible explanation, a problem formulation that corresponds to a solution and *vice versa*, and that their solutions cannot be strictly correct or false.[2] Tackling a problem of this type requires some initial insight, the exercise of some provisional set of rules, inference, or plausible strategy, in other words, the use of heuristic reasoning.[3]

Design is often guided by heuristic reasoning involving solution images, analogies, or restricted sets of form-giving rules that partially and provisionally define the "ends" or solution state of a problem, *i.e.,* what it should be like. Such heuristics, by virtue of the *a priori* knowledge that is incorporated, provide a framework for problem-solving behavior and exert a strong and dynamic influence over subsequent sequences of problem interpretation, solution generation, problem representation and solution assessment. During the course of designing one mode of heuristic reasoning may be found to be unproductive and give way to other kinds; co-mingling may even occur. As a result design appears to be essentially an emergent phenomenon where new information about a problem is generated, evaluated together with *a priori* knowledge, and solution strategies amended accordingly.

## TYPES OF HEURISTICS

Five classes of heuristics can be identified largely according to the kind of subject matter involved. They are: (1) the use of anthropometric analogies, (2) the use of

Source: Peter Rowe, "A Priori Knowledge and Heuristic Reasoning in Architectural Design," *Journal of Architectural Education,* 36, no. 1 (Fall 1982), pp. 18–23. Copyright © 1982 by the Association of Collegiate Schools of Architecture, Inc. (ACSA). Reproduced by permission of MIT Press Journals.

literal analogies, (3) the use of environmental relations, (4) the use of typologies, and (5) the use of formal "languages." These classes were based on protocol analyses of architectural designers at work, mixed with some speculation.[4] Each class is by no means exclusive of the characteristics of others, nor totally inclusive of the range of possible heuristics. Rather, the classification is one of practical convenience for grouping and discussing observations.

### Anthropometric Analogies

The use of an anthropometric analogy employs a construct describing the physical occupancy of a space, with relational and metric qualities, that guides further design activity. This form of reasoning is often used by naive designers with little or no experience with other forms of heuristics. In one such case a person without any architectural background was observed producing a staircase design based upon imagining someone ascending through a room in a certain manner. The result was a graceful form for which the subject appeared to have no prior reference. Architecture may be peculiarly suited to this sort of process by virtue of the close correspondence that seems able to be attained between the act of visualization and the 3-dimensional artifact itself. Less naive designers may also resort to using this approach when others fail to yield satisfactory results.

### Literal Analogies

This kind of heuristic incorporates borrowing of existing forms, or form-giving constructs, as a point of departure for structuring a design problem and for facilitating further information processing. They are literal analogies because in all cases the subsequent architectural forms that are derived match very closely the conformation of the physical analog. Here, a useful distinction can be made between iconic analogies and canonic analogies.[5]

The scope of references for the development of iconic analogies appears to be extremely broad. They can encompass objects from the natural world, or objects outside of architecture *per se,* such as LeCorbusier's admitted use of the shell of a crab for the roof of Ronchamp Chapel.[6]

They can also include imagery from some scene, painterly conception, or narrative account of real or imagined circumstances. Or they can incorporate, for their iconographic value, artifacts and elements squarely within the realm of architecture. The resulting analogy appears useful to a designer by virtue of its symbolic iconographical meaning, a meaning that is in some ways synonymous with an intention that, once realized, provides additional structure to a problem from which other information can be organized.

Canonic analogies have as their basis "ideal" proportional systems usually manifested in the form of abstract geometrical patterns or shapes. Such configurations, like cartesian grids or platonic solids, are often employed as guidelines to give shape to design problems and to help ease transition into the realm of 2- and 3-dimensional design. In a less palliative manner, canonic analogies may be pursued almost as ends in themselves for the purpose of exploring possibilities for spatial organization and order.

Specific analogs are apt to possess simultaneously both iconic and canonic properties. For example, the architectural rendering of an essentially canonic form may itself symbolize iconographically a particular aesthetic position.[7] For some observers the ubiquity of the urban grid may be strongly identified with decentralized, non-authoritarian organizations, whereas for others the uniformity may symbolize control and oppression. However, the final distinction as to whether a particular literal analogy is iconic or canonic is both a matter of degree and a matter of the purposes to which it is put by a designer.

**Environmental Relations**

Here, use is made of a principle or set of principles, often derived empirically, that represents what appear to be appropriate relationships between man and his environment and between components of the building fabric. Typically, special information about behavior as a determinant of form, or the influence of other environmental factors, such as climate, physiography, materials and resource availability, are incorporated. Also included are principles describing the expected behavior of the material substance of the building itself. In many ways the typological class of heuristic, described in the next section, fits the same definition. However, the heuristic principles involved with environmental relations are not necessarily drawn from past building practice, do not necessarily represent "ideals" in the same manner as typologies, and the arguments inherent in the principles invariably reflect highly problem-oriented interpretations of the design constraints.

The logical structure behind the application of this kind of heuristic for solution generation seems to be: if problem "X" is encountered, then take formal action "Y," under conditions "Z." The principle involved, in effect, creates a bridge between a perceived problem and an ensemble of form-giving characteristics representing its potential solution. Alexander's "patterns"[8] exhibit these properties, as do other relational constructs such as those dealing with structural behavior and "bubble diagramming" procedures that link preference information about physical adjacency to a 2- or 3-dimensional spatial arrangement.

**Typologies**

As heuristics, typologies allow one to make use of knowledge about past solutions to related architectural problems. Further, in a prototypical sense, they embody valid principles that appear to the designer to have exhibited constancy or invariance.

At least three sub-classes can be discerned.[9] First, there is the use of a building type as "a model," representing characteristics worthy of emulation, that seems to provide for the perceived needs, uses and customs found in a problem under consideration. Here, the symbolic meaning attributed to the type, as "a model," is of as much importance as the organizational principles that are incorporated. Second, there is the "organizational typology" used primarily as a framework for solving problems concerning distributions of uses or conformation of functional elements. Third, there are "elemental types" representing prototypes for solving particular classes of problems that recur in different design situations; for example, the problem of "entry," methods of handling vertical circulation, and the problem of rendering the transition between the ground plane and a building.

In some ways this subclassification is arbitrary. A single building may incorporate all three dimensions and be "a model," an organizational framework, and provide a repertoire of solutions for particular kinds of sub-problems. However, it is not the completeness of the building type that is at issue here, but rather the purposes to which the type is put in order to guide problem-solving activity.

Most typologies implicitly possess iconic and canonic qualities in the sense discussed earlier in connection with literal analogies. The difference is that the presence of these qualities in "types" is confined to the realm of architectural expression, and that one of the intentions presumably behind the use of a typology is its quality of being "tried and true" in the formation of architecture. This property is not necessarily found in all literal analogies.

## Formal "Languages"

In this classification the use of formal "languages" is a heuristic process where the content represents generalizations of the information inherent in other kinds of heuristics, particularly those involving typologies and man-environment relations. It is "language-like" at least to the extent that the process's guiding structure imposes an internal consistency that allows for the meaningful ordering and "correct" functioning of formal elements. Explicit treatises on the "classical language," for instance, provide a repertoire of architectural elements, rules for their combination and transformation, and a prescription of the purposes for which various ensembles of elements are deemed most appropriate.[10] The so-called "pattern language"[11] is a no less deterministic method of reasoning, although ostensibly concerned in a different manner with architectural composition. No less highly developed "languages" may be idiosyncratic and derived from a constant way of doing things over an extended period of time. More often than not, this form of heuristic reasoning is to be found in the work of experienced designers.

## HEURISTICS AND DESIGN BEHAVIOR

In practice each class of heuristic is by no means self contained. Some designers may habitually use a small variety of heuristics, while others may be broader in their applications. Heuristic search through a "problem space" essentially involves the generation of solutions in a sequence of stages with evaluations of interim solutions being made at each stage. The new information provided by the evaluation is combined with prior knowledge about the problem structure in order to guide further steps. It is unknown beforehand if a particular sequence of steps will actually yield a solution. Or, to put it another way, it cannot be ascertained if a proposed solution is really a solution until the complete proposal can be made.[12] The types of heuristics discussed so far were defined by the type of *a priori* knowledge that they furnished and, therefore, their generative role in solution generation and evaluation.

Designers may employ one kind of heuristic at the outset of a problem and later abandon it in favor of another type. A designer may also switch back and forth between heuristics at either end of an imagined "iconic–canonic" or "subjective–objective" spectrum. Essentially, the power of each class of heuristic derives from the ability of its application to furnish new information about problem constraints (e.g. criteria, goals, objectives, etc.), and for providing fruitful avenues for responding to constraints that are already established.

**New Information**

One positive outcome of an appropriate course of heuristic reasoning is that it throws the problem under consideration into a new light. Consequently, both the scope of the problem (*i.e.,* constraints) and the promise of various solution methods (*i.e.,* courses of action) can be reinterpreted. The process has the effect of providing valuable new information. The testing procedures involved, such as the use of a principle about man-environment relations or organization of "type," provide this new information in at least 3 ways.[13] First, the test provides information regarding the conformity of the proposed solution with known constraints. Second, it provides information regarding progress towards the overall goal of a solution, *i.e.,* does the strategy appear to be working. Third, the heuristic can provide new information regarding other constraints not so far considered, *i.e.,* problem re-definition.

**Referential Bases**

At a general level, the logical structure of solution generation within each heuristic class can be represented by a statement with three parts: condition, action and intent. The statement is usually expressed in the following manner: if condition "A" exists, then take action "B," for intent of "C."[14] Clearly, generative procedures of this form will automatically satisfy certain problem constraints. The constraints are incorporated within the generator itself. The same sort of thing can be said about each class of heuristic from which a particular generative procedure may be drawn. For example, certain kinds of constraints are implicit in the use of "typologies" that may not be present in the use of particular literal analogies, and *vice versa*. Further, essentially the same kind of constraint may be able to be articulated in a different way from one heuristic class to another, materially affecting the final outcome.

The referential basis for the form-giving power of a heuristic, like one involving an analogy, derives from the correspondence between an intention, prevailing conditions and formal action. Further, designers often seem to know of and express intentions through examples at hand. Actions and intentions may become intertwined to the point where they are largely synonymous, attention is shifted, and a transposition occurs in the logical structure of reasoning to: if action "X" (intention"Y") then condition "Z." Consequently, the process of solving the problem takes on a character of being "end-justified" *i.e.,* re-definition of the problem conditions to fit a proposed solution. This may appear to be perverse, but when one remembers the under-constrained nature of most design problems it can prove to be quite a plausible and even necessary approach.

Each class of heuristic so far described has its own self-referential rule structure, if logically extended, that is largely governed by the incorporated subject matter. For example, in a proportional device such as a 333 cube there are many implied patterns of orthogonal and diagonal subdivision that can be exploited to give rise to a building form.[15] However, the full implications, or consequences, of such a model may not be fully perceived by the designer when the heuristic is first employed. Discrepancies between the ultimate consequences and those initially perceived usually result in "back-tracking," where designers retrench their positions and change from one line of reasoning to another. It has been observed when such difficulties are encountered that switches among classes of heuristics are often made in directions most dissimilar from the type of heuristic last employed.

**Independent Qualities**

As mentioned earlier, the applications of all heuristics are at least partially problem oriented, *i.e.,* methods of generating and assessing solutions are assembled that incorporate and show cognizance of the problem constraints that are explicitly given. On the other hand, constraints may be independently supplied by the heuristic process.[16] Heuristic classes, in the manner described, latently incorporate *a priori* constraints consistent with the subject matter of the class. In a particular problem application some of these constraints are likely to be autonomous, or independent of the problem as given. Moreover, this may be very positive in helping to re-define the problem in a more complete manner. However, should the autonomy of the constraints be extremely pronounced, or superficial, there may also be a high risk of arbitrarily "end-justifying" the solution.

**Order of Application**

The sequential order in which various types of heuristics are brought to bear on the problem can materially affect the solution.[17] Given the fair degree of problem independence in the constraint structure of each heuristic class, it probably makes a difference which kind of heuristic is employed in what order, and therefore the manner in which the problem becomes structured. The use of a particular heuristic, such as a building "type," will reveal the problem in a certain light, suggest solution strategies and tests, and consequently order the priority and scope of further problem-solving efforts. Furthermore, the residual effect of a heuristic strategy, long-since abandoned, is often manifested in the final solution.

**Style**

In the production of architecture a major source of style is the design process itself,[18] and hence the heuristic reasoning processes involved play a central role. Style is usually attributed to a building according to its features. As shown in preceding sections, these features, or final manifestations of a design process, are strongly determined by the classes of heuristics adopted, their use and interpretation, and the sequence in which the overall process unfolds. It is not only the structure of the process that is important but also the "object qualities" of the *a priori* "models" that are used.

Consistency in style among the output of particular designers can then be understood as a habitual way of doing things, of solving design problems. The tendency for this consistency to be most pronounced during particular times in an architect's career, say towards the end of stylistic episodes or "periods," is also understandable. A fluency in a particular way of designing and the consistency that comes with it are only reached from experience and constant development.

By extension, architectural style in the broadest sense may be regarded, in large measure, as being congruent with the collective adoption of certain design practices and forms of heuristic reasoning. When dominant forms no longer prove productive they are replaced, and shifts in style can be observed. Undoubtedly social and cultural factors come into play and help shape and converge the design strategies that become commonly practiced. In fact, it is quite plausible that the individual discovery and sharing of information about the contemporary efficacy, or usefulness, of knowledge provided by commonly practiced heuristic types is the central mechanism in the process of stylistic change.

### Role of Representation

The role of representation in heuristic reasoning cannot be underestimated. For it is through the reciprocity between mind and image involved in rendering a solution that judgements or tests are applied and new information generated.[19]

In this context, the act of drawing can be either a deductive or an inductive process. It is deductive in the sense of: "if this element is here and that element is over there, then under the current line of reasoning, this other element must be placed so." Under a particular heuristic mode it essentially involves re-writing the problem from one state to the next. It is inductive in the sense of finding patterns among the "markings" representing an understood state of the problem that suggest other ways of looking at the situation, *i.e.,* that result in alternative lines of heuristic reasoning. Either way, drawing is a process of discovery: of becoming aware of formal arrangements and possibilities.

How a drawing is made, and therefore what view of the problem is offered, can either facilitate or inhibit the design process. Generally in the course of design, drawing proceeds from referential sketches through consolidation in a "parti," or diagram, to more definitive drawings. Of relevance to heuristic reasoning are the data compression and referential aspects of drawings as portrayals of design ideas. In many cases these characteristics have quite a personal meaning to an individual designer, as evidenced by the difficulty of interpretation encountered when others try to redescribe the drawing in other than mundane terms. Clearly the scale and precision of a drawing reveals certain qualities of a design solution that renderings of a different scale and precision might not accomplish. However, regardless of scale and precision, there is good reason to believe that the very type of rendering itself strongly influences its referential significance. For it is the choice medium, and what can effectively be represented in that medium, that governs the possible interpretations that might be given a rendering and, therefore, the type of reasoning that might be further exercised.

### A BRIEF CASE STUDY

A brief case study will serve to illustrate some of the prior discussion. This particular case study is drawn from a detailed protocol analysis of a student's work and is used because it seems to be typical of many such protocols. The program for the project called for a hotel and a comprehensive health facility. The site included a lake and an existing hospital to which the proposed building complex was to be connected. The following narration and accompanying illustrations are extracts from the protocol and describe some of the key attempts at solution generation and evaluation.[20] The quotations in the text indicate direct narration by the designer.

Initially, for planning purposes, the program was divided into the two parts just

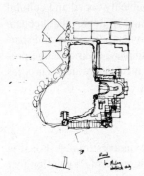

Figure 26–1 [Used by permission of Peter G. Rowe]

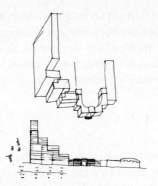

Figure 26–2 [Used by permission of Peter G. Rowe]

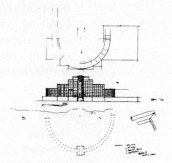

Figure 26–3 [Used by permission of Peter G. Rowe]

described, with a third element housing functions that were shared in common. The design process then proceeded from an intention . . . "to address and preserve the lake" . . . as a major part of the scheme. Here, the first move was to create a . . . "formal space next to the lake but to shield the space from the outside" . . ., or northern portions of the site. A "classical" building type served as the point of departure for linking the two major program elements (Figure 26–1). This choice was made because of the apparent correspondence between the example type and the problem at hand: location on the water, symmetry and formality, and the presentation of a rusticated façade protecting the public side of the site. However, there also appears to be simultaneity in the adoption of the type and the initial intention of enclosing a space by the lake. The intention derived as much from the example as it did from a study of the site without considering the type. In any event, the location of a formal space between the building and the lake edge, a continuing theme throughout the design, clearly derives from adoption of the particular reference.

Problems were immediately encountered with this arrangement due to the overpowering presence of the two program elements on either side of the central feature, namely the hotel and the health facility (Figure 26–2). As a consequence, the initial arrangement was generalized into a semi-circular plan for the hotel (Figure 26–3), with the structure . . . "defining a circulation path through the pieces at the base" . . . and thus, . . . "having an ordering impact" . . . on the scheme. An attempt was then made to reconcile the assymmetrical program with the symmetrical spatial conception by elongating one arm of the composition and shifting the inward lakefront focus around to the eastern edge of the site.

Evaluation of the scheme at this stage revealed problems with the scale and shape of the hotel, and a lack of formal difference between the major functions of the health facility and the service functions of the hotel . . . "It was a problem with thinking of it (the health facility) just as a piece in a formal arrangement, so I started analyzing it as a separate piece" . . . Here, a shift occurred in the type of heuristic employed from a pre-occupation with formal composition derived from a building type to a concern with the geometry of structural bays and abstract functional arrangement of required facilities. . . . "Basically I was trying to take a structural diagram, a functional diagram and a circulation diagram and combine them". . . (Figure 26–4). The result was an alternation of large and small structural bays within which circulation and functions of various sizes were appropriately accommodated. The same strategy was used to plan the service functions of the hotel . . . "Working strictly in sort of square bubbles, and (within) an understanding of proximity requirements, I tried to arrange the pieces so they would make sense" . . .

With these problems at least partially resolved attention was again shifted to the overall plan of the complex . . . "The problem of bringing in an axis (with the symmetrical form) was that it had no relationship to everything else that was going on. If I was going to do this I would have to develop a relationship . . . so then I went back to looking at the whole site" . . . An axial composition then evolved with a progression associated with arrival from the main entrance to the site through to the lake, with the buildings distributed symmetrically about the axis. This also led to a realization that a formal outdoor space was required where you arrived at the building complex for functional reasons and in order to complete the composition. Further, . . . "I now felt that I was dealing with 2 major spaces: one at the end of the axis when you arrive, the other facing the lake . . . (also) . . . now I have a formal public arrival space and a more informal space facing the lake" . . .

However, the problem now became one of deciding which way the curvilinear form of the hotel should face . . . "I realized I was trying to use the hotel piece to solve all problems and was having difficulty with this" . . . In fact, at this stage experiments were made with many different plan shapes (Figure 26–5). Ultimately, a decision was made to use a more simplified form of the hotel as a straight slab running across the site. This arrangement finally resolved the overall "diagram" or formal scheme of major elements in the project (Figure 26–6).

Within this framework, solutions to more detailed problems were developed. The open-ended expression of the primary circulation corridor, defined by the columns of the hotel slab, was terminated at the junction with semi-public functions in the form of a circular alcove (Figure 26–7). The columns themselves, and the section through the base of the hotel, were specially articulated to give a sense of grandeur to the public entry space (Figure 26–8), and to allow for a view through the building to the lake (Figure 26–9). The façade on the northern public face of the building was composed from a proportional grid with intervals determined so as to diminish the apparent

**Figure 26–4** [Used by permission of Peter G. Rowe]

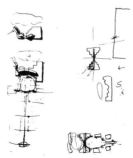

**Figure 26–5** [Used by permission of Peter G. Rowe]

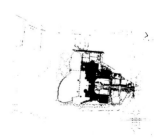

**Figure 26–6** [Used by permission of Peter G. Rowe]

scale of the building (Figure 26–10). This façade was further divided into a distinct "base," "middle" and "top," no doubt as an outgrowth of the classical references used earlier. A more informal textural treatment was proposed for the southern façade (Figure 26–11), although again the organization properties of a grid are evident . . . and so the process continued. The only subsequent modifications to the formal arrangement of the plan arose through attempts to extend the already successful strategy of expressing site constraints. Several lines at 45 degrees were introduced, relating the existing hospital with the building complex, with an additional purpose of helping organize a greater informality in the architecture along the lake front.

In this case several distinct lines of heuristic reasoning, where *a priori* knowledge was used to control the decision-making process, can be identified. The use of a specific building type, although more as an iconic analogy than as a typology, gave initial form to the project. Relational models dealing with structural organization and expressed preference for functional proximity were used effectively to solve specific layout problems. To some extent a formal language, or at least a set of "elemental typologies," was used in the sectional and façade developments. Canonic analogies were used throughout as organizational devices for resolving various functional and formal problems. The symmetrical compositional preoccupations engendered by the heuristic first employed seemed to persist, exerting an influence over subsequent lines of investigation. "Backtracking" could be seen on several occasions, and the independent qualities of constraints introduced with the adoption of different heuristics often seemed to provide fruitful points of departure for subsequent design.

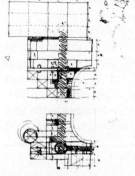

**Figure 26–7** [Used by permission of Peter G. Rowe]

**Figure 26–8** [Used by permission of Peter G. Rowe]

**Figure 26–9** [Used by permission of Peter G. Rowe]

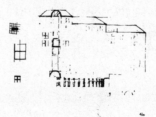

**Figure 26–10** [Used by permission of Peter G. Rowe]

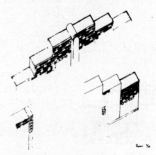

**Figure 26–11** [Used by permission of Peter G. Rowe]

## SUMMARY

The central thesis of this paper has been that the various classes of *a priori* knowledge incorporated in heuristic reasoning processes exert a strong influence over architectural design activity. Various types of heuristics were identified and described according to the type of knowledge that seemed to be most immediately involved in shaping form. This classification is by no means complete and probably represents an overly simplified and singular view of the matter. There is no reason to suspect, for instance, that many form-giving principles and analogies can't be multifaceted and used to support several different lines of reasoning. Nevertheless, the classification served a useful purpose in helping present the topic, and as a point of reference for later discussion.

The importance of *a priori* knowledge in heuristic reasoning appears to derive from the directness with which partial solution states of a problem can be provisionally specified, and thus orientation given to subsequent problem-solving activity, and from the independent qualities of the constraints that are offered. In architectural design a certain amount of "end-justification" and autonomy of constraints beyond the immediate problem context appear to be both unavoidable and necessary. However, misunderstanding of the consequent conditions implicit in following a particular line of reasoning can easily result in arbitrariness and capriciousness within the final design. Furthermore, superficiality in the number, type and complexity of constraints can also have disastrous effects on a solution. Finally, initial ideas, and the order in which various kinds of heuristics are employed, seem to have residual effects that strongly influence the degree to which "backtracking" and problem re-formulation can be satisfactorily accomplished.

## NOTES

1 Herbert A. Simon, Allan Newell, and J. C. Shaw, "The Processes of Creative Thinking," in *Contemporary Approaches to Creative Thinking,* ed. H. E. Gruber and M. Wertheimer (New York: Lieber-Atherton, 1962), pp. 63–119; and Herbert A. Simon, "Structure of Ill-Structured Problems," *Artificial Intelligence* 4 (1973),pp. 181–201.
2 C. West Churchman, "Wicked Problems," *Management Science* 4, no. 14 (1967), pp. 141–42; and Horst W. J. Rittel, "On the Planning Crisis: Systems Analysis of the First and Second Generations," *Bedriftskonomen,* no. 8 (1972), pp. 390–96.
3 G. Polya, *How to Solve It* (London: Anchor Books, 1957).
4 The protocol analyses involved extensive interviews with designers conducted at various stages during a design problem. During the interview, the subjects were asked to reconstruct, with the aid of sketches, how they solved the problem. This narration was tape recorded and copies made of the salient drawings.
5 Geoffrey Broadbent, *Design in Architecture: Architecture and the Human Sciences* (New York: John Wiley & Sons, 1973), chapter 2.
6 LeCorbusier, *The Chapel at Ronchamp* (London: Architectural Press, 1958).
7 Alan Colquhoun, "Typology and Design Method," *Perspecta* 12 (1967), pp. 71–74.
8 Christopher Alexander et al., *A Pattern Language* (New York: Oxford University Press, 1977).
9 This classification, although derived for different reasons, is very similar in the manner in which characteristics are distinguished to the five class system proposed by Vidler. Anthony Vidler, "The Idea of Type: The Transformation of the Academic Ideal, 1750–1830," *Oppositions* 8 (1977), pp. 95–115.

10 A. Palladio, *I Quattro Libri Dell'Architettura* (London: Ware Edition, 1738); Jean-Nicholas-Louis Durand, *Precis des Lecons d'Architecture Donnees a l'Ecole Polytechnique,* 2nd ed. (Paris, 1813); and John Summerson, *The Classical Language of Architecture* (Cambridge, MA: The MIT Press, 1979).

11 Alexander et al., *A Pattern Language;* and Christopher Alexander, *The Timeless Way of Building* (New York: Oxford University Press, 1979).

12 Polya, *How to Solve It.*

13 Simon, Newell, and Shaw, "The Processes of Creative Thinking."

14 Omer Aiken, "How Do Architects Design?" in *Artificial Intelligence and Pattern Recognition in Computer-Aided Design,* ed. Latombe (New York: North-Holland Publishing Co., 1978).

15 Peter Eisenman, "Aspects of Modernism: Maison Domino and the Self-Referential Sign." *Oppositions* 15/16 (1979), pp. 118–29.

16 Herbert A. Simon, "Style in Design," in *EDRA-TWO: Proceedings of the Second Annual Environmental Design Research Association Conference,* ed. John Archea and Charles Eastman (Pittsburgh, PA, October 1970), pp. 1–10.

17 Ibid.

18 Ibid. and E. H. Gombrich, *Art and Illusion: A Study in the Psychology of Pictorial Representation* (New York: Pantheon).

19 M. Graves, "The Necessity for Drawing: Tangible Speculation," *Architectural Design* 47, no. 6 (1977), pp. 384–94.

20 The case study used to illustrate this paper is drawn from the work of Elizabeth Rupp, a recent graduate student at the School of Architecture, Rice University.

*Author's Comment:* The term "a priori knowledge" as used in this essay simply means knowledge acquired before tackling a particular design problem, rather than an attempt to become involved in the broader philosophical question of innate or empirically derived knowledge.

## 选读 27
### "建筑设计工作室中艺术与应用科学的结合"
#### 唐纳德·艾伦·舍恩

舍恩一直致力于设计师如何解决复杂环境问题的研究。他把设计过程描绘成"在某种情况下与物质材料进行的反思式对话"。他认为建筑工作室的设计过程与文化同大学主导的教育模式相抵触，并展示这种抵触怎样为设计师参与行动导向型的教育进程提供了独特的机会。这篇文章研究了应用科学和建筑设计工作室之间的关系。舍恩认为设计师应该向科学家学习他们做了些什么，而不是解释他们的研究成果。

**READING 27**

# Toward a Marriage of Artistry and Applied Science in the Architectural Design Studio
## Donald Alan Schön

Schön has studied the ways that designers solve complex environmental problems and has described the design process as "a reflective conversation with the materials of a situation." He asserts that the processes and culture of the architectural studio are at odds with traditional university-based educational models and shows how this conflict presents designers with a unique opportunity by participating in action-oriented educational processes. This article examines the relationship between the applied sciences and the design studio setting. Schön suggests that designers should learn what scientists *do,* rather than interpreting their research results.

## INTRODUCTION

Contemporary professional education consists of two hermetic and disjoint systems. On one side, we find university-based schools of the professions that have adopted, in their pursuit of academic status, a curriculum modelled on the rigorous ideal of medical education. Following the positivist epistemology of practice that has shaped the modern research university, these schools adhere to a core of systematic, preferably scientific knowledge—first teaching the relevant science, basic and applied, then a practicum in the application of scientific knowledge to everyday practice.[1] On the other side, we find studios of visual and plastic arts and conservatories of music, dance, and drama. Here, the focus is on the artistry of performance or production and applied science occupies a controversial place, if present at all, on the margins of the curriculum.

Studio and conservatory tend to be freestanding institutions. When contained in a university, they are likely to be marginal, compartmentalized and low in status—the more prestigious the university, the lower their status. Hence, university-based studios and conservatories strive to establish a basis for their teaching in scholarly research. At the same time, because of the growing crisis of confidence in professional knowledge and education, educators are beginning to value the kinds of artistry professional schools are least equipped to teach. Some of them realize they have much to learn from the educational traditions of the studio and conservatory.[2]

In our bifurcated system of professional education, schools of architecture occupy a troubled, intermediate position.[3] Architecture is a hybrid, an occupation concerned with the design of usable structures and an art based on the forms of buildings and the experience of passage through their spaces. Architecture's reliance on older traditions of professional knowledge and education make the university uneasy. And even when architects try to gain a foothold in applied science, they cannot escape their profession's core of artistry; for they are designers and, although ancillary sciences may contribute to specialized design tasks, there is no general science of design. So architectural education still embraces, albeit with ambivalence, the studio traditions that might offer a basis for the renewal of education for artistry in the university-based professional schools.

The time is ripe for re-examining our two systems of professional education; we may be ready to imagine new ways of marrying applied science and artistry. With this end in view, I shall consider the architectural design studio from two points of view,

*Source:* Donald A. Schön, "Toward a Marriage of Artistry and Applied Science in the Architectural Design Studio," *Journal of Architectural Education* 41, no. 4 (Summer 1988), pp. 4–10. Copyright © 1988 by the Association of Collegiate Schools of Architecture, Inc. (ACSA). Reproduced by permission of MIT Press Journals.

exploring first how it might be taken as an exemplar for the university-based professional schools and second, how it might incorporate applied science.

## THE STUDIO AS A REFLECTIVE PRACTICUM IN DESIGNING

In *The Sciences of the Artificial,* Herbert Simon[4] claimed that designing is fundamental to all professions. But he saw designing as a form of problem solving—in its purest form, optimization—thereby ignoring situations of uncertainty, uniqueness and conflict where instrumental problem solving occupies a secondary place and problem *setting,* a primary one. In its most generic sense, designing consists in making representations of things to be built. In contrast to analysts or critics, designers put things together and make new artifacts. They juggle variables, reconcile conflicting values, and maneuver around constraints—a process in which, although some design products may be superior to others, there are no unique right answers and no moves that have only their intended consequences. With its webs of moves, discovered consequences, and implications, designing is a reflective conversation with the materials of a situation.

Artists make things and are, in this sense, designers. Indeed, the ancient Greeks used the term *poetics* to refer to the study of making things, treating poems as one category of things made. Professionals also make artifacts. Lawyers construct arguments, agreements, and laws; physicians make diagnoses and regimens of testing and treatment; planners construct spatial plans, policies and systems for orchestrating contending interests. More generally, professional practitioners frame problematic situations in accordance with their understandings and methods, and shape the very worlds of practice in which they live out their professional lives.[5] As makers of artifacts, they have much to learn from the architectural design studio which is, at its best, an exemplar of the process by which one learns to design.

The architectural design studio is a practicum, a virtual world that represents the real world of practice but is relatively free of its pressures, distractions, and risks. Here students learn, by doing, to recognize competent practice, appreciate where they stand in relation to it, and map a path to it. They learn the "practice" of the practicum, its tools, methods, and media. They do these things under the guidance of a studio master who functions less as a teacher than as a coach who demonstrates, advises, questions, and criticizes. They work with other students, who sometimes play the coach's role. As they immerse themselves in the shared world of the practicum, they unconsciously acquire a kind of background learning of which they will become aware as they move to other settings later on.

Our view of the practicum's work depends on our view of professional knowledge. If we focus on facts, rules, and procedures, we will see the practicum as a form of technical training. If we focus on professional knowledge as a way of thinking—"thinking like" a lawyer, teacher, or manager, for example—we will attend to the ways in which students learn to reason their way from general principles to concrete cases. If we focus on the kinds of inquiry by which practitioners sometimes make new sense of uncertain, unique, or conflicted situations—the process I call "reflection-in-action"—then we will notice how, in a practicum, students learn to construct and test new categories of understanding. It is important to add that the third kind of practicum may depend on the first two. Perhaps we learn to reflect-in-action by learning first to apply standard rules, facts, and operations; then to reason from general rules to concrete cases; and only then to develop and test new forms of understanding when familiar ways of thinking fail.

Practicums of the third kind exist in the second system of professional education, to which I have referred above: the deviant traditions of the studio and conservatory. Practicums are sometimes also found in apprenticeships or—less often, and usually without formal legitimacy—in the peripheral regions of the normative curricula of the professional schools. I call these practicums "reflective" because they aim at helping students learn to become proficient in various kinds of reflection-in-action and because, as we shall see, they depend for their effectiveness on coach and student entering into a kind of communication that is, at its best, a dialog of reciprocal reflection-in-action.

From my participation in studies of the architectural design studio, I have drawn a description of conditions and processes inherent in any reflective practicum.[6]

To begin with, I have observed that students must begin designing before they know what it means to do so. They quickly discover that their instructors cannot tell them what designing is, or that they cannot learn what their instructors mean by what they do say, until they have plunged into designing. Hence, in the early stages of the design studio, confusion and mystery reign. Yet in a few years or even months, some students begin to produce what they and their instructors regard as progress toward competent design. Coach and student finish each other's sentences and speak elliptically in ways that mystify the uninitiated. They seem to have achieved a convergence of meaning.

They make this transition—those who succeed in doing so—by joining in a dialog of words and actions. The student reflects on what he hears the coach say or sees the coach do, and evinces his or her understandings in further performance. The coach, in turn, interrogates the student's performance in order to discover what it reveals in the way of knowledge or ignorance, and considers what further demonstration, questioning, advice, or criticism might help the student.

A good coach must be able to demonstrate designing and describe it, particularizing what he or she does or says to fit the student's momentary confusions, questions, difficulties, or potentials. So the coach improvises, drawing on his repertoire, reflecting on his own spontaneous performance, conducting on-the-spot experiments in design and communication. In this process, the coach moves up or down the ladder of reflection, shifting from designing to description of designing, or from description to reflection on description, and back again to designing.

The student may reflect on her own spontaneous performances in order to discover what she already knows that helps or hinders her learning; and the student tries, through reflective imitation, to construct in her own action the features essential to the coach's demonstration. The student tries to strike a balance between taking responsibility for self-education in designing, and remaining open to the coach's help. For the student as for the coach, two kinds of practice are involved in the practicum: the substantive designing the student tries to learn and the reflection-in-action by which she tries to learn it.

The coach assumes that an initial instruction will be sufficient to get the student to do *something*. This initiative gets the dialog started and provides a first occasion for feedback, which the student is very likely to find confusing or ambiguous.

Nevertheless, however incomplete or mechanical the initial moves may be, the student begins to learn what it feels like to carry them out. In Wittgenstein's potent phrase, the student learns the meaning of the operation by performing it. And this performance allows the coach to give a new instruction or demonstration in order to correct the error discerned. Or the student, although unable to say why, does something that feels wrong, and the coach provides a way of explaining it. So the stage is set for a continuing dialog of reciprocal reflection in and on action, within which the student increasingly grasps what it means to design and thereby increases in ability to participate in the dialog.

The paradox of learning to design—that the student cannot be told about it ahead of time in any way she can understand, but must begin to do it in order to learn what it is—carries with it a predicament. The plunge into doing, without knowing what one needs to learn, provokes feelings of loss. The student feels a loss of control, competence, and confidence and, with this, a sense of vulnerability and dependency. The coach must accept being unable at first to tell the students about designing and must cope with their reactions to the predicament in which they have been placed.

Occasionally, a student reveals an ability to enter into an instructor's view of designing, secure in the ability to break it open later on. More often, the student's initial sense of vulnerability turns to defensiveness, and the learning predicament may become a learning bind. This may happen in the classroom, as well, but tends to be masked there by conventional habits of lecturing and note-taking. In a reflective practicum, the coach's ability to avoid or dissolve learning binds depends on the behavioral world he helps create with the students and his ability to foster a relationship open to inquiry.

In order to be credible, a reflective practicum must become a world with its own culture, including its own language, norms, and rituals. Otherwise, it risks being overwhelmed by the academic or professional cultures that surround it. But if it succeeds too well in establishing its own culture, isolated from the larger worlds of university and practice, then it may become, in Hermann Hesse's phrase, a mere "glass bead game."

A reflective practicum is unlikely to flourish as a second-class activity. The professional school must give it high status and legitimacy, or it falls prey to the dilemma of Nathan Glazer's "schools of the minor professions"[7] where students are forced to choose between low-status "relevance" or high status "rigor." Coaches must be first-class faculty members, and the process of coaching and learning must become central to the intellectual discourse of the school.

In the university-based professional schools, prevailing models of professional knowledge and classroom teaching are bound to be hostile to the creation of a reflective practicum like the architectural design studio, where overriding importance is attached to the process of coaching students in learning by doing. On the other hand, educators brought up in the tradition of the architectural design studio are likely to be hostile to the introduction of applied science and scholarship. In both cases, the challenge is to create a workable marriage of artistry and applied science, reflective practicum, and classroom teaching.

## APPLIED SCIENCE IN STUDIO EDUCATION

There is a very long list of applied sciences that at least some educators have thought architects needed to know. Among these are energy-related engineering and design, soil mechanics, structural engineering, building materials and technologies, geology, topography, solar engineering, acoustics, wind effects, earthquakes and earthquake hazard reduction, building economics, building finance, building diagnostics, urban development and design, law as it applies to architecture and building, the dynamics of groups and team functioning, the anthropology of architectural practice, urban politics, architectural history, the structure of the building industry, and computer science.

Each of these fields of study has been introduced into the curriculum of at least one architectural school, using one of several strategies of introduction. Some schools have established rudimentary courses in applied science as a preliminary to studio experience. Specialized studios, like Ralph Knowles's solar envelope studio at the University of Southern California, have focused on a particular body of knowledge. Larger studios, like the "Total Studio" taught in environmental design at MIT, have incorporated lectures or mini-courses in particular fields of knowledge at key points in the development of a studio project. Resources for technical assistance in special fields have been created and made available to students for their *ad hoc* use, with or without recourse to computer environments.

These strategies reflect different approaches to the problem of introducing fields of special knowledge into an architectural curriculum. Those who have introduced small chunks of applied science into studio projects, mindful of the problem of motivating students to learn, have tried to teach specialized knowledge when students are most likely to see the need for it. Those who advocate a base of applied science prior to studio experience argue that knowledge must be conveyed prior to its use, and that the extraordinary demands of the studio make students resist learning material whose utility is not immediately apparent to them. Those who have created knowledge-specialized studios want to present a critical mass of knowledge in a context where its implications for designing are immediately apparent.

Several of these problems—having too much to teach, making students receptive to knowledge—are by no means peculiar to schools of architecture. What is both more special to architecture and more fundamental is that the teaching of applied sciences tends to be accompanied in the studio by a view of knowledge and a mode of teaching that are very different from the epistemology and pedagogy built into the research university.

Students truly responsive to the messages of the design studio, immersed in learning to think like designers, are often puzzled by, skeptical about, or downright hostile to the various forms of applied science some of their professors feel they need to know. And these students' attitudes are, in many cases, mirrors of the attitudes of influential design instructors. The difficulty of making a productive marriage of applied science and studio education is not only a matter of overload and logistics, or quirks of style, habit, and personality; rather, it goes to the heart of the discrepancies between forms of knowledge and teaching honored in the university and the studio's epistemology of practice, its emphasis on the artistry of designing, and the basic assumptions that go into its version of the reflective practicum—discrepancies that underlie architecture's marginal position in the university. Hence, the problem of marrying studio education with the teaching of applied science is central not only for architectural education *per se* but for architecture's future role in the university.

I shall consider this problem in the light of four ideas which are as important to science teaching in general as they are to the special case of science in the architectural studio:

## 1. Science as a Body of Research Results vs. Science as a Method of Inquiry

When we teach science, not only to architects but to other kinds of students, we tend to present it as a body of facts, theories, and techniques, in short, as a product of scientific research. Moreover, the usual way of presenting science in the classroom, the one also favored by scientific journals and texts, presents it in the declarative mode, in sentences as unambiguous as possible, on the basis of a retrospective view that justifies what scientists have inferred from their observations, experiments, and analyses. The research processes that led up to these products tend to remain hidden.

Science looks like a very different enterprise depending on whether one encounters it in the form of its results—astringent, distanced, maximally objective—or in the form of its before-the-fact processes of inquiry.

These "retrospective" and "prospective" views might also be called "justificatory" and "exploratory," or "analytic" and "phenomenological," depending on the features chosen for emphasis. The important point is that the teaching of science, to architects and others, is always under a certain *view* of science—a view shaped by a wish either to meet prevailing standards of evidence, argumentation, and elegance of presentation, or to be faithful to the experience of before-the-fact discovery and invention.

## 2. Learning Theories about Phenomena vs. Getting a Feel for the Behavior of Phenomena

When science is presented in a retrospective way, the symbolic generalizations used to describe research results do not convey the feel of the phenomena they describe, and do little to help us recognize them when we see them. I am reminded of a story a chemist colleague of mine used to tell. He cited a journal article about the instability of certain hydrogen peroxide compounds. Reading that article with its equations and sparse analytic arguments, one would never guess that the data informing it had been produced when the back of a plant blew out.

The modes of experimentation peculiar to scientific inquiry are characterized by errors, anomalies, uncertainties, and confusions—all of which are masked by the neat, self-contained formulas and formal models typical of retrospective presentations of science. For this reason, some scientific educators are passionately devoted to the proposition that students must be exposed to hands-on laboratory experience. One of these educators, a physicist at MIT, put it this way:

> In physics, there is some kind of physical interaction going on, to which some instrument responds, and you have to understand every step of it. The instrument is part of it ... You take a curve with zero input, you get an experimental curve and you convolute those two together, point by point, and you know that you're doing the right thing! Or if you have doubts about it, you can understand. Maybe something different happens when you turn off the current. What could be happening?

If you make a fast Fourier Transform, where will you ever think about this? Typically, you're supposed to turn off the magnetic field to get your zero curve . . . and I say, "Are you sure the magnetic field is zero? Did you de-gauss the magnet?" Now, if you don't get through some kind of procedure like this, you're not going to see that.[8]

In this physicist's view, it is crucially important for students to experience how established scientific theories usually do *not* precisely fit experimental data.

When scientists occasionally describe their encounters with phenomena, their descriptions are often poetic. For example, this is Evelyn Keller's report of the geneticist, Barbara McLintock's, description of her encounters with the chromosomes of maize:

> I found that the more I worked with them, the bigger and bigger (they) got, and when I was really working with them I wasn't outside, I was down there. I was part of the system. I was right down there with them, and everything got big. I even was able to see the internal part of the chromosomes—actually everything was there. It surprised me because I actually felt as if I were right down there and these were my friends.[9]

When science is taught to students of the professions as a method of inquiry rather than as a body of research results, it can be clearly seen to resemble what skilled practitioners do in their own on-the-spot research. For skilled practice is, in its own right, a form of experimentation with its own discoveries of patterns in phenomena, and its own generation and testing of hypotheses.

Significantly, even when a professional practitioner makes practical use of scientific principles or techniques, he must conduct an intermediate form of on-the-spot research. So, for example, physicians speak of the large number of patients whose cases are "not in the book," and of the need to conduct experimental inquiry in the office or clinic in order to modify or combine standard diagnostic categories or even, on occasion, to evolve wholly new categories.

More specifically, the prospective view of science reveals a kind of inquiry that is close in spirit to designing. For design is itself a kind of experimentation, though one that bears only a family resemblance to experimentation in a scientific laboratory. In *The Reflective Practitioner* I described three functions of experimentation in designing: exploration, move-testing, and hypothesis-testing.[10] I proposed that whereas the logic of hypothesis-testing is the same in designing as in science, the overall logic of experimentation in design is unique. Here a single design move may bear the burden of all three functions. For example, a designer confronted with a "screwy site"—one that offers no initial coherence for design—may choose to impose on it an arbitrary geometrical discipline, testing whether the imposed geometry carved into the slope may then "work with" the contours of the buildings. In such a process, the designer explores the properties of the site, tests a move (the imposition of the geometry), and tests an underlying hypothesis about the potential fit of the modified contours of the slope to the shapes of the buildings.

It is when we see science and art only retrospectively, through their results, that art and science seem most disjoint. When we are exposed to their before-the-fact processes of inquiry, they seem much more like each other.

Many scientists readily admit that when they teach, they don't teach what they actually do. So most students "learn science" without learning what it is like to *do* science. This is true not only of the teaching of science in architecture but of science teaching in all professional education. For example, a former Dean of the Albert Einstein School of Medicine spoke with feeling, in our discussions together, about his vain attempts to get biologists on his faculty to teach medical students by engaging them in the process of doing biology.

Clearly, it is difficult to teach science in a prospective way. In order to do so, one would have to solve important intellectual problems of educational design—especially when students begin with relatively impoverished understandings of science and there is limited time available for teaching.

Yet there are interesting examples of science teaching on which to build. In some of these, teachers have made productive use of computer environments to create systems that short-cut the drudgery often involved in running experiments. In the Civil Engineering Department at MIT, for example, Professor John Slater has developed a computer program called GROWLTIGER as an aid to the teaching of structural engineering. With GROWLTIGER, a student can represent a structure like a truss or a frame as a pattern of lines on the computer screen. The student need only set the geometry of the structure and select the dimensions of its members and the type of steel to be used. The student can then ask the computer to analyze the forces, moments, and deflections, and the program will display them graphically on the screen. In order to provide a better feel for the impact of the loading, the stiffness of the structure, and the resulting deflections, the program can display deflections moving from zero to the full range. After one iteration of design and analysis, the student can check whether the structure behaves within the allowable tension in the beams and the allowable deflection of the structure. The student can then decide to change the geometry, the loading, or the stiffness, and redo the analysis in a matter of seconds.

When students speak of their experiences with GROWLTIGER, many of them refer, first of all, to their pleasure at being spared the drudgery of carrying out lengthy calculations. But they also speak of the experience of *visualizing* the effects of loads on the structures they design. They observe that their ability to make many quick iterations of design and analysis gives them a feel for the behavior of structures which they do not get from exposure to theory in lectures and texts. They sometimes adopt an apologetic tone when they note that, with GROWLTIGER's help, they have come to know structures in a very different way, even though they had passed an introductory course in statics and were supposed to know the basics already.

And some students, perhaps the more talented among them, spoke of the surprises they encountered and the puzzles they had to solve when the structures they drew turned out to behave contrary to their expectations. As one student put it:

> I applied a windload and I saw it lean a little and I noticed that because the building was so vastly cantilevered, only the middle third of it was supported. I saw the continuity of how it would have to behave if it weren't going to fracture and fail catastrophically . . . (I saw that) if I wanted to get rid of some deflection in a floor girder, I could stiffen up the wall columns . . . (then I saw that) by making some columns a little bit wider, *without necessarily changing the weight or even lowering the weight* I could make the building stiffer.[1]

Just such a feel for the phenomena (in this case, structural behavior) is what designers need, even more than they need to know the relevant equations—a feel, for example, for the behavior of solar envelopes, soils and foundations, the effects of wind on buildings, or the characteristic phenomena of building economics and finance. These are what scientists and engineers are fond of describing as their "intuitions." The approach to science teaching illustrated by GROWLTIGER seems to work by helping students develop such intuitions, while at the same time it exposes them to processes of design, experimentation, analysis, and observation that closely parallel the before-the-fact processes of science.

## 3. Prototypes, Exemplars, and Precedents in Scientific Inquiry and Architectural Designing

In his "Second Thoughts on Paradigms," Thomas Kuhn describes how students learn physics.[12] He believes that they do *not* proceed by first grasping symbolic generalizations, which they later learn to apply. Rather, he suggests,

> The student discovers a way to see his problem as like a problem he has already encountered. Once that likeness or analogy has been seen, only manipulative difficulties remain.[13]

> Typically, a beginning student of physics learns to solve many canonical problems—like figuring out the acceleration of a ball rolling down an inclined plane.

> In the course of their training a vast number of such exercises are set for them, and students entering the same specialty regularly do very nearly the same ones, for example, the inclined plane, the conical pendulum, Kepler ellipses, and so on.[14]

These concrete problems with their solutions are what Kuhn calls "exemplars," a community's standard examples. Starting with such exemplars, the student must learn to see other problems as analogous to them, and then, in Kuhn's words, "an appropriate formalism and a new way of attaching its symbolic consequences to nature follow."[15]

If Kuhn is right, beginning scientists learn their trade by familiarizing themselves with prototypes by reference to which they learn to see other problems as similar, before they can say "similar with respect to what." The implications of Kuhn's view for science teaching are clear: this is how students learn science regardless of the pedagogies by which they may be "taught" it. But Kuhn's thesis also has important implications for the connections between doing science and designing, learning to do science and learning to design.

In our studies of design knowledge and reasoning among practiced designers, William Porter and I have been struck by the importance of what we have labelled "types," though in a sense somewhat different from the ordinary use of this term in architecture. Like other observers of designers—for example, Arnheim[16] and Habraken[17]—we have noticed that types play a variety of crucial roles in architectural designing.[18] We have thought of types as particulars that function like general categories; or, to put the matter differently, general categories that have the fullness of particulars. Types, in this sense, seem to function both as holding environments for design knowledge and as generators of leading ideas that guide whole sequences of design moves.

As we have examined the protocols of practiced designers at work on the same design exercise, we have identified several sorts of types, each with its own characteristic functions. There are, for example, "functional building types," like "branch library" or "suburban site." These serve as reservoirs of commonplace knowledge on which designers draw to fill in the intermediate premises of their design reasoning. There are "references—for example, "a Richardson library"—that serve, often under special conditions, as generators of strings of design moves. There are spatial *Gestalts*—for example, a view of a particular geometry as "a middle area with three pods attached to it"—which form the essential background on which the designer works, and in relation to which he or she sets the problems to be solved. And there are, finally, what we have called "experiential archetypes." These might be illustrated by a cave, a movement from light to dark to light, an entrance "like a pair of outstretched arms," or a concave entrance that "captures you and draws you into the heart of the building." These also function as generators of leading ideas, but they are especially prominent in zones of designing that architects sometimes describe as "art" or "poetry." Experiential archetypes seem to be called upon when the mundane task of making a building work presents itself as one of particular difficulty.

It would take a good deal more space than is available here to do justice to our observations and speculations about the function of types in architectural designing. What I can suggest, however, is that it may be fruitful to think further about the roles of prototypes, exemplars, and canonical examples in learning both to design and to do science. If types do function as holding environments for design knowledge and generators of leading ideas, then studio instructors may wish to focus more explicitly on the processes by which students become familiar with types and their functions, and build up repertoires of types on which they draw in their designing. It may be interesting, then, to explore how architectural types may be enriched through linkages to exemplars in the applied sciences—how functional types of buildings, for example, may be enriched to include connections to the behavior of structures and their responses to environmental stresses.

Moreover, recognition of the functions of prototypes and canonical examples may help us to make more fruitful connections between students' ways of learning to do science and their ways of learning to design.

## 4. Kinds of Thinking Peculiar to Skilled Scientists and Skilled Designers

In the department where I now teach, there was in times gone by a social scientist—himself a renegade architect—who declared that "architects don't think" and refused to allow them in his classes. In my own early teaching at MIT, before I learned to appreciate the kinds of thinking that go into skillful designing, I was struck by how hard it seemed to be for many students of architecture to "think" as *I* understood thinking—to make clear verbal arguments and reflect critically on them in the light of internal consistency, evidence, and disconfirmability.

There do seem to be important differences in habitual patterns of thinking among graduate-level students of architecture and students of the physical or social sciences. Students who learn in the studio how to think like a good designer tend not to learn how to manage the verbal argumentation and criticism characteristic of social science and policy analysis. And the reverse tends also to be true: an ability to construct and criticize verbal arguments is no guarantee of skill in designing. To the extent that this observation holds true—and it is likely to be controversial—there are important implications for both studio education and the studio's place in the larger university. How robust is this difference in thinking? How widespread? How significant for students' later practice?

I have worked, on occasion, with architects engaged in writing theses for an advanced degree, and have tried to introduce them to what ought to be a helpful metaphor: designing and building an argument. For arguments must be constructed; one must discover, through exploration and testing, what their fundamental structure might be; it is helpful to sketch them before developing them fully; and as one tests them, one often gets a salutary surprise (as in architectural designing) which leads to reframing the argument and, sometimes, to a new idea.

But for all these tempting analogies, it seems to be extraordinarily difficult for architectural students to cross over. I am not sure why. Perhaps the difficulty has to do with differences in media, or with the powerful attraction of architectural fashions, or with the countervailing mystiques of architecture and social science.

However this may be, some students do manage to cross over, from either side of the divide, and make themselves formidable exceptions to the general pattern.

One student interviewed in a study of computers and education at MIT offers a particularly illuminating example of what it means to carry over skill in designing from architecture to other domains. Originally educated as an architect, this student has become proficient at computer programming. He believes that his way of programming owes a great deal to his architectural training, and he has reflected on the analogies between the two practices. "Now basically," he says,

> what I do first is . . . try to write [the program] as quickly as possible. Without thinking. When I do it, I just write the steps it would take my brain to do it. That's very crude. And I don't usually keep those programs for the most part, because those are very inefficient programs. And then I go swimming . . . or whatever, and while I do that, I think about the program: how can I make it more efficient? . . . and I say, "How can I make the computer do this in such a way that it will take me 10 minutes to write the program? How can I make sure the computer doesn't overwork? How can I drop possible bugs?"
>
> This is what I call catching the program. Just the first catch. I don't think anybody has told me that. But that is basically what I've learned by doing it. Most of the things I do are like that. I first do a sketch . . . I like architecture, and was pursuing a degree in architecture, and the first thing that you learn there is just draw a sketch. So that's what I do in everything . . . [19]

Might it be possible to help students make the sort of crossover this student made for himself (doing it, as he says, because "it comes to me naturally")? If it were possible, there might be important consequences for broadening studio education and bridging between it and education in physical and social science. Certainly it will be important, in pursuing this question, to explore new ways of helping students reflect on their designing . There may be some particular merit in exploring computer programming as a transitional activity linking architectural design to verbal argumentation.

## CONCLUSION

The themes mentioned in the preceding pages seem to me to support a single proposition: the apparent disjunction between science and architectural design, like the more general split between the sciences and the arts, has its roots in a particular view of science. If students of architecture come to experience science as a form of prospective inquiry—one that brings them into direct contact with the phenomena, methods of experimentation and canonical examples of before-the-fact science—then learning science is likely to have a very different and much more powerful meaning for them. Learning to do science, like learning to design, occurs most favorably in the context of a reflective practicum; the paradox, predicament and dialog of coach and student inherent in learning to design are mirrored in learning the skills of scientific inquiry.

What is more, if we learned this way of teaching applied science to architects, we would help them to make fruitful connections between characteristic exemplars and ways of thinking on both sides of the divide. For science itself is a design-like practice, though that fact is hidden from us when we see only the results of science and have no experience of its processes. The positivist epistemology of practice underlying the modern research university emphasizes a retrospective view of science as a body of facts, theories, and techniques which professions like architecture are meant to apply. But when we experience science and architectural designing as before-the-fact inquiries, we become aware of their deep similarities and potentials for reciprocal influence.

Of course, the boundaries between these design-like practices are by no means transparent; transfer of learning across them is neither easy nor inevitable. What is in question, rather, is an attitude toward learning which casts a very different light on the place of the arts, architecture among them, in the larger university, and bridges the chasm that separates our two systems of professional education.

Architecture's marginality in the university stems both from the retrospective view of science that has long prevailed in the university, and from the fact that architects themselves have tended to keep their own processes of inquiry private, tacit, and sometimes even mystical—have tried, perhaps defensively, in order to emphasize their differences from other fields, to protect themselves from reflection on their own skillful practice.

**NOTES**

1. See Edgar Schein, *Professional Education* (New York: McGraw-Hill, 1972).
2. I have developed this argument at some length in Donald A. Schön, *Educating the Reflective Practitioner* (San Francisco: Jossey-Bass, 1987).
3. See Donald A. Schön, *The Design Studio* (London: Royal Institute of British Architects Press, 1986).
4. Herbert Simon, *The Sciences of the Artificial* (Cambridge, MA: MIT Press, 1976).
5. The line of thought developed here owes a great deal to the "constructionism" advanced by the philosopher Nelson Goodman, in his *Ways of Worldmaking* (Indianapolis, IN: Hackett, 1978).
6. I refer here to the study of architectural education directed by Maurice Kilbridge of Harvard and William Porter of MIT in the mid-1970s, and to subsequent studies conducted at MIT in the Design Theory and Methods Group.
7. Nathan Glazer, "The Schools of the Minor Professions," *Minerva,* 1974.
8. Interview with an MIT physicist, quoted in a study conducted by Donald A. Schön and Sherry Turkle, *Project Athena at MIT,* MIT, mimeo, 1988.
9. Evelyn Fox Keller, *A Feeling for the Organism* (New York: W. H. Freeman and Company, 1983).
10. Donald A. Schön, *The Reflective Practitioner: How Professionals Think in Action* (New

11 Quoted from Schön and Turkle, *Project Athena at MIT.*
12 Thomas Kuhn, *The Essential Tension* (Chicago: University of Chicago Press, 1977).
13 Ibid., p. 305.
14 Ibid.
15 Ibid., p. 306.
16 Rudolph Arnheim, *Visual Thinking* (Berkeley, CA: University of California Press, 1969).
17 John Habraken, *The Appearance of the Form* (Cambridge, MA: Atwater Press, 1986).
18 Donald A. Schön, "Designing: Worlds, Rules and Types," in *Design Studies,* Milton Keynes (England, 1988).
19 Schön and Turkle, *Project Athena at MIT.*

选读 28
"纸板屋"
弗兰克·劳埃德·赖特

弗兰克·劳埃德·赖特以一种能够突显出其建筑原则的力量与信念的方式进行写作。他的著作由直截了当、铿锵有力的陈述句构成，强调了他关于人们应该如何体验建筑和如何生活在现代社会中的理论。在《纸板屋》一文中，他勾勒出了在美国适宜居住生活形式的种种特征，并且无论何种设计原则，只要他认为这些原则打断了那一进程，赖特均以口诛笔伐待之。他反对"盒子式"的建筑，因为这种建筑阻碍人们享受阳光，妨害了自然和运动的自由自在对人的有益影响。对于他自己在欧洲现代住房设计中发现的简易轻便的材料质地，赖特也没有放过，他指出："……大多数'现代主义'的新建住房都处心积虑地让自己看上去像是用剪刀从纸板上剪下来的一样。一块块纸板折叠弯曲做成矩形，偶尔还在表面点缀一张流线形纸板聊以慰藉。"

## READING 28
## The Cardboard House
### Frank Lloyd Wright

Frank Lloyd Wright wrote in a manner that reflected the strengths and convictions of his personal architectural principles. His texts are composed of direct and forceful declarative sentences that emphasize his theories about how people should experience architecture and live in the modern world. In "The Cardboard House" he outlines the characteristics that form a suitable form of residential life in America, and he verbally assaults any and all design principles that he feels interrupt that process. He argues against architecture that creates a "box" within which people are trapped from the beneficial effects of the sun,

*Source: The Future of Architecture* by Frank Lloyd Wright. Copyright © 1970 The Frank Lloyd Wright Foundation, Scottsdale, Arizona. Reproduced by permission of The Frank Lloyd Wright Foundation. Plume Books, pp. 149–62.

nature, and freedom of movement. He also takes aim at the insubstantial and flimsy material quality that he found in modern European housing designs. He states: "... most new 'modernistic' houses manage to look as though cut from cardboard with scissors, the sheets of cardboard folded or bent in rectangles with an occasional curved cardboard surface added to get relief."

"Inasmuch as the rivalry of intelligences is the life of the beautiful—O poet!—the first rank is ever free. Let us remove everything which may disconcert daring minds and break their wings! Art is a species of valor. To deny that men of genius to come may be the peers of the men of genius of the past would be to deny the ever-working power of God!" . . .

Now what architecture for America?

Any house is a far too complicated, clumsy, fussy, mechanical counterfeit of the human body. Electric wiring for nervous system, plumbing for bowels, heating system and fireplaces for arteries and heart, and windows for eyes, nose, and lungs generally. The structure of the house, too, is a kind of cellular tissue stuck full of bones, complex now, as the confusion of bedlam and all besides. The whole interior is a kind of stomach that attempts to digest objects—objects, "objets d'art" maybe, but objects always. There the affected affliction sits, ever hungry—for ever more objects—or plethoric with overplenty. The whole life of the average house, it seems, is a sort of indigestion. A body in ill repair, suffering indisposition—constant tinkering and doctoring to keep alive. It is a marvel we its infestors do not go insane in it and with it. Perhaps it is a form of insanity we have put into it. Lucky we are able to get something else out of it, though we do seldom get out of it alive ourselves.

But the passing of the cornice with its enormous "baggage" from foreign parts in its train clears the way for American homes that may be modern biography and poems instead of slanderous liars and poetry crushers.

A house, we like to believe, is *in statu quo* a noble consort to man and the trees; therefore the house should have repose and such texture as will quiet the whole and make it graciously at one with external nature.

Human houses should not be like boxes, blazing in the sun, nor should we outrage the machine by trying to make dwelling places too complementary to machinery. Any building for humane purposes should be an elemental, sympathetic feature of the ground, complementary to its nature environment, belonging by kinship to the terrain. A house is not going anywhere, if we can help it. We hope it is going to stay right where it is for a long, long time. It is not yet anyway even a moving van. Certain houses for Los Angeles may yet become vans and roll off most anywhere or everywhere, which is something else again and far from a bad idea for certain classes of our population.

But most new "modernistic" houses manage to look as though cut from cardboard with scissors, the sheets of cardboard folded or bent in rectangles with an occasional curved cardboard surface added to get relief. The cardboard forms thus made are glued together in boxlike forms—in a childish attempt to make buildings resemble steamships, flying machines, or locomotives. By way of a new sense of the character and power of this machine age, this house strips and stoops to conquer by emulating, if not imitating, machinery. But so far, I see in most of the cardboard houses of the "modernistic" movement small evidence that their designers have mastered either the machinery or the mechanical processes that build the house. I can find no evidence of integral method in their making. Of late, they are the superficial, badly built product of this superficial, new "surface-and-mass" aesthetic falsely claiming French painting as a parent. And the houses themselves are not the new working of a fundamental architectural principle in any sense.

They are little less reactionary than was the cornice—unfortunately for Americans, looking forward, lest again they fall victim to the mode. There is, however, this much to be said for this house—by means of it imported art and decoration may, for a time, completely triumph over "architecture." And such architecture as it may triumph over—well, enough has already been said here to show how infinitely the cardboard house is to be preferred to that form of bad surface decoration. The simplicity of nature is not something which may easily be read—but is inexhaustible. Unfortunately the simplicity of these houses is too easily read—visibly an attitude, strained or forced. They are therefore decoration too. If we look into their construction we may see how construction itself has been complicated or confused, merely to arrive at exterior simplicity. Most of these houses at home and abroad are more or less badly built complements to the machine age, of whose principles or possibilities they show no understanding, or, if they do show such understanding to the degree of assimilating an aspect thereof, they utterly fail to make its virtues honorably or humanly effective in any final result. Forcing surface effects upon mass effects which try hard to resemble running or steaming or flying or fighting machines, is no radical effort in any direction. It is only more scene painting and just another picture to prove Victor Hugo's thesis of renaissance architecture as the setting sun—eventually passing with the cornice.

The machine—we are now agreed, are we not—should build the building, if the building is such that the machine may build it naturally and therefore build it supremely well. But it is not necessary for that reason to build as though the building, too, were a machine—because, except in a very low sense, indeed, it is not a machine, nor at all like one. Nor in that sense of being a machine, could it be architecture at all! It would be difficult to make it even good decoration for any length of time. But I propose, for the purposes of popular negation of the cornice-days that are passed and as their final kick into oblivion, we might now, for a time, make buildings resemble modern bathtubs and aluminum kitchen utensils, or copy pieces of well-designed machinery to live in, particularly the liner, the airplane, the streetcar, and the motor bus. We could trim up the trees, too, shape them into boxes—cheese or cracker—cut them to cubes and triangles or tetrahedron them and so make all kinds alike suitable consorts for such houses. And are we afraid we are eventually going to have as citizens machine-made men, corollary to machines, if we don't look out? They might be face-masked, head-shaved, hypodermically rendered even less emotional than they are, with patent leather put over their hair and aluminum clothes cast on their bodies, and Madam herself altogether stripped and decoratively painted to suit. This delicate harmony, characteristic of machinery, ultimately achieved, however, could not be truly affirmative, except insofar as the negation, attempted to be performed therein, is itself affirmative. It seems to me that while the engaging cardboard houses may be appropriate gestures in connection with "Now What Architecture," they are merely a negation, so not yet truly conservative in the great cause which already runs well beyond them.

*Organic simplicity* is the only simplicity that can answer for us here in America that pressing, perplexing question—now what architecture? This I firmly believe. It is vitally necessary to make the countenance of simplicity the affirmation of reality, lest any affectation of simplicity, should it become a mode or fashion, may only leave this heady country refreshed for another foolish orgy in surface decoration of the sort lasting thirty years "by authority and by order," and by means of which democracy has already nearly ruined the look of itself for posterity, for a half century to come, at least.

Well then and again—"what architecture?"

Let us take for text on this, our fourth afternoon, the greatest of all references to simplicity, the inspired admonition: *"Consider the lilies of the field—they toil not, neither do they spin, yet verily I say unto thee—Solomon in all his glory was not arrayed like one of these."* An inspired saying—attributed to an humble architect in ancient times, called carpenter, who gave up architecture nearly two thousand years ago to go to work upon its source.

And if the text should seem to you too far away from our subject this afternoon—"THE CARDBOARD HOUSE"—consider that for that very reason the text has been chosen. The cardboard house needs an antidote. The antidote is far more important than the house. As antidote—and as practical example, too, of the working out of an ideal of organic simplicity that has taken place here on American soil, step by step, under conditions that are your own—could I do better than to take apart for your benefit the buildings I have tried to build, to show you how they were, long ago, dedicated to the ideal of organic simplicity? It seems to me that while another might do better than that, I certainly could not—for that is, truest and best, what I know about the subject. What a man *does, that* he has.

When, "in the cause of architecture," in 1893, I first began to build the houses, sometimes referred to by the thoughtless as "The New School of the Middle West" (some advertiser's slogan comes along to label everything in this our busy woman's country), the only way to simplify the awful building in vogue at the time was to conceive a finer entity—a better building—and get it built. The buildings standing then were all tall and all tight. Chimneys were lean and taller still, sooty fingers threatening the sky. And beside them, sticking up by way of dormers through the cruelly sharp, sawtooth roofs, were the attics for "help" to swelter in. Dormers were elaborate devices, cunning little buildings complete in themselves, stuck to the main roof slopes to let "help" poke heads out of the attic for air.

Invariably the damp sticky clay of the prairie was dug out for a basement under the whole house, and the rubblestone walls of this dank basement always stuck up above the ground a foot or more and blinked, with half-windows. So the universal "cellar" showed itself as a bank of some kind of masonry running around the whole house, for the house to sit up on—like a chair. The lean, upper house walls of the usual two floors above this stone or brick basement were wood, set on top of this masonry chair, clapboarded and painted, or else shingled and stained, preferably shingled and mixed, up and down, all together with moldings crosswise. These overdressed wood house walls had, cut in them—or cut out of them, to be precise—big holes for the big cat and little holes for the little cat to get in and out or for ulterior purposes of light and air. The house walls were be-corniced or bracketed up at the top into the tall, purposely profusely complicated roof, dormers plus. The whole roof, as well as the roof as a whole, was scalloped and ridged and tipped and swanked and gabled to madness before they would allow it to be either shingled or slated. The whole exterior was bedeviled—that is to say, mixed to puzzle pieces, with corner boards, panel boards, window frames, corner blocks, plinth blocks, rosettes, fantails, ingenious and jigger work in general. This was the only way they seemed to have, then, of "putting on style." The scroll saw and turning lathe were at the moment the honest means of this fashionable mongering by the wood-butcher and to this entirely "moral" end. Unless the householder of the period were poor indeed, usually an ingenious corner tower on his house eventuated into a candlesnuffer dome, a spire, an inverted rutabaga or radish or onion or—what is your favorite vegetable? Always elaborate bay windows and fancy porches played "ring around a rosy" on this "imaginative" corner feature. And all this the building of the period could do equally well in brick or stone. It was an impartial society. All material looked pretty much alike in that day.

Simplicity was as far from all this scrap pile as the pandemonium of the barnyard is far from music. But it was easy for the architect. All he had to do was to call: "Boy, take down No. 37, and put a bay window on it for the lady!"

So—the first thing to do was to get rid of the attic and, therefore, of the dormer and of the useless "heights" below it. And next, get rid of the unwholesome basement, entirely—yes, absolutely—in any house built on the prairie. Instead of lean, brick chimneys, bristling up from steep roofs to hint at "judgment" everywhere, I could see necessity for one only, a broad generous one, or at most, for two, these kept low down on gently sloping roofs or perhaps flat roofs. The big fireplace below, inside, became now a place for a real fire, justified the great size of this chimney outside. A real fireplace at that time was extraordinary. There were then "mantels" instead. A mantel was a marble frame for a few coals, or a piece of wooden furniture with tiles stuck in it and a "grate," the whole set slam up against the wall. The "mantel" was an insult to comfort, but the *integral* fireplace became an important part of the building itself in the houses I was allowed to build out there on the prairie. It refreshed me to see the fire burning deep in the masonry of the house itself.

Taking a human being for my scale, I brought the whole house down in height to fit a normal man; believing in no other scale, I broadened the mass out, all I possibly could, as I brought it down into spaciousness. It has been said that were I three inches taller (I am 5' 8½" tall), all my houses would have been quite different in proportion. Perhaps.

House walls were now to be started at the ground on a cement or stone water table that looked like a low platform under the building, which it usually was, but the house walls were stopped at the second story windowsill level, to let the rooms above come through in a continuous window series, under the broad eaves of a gently sloping, overhanging roof. This made enclosing screens out of the lower walls as well as light screens out of the second-story walls. Here was true *enclosure of interior space*. A new sense of building, it seems.

The climate, being what it was, a matter of violent extremes of heat and cold, damp and dry, dark and bright, I gave broad protecting roof-shelter to the whole, getting back to the original purpose of the "cornice." The undersides of the roof projections were flat and light in color to create a glow of reflected light that made the upper rooms not dark, but delightful. The overhangs had double value, shelter and preservation for the walls of the house as well as diffusion of reflected light for the upper story, through the "light screens" that took the place of the walls and were the windows.

At this time, a house to me was obvious primarily as interior space under fine *shelter*. I liked the sense of shelter in the "look of the building." I achieved it, I believe. I then went after the variegate bands of material in the old walls to eliminate odds and ends in favor of one material and a single surface from grade to eaves, or grade to second story sill-cope, treated as simple enclosing screens—or else made a plain screen band around the second story above the windowsills, turned up over on to the ceiling beneath the eaves. This screen band was of the same material as the underside of the eaves themselves, or what architects call the "soffit." The planes of the building parallel to the ground were all stressed, to grip the whole to earth. Sometimes it was possible to make the enclosing wall below this upper band of the second story, from the second story windowsill clear down to the ground, a heavy "wainscot" of fine masonry material resting on the cement or stone platform laid on the foundation. I liked that wainscot to be of masonry material when my clients felt they could afford it.

As a matter of form, too, I liked to see the projecting base, or water table, set out over the foundation walls themselves—as a substantial preparation for the building. This was managed by setting the studs of the walls to the inside of the foundation walls, instead of to the outside. All door and window tops were now brought into line with each other with only comfortable head clearance for the average human being. Eliminating the sufferers from the attic enabled the roofs to lie low. The house began to associate with the ground and become natural to its prairie site. And would the young man in architecture ever believe that this was all "new" then? Not only new, but destructive heresy—or ridiculous eccentricity. So new that what little prospect I had of ever earning a livelihood by making houses was nearly wrecked. At first, "they" called the houses "dress-reform" houses, because society was just then excited about that particular "reform." This simplification looked like some kind of "reform" to them. Oh, they called them all sorts of names that cannot be repeated, but "they" never found a better term for the work unless it was "horizontal Gothic," "temperance architecture" (with a sneer), etc., etc. I don't know how I escaped the accusation of another "renaissance."

What I have just described was all on the *outside* of the house and was there chiefly because of what had happened *inside*. Dwellings of that period were "cut up," advisedly and completely, with the grim determination that should go with any cutting process. The "interiors" consisted of boxes beside or inside other boxes, called *rooms*. All boxes inside a complicated boxing. Each domestic "function" was properly box to box. I could see little sense in this inhibition, this cellular sequestration that implied ancestors familiar with the cells of penal institutions, except for the privacy of bedrooms on the upper floor. They were perhaps all right as "sleeping boxes." So I declared the whole lower floor as one room, cutting off the kitchen as a laboratory, putting servants' sleeping and living quarters next to it, semidetached, on the ground floor, screening various portions in the big room, for certain domestic purposes—like dining or reading, or receiving a formal caller. There were no plans like these in existence at the time and my clients were pushed toward these ideas as helpful to a solution of the vexed servant problem. Scores of doors disappeared and no end of partition. They liked it, both clients and servants. The house became more free as "space" and more livable, too. Interior spaciousness began to dawn.

Having got what windows and doors there were left lined up and lowered to convenient human height, the ceilings of the rooms, too, could be brought over on to the walls, by way of the horizontal, broad bands of plaster on the walls above the windows, the plaster colored the same as the room ceilings. This would bring the ceiling surface down to the very window tops. The ceilings thus expanded, by extending them downward as the wall band above the windows, gave a generous overhead to even small rooms. The sense of the whole was broadened and made plastic, too, by this expedient. The enclosing walls and ceilings were thus made to flow together.

Here entered the important element of plasticity—indispensable to successful use of the machine, the true expression of modernity. The outswinging windows were fought for because the casement window associated the house with out-of-doors—gave free openings, outward. In other words the so-called casement was simple and more human. In use and effect, more natural. If it had not existed I should have invented it. It was not used at that time in America, so I lost many clients because I insisted upon it when they wanted the "guillotine" or "double-hung" window then in use. The guillotine was not simple nor human. It was only expedient. I used it once in the Winslow House—my first house—and rejected it thereafter—forever. Nor at that time did I entirely eliminate the wooden trim. I did make it "plastic," that is, light and continuously flowing instead of the heavy "cut and butt" of the usual carpenter work. No longer did the trim, so called, look like carpenter work. The machine could do it perfectly well as I laid it out. It was all after quiet.

This plastic trim, too, with its running "backhand" enabled poor workmanship to be concealed. It was necessary with the field resources at hand at that time to conceal much. Machinery versus the union had already demoralized the workmen. The machine resources were so little understood that extensive drawings had to be made merely to show the millman what to leave off. But the trim finally became only a single, flat, narrow, horizontal wood band running around the room, one at the top of the windows and doors and another next to the floors, both connected with narrow, vertical, thin wood bands that were used to divide the wall surfaces of the whole room smoothly and flatly into folded color planes. The trim merely completed the window and door openings in this same plastic sense. When the interior had thus become wholly plastic, instead of structural, a new element, as I have said, had entered architecture. Strangely enough, an element that had not existed in architectural history before. Not alone in the trim, but in numerous ways too tedious to describe in words, this revolutionary sense of the plastic whole, an instinct with me at first, began to work more and more intelligently and have fascinating, unforeseen consequences. Here was something that began to organize itself. When several houses had been finished and compared with the house of the period, there was very little of that house left standing. Nearly every one had stood the house of the period as long as he could stand it, judging by appreciation of the change. Now all this probably tedious description is intended to indicate directly in bare outline how thus early there *was* an ideal of organic simplicity put to work, with historical consequences, here in your own country. The main motives and indications were (and I enjoyed them all):

First: To reduce the number of necessary parts of the house and the separate rooms to a minimum, and make all come together as enclosed space—so divided that light, air, and vista permeated the whole with a sense of unity.

Second: To associate the building as a whole with its site by extension and emphasis of the planes parallel to the ground, but keeping the floors off the best part of the site, thus leaving that better part for use in connection with the life of the house. Extended level planes were found useful in this connection.

Third: To eliminate the room as a box and the house as another by making all walls enclosing screens—the ceilings and floors and enclosing screens to flow into each other as one large enclosure of space, with minor subdivisions only.
Make all house proportions more liberally human, with less wasted space in structure, and structure more appropriate to material, and so the whole more livable. *Liberal* is the best word. Extended straight lines or streamlines were useful in this.

Fourth: To get the unwholesome basement up out of the ground, entirely above it, as a low pedestal for the living portion of the home, making the foundation itself visible as a low masonry platform on which the building should stand.

Fifth: To harmonize all necessary openings to "outside" or to "inside" with good human proportions and make them occur naturally—singly or as a series in the scheme of the whole building. Usually they appeared as "light screens" instead of walls, because all the "architecture" of the house was chiefly the way these openings came in such walls as were grouped about the rooms as enclosing screens. The *room* as such was now the essential architectural expression, and there were to be no holes cut in walls as holes are cut in a box, because this was not in keeping with the ideal of "plastic." Cutting holes was violent.

Sixth: To eliminate combinations of different materials in favor of mono-material so far as possible; to use no ornament that did not come out of the nature of materials to make the whole building clearer and more expressive as a place to live in, and give the conception of the building appropriate revealing emphasis. Geometrical or straight lines were natural to the machinery at work in the building trades then, so the interiors took on this character naturally.

Seventh: To incorporate all heating, lighting, plumbing so that these systems became constituent parts of the building itself. These service features became architectural and in this attempt the ideal of an organic architecture was at work.

Eighth: To incorporate as organic architecture—so far as possible—furnishings, making them all one with the building and designing them in simple terms for machine work. Again straight lines and rectilinear forms.

Ninth: Eliminate the decorator. He was all curves and all efflorescence, if not all "period."

This was all rational enough so far as the thought of an organic architecture went. The particular forms this thought took in the feeling of it all could only be personal. There was nothing whatever at this time to help make them what they were. All seemed to be the most natural thing in the world and grew up out of the circumstances of the moment. Whatever they may be worth in the long run is all they are worth.

Now *simplicity* being the point in question in this early constructive effort, organic simplicity I soon found to be a matter of true coordination. And beauty I soon felt to be a matter of the sympathy with which such coordination was effected. Plainness was not necessarily simplicity. Crude furniture of the Roycroft-Stickley-Mission Style, which came along later, was offensively plain, plain as a barn door—but never was simple in any true sense. Nor, I found, were merely machine-made things in themselves simple. To think "in simple," is to deal in simples, and that means with an eye single to the altogether. This, I believe, is the secret of simplicity. Perhaps we may truly regard nothing at all as simple in itself. I believe that no one thing in itself is ever so, but must achieve simplicity (as an artist should use the term) as a perfectly realized part of some organic whole. Only as a feature or any part becomes a harmonious element in the harmonious whole does it arrive at the estate of simplicity. Any wild flower is truly simple, but double the same wild flower by cultivation, it ceases to be so. The *scheme* of the original is no longer clear. Clarity of design and perfect significance both are first essentials of the spontaneously born simplicity of the lilies of the field who neither toil nor spin, as contrasted with Solomon who had "toiled and spun"—that is to say, no doubt had put on himself and had put on his temple, properly "composed," everything in the category of good things but the cookstove.

Five lines where three are enough is stupidity. Nine pounds where three are sufficient is stupidity. But to eliminate expressive words that intensify or vivify meaning in speaking or writing is not simplicity; nor is similar elimination in architecture simplicity—it, too, may be stupidity. In architecture, expressive changes of surface, emphasis of line and especially textures of material, may go to make facts eloquent, forms more significant. Elimination, therefore, may be just as meaningless as elaboration, perhaps more often is so. I offer any fool, for an example.

To know what to leave out and what to put in, just where and just how—ah, *that* is to have been educated in knowledge of simplicity.

As for objects of art in the house even in that early day they were the *bête noir* of the new simplicity. If well chosen, well enough in the house, but only if each was properly digested by the whole. Antique or modern sculpture, paintings, pottery, might become objectives in the architectural scheme and I accepted them, aimed at them, and assimilated them. Such things may take their places as elements in the design of any house. They are then precious things, gracious and good to live with. But it is difficult to do this well. Better, if it may be done, to design all features together. At that time, too, I tried to make my clients see that furniture and furnishings, not built in as integral features of the building, should be designed as attributes of whatever furniture was built in and should be seen as minor parts of the building itself, even if detached or kept aside to be employed on occasion. But when the building itself was finished, the old furniture the clients already possessed went in with them to await the time when the interior might be completed. Very few of the houses were, therefore, anything but painful to me after the clients moved in and, helplessly, dragged the horrors of the old order along after them.

But I soon found it difficult, anyway, to make some of the furniture in the "abstract"; that is, to design it as architecture and make it "human" at the same time—fit for human use. I have been black and blue in some spot, somewhere, almost all my life from too intimate contacts with my own furniture. Human beings must group, sit, or recline—confound them—and they must dine, but dining is much easier to manage and always was a great artistic opportunity. Arrangements for the informality of sitting comfortably, singly or in groups, where it is desirable or natural to sit, and still to belong in disarray to the scheme as a whole—that is a matter difficult to accomplish. But it can be done now, and should be done, because only those attributes of human comfort and convenience, made to belong in this digested or integrated sense to the architecture of the home as a whole, should be there at all, in modern architecture. For that matter about four-fifths of the contents of nearly every home could be given away with good effect to that home. But the things given away might go on to poison some other home. So why not at once destroy undesirable things . . . make an end of them?

Here then, in foregoing outline, is the gist of America's contribution to modern American architecture as it was already under way in 1893. But the gospel of elimination is one never preached enough. No matter how much preached, simplicity is a spiritual ideal seldom organically reached. Nevertheless, by assuming the virtue by imitation—or by increasing structural makeshifts to get superficial simplicity—the effects may cultivate a taste that will demand the reality in course of time, but it may also destroy all hope of the real thing.

Standing here, with the perspective of long persistent effort in the direction of an organic architecture in view, I can again assure you out of this initial experience that repose is the reward of true simplicity and that organic simplicity is sure of repose. Repose is the highest quality in the art of architecture, next to integrity, and a reward for integrity. Simplicity may well be held to the fore as a spiritual ideal, but when actually achieved, as in the "lilies of the field," it is something that comes of itself, something spontaneously born out of the nature of the doing whatever it is that is to be done. Simplicity, too, is a reward for fine feeling and straight thinking in working a principle, well in hand, to a consistent end. Solomon knew nothing about it, for he was only wise. And this, I think, is what Jesus meant by the text we have chosen for this discourse—"Consider the lilies of the field," as contrasted, for beauty, with Solomon.

Now, a chair *is* a machine to sit in.

A home *is* a machine to live in.

The human body *is* a machine to be worked by will.

A tree *is* a machine to bear fruit.

A plant *is* a machine to bear flowers and seeds.

And, as I've admitted before somewhere, a heart *is* a suction pump. Does that idea thrill you?

Trite as it is, it may be as well to think it over because the *least* any of these things may be, *is* just that. All of them are that before they are anything else. And to violate that mechanical requirement in any of them is to finish before anything of higher purpose can happen. To ignore the fact is either sentimentality or the prevalent insanity. Let us acknowledge in this respect, that this matter of mechanics is just as true of the work of art as it is true of anything else. But, were we to stop with that trite acknowledgment, we should only be living in a low, rudimentary sense. This skeleton rudiment accepted, *understood,* is the first condition of any fruit or flower we may hope to get from ourselves. Let us continue to call this flower and fruit of ourselves, even in this machine age, art. Some architects, as we may see, now consciously acknowledge this "machine" rudiment. Some will eventually get to it by circuitous mental labor. Some *are* the thing itself without question and already in need of "treatment." But "Americans" (I prefer to be more specific and say "Usonians") have been educated "blind" to the higher human uses of it all—while actually in sight of this higher human use all the while.

Therefore, now let the declaration that "all is machinery" stand nobly forth for what it is worth. But why not more profoundly declare that "form follows function" and let it go at that? Saying "form follows function" is not only deeper, it is clearer, and it goes further in a more comprehensive way to say the things to be said, because the implication of this saying includes the heart of the whole matter. It may be that function follows form, as, or if, you prefer, but it is easier thinking with the first proposition just as it is easier to stand on your feet and nod your head than it would be to stand on your head and nod your feet. Let us not forget that the simplicity of the universe is very different from the simplicity of a machine.

New significance in architecture implies new materials qualifying form and textures, requires fresh feeling, which will eventually qualify both as "ornament." But "decoration" must be sent on its way or now be given the meaning that it has lost, if it is to stay. Since "decoration" became acknowledged as such, and ambitiously set up for itself as decoration, it has been a makeshift, in the light of this ideal of organic architecture. Any house decoration, as such, is an architectural makeshift, however well it may be done, unless the decoration, so-called, is part of the architect's design in both concept and execution.

Since architecture in the old sense died and decoration has had to shift for itself more and more, all so-called decoration is become *ornamental,* therefore no longer *integral.* There can be no true simplicity in either architecture or decoration under any such condition. Let decoration, therefore, die for architecture, and the decorator become an architect, but not an "interior architect."

Ornament can never be applied to architecture any more than architecture should ever be applied to decoration. All ornament, if not developed within the nature of architecture and as organic part of such expression, vitiates the whole fabric no matter how clever or beautiful it may be as something in itself.

Yes—for a century or more decoration has been setting up for itself, and in our prosperous country has come pretty near to doing very well, thank you. I think we may say that it is pretty much all we have now to show as domestic architecture, as domestic architecture still goes with us at the present time. But we may as well face it. The interior decorator thrives with us because we have no architecture. Any decorator is the natural enemy of organic simplicity in architecture. He, persuasive doctor-of-appearances that he *must* be when he becomes architectural substitute, will give you an imitation of anything, even an imitation of imitative simplicity. Just at the moment, he is expert in this imitation. France, the born decorator, is now engaged with Madame, owning to the good fortune of the French market, in selling us this ready-made or made-to-order simplicity. Yes, imitation simplicity is the latest addition to imported "stock." The decorators of America are now equipped to furnish *especially* this. Observe. And how very charming the suggestions conveyed by these imitations sometimes are!

Would you have again the general principles of the spiritual ideal of organic simplicity at work in our culture? If so, then let us reiterate: first, simplicity is constitutional order. And it is worthy of note in this connection that $9\times9$ equals 81 is just as simple as $2+2$ equals 4. Nor is the obvious more simple necessarily than the occult. The obvious is obvious simply because it falls within our special horizon, is therefore easier for us to *see;* that is all. Yet all simplicity near or far has a countenance, a visage, that is characteristic. But this countenance is visible only to those who can grasp the whole and enjoy the significance of the minor part, as such, in relation to the whole when in flower. This is for the critics.

This characteristic visage may be simulated—the real complication glossed over, the internal conflict hidden by surface and belied by mass. The internal complication may be and usually is increased to create the semblance of and get credit for—simplicity. This is the simplicity-lie usually achieved by most of the "surface and mass" architects. This is for the young architect.

Truly ordered simplicity in the hands of the great artist may flower into a bewildering profusion, exquisitely exuberant, and render all more clear than ever. Good William Blake says exuberance is *beauty,* meaning that it is so in this very sense. This is for the modern artist with the machine in his hands. False simplicity—simplicity as an affectation, that is, simplicity constructed as a decorator's *outside* put upon a complicated, wasteful engineer's or carpenter's "structure," outside or inside—is not good enough simplicity. It cannot be simple at all. But that is what passes for simplicity, now that startling simplicity effects are becoming the *fashion.* That kind of simplicity is *violent.* This is for "art and decoration."

Soon we shall want simplicity inviolate. There is one way to get that simplicity. My guess is, there is *only* one way really to get it. And that way is, on principle, by way of *construction* developed as architecture. That is for us, one and all.

## THE DESIGN PROCESS—SUGGESTED READINGS

Alexander, Christopher; Sara Ishikawa; and Murray Silverstein. *A Pattern Language: Towns, Buildings, Construction.* New York: Oxford, 1977.

Anthony, Kathryn. *Design Juries on Trial.* New York: Van Nostrand Reinhold, 1991.

Broadbent, Geoffrey. *Design in Architecture.* New York: Wiley, 1973.

Fitch, James Marston. *Historic Preservation: Curatorial Management of the Built World.* Charlottesville: The University Press of Virginia, 1990.

Habraken, N. J. *Supports: An Alternative to Mass Housing.* New York: Praeger, 1972.

Koestler, Arthur. *The Act of Creation.* New York: Macmillan, 1967.

Leatherbarrow, David. *The Roots of Architectural Invention.* New York: Cambridge University Press, 1993.

Preiser, W. F. E.; Jacqueline Vischer; and Edward T. White, eds. *Design Intervention.* New York: Van Nostrand Reinhold, 1991.

Pye, David. *The Nature and Aesthetics of Design.* New York: Van Nostrand Reinhold, 1978.

Rudofsky, Bernard. *Architecture without Architects.* New York: Museum of Modern Art, 1964.

Sanoff, Henry. *Integrating Programming, Evaluation and Participation in Design.* Brookfield, VT: Avebury, 1992.

Schön, Donald. *The Reflective Practitioner: How Professionals Think in Action.* New York: Basic Books, 1983.

# 8

# THE SOCIAL IMPLICATIONS OF ARCHITECTURE
# 建筑的社会含义

选读 29
"为无居可归者提供居所"
理查德·布克敏斯特·富勒

理查德·布克敏斯特·富勒是一位设计哲学家,他认为设计行业的主要目标是通过预估未来需求来提高人们的生活现状。他的著作从本质上来说往往带有推测性的意味;他对设计过程与建成形式的社会成果的关注要高于他对建筑的关注。在这篇文章中,他回顾了1976年联合国人居大会进程,并将其置于自己50余年环境设计师的职业生涯背景之下。他认为,设计回应就是为绝大多数生活在有害健康的不人道环境中的人们解决住有所居的问题。富勒支持现代设计原则,通过"少投入、多收获"来改善环境。他也是倡导这一原则的人群中最富有干劲、充满创造性的学者之一。读毕全文,读者不禁会为他的"技术是为了给人类提供更好服务"的乐观观点所打动。

## READING 29

## Accommodating Human Unsettlement
### Richard Buckminster Fuller

R. Buckminster Fuller was a design philosopher who viewed the primary purpose of the design professions as a means of improving the human condition

*Source:* "Accommodating Human Unsettlement" by R. Buckminster Fuller in *Town Planning Review* 49, no. 1, pp. 51–60. Copyright © 1978 Allegra Fuller Snyder. For more information on the work of Buckminster Fuller please contact:
  The Buckminster Fuller Institute
  2040 Alameda Padre Serra, #224
  Santa Barbara, California 93103
  805-962-0022, fax 805-962-4440
  e-mail BFI@aol.com
Reproduced by permission of The Estate of R. Buckminster Fuller.

through a process of anticipating future needs. His writings tend to be speculative in nature; his discussions of architecture are secondary to his concerns with the design process and the social outcomes of built forms. In this article he reviews the proceedings of the 1976 United Nations Habitat Conference and places it in the context of his own 50-year career as an environmental designer. He argues for design responses that address the problems of housing the vast majority of humankind who must exist in unhealthy and inhumane conditions. Fuller was one of the most forceful and innovative advocates of the modern-design principle of improving the environment by doing "more with less." The reader cannot help but be affected by his sense of optimism in placing technology at the service of humanity.

The United Nations Conference on Human Settlements *Habitat* 1976 occurred in the penultimate year of my 1927 conceptioning of and all-out commitment to a fifty-year gestation period of economic initiatives, philosophic formulations, artifact inventions, their physical realisations, practical proving, progressive development and integration with general evolutionary events, all planned to culminate in the 1977 birth of a new World-around industry: that of an air-deliverable, air-serviceable and air-removable dwelling machine and environment controlling mass manufacturing and renting industry, which would employ humanity's maximumly informed and performing sciences and technologies and most advanced production techniques, to comprehensively and adequately accommodate all human living and development needs with the dwelling machines also serving as effective harvesters and conservers of all local income energies of the vegetation, sun and wind as well as of the energies in human and food wastes—and most importantly of all, to serve as spontaneous, comprehensively effective, self-teaching devices of both the young and the old children therein dwelling.[1]

All of the Dymaxion artefacts which I have developed have come into socio-economic use only in emergencies when all customary means of solving problems were either physically inadequate or prohibitively expensive and there were no alternatives but to use my more efficient high performance developments, reducing materials, energy, labour and overhead input costs.

When I commenced my project in 1927 at the age of 32, I was moneyless, jobless with a dependant wife and new born daughter. Despite the self-discipline of never asking anyone to listen to me, nothing could be in more marked contrast to my then unknown unlistened to sociological state than the 150 world-around audiences who have asked me to address them in each of the last five years, or the US Senate Foreign Policy Committee's invitation to me to speak to them on world political trends,[2] nor the plurality of invitations that I received to speak at the United Nations Vancouver Habitat Conference in 1976. This interest seems powerfully to suggest the relevance of my fifty-year program and the extent to which it has developed.

## HABITAT AND HUMAN SETTLEMENT

I was invited to Habitat under four prime auspices: as the special guest of Habitat itself, as a member of Barbara Ward's pre-Habitat Vancouver Symposium, as President of the World Society for Ekistics and as a guest of Vancouver's Habitat Committee. I also went as leader of the combined World Game students and Earth Metabolic students *Now House* project.

On the day the Habitat conference opened, front page photographs in newspapers around the world showed the acrylic skin of my USA 275-foot diameter, geodesic dome of Montreal's Expo 67 being completely burnt out. First reports that the dome had burned to the ground were untrue; the steel structure was undamaged. Since the invisible acrylic skin had been mounted inside the spherical structure, the structural appearance had not been changed. No one was inside and no one was hurt. Within ten days, even before Habitat closed, Montreal announced its intention to rehabilitate the dome.[3]

It almost seemed as though the non-structural skin of the great unharmed geodesic dome had been set afire by some mystical evolutionary wisdom to remind the world of geodesics' very high structural performance, accomplished with only three per cent of the weight of any given material necessary to produce equivalent structural and functional capabilities by any other known alternate engineering systems. Apparently, the one hundred thousand geodesic domes built around the world in the last thirty years have proven their economic value, reliability and economy to such an extent that this fire brought no charges of inadequacy of geodesic dome principles. The Expo 67 dome event and the progressively increasing magnitude of human numbers interested in listening to me—as the protagonist of a design science revolution by which to accommodate physically the now evident evolutionary insistence on world-around *unsettlement of humanity*—seemed in marked contrast to related aspects of Habitat and its technological focus almost exclusively upon nationalistically emphasised, local, immobile, and "one-off" tailoring of *human settlements*.

At the opening press conference of the Barbara Ward Group at Habitat I reported that in April 1976 the Club of Rome had issued a public reversal of its 1972 *Limits to Growth*[4] concept. There had been so many contradictions of the Club's 1972 pronouncement on the limits to growth that they had reconsidered their position. I said that I felt that the Club of Rome's first statement was funded by interests that were continuing to do what money had done in the past, that is, to rationalize selfishness. Assuming the political concept of fundamental inadequacy of life support for all humans on our planet, selfishness had been able to say "I have those for whom I'm responsible and because there is not enough life support for all, I am obliged to do various things that are utterly and completely selfish." I felt that the Club of Rome's *Limits to Growth* pronouncement represented the last attempt on the part of organised capitalists' selfishness to justify to the world public why their wealth would be unable to do anything about the third world. The initial Smithsonian announcement of the *Limits to Growth* was based on work done by an MIT professor of computer sciences who was given his input data by other MIT specialists.[5] I and many others were able to make well documented and fortunately effective public announcements that the Club of Rome's *Limits to Growth* pronouncement was a sadly ignorant statement. For instance, its authors cited only the very small remaining percentages of the world's unmined metal ore reserves, and were manifestly unaware that the metals on our Earth are continually being melted out of their last use and being recirculated in amounts greatly exceeding the tonnage of metals being newly mined and added into the cumulative circulatory system approximately 3 to 1, while the interim gains in technological "know-how" take care of ever greater numbers of humans per each pound of recirculating metal or other chemical substance into which technology invests its ever improving know-how, with the result that it is now engineeringly feasible to take care of all humanity at an unprecedentedly high standard of living without mining any more metals. In my view the Club of Rome's ignorance was occasioned by the over specialisation of scientists. I told the Habitat press conference that I thought the Club of Rome had manifested extraordinary courage and integrity in changing their public position, when they announced in Philadelphia that they had found on reinspection that their data was inadequate, ergo, their resulting conclusions were wrong. Later that week I received an invitation to lunch with Mr Peccei, President of the Club of Rome, when he personally verified their new position. I applauded his integrity.

## THE INFLUENCE OF FINANCE ON DEVELOPMENT

Though it produced almost no world-around newspaper reportage, the Vancouver conference was an historical watershed event.[6] The old established building world was conspicuous by its absence, though there were many other powerful lobbies present such as those of the Sierra Club, Audubon Society, World Population Institute and other foundations concerned with environmental subjects. It became clear that the great banks, confronted with escalating building costs which had passed the point of no return had withdrawn all support of real estate exploiters and of obsolete building technology in general. The "big money" of the world which has gone entirely transnational had found that whereas "you can't take it with you" into the next world, you also can't take it with you *around* the world: ownership has now become onerous. Big money has left all the sovereignly locked-in, local-property-game-players "holding the unmovable bags" of "real estate." Machinery becomes obsolete almost overnight, is unattractive as a continuing property and must be written off the books in five years. But machinery can be melted and reworked, to ever higher earning effectiveness only by ever improving know-how. "Know-how" has become the "apple" of transnational capitalism's eye.

As a consequence of the great 1929 crash, the monopolistic control of America's prime industrial establishments was broken. During the gradual recovery of the corporations under the aegis of the New Deal, the directors and executives of the corporations found that whether they were going to keep their jobs now, for the first time, depended entirely on the voting by stockholders to re-elect the directors and their managements, which depended entirely on whether the corporation management made profits. It was exactly at this time in history that the metals of World War I, which had been mined in such enormous profusion, began increasingly to reappear in the form of scrap. Suddenly the unforeseen recirculation of scrap began to break up the control of metal prices by the mine owners, who objected to this new development. However the new self-perpetuating managements realised that remelted metal was as pure as new metal, and more desirable because it cost less. They could make as much money by recirculation as they could by new production.

What they next found was that every time they developed a more desirable product, the sales increased. This made the wheels go round even faster so that management began to look for know-how to improve products. This brought about a completely new volition on the part of capitalism of our world. The post-1933 search for new know-how is why you see in the Sunday newspapers page after page of advertisements of great corporations looking for highly specialised, scientific and technical men with the experience generated know-how to produce new, improved and more desirable products.

All the great American corporations of yesterday have now moved out of America and their prime operations have become transnational and conglomerate and are essentially concerned with the game of selling their corporation's very complete, technical, managerial and vast credit handling and money making know-how. For this reason they are not interested in the older kind of properties. This set of unpredicted changes of volition explained the lack of concern of transnational conglomerate capitalism and their lack of opposition to the United Nations' Vancouver Conference's pre-occupation with human settlements, which they regarded as "peanuts."

The new capitalism is only mildly interested in trailers or mobile homes which are simply weather boxed platforms on which are mounted beautyrest mattresses, shower baths, washing machines, television, radio, air conditioning, lighting, cooking, refrigerating, bottled gas, tableware, toiletries, wardrobes and so forth. Mobile homes take the shape of a shoebox because they have to go through highway or railroad bridges. It is like living in the narrow shoebox shape of a railroad car.

Such mobile homes provide a place to live near jobs without having to buy a fixed home or a fixed piece of land. Because they are assemblies of mass production items their costs are low, but nowhere nearly as low as they could be if uncompromisingly designed for rental and easy maintenance rather than for sale and early replacement.

The new transitional capitalism's grand strategies are primarily formulated by international lawyers in their endeavours to vault legal barriers and avoid taxes. From 1800 to 1929, world economics were mastered by "Finance Capitalism" of the J. P. Morgan brand. From 1932 to 1952, we had "Federally Socialised Corporate Management Capitalism." Since 1952 we have had "Lawyer Desocialised and Strategied Supranational Managerial Capitalism." The grand strategy of the lawyer-managed supra-national capitalism is to keep governmental power widely deployed, ergo "conquered." Much of their media *news* has been a smoke screen diverting attention from what they were doing. For instance, while world news was spot-lighted on the Korean and Vietnam wars, the great USA corporations and banks were conglomerating and moving out of America into a world theatre of operations. In 25 successive annual appropriations of Foreign Aid, totalling 100 billion dollars, "riders" required that where a USA corporation was present in the country being aided, the aid funds had to be spent through those USA corporations. In this manner, the building of the supra-national corporations' foreign manufacturing plants took all the gold out of America. When all the "gold" was gone, the USA dollar was cut loose from gold, which lowered its world purchasing equity to a quarter of its pre "floated" value. This multiplied the supra-national corporations' gold backed relative monetary equity four-fold its previous value.

The world news media is controlled by transnational capitalism through advertising, its main source of income. The great corporations control that advertising. The amount of advertising placed in the media and the rates paid for it is predicated upon the size of the audience reached. Media management finds that the public appetite is for bad news. Whatever the psychological explanation may be the fact is that the media looks mainly for malignant news, rarely for benign news.

## RECOMMENDATIONS FROM VANCOUVER

Though they will probably be disregarded by many nations and probably much of the world press, four noteworthy recommendations emerged from the various meetings of Barbara Ward's invited group at the Habitat Symposium, from the Non-Governmental Organisations at the Habitat Forum and from the official delegates at the UN Habitat Conference itself.[7] The four, and the conditions to which there was a response, were as follows:

First, all around the world there are large squatter settlements, as for instance in Puerto Rico, Caracas and Bombay. These squatter settlements, which may increase by as many as a million people a year, are referred to formally as "self-help" groups because they improvise something to sleep under that sheds off the rain, whether it's three ply, corrugated paperboard or rusty corrugated iron. They are invariably on land that by law "belongs" to somebody else. The squatters are continually approached by racketeers who tell them, secretly, that the police are going to evict them, but if the squatters will pay the racketeers, arrangements can be made for them to remain. In this way the racketeers skim off all the money the squatters earn.

In order to cope with this phenomenon, the UN Vancouver Congress passed a very extraordinarily wise and humanly considerate resolution. In travelling around the world and visiting such squatter settlements, I have observed their beautiful community life. People in trouble co-operate in a thoughtful and loving way. Their way of life is so beautiful that I have always said that if I ever have to retire, it will be into one of those squatters' settlements. It was also observed by the majority of the UN delegates that the people coming to squat are very ingenious in the way they employ the limited available materials to provide shelter. Therefore, one of the first resolutions passed at the Vancouver conference and one also forged by the Barbara Ward Symposium recommended that all nations decree that all the land which these squatters occupy be made public lands, on which the people are allowed to remain. It was part of the same motion that the squatters be given much better materials with which to accomplish their environmental controlling.

The second resolution of note passed at Vancouver would remove the profit motive of real estaters who persuade farmers to give them an option on their land and then borrow government guaranteed funds to put in sewers, water and streets, thereby escalating land prices. The meeting recommended that all the nations individually arrange that whatever the increase in the value of land at the time of sale, it shall be taxed at a hundred per cent.

The third resolution that I want to draw attention to was the Barbara Ward Symposium recommendation that there should be a world moratorium on the further development of atomic energy.

The fourth resolution was one which Barbara Ward had herself conceived and introduced. It recommended that all around the world, by 1985, it be made physically and practically possible for all human beings to have fresh, safe, potable drinking, bathing and washing water. Around the world there are as yet many places where people are dying or suffering because of infected water. It is highly feasible within the present technology to make pure, safe water available to everybody anywhere.

## THE WORLD GAME DOMES AND THE *NOW HOUSE*

A mushroom group of foldable and moveable geodesic domes and modernised Indian tepees at Vancouver's Jericho Beach conference site demonstrated a young world's ability and inspiration to do something positive about its own future.

The World Game staff, from the Universities of Pennsylvania and Yale, called their exhibit the *Now House* because everything they had on display could be purchased *now* from industrial mass production sources.[8] All the labour of their production occurred under the controlled environment conditioning of factories: no rain, cold, heat, snow, ice and wind. The World Gamers exhibited four 14 foot 5/8 sphere polyester fibreglass geodesic domes with alternate translucent or opaque fibreglass hexagon or pentagon panels. These domes had no more need for old building technology than has the opening of an umbrella—a mobile, environment-controlling artefact. The World Gamers brought their exhibit from Philadelphia to Vancouver in one camper truck pulling one trailer.

The World Gamers first dug circular trenches slightly larger in diameter than the domes' circular bases. As they trenched they threw the earth into the enclosed circle and levelled it to form an elevated base for each dome. On top of the earth they laid overlapping corrugated aluminum panels which were surmounted first by aluminised foamboard to reradiate heat and next by plywood and again by indoor-outdoor carpeting. This made a very comfortable, springy and dry floor. They anchored the domes so that they could not blow away, for each one weighed only 225 pounds.

Three of the domes were positioned in a triangular pattern with ten feet between them. A high pole was mounted at the centre of the triangular area which in turn supported a watertight translucent canopy. The large triangular area between the domes and below the canopy was covered with the indoor-outdoor carpeting. The fourth dome stood mildly apart and could have been connected by a canopy but was not.

Ten of the World Gamers lived very comfortably and happily in the *Now House* installation. In the kitchen-bathing dome compact, economic but adequate shelving was provided on which to mount the kitchen equipment. They had a toilet which converted human waste into high-grade fertiliser. The heat necessary for this odourless process was provided alternately by electricity from the windmill hookup and by heat from the solar panel water-heating device. The toilet system produced fertiliser as a rich, dry, manured, loam-like substance which needed to be taken out of the system only once a year.

The windmill was equipped with a synchronous inverter, embodying new advanced efficiency, electronic circuitry for converting the direct current produced by the windmill into 110 volt alternating current required by most electrical equipment, making it possible to feed their alternating current directly into the public power lines. When windpower generated electricity is fed into batteries and an electric charge is later taken from the batteries for final light or power use, approximately half the energy is lost.

Feeding the unscheduled wind energy harvest directly into the power grid avoids this transfer loss. This innovation has now been accepted by the public utilities in twenty of the fifty United States. The utility companies pay the local windmill owner at wholesale rates for the energy he puts into the system and charge him at retail rates for the energy he takes out. This increases the economic advantage of both the private windmill owner and the public utilities. It is a fundamental energy "income gain" by humanity over and above dollar considerations. It is found that somewhere within a one hundred mile radius, the wind is always blowing, that is, within a two hundred mile diameter circle of 31,000 square miles. With the proliferation of such local windmills, the public utilities can progressively retire significant amounts of stand-by generating capacity, while also reducing their fossil fuel burning.

Arrayed between two of the three domes under the translucent canopy were banks of tomatoes and other food vegetation in hydroponic tanks, with noticeable growth accomplished during the short two-week period of the installation. The domes could be rotatingly rearranged with the translucent side south to impound enormous amounts of sun radiation. With the translucent panels north they remained cool and let in only the north light so desirable to artists.

The domes that were exhibited were priced at $750 each and the cost of the total package, including $17,000 worth of equipment, amounted to $20,000. Approximately 1,000 people a day visited the *Now House* and seemed genuinely enthusiastic. Their comments and the World Gamers' experience of the operation of the complex were invaluable in consideration of future improvements.

At Habitat, as elsewhere, I pointed out that the general principle of aiming for an ever higher performance with ever less inputs of energy, time and weights of material per given level of accomplished functioning, produced sumtotally a trend toward doing so much with so little that we have now arrived at a condition where performance is approximately invisible. Form is no longer following function. Functions have become formless. World humanity's *reality* of 1900 consisted of everything people could smell, see, touch and hear. Now, three quarters of a century later, 99.9 per cent of all humanity's practical everyday, worked-with realities are only instrumentally (non-sensorially) apprehendable and employable by humans. Therefore I emphasised to the Habitat audiences that they should disregard their conditioned reflexes which spontaneously look only for immediately visible manifestations of new improved ways of living.

In 1928, at the start of my fifty-year program, the structural mast of the 1928 Dymaxion House contained all its service mechanics, as did also the first full scale prototype produced at Beech Aircraft, Wichita, Kansas, in 1944–45.[9] Following general news publication of the latter, over 37,000 orders for the house were received by mail, many with cheques, all of which had to be returned because there was then no industry to manufacture or install these air-deliverable dwelling machines. Many distributors applied for sale franchises, but the electricians and plumbers who are everywhere exclusively licensed to connect houses to the water and electricity mains said that in order to survive they would have to take apart all the Dymaxion Houses' pre-assembled plumbing and electricity manifolds and re-assemble them. This would have tripled the costs and have been as illogical as would be local electricians and plumbers taking each purchased automobile apart in the owner's frontyard and reassembling it before finally permitting its use.

To avoid this nonsensical and wasteful situation, in 1947 after twenty years of my fifty year program had passed, the grand strategy was changed to one of concentration on the improvement of the shell structures themselves. Thus was I led to design the geodesic dome. And now at the completion of my fifty year campaign the No. 2 *Now House* is becoming publicly available as the air-deliverable, only-rentable, world-around dwelling machine, right on its fiftieth birthday.

## CHANGING CONCEPTIONS OF GLOBAL SCALE

World War I was so called because the stage on which it was acted was an historically unprecedented and entirely new world-around involvement. All the world's metal ore lands were involved in the production of new inanimate energy powered production machinery. When World War I was over the copper in the electric generators and motors did not rot as did the pre–World War I farm produce, nor did the copper return to the mines. The electric generators hooked up to the waterfalls kept producing electricity and the overland wires kept distributing that electricity to mass production factories and people's homes. Energy is the essence of wealth, wealth being the organised capability to support life.

When World War I was over, all the metal producing capability and energy generation persisted, with an enormous wealth gain by humanity. This high producing capability went not only into automobiles, but into farm machinery. It reduced the 90 per cent of humanity necessary on the farms to six per cent. Those not needed on the farms migrated to the cities, for food could now reach them anywhere. The new technology and its mass production under controlled environmental conditions made the old building craft technology operating under non-controlled environmental conditions, fundamentally obsolete; but society's preoccupation with accepted ways of earning its living obscured the fact. World War II took humanity's technology into the sky, deep into the ocean and eventually into outer space. These latter arts required an enormous step-up in doing more with less in order to make all logistics flyable, rocketable or electro-magnetically transmittable.

Subsequent to World War II it was found that all metals involved in the general technology of humanity were being consistently melted out of their old use forms on average every 22 years to become re-employed with an interim gained know-how to accomplish a far higher performance per pound, erg and hour technology for many times the numbers of humans served on the previous round. Japan became one of the world's greatest industrial countries employing only recirculating metal scrap.

In pre-automobile American cities and factory towns only the rich moved house, on fall and spring moving days, to bigger or smaller homes, as their changing means dictated. With the advent of the automobile, workmen could travel to better paid employment and factories could be located on new out-of-town sites. In 1950 the average American family moved out of town every six years. In 1975 the average American family was moving out of town every three years. When World War I began, the average American walked 1,100 miles a year and rode 300 miles in a vehicle of some type. In 1976, the average American still walks 1,100 miles each year but travels 20,000 miles by vehicle. And while there were no aircraft in 1900, by 1976 airport traffic was greater than that by rail when the century began. Humans with legs to move are freeing themselves from rooted dwelling patterns of earlier eras. Human settlements were inherent to agrarian and mill town ages: now human *unsettlement* is occurring.

That is why Vancouver's Habitat was an historical watershed. It marked the end of human settlement, in exclusively local geography and in major poverty. It was the beginning of the era of local geographical unsettlement and transition into the historically unprecedented and utterly unexpected condition of all humans—successfully-at-home-in-universe.

## PAYING FOR THE HOMES WE NEED

At the time of the 1929 crash and following depression and the beginning of the New Deal in 1933, the United States government took over the underwriting of the obsolete building industry.[10] Cutting loose from the historical earned savings, purchasing capability and instituting purchasing capability based on future earnings of the people, the US government instituting purchasing capability based on future earnings of the people, the US government instituted 20, 30 and 40 years mortgages that need, in effect, never be reduced so long as the periodically renegotiated interest was being paid. Had the buildings been as efficient and effective as air-space technology could render them, they would have paid for themselves in five years or better, as does all good machinery. What the government financed was the continuation and multiplication of inefficiency, manifest today in the fact that out of every 100 units of energy consumed in the US only five units of effective life supporting physical work is realised, that is, our "system" has an overall techno-economic efficiency of only five per cent. People can only have incomes through employment but 70 per cent of all the jobs in the USA are invented and produce no life support whatever. The last quarter century's vast transformation of cities all around the world to skyscraper clusters has produced space within which no life support is produced, only to accommodate job and money making. All the money making drives towards omni-automation and complete unemployment. Politics keeps inventing the jobs.

Post-1933 housing finance has shown that when the price of the median house goes above three times the median annual family income the family cannot demonstrate creditable house purchasing capability. A general condition of such inability has now been reached. Since the median family's life expectancy is 70 years and since the age of the median family's earners is 35 years, they have 35 years of life ahead but only 25 years before mandatory retirement, ergo, no more future earning years to hypothecate for home "buying" on the instalment plan which (theoretically) leads towards ultimate but rarely realised "owning." To continue underwriting the inefficiencies of miniature castle building of the building and real estate enterprise system, their governments would now have to give the housing to the median class and "forget about the lower half of humanity as unhouseable." Furthermore, examination of private individual homes shows that they are only superficially individual, for the hydraulic wash away of the earth surrounding their foundations discloses the private houses to be only fancy terminal boxes mounted on the ends of pipes with the whole community functionally a unit mechanical organism.

Not only has the progressive unsettlement of humanity completely upset all historical expectancy, but as with the individual median family's inability ever again to buy its home so, too, have we exhausted the possibility of our nation's people and its businesses paying for further government underwriting of our obsolete building industry. We still have an obsolete building industry: we must try a completely new approach and our Vancouver experience indicates one that may be appropriate.

## SETTLEMENTS ON THE NORTH FACE

Those who are world travellers are familiar with the scene at the airport baggage delivery turntables: along come well-strapped bundles of tubes and blue nylon which are picked up by young people and strapped on their backs. These packs open out into very small homes, but homes they are, and very satisfactory to youth in a world where there are so many satisfactory technological complementations of such world-around living in the form of electrified and plumbinged campsites and hostels.

At our Vancouver site, in addition to the four Turtle domes there were two smaller North Face domes. The name "North Face" derives from the north face of Mount Everest, for these domes were developed by successful Everest climbers for their high altitude, advanced base, dwelling devices—designed for environmental conditions far more formidable than those with which humans anywhere had ever before swiftly and effectively coped. The North Face domes are oval in plan and are geodesic. They are made with the highest tensile strength aircraft aluminum struts and have inner and outer skins of nylon with a double skin floor. They disassemble and roll into a pack two feet long by eight inches in diameter and weigh only eight pounds. An eight-pound home compounded with a sleeping bag permits human beings to be very intimate with nature under most hostile conditions.

Despite that they were going to have to move out of town and then out of state within only five years and would have preferred to be allowed to rent acceptably built and furnished homes in acceptable localities, those humans necessitous of getting to and holding their jobs while providing their families with favourable living, learning, playing and growing conditions have been forced to buy the acceptable homes by the speculative builders, at figures that would require a minimum of 30 to 40 years to pay off. Humanity in the non-socialist world is now being propagandised, coerced and often even forced to purchase all the immobile home properties, which gave rise to condominium or co-operative offices, apartment houses and owned single family dwellings. Yet the great industrial corporations have found such immobility to be untenable, and having now become transnational, they are concerned only with investments in service industries which rent rather than sell telephones, computers, cars, world hotelling, etc. and sell only armaments.

Eventual and perhaps even imminent, world disarmament will release the vast weapons industries for the production of air-deliverable dwelling machines. With general disarmament and the release to life promoting account of the fabulous production capacity of the world's industrial complexes will come the one day air delivery of whole cities wherein the operating energy efficiencies will be significantly multiplied and the social conditions provided by the omni-visible central community and the completely private, deployed dwelling areas; or the air delivery of single family dwelling machines to the remotest of sites; or of whole clusters of single family dwelling machines to near or far sites.

Before 1985 we will have abandoned the concept of having to earn a living. We will have given life long scholarships to everyone. We will have converted all the big city buildings to apartments and have eliminated 70 per cent of local commuting while vastly increasing long distance travel.

## MORE WITH LESS: THE HOPEFUL FUTURE

In Vancouver in June 1976, the young world in its own right opened a new chapter for human society by itself becoming committed realistically to doing more with less. Before the end of the century we will find all of humanity doing so much more with so much less that it will be enjoying a higher, legitimately richer and ethically more decent standard of living than has ever been experienced by any humans before us. With economic, physical and environmental success for all will come completely new economic accounting. We now have the metals comprehensively recirculating, and the know-how to accomplish all these tasks within the limits of already mined metals.

Since all political systems are predicated upon the misconception of fundamental inadequacy of human life support on our planet, their premise will have been proven invalid. We know how to live entirely within the scope of our daily star emanating radiation and gravity, energies income, ergo within a ten year world program we can provide all humanity with an amount of energy annually equal to that enjoyed exclusively by North Americans in 1972, while concurrently phasing out all use of fossil fuels. Nor need we longer have recourse to burning up our spaceship Earth's capital inventory of atoms.

The time-energy cosmic accounting and maximum efficiency alternative technologies as exclusively employed by scenario universe and spoken of by us as "nature" will be instituted in all human affairs and will be integratively operated by world-around satellite interlinked computers. With the computers' integrative examination of the physical and metaphysical resources available to human beings, it will be discovered that we are incredibly wealthy. Wealth, as stated before, being predicated on the degree of organised competence to nurture, protect and accommodate today's and tomorrow's lives. It will be clearly manifest that we have aboard spaceship Earth four billion, billionaire, heirs-apparent who have never been notified of their magnificent inheritance which has been over long hidden within the world's obsolete laws, customs and administrations whose divorcement of money from real wealth has hidden from the whole world the late twentieth century realised existence of omnihumanity sustaining inexhaustible wealth.

## NOTES AND REFERENCES

1. On the Dymaxion House see *The Harvard Crimson,* May 22, 1929, pp. 24–32.
2. *Hearings before the Committee on Foreign Relations,* US Senate, Ninety Fourth Congress, First Session, May/June 1975 (Washington, DC: US Government Printer, 1975), pp. 181–202.
3. R. Sheppard, "Buckminster Fuller Has New Dome Plan," *Montreal Star,* July 10, 1976.
4. D. H. Meadows, D. L. Meadows, J. Randers, and W. W. Behrens, *The Limits to Growth* (London: Earth Island; New York: Universe Books, 1972).
5. See J. W. Forrester, *World Dynamics* (Cambridge, MA: Wright-Allen Press, 1971).
6. For a discussion of Habitat 1976, see Humphrey Carver, "Habitat 1976: The Home of Man," *Town Planning Review* 248, no. 3 (July 1977), pp. 281–86.
7. The resolutions and the roles of the various meetings are discussed in *Ekistics* 42, no. 252 (November 1976).
8. The domes are manufactured by the Molded Fiberglass Company, Ashtabula, Ohio, and the windmill by Kedco of Inglewood. California. The synchronous inverter came from Windworks, Mukwanago, Wisconsin. In all fifty firms voluntarily equipped the

9   See *Fortune* magazine, April 1946, for an article " Fuller's House: It Has a Better Than Ever Chance of Upsetting the Building Industry."
10  On this and associated issues of efficiency and attitudes in the building industry, see "The Rebirth of the American City," *Hearings before the Committee on Banking, Currency and Housing.* US House of Representatives, Ninety-Fourth Congress, Second Session, September 30, 1976 (Washington, DC: US Government Printer, 1976), pp. 1011–16.

选读 30
"设计：探究与含义"
德隆·林顿

林顿在这篇文章中针对设计教学提出了一种相应的语言学类比方法，并且讨论了在设计过程中会遇到的各种问题。他为建筑设计师教育提供了一种正规范式。林顿强调设计师需要深入探究建筑物存在的最基本理由，特别是对于那些聚焦于人类居住行为的理由更要仔细研究。他认为设计过程"尤应包括一种对居住者的感知，因为他们不仅穿梭于此，而且栖息于此"。

## READING 30

# Design: Inquiry and Implication
**Donlyn Lyndon**

Lyndon presents in this article a linguistic analogy to the teaching of design and discusses a range of concerns encountered in the design process. He presents a normative model for the education of architectural designers. Lyndon emphasizes the need for the designer to inquire deeply into the fundamental reasons that buildings exist, specifically those concerns that are centered on the act of human habitation. He states that the design process "should include most especially a sense for inhabitants, not only as they pass through, but as they dwell in places."

In his study, "Learning a Language, Learning to Design," Donald Schön observes that learning to design as evidenced in the studio is similar to learning to use a language. More specifically it is what he calls, after Wittgenstein, a "language-game," where speaking and acting are closely related; where "language and the actions into which it is woven are considered a whole." The child learns not only the names of things and how to construct descriptions, but also the uses and values and peculiarities of the things themselves. In this active explorative mode of learning the learner comes to associate given elements of the language with what Schön calls "moves," with the approved or

*Source:* Donlyn Lyndon, "Design: Inquiry and Implication," *Journal of Architectural Education,* 35, no. 3 (Spring 1982), pp. 2–8. Copyright © 1982 by the Association of Collegiate Schools of Architecture, Inc. (ACSA). Reproduced by permission of MIT Press Journals.

disapproved, workable or unworkable actions that may attend the use of words or, in this case, designing lines on paper. Designers choose moves, graphic or verbal utterances, on the basis of their appreciation of a present design situation. Each move has consequences which reverberate through a range of what he calls "design domains." Each move changes the situation a designer confronts. Schön contends that to proceed, the designer must have a "repertoire of Domain Specific and Cross Cutting Terms." The designer must be able to see the implications of any move in relation to many domains; to have what Schön calls "multiple vision." To proceed through a design process, Schön suggests, is to conduct a series of drawn experiments, in which a given move is tested for its implications in several domains and for the suggestions it offers for further moves. Schön differentiates "design domains" (the clusters of factors and concerns that form in some sense the substance of the building enterprise; that is, the many ways in which design acts have consequence) from those procedures that are useful or necessary for the designer to wend their way through the choices presented. For the case study in question (conducted some years ago), Schön identifies twelve design domains as having been present in the discussion. They are quite predictable: program/use, siting, building elements, organization of space, form, structure/technology, scale, cost, building character, precedent, representation, explanation.

Schön makes no claim for the generality of these domains or for the case process he describes. But it closely matches my own experience with teaching design in studios. He points out very usefully that in many cases what appears to be conflict in the pattern of teaching between teachers is the very real question of what sets of domains people are concerned with and bring forward in a studio, while there is frequent agreement on the way in which the designer moves through the process. In the case Schön examines, during a critique the teacher models this movement through talking about the project and making sketches and diagrams.

The model involves using both drawings and words and in the exchanges recorded there seem to be several characteristics that bear noting. First, there is extensive use of what Schön calls "spatial action language." These are phrases that attribute action to a design element. "The upper level could drop down two ways" is the example that is noted. Such phrases, Schön notes, intertwine the designer's actions with the building's form and with the potential for experience within the building. Similarly in the transcribed critique, Schön detects four distinctly different uses of the word "form." Form 1 is the shape of the building or the building element, as in "hard-edged bench." Form 2 is geometry as in "geometry of parallels." Form 3 is the visible manifestation of the organization of space as in "marking a level difference." Form 4 is the "active sequential felt-path experience of one who travels through the organized space, apprehending the figures of design, taking in the sensing and feeling qualities of the building, and detecting the relations and order of the whole as reflected and realized through the parts." (This is a wonderfully articulated and ambitious description of a word that we often use.) Finally, Schön notes the importance to the process as modeled and described of "cycling back and forth between the unit and the total," which is also a cycling back and forth between deep involvement and distancing. Bill Kleinsasser, at the University of Oregon, has characterized this as an alternation between commitment and response.

This model of design as it is experienced in the studio format presents us with a number of useful insights. First, the language pertinent to designing is in this view inherently and usefully ambiguous: Words are recipient of many meanings, just as the forms which they describe are intended to be active in many domains. Both the multiplicity of implications in spatial action language and the evocative ambiguity of using the term "form" in several senses are clearly consonant with the design situation in which a given set of things placed in space is expected to have consequences in multiple domains and to be evaluated with respect to quite varied criteria. These attributes of the terms used seem to be residual consequences of the language accumulated by designers who have been striving to maintain some hold on the synthetic whole while examining diverse domains of consequence.

Throughout the case studies, single terms develop multiple reference in a way that seems peculiarly useful to the multiple vision required of designers. Nonetheless, they are bewildering for an outsider and amplify the opportunities for obfuscation. This also does not make it any easier for those who wish to enter the dialogue, be they new designers setting out to learn or interested observers or users (clients perhaps) anxious to watch over or participate in the process at work.

We should note, however, that extensive efforts at clarification and precision in the terms used in design may often be misplaced; especially if that clarification is directed to a reduction of the possible terms of reference for any set of actions, rather than to the clarifying of parts and the developing, in a structured way, of multiple frames of reference. What is more likely called for is a greater participation in the language game itself—shared experiences that can serve as useful references for both designers and newcomers. Concrete means must be devised through which new domains of concern can enter into the designer's consciousness, and useful analogs must be devised that allow newcomers to develop the capacity for multiple vision without being bound only to those possibilities that the designer has learned to imagine and refer to. The construction of useful references is a primary purpose of the university and a necessity for the field. Fortunately, much has happened in recent years that makes real progress in this regard, albeit in various ways, ranging from the study of typologies on the one hand, to the publication, for instance, of Alexander's *A Pattern Language*.

This brings us to the central question. Does design have to do with mobilizing established formulas, or is it acting out an internally programmed pattern of response to stimuli present in the design situation. Or is it instead a form of sustained inquiry; or is it some of each and then more? How does our model of teaching and the realm of discourse in which it exists address these various conceptions of the design process?

## FORMULA

In what ways does design have to do with established formulas? What is the place, that is to say, of codified routines or prescribed forms that suit specific situations, usually of singular purpose. "Formula" in architecture is used differently than in engineering or chemistry; it tends to be a pejorative word. You would probably look for quite a long time in the literature of architecture before you find critics saying with admiration, "this is a great formula building" or even, "the architect of this building has used all the most advanced formulas for design." (On this latter point, however, you may have to only wait for another couple issues of *Progressive Architecture*.) Formula has been anathema to most conceptions of architectural design—the symbol of a mindset unconcerned with people or place, or for that matter, learning. In the

Schön description, Architectural Design is inherently concerned with the evaluation of form in multiple domains. The process involves an extensive network of branching possibilities and this makes it clear why we shun the use of formula as a norm. Especially in education, we are still trying to expose the student to a language game that is meant to help them develop facility in imagining the interconnectedness of form, space and consequence. If we see architecture as entering into human affairs in complex ways (and there is little to suggest that we can do otherwise) then the uses of formula can be expected to be few. In reality, of course, we use formulas all the time and not only for beam size and tread-riser ratios, but for fire stairs, and all too often for ceiling heights and kitchen windows or the relation of coat closets to the front door. Such formulas are extremely useful for getting in order some of the parts of a project.

As a young instructor, I was much moved by a discussion with Freidrich Keisler in which he contended that one of the possibilities of his endless house was that you would constantly improvise. For example, you would invent each day where to put your hat when you came home. It's a lovely vision, I suppose, but now in the midst of middle age I find that I'm tired of trying to figure out where to put things. While his septuagenarian energy and intensity and the formal invention that flowed from it remain exemplary, the described goal is a bit fatuous. Formula is anathema when it obscures the capacity for multiple vision, or when its use creates buildings that deny the opportunity for multiple experience and interpretation; or when, for instance, the established codes for fire stairs thoughtlessly become the norm for how we move through space vertically, as they so often do. It is as important that our students should know well as that they should view critically the conventions that make buildings easy and predictable. Fault lies not in codified information, but in its misapplication.

Recent discussions of typology in architecture have refocused our attention on formula solutions although with a certain amount of rhetorical grandeur. This should be a cause for considering once again how and where we spend design energy. The great advantage of building types is that they let you say "Oh, it's one of those," and go on to see either how it fits in the larger context, or better yet, how it's made close at hand. Part of the enthusiasm for working with the idea of building types as described by Rossi, Krier and others is that designers can use known units to examine larger scale relationships without leaving a void in their imagining. Type, when it is not frozen and left unconsidered (as much too often it is), can become at one moment a means for examining larger contextual relationships (something to draw that is grounded in previous experience), and at another moment the vessel into which the designer, or later the user can pour specific individual imaginings. When transformed, as Dean Robert Harris would have it, into archetype—into the essential coincidences of form, action and purpose, then we are dealing with the stuff that makes places momentous.

Here again, the danger lies not in the study of types but in the doctrine of types. What is a strength can become a weakness. A house type, for instance, as a means to hold an organizational level in mind, can become (often does become) a substitute for considering the lives and experiences of those who will live there, rather than a basis for gathering understanding and investing care. Type form must be understood as a way of summarizing precedents not a decree of appropriate form. Forms become appropriate when they can be taken over by their users.

## SHIFTING SCALES

The nature of the design problems we give and the balance of concerns they represent are essential elements of a curriculum strategy. An oscillation of attention to scale similar to that which Schön describes as essential to the study of a given design problem is also essential to the overall pattern of the curriculum. We should not, unless we believe the individual to be principally a tool of social organization, allow a study program to be based solely on the use of type dwelling units conceived as tools for structuring urban space, any more than we should deal only with individual house forms that are not considered for their consequences in the public domain. Urban form, urbanist typology, is heady stuff—as we all learned in the late 1950s. The grand urban gesture, once accepted, leaves little room for "trivial" differences; differences that may be of enormous consequence to the individuals.

One of the most positive aspects of recent building-type studies has been the multiplicity of scales that they typify. The existence within the vocabulary they propose of building forms that assemble units and conform to a public presence as well as to a private one, is a great improvement on the norms that dominated during the last several decades. In the marketplace, it still remains that a limited number of unit types is simply multiplied to make a whole building, or in suburbia to form a development tract, with little or no intermediate recognizable form. The expression of collective scale and the identity of recognizable forms that represent some kind of a grouping of people beyond the individual, yet less than the system as a whole, is a central cultural issue. While on one hand, recognition of sanctity in the person has in many ways enriched our sense of what a society and its built form can and ought to be, there is now and often has been a parallel impoverishment that has been brought about by the absence of public forms that can be appropriated. If we seem now to be recognizing again a positive value in such things as markers, gates, porticos, courts, passages and collonades, and the full paraphernalia of building dress, with pediments, wreaths, windows of appearance and (watch for them soon) door guardians, it indicates among other things that we have added (or rather dusted off) another domain in Schön's terms: the domain of concern for how buildings signify as they are experienced.

That specific elements of what might be called civic art play such an important figural role in these studies, indicates further an interest in the presence of government or at least of authority to an extent that is sometimes alarming to those of us who took part in the 1960s. This apparent cultural transformation is perhaps similar to the evolution of thought in someone like Richard Sennett whose first major work in the late 60s was titled *The Uses of Disorder,* and traced the opportunities for personal differentiation and creative renewal among social groups in situations where traditional bonds were loosening, or did not apply. In a more recent book, *The Fall of Public Man,* Sennett examines in various ways what he describes as a "tyranny of intimacy"; the ultimate impoverishment of relations between people when they are devoid of supporting social conventions. Forms of behavior, dress and architecture, for instance, can be used for their effectiveness in public discourse without making insistent demands on one's own personality; the absence of codes is finally limiting. Social conventions free individuals from being always on the line with their true feelings; true feelings that Sennett claims have to become trivialized if they are to be in view constantly. This varies quite a bit from the earlier vision.

We need conventions and we should learn to work with them, critically and creatively. Sennett, like many architects, looks back at the 18th century as the time when social structure was buttressed by a thoroughly developed set of cultural conventions. These conventions, in architecture at least, were based on centuries of development and reinterpretation, within a heritage of classical form. They were as yet undisturbed by the massive shifts in production, capital distribution and labor specialization that characterized the 19th century and which in everchanging guise continue to shape our lives and our fortunes.

## GUISES

The 18th century, or rather its architects, created guises right and left, attempting to represent what was happening to them in terms drawn from previous experience, sometimes live—often stilted. The modernist sensibility, like early Marxism, rooted its confidence in the means of production, attempting to capture the dynamic energies of technological change and turn them to social purpose, to the building of new societies. Societies as it turns out, do not like to be built. Or, put more malevolently, those who maintain control of production prefer not to have it be captured. And those who are excluded from significant power, who doubt their capacity for change, tend to prefer that which is tangible and known to that which is more difficult and risky. This is especially so when they are constantly confronted with apparent choices of great richness. The social structure in which we now find ourselves has served effectively to balance these concerns and prevent radical change. As small investors the majority can identify with the pecuniary interests of the controlling minority, and as consumers in a society rich with apparent choice, we can exercise our need for differentiation without taking on any very demanding or threatening tasks of reconstruction. The state of our society is eloquently and sadly represented on the discordant and commercialized edges of every town. We have come to accept spurious choice as the staple of our diet, both literally and figuratively.

The extent to which we have been cast in a consumptive role, a fundamentally dependent role, was evident the other day on the radio as it carried a commercial and a news item in succession. The commercial (whose catchy lingo I cannot possibly recapture) advertised the exciting opportunities provided by the largest chain of dinette stores in California. Yes, that's right. A *chain* of stores specializing in the selling of dinette sets, and you can personalize a set of your very own by mixing styles or finishes and fabrics. The news item, following immediately, was a description of a Ralph Nader call for improvement in the educational system of America. He was reported to advocate that we should spend more time in the schools educating the students to be better citizens *and better consumers*. Nader, no doubt, meant to call attention to the intelligent and informed use of shopping choices as a form of political influence on the nature of the choices we are offered and on the way that things are produced. But in so doing he dramatized, as well, the terms of our social contract. Responsible consumption of goods is akin to playing one's appropriate role as a citizen.

## RESPONSE SYSTEM

In contrast to the formula, or type approach to design, lies the view of design as the activator of complex responses: the designer as the creative agent of forces inherent in the design situation. The architect in this view is a master coordinator. Sensitivity and responsiveness are likely to become major qualitative terms in this context and the operative rules for designing are generally internalized. Architects, or more particularly for our purposes, design students, are encouraged to enter the situation with blank paper, to serve principally as transcribers of that which is inherent in the situation. Seldom if ever is this true. In one way or another the architect brings to the situation biases and preconceptions. To presume that they are not there is to mask their influence. Its rather akin to Victorian silence about sex.

In a more recent socially conscious mode, architects have been prone to consider themselves simply as facilitators of the concerns and desires of those for whom they build. Alex Tsonis and Liane Lefaivre in a paper "In the Name of the People" criticized sharply what they term a Populist approach to architecture—of "giving people what they want." This creates an architecture that does not alter fundamental conditions, is at best marginally liberating and at most supports a situation where "the structure of dependencies is hidden behind the phenomena of possession."

**INQUIRY**

Robert Harris has espoused a different conception of possession—a deeper more knowing form of possession. In discussion of the paths to appropriate form, he has suggested something much more like appropriation: a process concerned with the act of taking possession through knowledge, not with the receiving or dispensation of experience. Taking possession through inquiry and association, through making things fit our sense of self, nature and society, is a line of approach which is open to us all, as designers and as people who use things. We are, all of us, both. What should be sought are places that are open to an active role on the part of their inhabitants; not by being neutral, the built equivalent of the blank sheet, but by being invested by their architect with imagination and care. Such places have been made with concern for the peculiarities of a dwelling presence, for what it's like to be there and to attend to the place around you. Care, I suspect, begets care. Obviously this is not always so, but often enough to be worth the investment.

Human consequence fully imagined is the essence of the Humanist tradition. In this lies the most fundamental task of education (or perhaps, we should say of "living"). Harris has enjoined us to do this in the way of a designer, full of "what ifs," open, exploratory, seeking what things might mean, could mean, may mean, to others—attempting to find structural relations to other places or things, and to the actions of people in private and together. He has talked about how to transform dwelling in a place into something that might be used in understanding how to make other places. In this he has drawn a good bit of format from Christopher Alexander's studies in Pattern Language, where there is a fundamental commitment to building knowledge and to developing ways that people can share experience. What is involved is behaving as an active co-imaginer of the places that we describe.

There is another, more dominant Humanist perspective; one that starts from the critical examination of propositions on which evaluation might be based. It deals much more skeptically, and to my mind somewhat less usefully, with the specific places at hand. This perspective tempts us to consider the places we examine (Piazza d'Italia, for instance) not as something to try to imagine oneself in, to understand and experience, but as instances of a category of concern that are useful to an argument. Observations then are limited to the confirmation of hypothesis and with the specific as an instance of a predetermined class. This may lead to interesting points and identify important issues to consider but it does not sustain design inquiry.

## LISTENING

We must attend to the education of the imagination. This involves in Constance Perin's phrase, using ideas to "stoke the imagination." We must attend to the culture as it is lived and experienced by others and not only by those who live within our own cultural framework. We must learn to listen, in the full sense of being willing to engage in a dialogue with others and to change our minds. To listen carefully we must risk argument. For few of us speak so clearly that we mean always only what we say. Challenge is essential to clarification. We must nurture that constantly in our students. There is no reason to suppose that this is less true of clients than of architects, or less true of poor clients than of rich ones. Listening is hard work and inquiry is, after all, not a method but a state of mind.

There is also another listening to do. As William Porter has observed in discussing Giancarlo de Carlo's Education Building in Urbino, design involves both a public and a private conversation. In de Carlo's case, the discussion with himself was traced in the plans; a discussion about making things with circles, in the most simple sense, and on a larger scale, about making new buildings that do not disrupt the environmental structure of a town in which he has invested much of his professional life during the last decade. Urbino is a place in which de Carlo has developed a community of concern and it is a context in which consequences of each possible move in the language game of design can be richly imagined. To educate our own imaginations, we must be very specific and very knowing and in some mysterious way both open and purposeful. We need to be willing to entertain the improbable and episodic and yet intent on some inner abiding purposes.

Imagination is fueled not only by ideas and propositions, but by study. The suggestiveness of gestures inherent in the design process itself is essential fodder for invention. The language game of design, as Schön would have it, involves a continual spinning out of consequences as evident in multiple domains. Each move in the tentative evolution of a design is predicated on the conditions established by its predecessors. Predicated, that is to say, on an assessment of the work done so far as viewed from multiple vantage points, from the various design domains that the architect has taken to be pertinent to the problem. How the imagination is directed is then, in part, a matter of what design domains are invoked. In this, education plays a great role, and the settings in which it takes place can be decisive. Design as Formula takes these domains to be immutable and pertinent one at a time. Design as a Response System tends to consider that the domains are all externalized, established by the society, but that their resolution is a matter of integrative private imagining: the simultaneous achievement of fit in separate domains. Design as inquiry expects the designer to play a part in establishing the domains and to do that openly and with others.

The architect, in this latter model, brings into the design situation an agenda of concerns, a set of problems to explore. We have every reason to expect that agenda to be serious. Indeed it is the great purpose of design education in a university setting and of the alignment of architecture with the humanist traditions to demonstrate that inquiry can be serious and committed, independent in some measure of the pressures of the moment.

## INHABITING

We need to examine critically not only the design domains, but our ways of moving through the design process. For instance, as we infuse design learning with spatial action language, we must attend to the nature of inhabitation that we invoke, that we model. Generally, the kind of terms that get used have to do with moving through spaces, experiencing them in sequence, recognizing the interaction of the space and the person in a momentary encounter with the building. Spatial action language, at least as I have experienced it, tends to have to do with a tourist view of the buildings that we design. But we also need to identify with the design process language that talks about being in the place, staying with it, living in it, working at it; that deals with things such as dwelling, changing and being. Louis Kahn tended to do this, though his spatial action language had more to do with passivity than with actions; he didn't often talk about people changing things which they were in.

We need to see, too, that the language that is used in the studios helps to identify the first excitement of seeing the world in a different way—of seeing places as invested with care, acting as a repository of inventiveness and discrimination. With the thoughtful fashioning of forms the imaginative energies that are invested in a place are there waiting to be recognized and appropriated by the people who dwell in it; sitting as it were, like words on a shelf. Glenn Lynn, in discussions at Berkeley, has observed that while the making of architecture in all its complexity is best learned in the office, in the practice of architecture, the *idea* of architecture is transmitted, now at least, through the universities. Why? Because the university as a setting encourages, even demands, that we consider questions to be of central importance. In the university we are meant to be concerned with what might be as well as with what is, or at the moment, can be.

The idea of architecture has to do with the potential for congruence between the way things in the world are made, and the sense of ourselves most fully imagined. It has to do with columns to be upright with, walls to be bounded by, flat places under foot, and cover from the sky. The idea of architecture has to do with essences, or with what Alexander calls "the quality without a name." The idea of architecture has to do with care invested; with offerings made to a place that can be shared, there to be kept until they are appropriated, made a part of someone else's world not by the architect, but by the person dwelling there. "Man's mind" Rabindranath Tagore has written, "longs to be the playmate of things." The idea of architecture may have to do with emptiness, it has nothing to do with vacuousness.

## THE SETTINGS

We should recognize and think carefully about the ways in which schools establish situations for learning that can serve as exemplary in later life. Probably one of the most important things that the school does is create moments that we remember and use to guide our future actions. Schools establish exemplary situations in a variety of ways. First, and we pay most attention to that, is the role of the teacher. Teachers, and the way in which they respond to us, the moments we remember of exchange with them, form exemplary situations of lasting power. And then of course there are the students, the discourse that is established in a community of people pursuing common goals. Third, there are the occasions, the lectures, the seminars, the coffee breaks, the charrettes and reviews. The concerns expressed in these situations model how we might continue to behave in the society beyond. The physical setting itself, the place in which things happen, also has consequence for us. The studio, the laboratory, the seminar room, the office; how we use these, how we care for them, what we do with them and to what extent they are seen by us as part of the world we can change, are all important considerations.

Peter Collins, in his article in the November 1979 issue of the *JAE* on the history of architectural education, describes the first Academie Royale and its operation under J F Blondell.

> These different lessons were given in several rooms which looked out onto a large garden. One room was used by the junior students designing projects, in both of these rooms, sets of finished drawings to large scale were exhibited. Next to it was a room used to display various techniques of drawing, including a number of originals, with specimens of sculpture in the round and low relief. The fourth room was for lectures in mathematics, perspective, fortifications, quantity surveying and theoretical stereotomy. Finally, there was a large room which contained books, instruments, all kinds of models and a fine collection of framed drawings. It was here that lessons were given in experimental physics.

This place in which architecture happened was surrounded with the works of the past. There is in this an admirable sense of care and orderliness, and a reverence for the tools of inquiry. Often the settings in which architectural education now takes place are not that way. Often now the settings in which architectural education takes place are bounded only with images of the present, and the places we work in generally lack a sense of cultural presence. The studio culture is present to be sure, but often only one generation deep. Seldom is there the abiding sense of serious pursuit, of standards accumulated, of the stuff of architecture lived with. It makes one long for a few good plaster casts.

Plaster casts, of course, went out (not so long ago, in fact) with the discarding of classical precedent, with the last vestiges of the view of the world as a homogenous whole with eternal verities common to us all. They may soon be back to haunt us. As the first blush of enthusiasm for diverse environmental (that is to say territorial) claims has come to seem less compelling, fundamentally less provocative than originally supposed, we have come to see that what has been missing from architecture was not the specific needs for identification of a group, but the fundamental needs for identification that we all share . . . *the fond hope that the places around us might, after all, fit and be home.*

At the core of the problem that most university settings face, is the unresolved conflict between work and academic class. What we have created in most cases, are surrogate work places in which no one can make a very great investment. They are ruled usually by the general university administration scheduling office over which people have little control. Because often both the student and faculty relation to the place is temporary it is difficult to build up a kind of sustaining inquiry around us in the place itself. What we need is real places of work and conducive places for exchange about the nature of that work.

Generally, architecture education is caught somewhere in the middle between two educational visions. On the one hand there is the example of the first Law School at Berkeley where there were classrooms, a club room, a library, and a very small dean's office, but no faculty offices; where indeed the point was for those who practice law somewhere else, to tell about it there and to use perhaps the resources of the library. And on the other hand there is the vision of education as embodied in specialized laboratories in science where the university is the indispensable setting for the professor's work, and all of it takes place surrounded by the apparatus of that particular place. Perhaps we have the best of both worlds, but in most cases, I think we have the worst. If we are to take the making of exemplary situations to be the core of our business, then we must take the influence of our environment seriously. We had best give serious thought to the ways in which we use our buildings and the qualities of work and dwelling that are possible there and/or we should continue to question the focus of our studies—perhaps more should happen elsewhere.

## KNOWLEDGE AND CRAFT

We also need to be sure that schools provide students with the full sense of the achievements and the limits of their craft, and a spirited way of working at it. This includes a deep involvement in history as well as direct observation of forms and uses from both an architect's and an anthropologist's, if you will, vantage point. It involves a real sense for technical achievements. For knowing that craft is required to make a building come true completely. It involves exposure to both the public and the private conversations that are necessary to make an architecture that is worthwhile. We must give evidence in our teaching of the process of staying with a problem; of making it live through constant inquiry. Many faculty are much too prone to think of the new problem as interesting, to feel a necessity for constantly shifting attention from one set of conditions to another rather than staying with, abiding with, following through on a problem that has been established over a long period of time. We must be concerned that the design domains as they are evidenced in the design studios and in our teaching, in the seminar and lecture courses that we offer, establish in the designer a full sense of the way in which buildings have consequence. That should include most especially a sense for inhabitants, not only as they pass through, but as they dwell in places.

Finally, as I have written before, we should be sure that our schools are understood to be building, and not just transmitting, knowledge. They should be organizing information about the built environment, the natural ecologies into which it fits, the acts of inhabitation within it, and the opportunities for thoughtful speculation it provides. And they should do so in ways that are useful to designers and to the public—expanding and clarifying the opportunities for action as well as articulating domains of evaluation. Through the structure of knowledge that we build up and the way in which we tell things to ourselves we should be inviting others to participate.

Inquiry, in its fullest sense, is a precious resource and the university has a mandate to sustain it.

选读 31

"设计的责任：五个神话和六个方向"

维克多·帕帕奈克

维克多·帕帕奈克将设计行业置于人类文化的背景之下进行讨论。他认为建筑是众多与设计有关专业里唯一需要对环境问题做出响应的专业。这篇选读出自他的著作《为真实世界而设计》一书。在这篇选读中，帕帕奈克回顾了将工业品设计当作社会、经济和美学因素产物的历史。帕帕奈克认为设计师必须把建成对象的创造当作一个社会责任问题来充分考虑，同时必须视其为一种技术能力或艺术方式。他概述了一种将两方面内容置于平等地位的设计方法，其中一方面内容是生态敏感性和社会必需性，另一方面则是正式的秩序、先例、专业主义这些更为传统的原则。帕帕奈克将设计师的多重身份定义为：他致力于解决与贫穷困苦和老弱病残的人群相关的问题，为他们提供人文环境；他研究人的需求，也创造建筑的形式；他倡导营造可持续的全球环境。

READING 31

## Design Responsibility: Five Myths and Six Directions

**Victor Papanek**

Victor Papanek writes about the design professions in the context of human culture and sees architecture as only one of a variety of design responses to environmental problems. In this selection from his book, *Design for the Real World,* Papanek reviews the design of industrial objects as the products of social, economic, and aesthetic forces. Papanek argues that the designer must view the creation of built objects as much a question of social responsibility as one of technical competence or artistic license. He outlines a design approach that places ecological sensibilities and social necessities on an equal footing with the more traditional principles of formal order, precedent, and professionalism. Papanek defines the designer's multiple roles as: one who addresses the problems of providing humane environments for the economically and physically disadvantaged; as a researcher of human need as well as a creator of built form; and as an advocate for a sustainable global environment.

*One cannot build life from refrigerators, politics, credit statements and crossword puzzles. That is impossible. Nor can one exist for any length of time without poetry, without color, without love.*

Antoine de Saint-Exupéry

*Source:* "Design Responsibility: Five Myths and Six Directions" in *Design for the Real World: Human Ecology and Social Change* by Victor Papanek. Copyright © 1984 by Victor Papanek. Reprinted courtesy of Academy Chicago Publishers. pp. 215–47.

Industrial design differs from its sister arts of architecture and engineering. Whereas architects and engineers routinely solve real problems, industrial designers are often hired to create new ones. Once they have succeeded in building new dissatisfactions into people's lives, they are then prepared to find a temporary solution. Having constructed a Frankenstein, they are eager to design its bride.

One basic performance requirement in engineering hasn't really changed too much since the days of Archimedes: be it an automobile jack or a space station, it has to work, and work optimally at that. While the architect may use new methods, materials, and processes, the basic problems of human physique, circulation, planning, and scale are as true today as in the days of the Parthenon.

With accelerating mass production, design has become responsible for all of our means of communication, transportation, consumer goods, military hardware, furniture, packages, medical equipment, tools, utensils, and much else. With a present worldwide need of 650 million individual family living units, it can be safely predicted that even "housing," still built individually by hand, will become a fully industrially designed, mass-produced consumer product by the end of the century.

Buckminster Fuller made an early start toward mass-produced housing with his Dymaxion House (experimentally produced by the Beech Aircraft Company in Wichita, Kansas) in 1946. Later came his Domes, which started a whole generation of "Dome freaks" busily building geodesic carbuncles with a dismaying capacity for leaking covers. Other attempts came through an intelligent reappraisal of trailers stacked vertically three units high. These experiments were carried out under grants from Housing and Urban Development in Lafayette, Indiana, in the mid-sixties. The most promising mass-produced house now is manufactured in Japan by Misawa Homes. These buildings can be put together in hundreds of different configurations, are inexpensive and quickly built, and are made with a new kind of concrete.

Even now the contemporary architect is frequently no more than a master assembler of elements. *Sweet's Catalogue* (twenty-six bound volumes that list building components, panels, mechanical equipment, and so forth) occupies an honored place on the shelves of an architect's working library. With its help, he fits together a puzzle called "house" or "school" or whatever by plugging in the components—designed, for the most part by industrial designers, and listed conveniently among the 10,000 entries in *Sweet's*. Quite naturally architectural offices use computers and merely feed all of *Sweet's* pages, as well as the economic and environmental requirements of the job, into the computer. The computer assembles all the bits, relates all the information to square-foot costage, and comes up with the solution. With endearing candor, some architects have taken pains to explain that "the computer does an excellent job."

By contrast, as in the case of the TWA Terminal at Kennedy International Airport, the architect may create a three-dimensional trademark, an advertisement through which people are fed, but whose function it is to create a corporate image for the client, rather than provide comfort and facilities for passengers. Having myself been trapped at the TWA Terminal during a fifteen-hour power blackout, I can vouch for the inappropriateness of this sculptural environment to process people, airplanes, cars, food, water, waste, or luggage.

The lacy mantles and Gothic minarets of Edward Durell Stone and Yamasaki are little more than latter-day extensions of the Chicago Fair of 1893. Frothy trifles, concocted to reinject romanticism into our prefabricated, prechewed, and predigested cityscape, can nonetheless be revealing. For who could see Yamasaki's soaring Gothic arches at the Seattle Science Pavilion without realizing that here science was at last elevated through glib design clichés to the stature of religion? One almost expected Dr. Edward Teller to appear one Sunday morn, arrayed in laboratory vestments, and solemnly intone "$E = mc^2$."

One of the difficulties with design by copying, design through eclecticism, is that the handbooks, the style manuals, and floppy disks continuously go out of style and become old-fashioned and irrelevant to the problem at hand. Furthermore, it is not just aesthetics that is eliminated in designing via *Sweet's* and/or the computer. "The Concert Hall and the Moonshot Syndrome," by William Snaith in his *Irresponsible Arts,* gives an excellent example of how design fails when it relies exclusively on copying and computer-generated models.

If the need for some 650-million housing units around the world is to be met, surely the answer lies in rational rethinking of what housing means—or can mean—and the developing of totally new processes and concepts.

The architect as heroic master builder and the architect who defiles this fair and pleasant land with gigantic sterile file cabinets ready to be occupied by interchangeable people are both anachronisms.

When Moshe Safdie designed and built Habitat, an example of a radically new type of shelter, for the Montreal Exposition of 1967, he was among the first architect-planners who attempted to use a modular building system intelligently. Habitat has often been faulted for being both too expensive and too complex. In reality Habitat is probably the least expensive and at the same time most varied *system* that can be devised, and it is instructive to note that the Canadian Exposition Board made it impossible to build more than one-third of the units. The strength of Habitat lies in the fact that once a large amount of money has been invested in basic building and handling equipment, the system then begins to pay for itself as more units are built. For a fuller understanding of the Habitat system, see Safdie's two newer projects in Puerto Rico and Israel (see also R. Buckminster Fuller's *Nine Chains to the Moon,* p. 37).

In clothing design, as in architecture, the industrial designer has entered the field through the back door, creating disposable work gloves (2,000 to a roll), ski boots, space suits, protective throwaway clothing for persons handling radioactive isotopes, and scuba gear. Lately, with the introduction of "breathing" and therefore usable leather substitutes, much of the boot, belt, handbag, shoe, and luggage industry, too, is turning to the product designer for help. New techniques in vacuum forming, slush molding, gang turning, and so forth, make mass-production design possible for products traditionally associated with handcrafted operations.

The lesson of this book, to design for people's *needs* rather than their *wants,* can be applied to clothing design as well. Fashion design is much like automotive styling in Detroit: applying Band-Aids to cancerous sores. Women have been permanently disabled by wedgies, elevator shoes, stiletto heels, and pin heels. The influence of girdles on women's diaphragms, digestive systems, and pulmonary abilities could lead to a book by itself. But there are genuine needs here as well: the design of clothes for handicapped children and adults making it possible for them to dress or undress themselves—resulting in greater pride and self-confidence. Most fashion is designed for people who are seventeen years old or, more disastrously, their middle-aged brothers or sisters fancying themselves as teenagers. Little or no clothing is designed for the elderly, the obese, people who are unusually short or very tall.

Satisfying the need for tools, shelter, clothing, breathable air, and usable water is not only the job and responsibility of the industrial designer but can also provide enormous new challenges.

Mankind is unique among animals in its relationship to the environment. All other animals adapt *autoplastically* to a changing environment (by growing thicker fur in the winter or evolving into a totally new species over a half-million-year cycle); only mankind transforms earth itself to suit its needs and wants *alloplastically*. This job of form-giving and reshaping has become the designer's responsibility. A hundred years ago, if a new chair, carriage, kettle, or a pair of shoes was needed, the consumer went to the craftsman, stated his wants, and the article was made for him. Today the myriad objects of daily use are mass-produced to a utilitarian and aesthetic standard often completely unrelated to the consumer's need. At this point Madison Avenue must be brought in to make these objects seem desirable.

How the smallest change in design can have far reaching consequences can be explained through example. Automotive designers in Detroit might set themselves the goal to make car dashboards more pleasing through a symmetrical arrangement of all control knobs, and by relocating ashtrays, air conditioning controls, and wiper and heater switches. The results? *As many as 20,000 people killed outright and another 80,000 maimed on our highways during any given five-year span.* These 100,000 deaths and accidents would be caused by the driver having to reach only eleven inches further, diverting attention from the road for an extra second or two. These figures are an extrapolation of the Vehicular Safety Study Program at Cornell University. In 1971 a General Motors executive said: "GM bumpers offer 100 percent protection from all damage if *the speed of the car does not exceed 2.8 miles per hour.*" (Italics supplied.) Meanwhile the president of Toyota Motors has built a $445,000 shrine to "honor the souls of those killed in his cars" (quoted in *Esquire,* January 1971). By 1982 I saw many small shrines and memorial tablets built by the president of Honda in Japan to victims of accidents in their cars.

In late April 1983 the National Highway Traffic Safety Administration stated that General Motors might have to recall 5 million midsize cars and trucks made in 1978–1980. Should this recall be ordered, General Motors would have the distinction of having been involved in the three biggest recalls in history: in 1971 6.7 million GM cars and light trucks had to be recalled, followed by a recall of 6.4 million midsize cars in 1981. This would mean that GM had to recall a total of nearly 19 million vehicles—*or nearly half of its entire production*—due to design and engineering mistakes.

Consider the home appliance field. Refrigerators are not designed, aesthetically or even physically, to fit in with the rest of the kitchen equipment. Rather, they are designed to stand out well against competing brands at the appliance store and scream for the consumer's notice. Once bought, they still shrilly clamor for attention in the user's home—destroying the visual calm and unity of the kitchen.

Through wasting design talent on such trivia as mink-covered toilet seats, chrome-plated marmalade guards for toast, electronic fingernail-polish dryers, and baroque fly-swatters, a whole category of fetish objects for an abundant society has been created. I saw an advertisement extolling the virtues of diapers for parakeets. These delicate unmentionables (small, medium, large, and extra large) sold at one dollar apiece. A long-distance call to the distributor provided me with the hair-raising information

FIGURE 31-1  Advertisement for diapers to be used for parakeets. Author's collection. [Used by permission of Victor Papanek]

that 20,000 of these zany gadgets were sold each month in 1970.

In all things, it is appearance that seems to count, form rather than content. Let's unwrap a fountain pen we have just been given. At first there is the bag provided by the store. Nestled in it is the package, cunningly wrapped in foil or heavily embossed paper. This has been tied with a fake velvet ribbon to which a pretied bow is attached. The corners of the wrapping paper are secured with adhesive tape. Once we have removed this exterior wrapping, we come upon a simple gray cardboard sleeve. Its only function is to protect the actual "presentation box." The exterior of this little item is covered with a cheap leatherette that looks (somewhat) like Italian marble. Its shape conjures up the worst excesses of the Biedermeier style of Viennese cabinetry during the last and decadent stages of that lamentably long period. When opened, the vistas thus revealed would gladden the heart of Evelyn Waugh's *The Loved One,* for they match the interior appointments of a Hollywood-created luxury coffin to a nicety. Under the overhanging (fake) silk lining and resting on a cushion of (phony) velveteen, the fountain pen is at last revealed in all its phalliform beauty. But wait, we are not yet done. For the fountain pen itself is only a further packaging job. A recent confection of this type (selling for $150.00) had its outer casing made not of *mere* silver, but of "silver obtained by melting down ancient 'pieces of eight'" recovered, one must assume, at great expense from some Spanish galleon fortuitously sunk near the Parker Pen factory three centuries ago. A (facsimile) map, giving the location of the sunken ship and tastefully printed on (fake) parchment, was enclosed with each pen. However, whatever the material of the pen-casing, within it we find a polyethylene ink-cartridge (manufacturing cost, including ink, 3¢) connected to a nib.

In the case of the silver pen cited above, the retail price of the silver pen in its package is approximately 145,000 percent higher than the cost of the basic writing tool. We may say that inexpensive pens are, after all, available and that the example mentioned merely illustrates "freedom of choice." But this freedom of choice is illusory, for the choice is open only to those to whom the difference between spending $150 or 39¢ is immaterial. In fact, a dangerous shift from primary use and need functions to associational areas has taken place here, since in most ways the 39¢ ball-point pen outperforms the $150 one. Additionally the tooling, advertising, marketing, and even the materials used in packaging represent such an exercise in futile waste-making that it is not acceptable except to a pampered elite.

This is *not* an argument against comparatively high prices that are a result of outstanding quality. My own fountain pen (a German *Mont Blanc*) was given to me by my father on my tenth birthday; it has given me excellent service—with two minor repairs—for nearly forty-four years and is still working well and is unusually handsome.

The example of pens could be easily duplicated in almost any other area of consumer goods: the packaging of perfumes, whisky decanters, games, toys, sporting goods, and the like. Designers develop such trivia professionally and are proud of the equally professional awards they receive for the fruits of such dedicated labor. Industry uses such "creative packaging"—this, it is useful to note, is also the name of a magazine addressed to designers—in order to sell goods that may be shabby, worthless, or just low in cost, at grossly inflated prices.

In 1981 Americans for the first time paid more for the packaging that contained their food than was paid to farmers as net income, according to the Department of Agriculture's Economic Research Service. Twenty-three billion dollars was paid for food packaging by consumers in 1981 compared with a net farm income of nineteen point six billion. This is expected to rise year by year. Here are some examples:

A beer can (or bottle) costs five times as much as the beer it holds.

A potato chip bag, table syrup bottle, chewing gum wrapper or soft drink bottle cost twice as much as the foods they contain.

A breakfast cereal package, soup can, frozen food box, baby food jar, or dessert box costs one-and-one-half times as much as the foods inside. (Associated Press, 20 September 1982; Department of Agriculture *National Food Review,* 7 July 1981).

In communication and transport, other new challenges emerge globally. Nearly twenty-two years ago I was approached by representatives of the United States Army and told of their practical problems concerning parts of the world (like India) where entire village populations were illiterate and unaware that they lived in, and were part of, a nation-state. Unable to read, and without enough power for radios or money for batteries, they were effectively cut off from all news and communication. In 1962 I began to design and develop a new type of communications device.

An unusually gifted graduating student, George Seegers, did the electronic work and helped to build the first prototype. The resulting one-transistor radio, using no batteries or current and designed specifically for the needs of developing countries, consisted of a used tin can. (As illustrated in this book, a used juice can is shown, but this was no master plan to dump American junk abroad: there is and was an abundance of used cans all over the world.) This can contained wax and a wick that burned (just like a wind-protected candle) for about twenty-four hours. The rising heat was converted into enough energy (via thermocouples) to operate an earplug speaker. The radio was, of course, nondirectional, receiving any and all stations simultaneously. But in emerging countries, this was then of no importance: there was only *one* broadcast (carried by

**FIGURE 31-2** The same radio as described on the previous page but decorated with colored felt cutouts and seashells by a user in Indonesia. The user can embellish the tin-can radio to his own taste. Courtesy UNESCO. [Used by permission of Victor Papanek.]

relay towers placed about fifty miles apart). Assuming that one person in each village listened to a "national news broadcast" for five minutes daily, the unit could be used for a year until the original paraffin wax was gone. Then more wax, wood, paper, dried cow dung (which has been successfully used as a heat source for centuries in Asia), or for that matter anything else that burns could continue to keep the unit in service. All the components: earplug speaker, hand-woven copper radial antenna, an "earth" wire terminating in a (used) nail, tunnel-diode, and thermocouple, were packed in the empty upper third of the can. The entire unit was made for just below 9¢ (1966 dollars).

It was much more than a clever little gadget, constituting a fundamental communication device for preliterate areas of the world. After being tested successfully in the mountains of North Carolina (an area where only *one* broadcast is easily received), the device was demonstrated to the Army. They were shocked. "What if a Communist," they asked, "gets to the microphone?" The question is meaningless. The most important intervention is to make information of all kinds freely accessible to people. After further developmental work, the radio was given to the U.N. for use in villages in Indonesia. No one, neither the designer, nor UNESCO, nor any manufacturer, made any profit or percentages out of this device since it was manufactured as a "cottage industry" product.

In 1967 I showed color slides of the radio at the *Hochschule für Gestaltung,* at Ulm in Germany. It was viewed with dismay because of its "ugliness" and its lack of "formal" design. Of course, the radio *was* ugly. But there are good reasons for this. It would have been simple to paint it (gray, the people at Ulm suggested). But painting it would have been wrong: I felt that ethically I had no right to make aesthetic or "good taste" decisions that would affect millions of people in Indonesia, members of a different culture.

The people in Indonesia decorated their tin-can radios by pasting pieces of colored felt or paper, pieces of glass, and shells on the outside and making patterns of small holes toward the upper edge of the can. In this way it has been possible to bypass "good taste" and to design directly for the needs of the people by "building in" a chance for them to make the radio truly their own through design participation.

It is more than twenty years now since the tin-can radio was first used. Two decades later, the people of Indonesia use normal broadcasting channels; in Bali and Java ordinary stereo AM-FM radios are used by nearly everyone—much as anywhere else. One of the original tin-can radios is still on view as a sort of historical artifact in the museum at Jakarta. However I am told the radio is still used in West Irian (the Indonesia-ruled western half of Papua, New Guinea). West Irian is at a stage of development comparable to that of the rest of Indonesia two decades ago.

The story of the tin-can radio shows that it is possible—or at least was possible—to practice decent and ethical design intervention in a developing country. But it must be emphasized that the intervention was small and on a village level. Large-scale design in the Third World by outsiders has never worked. During the fifties large design offices, such as Joe Carreiro of Philadelphia, Chapman and Yamasaki of Chicago, and others, performed design development in Third World countries at the request of the State Department. But most of their work was a sort of "win the minds and hearts of the countryside" operation: they helped to design and manufacture craft-based objects that would appeal to American consumers. In other words, they did not design for the needs of people in India, Ecuador, Turkey, or Mexico; instead they worked for the fancied wants of American consumers. The fallacy of this approach has been shown in an earlier chapter of *Design for the Real World*. During the seventies and early eighties similar large-scale designs have been carried out in developing countries, this time predominantly by architects. When a developing nation is cluttered up with large buildings and consumer objects all designed and developed somewhere else, the effects tend to be disastrous. The verdict is already in for Iran; it is about to be pronounced in the Philippines; for most of Latin America the jury is still out.

If we turn from the real and fancied needs of developing countries to our own cities, we see a similarity between expectations that constantly rise and a decaying reality.

Our townscape bears the stamp of irresponsible design. Look through the train window as you approach New York, Chicago, Detroit, Los Angeles. Observe the miles of anonymous tenements, the dingy, twisted streets full of cooped-up, unhappy children. Pick your way carefully through the filth and litter that mark our downtowns or walk past the monotonous ranch houses of suburbia where myriad picture windows grin their empty invitation, their tele-vicious promise. Breathe the cancer-inducing exhaust of factory and car, watch the strontium-90 enriched snow, listen to the idiot roar of the subway, the squealing brakes. And in the ghastly glare of the neon signs, under the spiky television aerials, remember: this is our custom-designed environment.

How has the profession responded to this? Designers help to wield power to change, modify, eliminate, or evolve totally new patterns. Have we educated our clients, our sales force, the public? Have designers attempted to stand for integrity and a better way? Have we tried to push forward, not only in the marketplace, but by considering the needs of people?

Listen in on a few imaginary conversations in our design offices:

"Boy, wrap another two inches of chrome around that rear fender!"

"Somehow, Charlie, the No. 6ps red seems to communicate freshness of tobacco more directly."

"Let's call it the 'Conquistador' and give people a chance for personal identification with the sabre-matic shift control!"

"Jesus, Harry, if we can just get them to PRINT the instant coffee right on to the paper cup, all they'll need is hot water!"

"Say, how about roll-on-cheese?"

"Squeeze-bottle martinis?"

"Do-it-yourself shish-kebab kits with disposable phenolic swords?"

"Charge-a-plate divorces?"

"An aluminum coffin communicating 'nearness-to-God' (nondenominational) through a two-toned anodized finish?"

"A line of life-sized polyethylene Lolitas in a range of four skin shades and six hair colors?"

"Remember, Bill, the corporate image should reflect that our H-bombs are always PROTECTIVE!"

These imaginary conversations are quite authentic: this is the way designers talk in many offices and schools, and this is also the way in which new products often originate. One proof of authenticity is that of the eleven idiocies listed above, all but two—charge-a-plate divorces and protective H-bombs—have by now become available.

Is this just a hysterical outburst, directed toward some of the phonier aspects of the profession? Aren't there designers working away at jobs that are socially constructive? Not enough. Few articles in the professional magazines or papers presented at design conferences deal with professional ethics or responsibilities going beyond immediate market needs. The latter-day witch doctors of market analysis, motivation research, and subliminal advertising have made dedication to meaningful problem-solving rare and difficult.

The philosophy of most industrial designers today is based on five myths. By examining these, we may come to understand the real underlying problems:

**1** *The Myth of Mass Production:* In 1980, 22 million easy chairs were produced in the United States. Dividing this number by the 2,000 chair manufacturers, we find that, on the average, only 11,100 chairs could have been produced by each manufacturer. But each manufacturer has, on the average, ten different models in the line; this reduces our number to only 1,000 or so chairs of one kind. Since furniture manufacturer's lines change twice a year (in time for the spring and autumn market showings), we see that, on the average, only 500 units of any given chair were produced. This means that the designer, far from working for 235 million people (the market he is trained to think about), has, on the average, worked for 1/5,000 of 1 percent of the population. Let's contrast this with the fact that in underdeveloped areas of the world there exists a present need for close to two *billion* inexpensive, basic seating units in schools, hospitals, and houses.

**2** *The Myth of Obsolescence:* Since the end of World War II, an increasing number of responsible people at the top levels of management and government have voiced the myth that, by designing things to wear out and be thrown away, the wheels of our economy can be kept turning *ad infinitum.* This nonsense is no longer acceptable. Polaroid cameras, even though new models routinely replace earlier ones, don't become obsolete since the company continues manufacturing film and accessories for them. The German Volkswagen has moved into a leading position in supplying the transportation needs of the world by carefully refraining from major style changes or cosmetic jobs. The Zippo lighter sells far better than all other domestic lighters combined, even though (or could it be because?) the manufacturer guarantees to repair or replace its case and/or guts for life. There is ironic justice in that. For it was in 1931 that George Grant Blaisdell, a nonsmoking American, noticed that some of his friends carried windproof, dependable, Austrian cigarette lighters that sold in chain stores for twelve cents. He tried importing them directly and selling them at one dollar apiece, but, with a public unwilling to pay that much during the Depression, he stopped. He waited for the expiration of the Austrian patent and began producing it in 1935 with a lifetime guarantee. The Zippo lighter has moved from an item made on $260 worth of secondhand tools, in a $10 room in Brooklyn, to a production level of 3 million units per year. Since so many of our products are made obsolete by technology, the question of *forced* obsolescence becomes redundant and, in terms of scarce raw materials, a dangerous doctrine.

**3** *The Myth of the People's "Wants":* Never in recent times have the so-called wants of people been investigated as thoroughly by psychiatrists, psychologists, motivation researchers, social scientists, and other miscellaneous tame experts, as in the case of the ill-fated Edsel. That mistake cost $350 million and led one comedian to quip that the mistake "was being handled by the Ford Foundation."

"The people want chrome, they like change," except that Volkswagen, Honda, Renault, Volvo, Saab, Mercedes Benz, Datsun, Toyota, and Fiat have exploded that idea thoroughly. So thoroughly, in fact, that over the last twenty years, Detroit has had to start producing compact cars whenever foreign imports began to seriously affect American sales figures. As soon as foreign imports began to drop off, compact cars were again advertised as "the biggest, longest, lowest, most luxurious of them all." This stylistic extravaganza has now again increased the number of small Japanese and European imports coming into this country.

The myth of the people's wants continues to be used by industry and some designers. With the oil crises of 1973 and 1978 behind us, we are early in 1984 facing the possibility of still another shutoff due to the Iraqi-Iranian war. The people have demonstrated their wants by buying subcompacts and—insisting on quality—buying many of these cars from Japan. But in the real world of multinationals, mass unemployment in Detroit, and economic downtrends, other facts assume greater importance. Three of the four big car makers in America have now associated themselves with European or Japanese companies to produce high-quality subcompacts cooperatively. Meanwhile Japan has limited itself to exporting a smaller number of cars to the United States, and this has directly led to a fine irony: the Japanese now export larger and more luxurious cars to the United States, so that their profits can increase with a smaller number of units exported.

**4** *The Myth of the Designer's Lack of Control:* Designers often excuse themselves by explaining that it's "all the fault of the front office, the sales department, market research," and so forth. But of more than 200 mail-order, impulse-buying items foisted on the public in 1983, a significantly large number were conceived, invented, planned, patented, and produced by members of the design profession.

In the magazine *Products That Think* (no. 12, JS&A Corporation), an electronically heated ice-cream scooper designed in France is offered for $24. The "Electronic Burger" (in the same issue) is, according to the description, "an AM radio shaped like a hamburger with its speaker at the bottom of the bun." One assumes the speaker has been placed there so that the sound is barely discernible. As mentioned earlier, a $30,000 solid gold telephone was available from Diners Club for the 1983 Christmas market. For $149 an overnight electric trouser-warmer is available under the title "Hot Pants."

These 1983 items recall one of my all-time favorites: "Mink-Fer," a tube of deodorized mink droppings sold at $1.95 in 1970 as a Christmas fertilizer for "the plant that has everything."

**5** *The Myth That Quality No Longer Counts:* While Americans have for years bought German and, later, Japanese cameras, Europeans now line up to buy Polaroid cameras and equipment. American Head skis are outselling Scandinavian, Swiss, Austrian, and German skis around the world. Sales of Schlumbohm's Chemex Coffee Maker are diminished only somewhat by a recent German copy of it. The United States Universal Jeep designed by Willys in 1943 (since modified, and sold by American Motors) is still a desirable multipurpose vehicle; the only competition to its descendants comes from the British Land Rover and the Japanese Toyota Land Cruiser, both updated and improved versions of the Jeep.

The one thing these and some other American products that still command world leadership hold in common is a radical new approach to a problem, excellent design, and the highest possible quality.

Something can be learned from these five myths. It is a fact that the designer often has greater control over his work than he believes he does, that quality, new concepts, and an understanding of the limits of mass production could mean designing for the majority of the world's people, rather than for a comparatively small domestic market. Design for the people's *needs* rather than for their *wants*, or artificially created wants, is the only meaningful direction now.

Having isolated some of the problems, what can be done? At present there are entire areas in which little or no design work is done. They are areas that promote the social good but call initially for high risk and, to begin with, low return. All that is needed is a selling job, and that is certainly nothing new to the industrial design profession.

Here are some of the fields that design has neglected:

**1. Design for the Third World** With the global increase in population over the last twenty years, nearly three billion people stand in need of some of the most basic tools and implements.

In 1970 I said that more oil lamps were needed globally than ever before. By 1984 this lack has become even more acute. *There are more people without electric power today than the total population of earth before electricity was generally used.* In spite of new techniques, materials, and processes, almost no radically new oil or paraffin lamps have been developed since Thomas Edison's day. In northeastern Brazil the local population began adapting used electric light bulbs to burn oil for illumination in the late seventies. The *Nordestinos* have difficulty understanding why light bulbs have to go through an electric cycle before being cut down into a container for oil. And it is a fact that Brazil now has to import used bulbs into the northeastern states where oil lamps exceed electric bulbs.*

Eighty-four percent of the world's land surface is completely roadless terrain. Often epidemics sweep through an area: nurses, doctors, and medicine may be only seventy-five miles away, but there is no way of getting through. Regional disasters, famines, or water shortages develop frequently; again aid can't get there. Helicopters work but are far beyond the money and expertise available in many regions. Beginning in 1962, a graduate class and I developed an off-road vehicle that might be useful in such emergencies. We established the following performance characteristics:

a The vehicle would operate on ice, snow, mud, mountain forests, broken terrain, sand, certain kinds of quicksand, and swamps.

b The vehicle would cross lakes, streams, and small rivers.

c The vehicle would climb forty-five-degree inclines and traverse forty-degree inclines.

d The vehicle would carry a driver and six people, or a driver and a 1,000-pound load, or a driver and four stretcher cases; finally it would be possible for the driver to walk next to the vehicle, steering it with an external tiller, and thus carry more load.

e The vehicle could also remain stationary and, with a rear-power takeoff, drill for water, drill for oil, irrigate the land, fell trees, or work simple lathes, saws, and other power tools.

We invented and tested a completely new material, "Fibergrass." This consists of conventional chemical fiberglass catalysts but with dried native grasses, hand-aligned, substituting for expensive fiberglass mats. This reduced costs. Over 150 species of native grasses from many parts of the world were tested. By also inventing new manufacturing logistics, it was possible to reduce costs still more. Various technocratic centers were to build components: heavy metal work was to be done in Egypt and Libya, Central Africa, Bangalore (India), and Brazil. Electronic ignitions were to be made in Taiwan, Japan, Puerto Rico, and Liberia. Precision metal work and the power train were to be done in the Chinese Democratic Republic, Indonesia, Ecuador, and Ghana. The Fibergrass body would be made by users on a village or cottage level, all over the world. Several prototypes were built (and are illustrated), and it was possible to offer the vehicle to UNESCO at a unit price of less than $150 (1962 dollars). But this is the point where ethical considerations became important: Although the prototypal vehicle worked well, and computer analysis by the U.N. told us that close to ten million vehicles could be used initially, we realized that we were conniving at ecological disaster. The net result of going ahead would have meant introducing ten million internal combustion engines (and consequently pollution) into hitherto undefiled areas of the world. We decided to shelve the off-road project until a low-cost alternative power source was available, which still has not happened.

(Historical note: Since I do not believe that patents work toward the social good, photographs of our vehicle were published in a 1964 issue of *Industrial Design* magazine. Since then, more than twenty-five brands of vehicles of this type, priced between

---

*See "Papanek 1983," pp. 148–49.

$5,500 and $8,000, have been offered to wealthy sportsmen, fishermen, and (as "fun vehicles") to the youth culture. These vehicles pollute, destroy, and create incredible noise problems in wilderness areas. The destructive ecological impact of the snowmobile is detailed in Chapter Ten [of *Design for the Real World*].)

General Motors, Mercedes Benz, Volvo, and others are now manufacturing off-road vehicles for many developing countries. While these vehicles bring about some benefits to the countries involved, they also violate some of the ecological standards that made us withdraw our own vehicle. Furthermore they tie the economy of Third World countries to corporations from the rich countries through direct import or franchise deals. There are notable exceptions to this: Volkswagen production in Mexico, Brazil, and other developing countries operates with reassuring autonomy.

As a result of our concern for pollution, we began exploring muscle-powered vehicles together with a group of Swedish students at *Konstfackskolan* in Stockholm. The Republic of North Vietnam moved 1,100 pound loads into the southern part of that country by pushing such loads along the Ho Chi Minh trail on bicycles. The system worked and was effective. However, bicycles were never designed to be used in just this manner. One of our student teams was able to design a better vehicle made of bicycle parts. The new vehicle is specifically designed for pushing heavy loads; it is also designed to be pushed easily uphill through the use of a "gear-pod" (which can be reversed for different ratios, or removed entirely). The vehicle will also carry stretchers and, because it has a bicycle seat, can be ridden. Several of these vehicles plug into each other to form a short train.

When students suggested the use of old bicycles or bicycle parts, they regretfully were told that old bicycles also make good transportation devices and that parts are always needed for replacement or repair. (The students may have been influenced somewhat negatively by the fact that a design student won first prize in the Alcoa Design Award Program by designing a power source, intended for Third World use, made of brand-new aluminum bicycle parts.)

Consequently we designed a new luggage carrier for the millions of old bicycles all over the world. It is simple and can be constructed in any village. It will carry more payload. But it will also fold down in thirty seconds and then can be used in its other capacity for generating electricity, irrigation, felling trees, running a lathe, digging wells, and pumping for oil. Afterward the bicycle can be folded up again and returned to its primary function as a transportation device. Except that it now has a better luggage carrier.

A Swedish student built a full-size sketch model of a vehicle that is powered by the arm muscles and can go uphill. This in turn led us at Purdue University to design an entire generation of muscle-powered vehicles that are specifically designed to provide remedial exercise for handicapped children and adults Figure 31-3.

**2. Design of Teaching and Training Devices for the Retarded, the Handicapped, and the Disabled** Cerebral palsy, poliomyelitis, myasthenia gravis, mongoloid cretinism, and many other crippling diseases and accidents affect one-tenth of the American public and their families (20 million people) and approximately 400 million people around the world. Yet the design of prosthetic devices, wheelchairs, and other invalid gear is by and large still on a Stone Age level. One of the traditional contributions of industrial design, cost reduction, could be made here. At nearly every

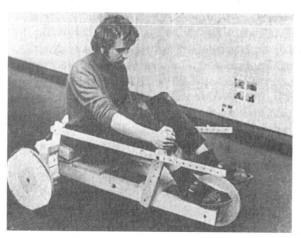

**FIGURE 31-3**  A version of a muscle-powered experimental vehicle designed by a student in Stockholm. [Used by permission of Victor Papanek.]

drugstore one can buy a transistor radio for as little as $8.98 (including import duties and transportation costs). Yet as mentioned previously, pocket-amplifier-type hearing aids sell at prices between $300 and $1,100 and involve circuitry, amplification elements, and shroud design not radically more sophisticated than the $8.98 radio.

Hydraulically powered and pressure-operated power-assists are badly in need of innovation and design.

Robert Senn's hydrotherapeutic exercising water float is designed in such a manner that it cannot tip over. There are no straps or other restraint devices that would make a child feel trapped or limited in his motions. At present hydrotherapy usually consists of having the child strapped to a rope attached to a horizontal ceiling track. In Robert Senn's vehicle all such restraints are absent. Nonetheless, his surfboardlike device is safer (it will absorb edge-loading of up to 200 pounds), and the therapist can move in much more closely to the child. Later, I explain further ideas we have developed in this field.

**3. Design for Medicine, Surgery, Dentistry, and Hospital Equipment**  Only recently has there been responsible design development of operating tables. Most medical instruments, especially in neurosurgery, are unbelievably crude, badly designed, very expensive, and operate with all the precision of a steam shovel. Thus a drill for osteoplastic craniotomies (basically a brace and bit in stainless steel) costs nearly $800 and does not work as sensitively as a carpenter's brace and bit available for $7.98 at any hardware store. Skull saws have not changed in design since predynastic times in Egypt. The radically new power-driven drill and saw for osteoplastic craniotomies was tested in vet labs devoted to experiments with animals. It promises to revolutionize methods in neurophysiology.

The cost of health care is rising astronomically. Regardless of who absorbs these costs in the long run, the fact remains that a great deal of the high expense can be attributed directly to bad design.

From time to time, illustrations of new biomedical equipment appear. Almost invariably these are "hi-style modern" cabinets, in nine delicious decorator colors, surrounding the same old machine. Hospital beds, maternity delivery tables, and an entire host of ancillary equipment are almost without exception needlessly expensive, badly designed, and cumbersome.

**4. Design for Experimental Research**  In thousands of research laboratories, much of the equipment is antiquated, crude, jury-rigged, and high in cost. Animal immobilization devices, stereo-encephalotomes, and the whole range of stereotactic instruments need intelligent design reappraisal.

Companies routinely overcharge governmental purchasing agencies by percentages that are incredibly high. Before a Senate Subcommittee investigating overcharges by manufacturers to Air Force purchasing agents, a simple hexagon Allen wrench was shown (a three-inch-long piece of six-sided wire bent to a right angle at one end). This sells to the public for 12¢. With one-eighth of an inch cut off and a 1¢ rubber grip slipped on, the same vendor sells the tool to the United States Air Force at $9,602 each! A piece of thin steel wire, about three inches long, was also shown. This wire sells for 1¢ per yard, hence the retail price of a four-inch chunk is about one-twelfth of a penny each. However industry sells this plain wire to the Air Force for $7,417 each under the formidable title Antennae Motor Safety Alignment Pin! Senate hearings have established that similar overcharging sometimes amounts to price increases as large as 230,000 percent—a practice that costs American consumers an estimated total of eighteen billion dollars annually. (All figures from U.S. Senate Subcommittee hearings on defense spending and the *MacNeil-Lehrer News Hour*, November 2, 1983.)

A simple electric laboratory timer made in upstate New York sells to amateur photographers for $89.50. Research laboratories pay $750 for the same device. An electric kitchen mixer is offered to consumers for $49.95 in white enamel or in stainless steel for $79.95. For lab use the *same* unit by the *same* manufacturer lists for $485. Value engineering is a sub-branch of design that has to do with cost reduction and assessing the value of specific parts in a machine. These value engineering techniques could play an important part to change the pricing of machines and devices for laboratories. After enough Senate investigations, manufacturers might even decide to sell laboratory apparatus at an *honest* profit, instead of defrauding the public and research establishments alike.

**5. Systems Design for Sustaining Human Life Under Marginal Conditions**  The design of total environments to maintain men and machines is becoming increasingly important. As mankind moves into jungles, the Arctic, and the Antarctic, new kinds of environmental design are needed. But even more marginal survival conditions will be brought into play as sub-oceanic mining and experimental stations on asteroids and other planets become feasible. Design for survival in space capsules has already become routine.

The pollution of water and air and the problems of toxic and atomic waste disposal also make a re-examination of environmental systems design necessary and are explored in Chapter Ten [of *Design for the Real World*].

**6. Design for Breakthrough Concepts** Many products have reached a dead end by now in terms of further development. This has led to "additive" design: more and more features or extra gadgets are added instead of reanalysing the basic problems and evolving new and innovative answers. Automatic dishwashers, for example, waste billions of gallons of water each year (in the face of a worldwide water shortage), even though other systems like ultrasonics for "separating-dirt-from-objects" are well within the state of the art. The rethinking of "dishwashing" as a system might make it easier to clean dishes, as well as solving one of the basic survival problems: water conservation. To this add: industrial water waste, toilets, showers.

Humidity control in homes and hospital rooms is important and can sometimes become critical. In many regions of the United States humidity levels are such that both humidifiers and dehumidifiers are needed. Such gadgets are costly, ugly, and ecologically extraordinarily wasteful of water and electricity. Researching this problem for a manufacturer, Robert Senn and I were able to develop a theoretical humidifier/dehumidifier without moving parts, using no liquids, pumps, or electricity. By combining a mix of deliquescent and antibacteriological crystals, we were able to develop a theoretical surface that would store twelve to twenty-four atoms of water to each crystal atom and release it again when humidity was unusually low. This material could then be sprayed onto a wall or woven into a wallhanging, eliminating the drain on electric power as well as noise pollution and expense of present-day systems. Experiments have continued for several years, and the device now works well. In 1982 test marketing was begun.

Problems are endless, and not enough breakthrough thinking is done. Consider the heating of rooms and houses. With heating costs rising many people have been forced to close off some rooms in their homes—especially in the Northeast of the United States—and install paraffin heaters, electric fires, or other space heaters that are only marginally safe. Add to this group of people those living in southern California, parts of Florida, Australia, and other areas where room heating is only temporarily needed. Basing my thinking on Frank Lloyd Wrights's "gravity heat," that is the fact that a warmed floor will reduce temporary and permanent heat requirements in a room, I began research in 1981 for another breakthrough answer to this. Using techniques borrowed from electric blankets, which take very little current to operate, I developed a system for modular electric rugs. Each electric rug is shock-proof and measures 39 by 39 inches, and they're easily plugged together. With very low energy usage they heat a room to comfortable temperatures. They are now being experimentally worked with by one of my clients in Australia.

Breakthrough concepts are also related to people's expectations and wants, as discussed earlier in this chapter. Catheryn Hiesinger describes such a change in people's consciousness and its impact on manufacturers: "While in 1964 visitors to the New York World's Fair were presented with an exhibit of model homes called The House of Good Taste, the 1982 World's Fair at Knoxville, Tennessee, displayed TVA conservation techniques, a Victorian house remodeled with energy-saving devices and appliances, and a factory-built home with a solar-heating system, approaching if not attaining the utopia of Victor Papanek's *Design for The Real World"* (*Design since 1945.* Philadelphia: Museum of Art, 1983).

These are six possible directions in which the design profession can and must go if it is to do a worthwhile job. Few designers have realized the challenge so far or responded to it. The action of the profession has been comparable to what would happen if all medical doctors were to forsake general practice and surgery and concentrate exclusively on dermatology, plastic surgery, and cosmetics.

## THE SOCIAL IMPLICATIONS OF ARCHITECTURE—SUGGESTED READINGS

Critchlow, Keith, and Jon Allen. *The Whole Question of Health.* London: The Prince of Wales' Institute of Architecture, 1994.

Day, Christopher. *Places of the Soul.* Wellingborough: Aquarian, 1990.

Fathy, Hassan. *Architecture for the Poor.* Chicago: University of Chicago Press, 1973.

Fuller, R. Buckminster. *Earth, Inc.* Garden City, NY: Anchor Press, 1973.

Harries, Karsten. *The Ethical Function of Architecture.* Cambridge, MA: MIT Press, 1997.

Huxtable, Ada Louise. *Goodbye History, Hello Hamburger.* Washington, DC: Preservation Press, 1986.

Papanek, Victor. *The Green Imperative.* New York: Thames and Hudson, 1995.

Rybcynski, Witold. *Taming the Tiger.* New York: Viking, 1983.

Soleri, Paolo. *Arcosanti.* Arcosanti, AZ: Cosanti Press, 1993.

Van der Ryn, Sim, and Peter Calthorpe. *Sustainable Communities.* San Francisco: Sierra Club, 1986.

Wright, Gwendolyn. *Building the Dream: A Social History of Housing in America.* New York: Pantheon, 1981.

Zukin, Sharon. *Landscape of Power: From Detroit to Disney World.* Berkeley: University of California, 1991.

# 9

## THE ARCHITECTURAL PROFESSION
## 建筑行业

选读 32
"建筑业与代达罗斯[译注]的风险"
朱迪思·R. 布劳

这篇选读出自《建筑师和事务所》一书。这是一本从社会学角度研究纽约市建筑事务所的书籍,它关注塑造建筑师职业生涯的社会、经济和技术因素。从这本书中衍生出一个重要的观点,建筑行业充满活性和极不稳定的前景就教育和历练建筑师而言却创造出了一种试验和创新的文化。与其他主要行业不同,建筑业如果想要正常运转,必须要建筑师对各执一词的设计理论、建设进程和客户要求做出回应,丝毫不能停歇。建筑师必须能够一人胜任许多角色才能在实践中取得成功。这些角色包括艺术家、商务专家、官员、社会改革者、用户代言人及技术人员。布劳最后得出结论,当建筑师有能力接受这些相互矛盾的事物时,他们也会获得独特的机遇,让其创造性得以发挥,个人的精神得以满足。

### READING 32

### Architecture and the Daedalean Risk
**Judith R. Blau**

This is a selection from *Architects and Firms,* a sociological study of architectural firms in New York City that focuses on the social, economic, and technical factors that shape the professional lives of architects. A major thesis that emerges from the study is that the highly dynamic and unstable professional expectations in architecture create a culture of experimentation and innovation

*Source: Architects and Firms: A Sociological Perspective on Architectural Practice* by Judith R. Blau. Copyright © 1984 by The Massachusetts Institute of Technology. Reproduced by permission of The MIT Press. pp. 133–45.

[译注] 代达罗斯,希腊神话人物,建筑师,雕刻家。

on the part of architects in their education and practices. Unlike the other major professions, architecture must operate in ways that require architects to constantly respond to contradictory design theories, construction processes, and client demands. The architect must be able to play a host of roles to succeed in practice, including that of the artist, business expert, bureaucrat, social reformer, user advocate, and technician. Blau concludes that the ability to come to terms with these contradictions gives the architect unique opportunities for creativity and personal satisfaction.

There is an increasing awareness that advanced capitalism has gone haywire in building markets, which precludes the efficiency, the integrity, and the assumed morality of rational approaches to environmental design. This is significant at several levels. It means in urban design a disillusionment with systems theory and rational decision models. Theorist Manfredo Tafuri (1979:173) writes, "Indeed the present efforts to make equilibrium work, to connect crisis and development, technological revolution and radical changes of the organic composition of capital, are simply impossible. To aim at the pacific equilibration of the city and its territory is not an alternative solution, but merely an anachronism." The growing awareness that market mechanisms come into conflict with building and environmental design also entails the recognition that political potentialities underlie the aesthetics of architecture (see Fuller 1983), and they are more manifestly evident in its praxis. The politicization of the aesthetics of architecture is evident in the recent emphasis on buildings that appeal to a greater diversity of taste through their complexity, pluralism, and multiformity—that is, an architecture that gives expression to countervailing tastes, if not countervailing social and political interests (Jencks 1981; Hubbard 1980). The political potential of architecture is recognized by, among others, Friedman (1983), who would turn building over to the arena of direct democratic decision making. To be sure, architecture is not moving in one uniform direction, but what it is moving away from is clear: a model of rational equilibrium, purism, and functional orthodoxy.

On the whole social scientists too have jettisoned their models of equilibrium—the theories that posit functional integration—and now pay greater attention to understanding the sources of cleavages and dysfunctions in modern society. Current theoretical and empirical investigations focus on such issues as the ways in which advanced countries achieve dominance and keep other countries economically and militarily dependent, the segmentation in labor markets that reinforces initial differences in training and rewards between blacks and whites and between women and men, the confounding of class and minority status that produces a seemingly intractable source of inequality, and the concentration of capital and power in the monopolistic sector of the economy at the expense of the competitive, entrepreneurial periphery.

Not only do these forms of cleft structure rest on contradictions in society, they generate new antagonisms over time—for example, when women make gains in improving their educational qualifications only to find their degrees are worth less than men's and that the division of household labor is not significantly altered either.

Contradictions in professional practice have been the subject of this book. Although architecture alone was used to make the argument, and thus the forms the contradictions take are more or less specific to architecture, I assume that the underlying processes are generalizable. In chapter 1 [of *Architects and Firms*] I pointed out the features that architecture shares with other established professions. But it is also important to remember the ways in which architecture differs from them.

With the exception of engineering, architecture is more fully involved with technocratic and corporate elites than other professions, which creates unusual difficulties. One is ethical; a worthwhile objective in any profession is that it provide services to all clients, not just to the very rich and powerful. A second difficulty of its dependence on elites in the private sector is that architecture enjoys few of the legal and monopolistic protections accorded, for example, the health professions, which are closely bound to public and quasi-public sectors. Moreover architecture's commercial standing is not very secure since it must compete with a large and efficient building industry to capture a corner of the market.

Whereas most other professional fields depend on a given theory that changes relatively slowly (the theory of genetic transmission, learning theory, or Newtonian physics, which suffices for most engineering fields), architecture (not unlike art and psychiatry, however) undergoes frequent apostasies as conceptions about design, function, and scale are routinely reevaluated. Not all members of the profession are equally affected by this process of reevaluation and change in theory, but its consequences for practice are far-reaching. To mention just a few aspects of change: buildings can be outdated before they are completed; the possibility of introducing a new approach is open to anyone; one problem typically has many design solutions; and the relation between design theory and technological means is highly variable. Such flux implied by these uncertainties is an important reason why members of the profession are not closely integrated and often unable to act in concert on major issues facing the field.

Architecture is also far from uniform with respect to its educational and certification standards. In part this is because of different emphases placed on the various roles of the occupation: artist, business expert, bureaucrat, social reformer, user advocate, technician. One role, artist, is conceived to lie somewhere beyond the educational and credentialing process.

Finally, the integral importance of architecture to the public is far less than that of medicine and law. Medicine stakes its legitimacy on life and health, law on personal legal protection and public safety. Architecture has no comparable domains; in fact it cannot even lay exclusive claims on building, except on its design or artistic value inasmuch as most building is done without any architects. (To paraphrase Gropius, after engineers have finished the building, architects may be called in too.)

These features combine to make architecture a profession that is less organized than most others and relatively weak. Yet they also contribute to a greater vitality and openness, the lack of which makes most prestigious occupational groups opaque to outsiders. Controversies are widely aired, there is considerable variation in forms of practice, and individuals differ in the degree to which they adopt one or a combination of occupational models. From one angle this means there is ample range for addressing a number of questions of sociological interest.

In the remainder of this chapter, I review the main conclusions of this study from a perspective that emphasizes that these and other contradictions are important for they are responsible for change and transformation in the profession and practice of architecture. Social and economic structure figures predominantly in this process, but ideas play more than a casual role.

## IDEAS AND STYLE

The question concerning the linkages among societal conditions, ideas, and architectural style is complex but can be addressed by the findings of this study. A brief summary of the contrast between architecture in premodern times and in contemporary societies helps clarify the issues.

In traditional Western and Oriental societies the values of the ruling class were easily translated into secular or religious buildings, as "architects appeared to mediate the 'affinity' between elites and stylistic conventions" (Larson 1983:53), and in relatively homogeneous cultures, without elites (and without architects), vernacular buildings reflected a "community of experience" (Rudofsky 1964). Such a close relationship between societal conditions and architecture, however, is undermined and often completely breaks down in contemporary Western societies.

There are three main explanations for why buildings appear now to be outside their historical times. One hinges on the assumption that social conditions govern or stimulate artistic ideas, which in turn find expression in building style. Although this chain of influence is clear enough, the process can be slow, in which case built architecture lags behind historical circumstances, or ideas may be reactive, in which case built architecture represents opposition to its historical circumstances. The second explanation assumes that architectural ideology is independent of circumstances and thus can engender changes in buildings with results independent of and unrelated to its context. The third explanation is that architectural products are not congruent with their times because they are capriciously influenced by social and economic conditions that are themselves in contention and flux; ideology may provide justification for architecture, but ideas themselves have no consequence.

To provide a concrete example, it is of interest to architectural historians that Beaux Arts' impractical romanticism took such deep roots in the later part of the nineteenth century, just at the time, Saint (1983:80) notes, that the "unsentimental priorities of capitalism were starting to gnaw at the cities." One view is that the Beaux Arts style was a reaction against commercialism and the rationality of capitalism (Richards 1970:36) while the international style represents acquiescence to it (Condit 1964). Ideology is thus not trivial, but it is more controlled by conditions than controlling them. The second explanation provided is that ideas about design respond, sympathetically or antithetically, to prior styles. In this vein Jordy (1976:349) argues that the Beaux Arts style was a reaction against the more naturalistic conceptions that had conditioned design in the earlier part of the century. Thus architectural intentions, not socioeconomic conditions, are the driving forces resulting in change. The third explanation of why buildings do not correspond to prevailing conditions is that architecture is controlled by contradictory and seemingly irrational, material forces in society (Colbert 1966; Tafuri 1980). To put the question in its most elementary form, Is architectural style largely the translation of ideas that respond to an objective reality; is it the product of ideas that themselves are autonomous; or is it driven by material conditions and beyond ideology?

What this study can lend to the controversy comes in the way of some empirical data, information obtained from architects and the heads of firms. I find that the ideas of rank-and-file architects are at fundamental odds with the prevailing style of built architecture and with the objectives evident in practice. That is, their ideas are altogether contrary to the monumentalism and stylistic poverty of the late versions of modernism and do not mesh either with the pragmatic priorities that govern the practice of architecture, at least as these priorities are stated by principals. The responses of architects may imply that they interpret the international style as the symbolic embodiment of monopoly capitalism and commercial imperatives and in its stead argue for an architecture based on humanistic values, for buildings that are less pretentious and more informal. In the absence of the organizational and economic constraints with which firm heads must contend, rank-and-file architects pay little heed to the interests of the paying client and say they would like to be more concerned with the users of buildings. It was only at the end of the 1970s that these revisionist ideas were clearly articulated by architectural critics and slowly and timorously incorporated into built architecture. This might suggest that a genuine avant-garde—one in full accord with its accomplished ideas—is impossible in architecture, and it might also suggest that ideas are not autonomously achieved or very effectual. These conclusions are not quite correct for they ignore the momentum required for new ideas to be consequential. Initially social and economic conditions and the nature of organized practice impede the realization of new design conceptions. Yet something of a snowballing process takes over as ideas become organized, as critics articulate and thereby strengthen these ideas, and as experimentation occurs.

The analysis of the ideas of the firm heads—I call them agendas—clarifies the role of ideas in this process of change. The agendas that principals have about practice are found to be efficacious, but only provided that they are logically related to particular conditions of practice. The conditions to which agendas must be logically related to be effective are in fundamental opposition to other conditions; ideas themselves are inert, and although structural affinities give them potential power, it is the simultaneity of structural alternatives that empowers ideas. If an agenda of aesthetics is emphasized, it will be translated into quality, but only if the firm is small, highly professionalized, and nonbureaucratic. An agenda of profitability is not inherently inconsistent with attaining quality design, but it is inconsistent with it if the office is structured along the lines of an efficient, bureaucratic enterprise. Agendas, in short, must take root in appropriate social forms if they are to be potent and operative.

Yet the contrarieties of social forms themselves are just as important. That is, elements that make up unusual forms to provide the fertile soil for ideas to become translated into practice are themselves contradictory. They are in the sense that they represent combinations of firm characteristics that are atypical and anomalous themselves, and especially so for the production of merited buildings. For instance, the small professionalized firm faces deficits in staff support and marketing services and lacks the capability of doing large-scale work. It exhibits, in other words, features antithetical to those of well-established practices that quite naturally produce high-quality work. Even so, an achieved coupling between values of excellence and the features of the small professionalized firm is realized in recognition for merit, whereas it is not so realized in well-established forms of practice, which are otherwise normally advantaged in competition for recognition.

Ideas appear thus to play a role in affecting change, but they must be sown at the right time and the right place, which is to say a time and place of contrast or of contradiction with prevailing institutional forces. A well-documented historical example from the visual arts helps to substantiate this conclusion. Antal (1966), among others, has documented how the progressivism and humanistic ideologies of pre-Revolutionary

French artists, of whom Jacques-Louis David is the most prominent example, became increasingly regressive as they continued to cater to bourgeois taste. They served, in fact, the very nationalism and elitism that courtly art had fostered for the ancien régime, an art against which they had ostensibly rebelled. It was not until the traditionally powerful Academy of Fine Arts began to collapse, in part by its own antiquated structure and in part by increasing competition from dealers, that ideological and stylistic revolt became possible (White and White 1965). But if it had not been for the ideas nurtured from the times prior to the Revolution and then sustained by Gericault and Courbet, the impressionists would not have been ideologically prepared to exploit the opportunities created by institutional contradictions.

The broad conclusion is that ideas have ontological status, yet paradoxically they are powerful when they are logically validated through being consistent with social structures that are themselves composed of antagonistic, or incompatible, elements. In other words, institutionally discordant conditions provide ideas with a greater chance to flourish and to be operative than they have under uniform and stable conditions. However, particular antagonistic elements must themselves be compatible with ideas; otherwise they would be mere pipe dreams or mere ideology in the original sense in the theory of historical materialism. Certain contrasting forms of organization empower the agendas of firm heads to be transformed into practice. In a broader historical context the polarities of institutional arrangements in France similarly created opportunities for painters. As for the convictions of rank-and-file architects, there is fairly good indication that progressive conceptions of design have acquired critical and public support through the snowballing process. Many chinks in the configuration of architecture production make it likely that with the accumulation of contradictory circumstances, dramatic changes can occur too in design practice that would make buildings more consistent with architects' convictions.

## ARCHITECTURE AND PROFESSIONALISM

Small and incompletely rationalized offices, though handicapped by diseconomies of scale, are the seedbed of ideas that produce high-quality work. And when they venture beyond the limits of their scale and flaunt their diseconomies, they do exceptionally good work. A complex structure, a large technical component, and bureaucratic formalization contribute to the likelihood of winning awards among smaller firms, in which these characteristics are least likely to be found. In contrast large firms usually have all of these features, and for that reason are normally advantaged.

The explanation for why small, peculiar firms excel is that eccentric surroundings are a spur to innovation. It appears that these firms retain many of the features of the small professional office while they undergo rationalization and bureaucratization at an uneven rate. For example, they have relatively open communication and a broad definition of employees' responsibilities, and they nurture a feeling of collegial camaraderie, while at the same time they are acquiring the organizational resources to obtain good clients and go after venturesome commissions. Firms that dare to take risks, act in a brazen manner in the marketplace, and at the same time preserve a congenial climate for the staff are often successful in closing the rift between professionalism and entrepreneurship.

Yet these are firms with internal contradictions. Over a five-year time span, the evidence suggests, meritorious offices with eccentric structures take the following trajectory: they usually survive as they become large and more completely rationalized (though sometimes they fail as they overextend themselves in the market). Thus what makes eccentric firms more likely to survive is their evolution in the direction of large, normally advantaged firms. In this way the risks they incur by their internal contradictions lead them to increasing normalization, at the cost that this undercuts their capabilities of continuing meritorious work. This is but one instance of the Daedalean risk.

## ARCHITECTURE AND BUREAUCRACY

For the reason that architecture requires mastery of techniques and a complex technology and because its practice is also a business, the organization of architecture is accompanied by bureaucratic features. The main hallmark of bureaucratic organization is differentiation, not only of individuals in terms of specialties and jobs but also of groups of individuals by hierarchical levels, departments, teams, and work groups. Coordination is achieved through written procedures, operating rules, hierarchy, and an administrative staff.

Architecture requires a bureaucracy to which it is, in part, discrepant. For example, there are incongruities between the expectations of those who become architects and the restraints bureaucracy imposes on them. The notions of the creative genius and of architecture as high art are preserved and reinforced in many quarters, not only by professional schools but also by architectural historians and by critics.

Imagination, however, is not long left unfettered in the places of organized practice. I discovered a vast gap between architects' expectations about what their careers would offer—the chance to be creative—and the realities of organized practice—specialization, banal assignments, inequalities in voice, and few opportunities to have design responsibilities.

Another contradiction arises from a combination of one usual and one unusual circumstance. There is, first, the usual and inevitable specialization of individuals that accompanies large size, and there is, second, an unusual tendency of large firms not to have specialized practices. For both economic and organizational reasons, a large organization of any kind has more minute internal specialization than a small one. This means that people have specialized responsibilities and that the organization itself is divided into many specialized subdivisions. This is true for architecture firms as well. Unlike other enterprises, however, larger architecture firms are more generalist in character than small ones with respect to the kinds of activities in which they engage and the kinds of buildings they produce. The large firms in this study are much less specialized on major dimensions than small ones. They offer a wider array of client services, including, for example, mortgage services, landscape design, and prefabrication, and they work on more diversified project types (although they do not have a wider range of clients than smaller offices). In all other professions the emblem of a complex base of knowledge is specialization and narrow expertise; there is a strong push among law firms, medical and dental group practices, accounting offices, veterinarian clinics, and brokerage houses to reduce their scope of services (while, presumably, improving their quality). Architecture, on the other hand, remains wedded to the traditions of the liberal professions and has tended to resist such specialization, at least at the level of the organization.

The wide range of services and projects that the larger office undertakes affects the work of all of its individual members, and their involvement in diverse services and projects is translated into greater personal power. Nevertheless the extensive specialization with respect to project responsibilities in the larger office results in less personal power. To illustrate, an architect who works in a large eclectic practice has greater knowledge about many kinds of projects compared with the architect who works in a small office, but he or she is assigned narrowly circumscribed tasks—for example, financial analysis or specification development. This coupling of eclectic organizational practice with respect to services and projects with specialization of individual tasks has contradictory effects on individual voice in the affairs of the firm. The first tends to widen the opportunity for exercising voice, whereas the second tends to reduce it. Because the negative effects of individual specialization on power are far greater than the positive effects of firm comprehensiveness, architects employed in larger offices tend to have less power than those employed in smaller offices.

These findings help to sharpen our understanding of why architects disparage, in whatever vague terms, the big production firm, and additional findings further clarify the issues involved. The data show that the reduced commitment to work and the relatively weaker professional identification usually found in large firms are the result of the way in which jobs are structured in them and the lesser voice in the large firms' affairs that its architects have.

What mainly underlies high commitment is a wide scope of responsibilities and voice in organizational matters and the autonomy that such voice implies. Both are more likely to be found in smaller offices, although occasionally they are also found in specially structured large ones—those with relatively independent project teams rather than a departmentalized organization. It should be emphasized again that large firms are advantaged in multiple ways, in spite of their lower levels of participation and of morale. In ordinary times their advantages—high-quality work, high profitability—are not affected by liabilities of low morale. In extraordinary times they grapple most with the dilemmas involving bureaucracy and professionalism, and architecture and business.

## ILLUSIONARY MYTHS

"Architects were not meant to design together; it's either all his work or mine."[1] Only the solipsistic conventions in architecture make such a statement comprehensible. This romantic conception of the designer-artist, still a current of thought in contemporary architecture, is strikingly at odds with what I find to be the basis of career commitment and vitality of architects' professional identity: voice and shared responsibilities.

Yet this is but one instance that traditionally rooted beliefs are at variance with organizational realities. Another finding contrary to this romanticization of the individual in architecture is that the more participatory the office, the more effective and efficient it is as an organization and as a business enterprise. Moreover, the democratically organized office is also most likely to design projects of superior merit.

Still another conception that is found to have little empirical support concerns the beliefs about clientage. It is often stated that having loyal clients is a mark of professional success. Also, having them, it is thought, makes it possible to dispense with the unsavory business of marketing one's wares. Loyal clients, my data indicate, have no clear benefits for practice, and they can be risky in times of an economic recession. Correspondingly a strong client orientation in the firm—that is, client satisfaction is a main component of the organizational agenda—will reduce the quality of the firm's projects when accompanied with a bureaucratized practice.

## ARCHITECTURE AND BUSINESS

In times of prosperity and growth, the commensuality between corporate capitalism and architecture is clearly evident. Under such conditions normally advantaged firms—those that are large, corporate, and fully rationalized as organizational entities—dominate core markets, markets that are highly organized and profitable. At the same time it is apparent that they sow many seeds that potentially enfeeble them. Even before a serious economic recession, their capacities as bureaucratic organizations begin to undermine their base of professionalism: departmentalization, a decline in the professional component and the routinization of work, a reduction in staff participation in decisions, and the consequent creation of social conditions that are not congenial to ideological progressivism. Most firms like these did survive the economic recession, but just barely; small entrepreneurial firms with a more intact professional focus tended to surpass them—and surpass them on their own terms, in productivity and profitability. The logic of the relative decline of the erstwhile core firms—modern offices integrated in corporate markets—is part of the dialectic of professionalism and commerce.

Another phase of the dialectic process is discernable when the contradiction between professionalism and commerce is resolved, even though the resolution is illusory. Some firms internally retain the contradictory features of high professionalism and successful integration in corporate markets. Yet they are more likely to fail than remain stable, just as they are more likely to expand and grow than remain stable. As such internalized contradictions are the very substance of dialectical transformation, it is to be assumed they are the immature predecessors of more completely rationalized offices that in time sheer off much of what becomes uncommon professionalism in big business.

The paradox is that firms that are likely to surpass the highly established and previously successful firms are those that are more subject to bankruptcy or ruin. They are governed by a potent structure of risk that contains within it the possibility of failure or great success. This is the principle of the Daedalean risk, and I have described the fates of the small professional firms as being suspended on a fulcrum: they cannot stand still for they lack the resources of the large, core firms to do so, and unless they seize the opportunities and thereby surpass core firms, they fail. Thus the dialectic of change in the practice of architecture can be seen from one perspective as the replacement of particular types of firms by others, but it also can be viewed as both resolution and creation of a series of contradictions over time. Economic forces that govern successful enterprise are counter to other forces that make architecture a successful profession, and this is what creates a dynamic of change and transformation.

## MATTER OF CHOICE

An attempt was made here to consider the processes that affect a segment of the urban market, composed of a set of professional architecture firms and the architects who work in these firms. A series of contradictions were identified from the data analyses from which inferences were derived concerning dialectic change and principles that explain such change. These principles are summarized as follows: when two opposing conditions are internalized in a structure, one asserts itself over the other, and which does depends primarily on the prevailing economic climate; the economic forces that pull firms into the orbit of successful enterprise are counter to the forces of professionalism; yet the forces that pull firms into that orbit are weakened during economic recessions, creating opportunities for small, professional firms; although ideas do not in and of themselves instigate change, when they are logically connected with structured elements that themselves are in antagonistic relations with other elements, ideas help to precipitate change and establish the substance of the newly emergent forms.

Clearly implicit in these principles is that there are no predetermined outcomes. The piling up of contradictions is redolent with vast indeterminacy, which in turn creates the opportunities for experimentation and innovation. This conclusion is itself the logic of Marx's conception of the dialectic process. There is, moreover, another reason for expecting choice to matter, and this fortifies the first. Ideas are found to play a role in change, and although it is a conditional role, they are instrumental just at the point of potential impasse, specifically in the midst of structural contradictions. It can be said that although architecture—as profession, as theory, as practice—faces now fulsome predicaments, its future is in no sense a forgone conclusion.

## NOTES

1 Paul Rudolph, quoted in Cranston Jones (1961:175). Such a statement is not atypical; see discussions by Gehry (Diamonstein 1980:40), Paolo Soleri (Heyer 1966:81), and Victor Lundy (Cranston Jones 1961:175). As Saint (1983:14) notes, the role of the individualistic designer is gradually whittled away with the increasing complexity of practice, but the concept thrives, attached to the charismatic basis of high art.

## REFERENCES

Antal, Frederick. *Classicism and Romanticism.* New York: Basic Books, 1966.
Colbert, Charles. "Naked Utility and Visual Chorea." In *Who Designs America?* ed. Laurence B. Holland. Garden City, NY: Anchor, 1966.
Condit, Carl W. *The Chicago School of Architecture.* Chicago: University of Chicago Press, 1964.
Diamonstein, Barbarlee. *American Architecture Now.* New York: Rizzoli, 1980.
Friedman, Yona. "Architecture by Yourself." In *Professionals and Urban Form*, ed. Judith R. Blau, Mark La Gory, and John S. Pipkin. Albany: State University of New York Press, 1983.
Fuller, Peter. *The Naked Artist.* London: Writers & Readers Publishing Cooperative Society, 1983.
Heyer, Paul. *Architects on Architecture.* New York: Walker and Co., 1966.
Hubbard, William. *Complicity and Conviction.* Cambridge, MA: MIT Press, 1980.
Jencks, Charles A. *The Language of Post-Modern Architecture.* Rev. ed. New York: Rizzoli, 1981.

Jones, Cranston. *Architecture Today and Tomorrow.* New York: McGraw-Hill, 1961.
Jordy, William H. *American Buildings and Their Architects.* Vol. 3. Garden City, NY: Doubleday, 1976.
Larson, Magali Sarfatti. "Emblem and Exception: The Historical Definition of the Architect's Professional Role." In *Professionals and Urban Form,* ed. Judith R. Blau, Mark La Gory, and John S. Pipkin. Albany: State University of New York Press, 1983.
Richards, J. M. *An Introduction to Modern Architecture.* Harmondsworth: Penguin, 1970 [1940].
Rudofsky, Bernard. *Architecture without Architects.* Garden City, NY: Doubleday, 1964.
Saint, Andrew. *The Image of the Architect.* New Haven: Yale University Press, 1983.
Tafuri, Manfredo. *Architecture and Utopia.* Cambridge, MA: MIT Press, 1979 [1973].
——— *Theories and History of Architecture.* New York: Harper & Row, 1980 [1976].
White, Harrison C., and Cynthia White. *Canvases and Careers.* New York: Wiley, 1965.

选读 33
"建筑业的挑战"
罗伯特·古特曼

古特曼长期从事建筑行业历史的研究，曾因《建筑实践》一书荣获 1988 年《进步建筑》杂志主办的研究大奖。这本书研究了建筑行业面对的众多挑战，而诸如以下这些因素则是造成这些挑战的根源：建筑师之间为争夺客户愈发激烈的竞争、新建建筑的复杂性、政府的要求以及每年向劳务市场输送的应届毕业生。这篇选读即出自《建筑实践》一书，古特曼在文中讨论了他认为对建筑师观念意识格外重要的五种挑战：为行业输送的建筑师需要满足建筑行业的要求；建筑师需要形成自己的实践哲学；建筑师需要对服务市场的实时情况了然于胸；建筑师需要找到维持收益以及偿付能力的手段；建筑师需要加入一个富于竞争力、斗志昂扬的行业组织。古特曼提出关于建筑行业的议题至今已有将近十年的时间，但它与最后一篇选读中博耶和米特冈对当下建筑行业的观察十分相似，它们二者的相似性还是非常引人关注的。

## READING 33

## Challenges to Architecture
### Robert Gutman

Gutman, a long-time student of the history of the architecture profession, received the 1988 *Progressive Architecture* Research Award for *Architectural Practice.* The book examines challenges to the profession stemming from such factors as the increased competitiveness among architects for clients, the complexity of new buildings and the demands of the state, and the supply of new

*Source:* "Challenges to Architecture" in *Architectural Practice: A Critical View* by Robert Gutman. Copyright © 1988 Robert Gutman. Princeton Architectural Press. pp. 97–111.

graduates entering the labor market. In this selection from the book, Gutman discusses five challenges that he viewed as especially salient to the consciousness of architects: the need to match the demand for practitioners to the supply of architects; the need to develop a philosophy of practice; the need to maintain a secure hold on the market for services; the need to find ways to maintain profitability and solvency; and the need to have a competent organization exhibiting high morale. It is quite striking to note the similarity between the issues raised about the architecture profession by Gutman almost 10 years ago and the current observations of Boyer and Mitgang in Reading 36.

I have described ten major conditions that form the context for architectural practice, and that have been undergoing significant transformations. They include (1) the extent of the demand for services; (2) the structure of demand; (3) the oversupply or potential oversupply of entrants into the profession; (4) the new skills required as a consequence of the increased complexity and scale of building types; (5) the consolidation and professionalization of the construction industry; (6) the greater rationality and sophistication of client organizations; (7) the heightened intensity of competition between architects and other professions; (8) increased competition within the profession; (9) the difficulties of achieving profitability and obtaining sufficient personal income; and (10) greater intervention and involvement on the part of the state and the wider public in architectural concerns.

These changes have been a source of anxiety and strain to architects. To the degree that individuals and firms have begun to acknowledge the new reality, the profession has been able to redefine the changes as challenges that should be confronted forthrightly. Some methods architects have developed have been very effective. Other challenges the profession has ignored. Still others, even when addressed responsibly, induce new conflicts and problems. Five challenges that are especially salient now in the consciousness of architects are discussed below, along with comments on the profession's response to them.

### Challenge I. The Need to Match the Demand for Practitioners to the Supply of Architects, and to Adjust the Number of Architects to the Potential Demand for Their Services

Architects are in the fortunate position of being able, like other service professionals, to convince prospective clients of the need for their services. To this extent, there is a system in place that helps to make demand responsive to supply. On the other hand, it is important to realize that this process is effective to only a limited degree. I have discussed many other factors, besides the marketing efforts of architects themselves, that have an impact on demand.

There is an interesting point here to consider. Compared, say, to physicians or lawyers, architects are in a weaker position to generate demand for their services. They have nowhere near the authority over the building industry that physicians enjoy in the medical care system; and they do not have as much influence over building rules and decisions as lawyers can exert through the courts and the legislative system. Furthermore, when architects are effective in helping to pass laws or encourage the adoption of public policies that stimulate construction, the design work often goes to other professions. Despite their relatively greater power, lawyers and physicians, especially physicians, have taken measures over the past fifty years to limit the number of students who study for their professions. Meanwhile, the architecture schools are admitting as many students as university budgets allow. Perhaps the absence of measures by architects to restrict enrollment and the limited control they are able to impose on forces influencing the demand for services, are connected events. In both cases, the evidence points to a profession that has difficulty regulating its position in society.

Another aspect of the demand–supply relationship is worth noting. Principals who run firms find it to their advantage to maintain a substantial flow of architects through the schools. It provides offices of all sizes with an inexpensive supply of young graduates who are well educated but nevertheless are prepared to do low-skilled work. Although architects are embarrassed by their pay levels compared to other old and established professions, without the reserve supply of cheap labor the profitability of firms would be below the current rate. Physicians have dealt with the problem by setting up a system of internships and residencies, and by overseeing the establishment of other licensed professions, including nursing and pharmacy. These subsidiary professions are arranged in a hierarchy, which doctors regulate and which do not threaten their hegemony. Lawyers have been moving in this direction through recognizing a cadre of paralegals. Architecture once approximated such a system when practitioners controlled education by means of the apprenticeship and pupilage programs. Now that education is handled independently through university-based schools, the profession appears to be less, rather than more, capable of forming this kind of clearly stratified hierarchy. If the profession were able to establish such a system, it could go a long way toward reducing the dangers inherent in a prospective oversupply of graduates, raise salaries and wages throughout the profession, and also probably stem the downward tide of profits. However, my guess is that architects will never be able to introduce this degree of rationality into the organization of the profession's work. The nature of architecture as an art, along with the architect's image of individual creativity, inhibits the development of the division of labor associated with such an undertaking. The artistic identity is also a reason why the unionization movement is almost totally absent in the profession. Unionization is proving a very effective method for adjusting the supply and remuneration levels of professionals to the demand for their services, but architects resist it. Even during the worst years of the Great Depression, when most architects were without regular work, fewer than 1,000 of the nation's 20,000 architects joined the Federation of Architects, Engineers, Chemists and Technicians. A leader of the Federation remembers them: "The architects were a pretty good group but they were pretty badly organized, pretty badly fragmented. They were individualists. They were a good group because they had a social consciousness."[1] At the present time the class-consciousness of salaried architects is unawakened. Only 1 percent of architects working in the private sector are union members, and these are professionals working for non-architectural organizations. The percentages are larger among physicians and lawyers.[2]

---

[1] Tony Schuman, "Professionalism and Social Goals of Architects, 1930–1980."
[2] Richard B. Freeman and James L. Medoff, "New Estimates of Private Sector Unionism in the United States," p. 162.

## Challenge II. The Need to Develop a Philosophy of Practice That Is Consistent but That Also Corresponds to the Expectations, Requirements, and Demands of the Building Industry

The concept of a philosophy of practice covers a range of concerns: methods by which jobs are obtained; types of jobs undertaken; the division of responsibility between the office and other firms and organizations in the building industry; modes of organizing work in the office; and the place assigned to esthetic and formal issues in building design.

A reasonably consistent philosophy is important at two levels. First, with respect to the individual firm, a philosophy of practice functions as a guide for dealing with recurring problems, such as forging a plan for the firm's development, acquiring a distinctive image and attaining a specific niche in the market for services. It also smoothes over problems that arise within practices around questions of management, recruitment, and employee incentives. Second, looking at the profession as a community of firms, it can be argued that a shared viewpoint is also an advantage. It can assist architects in thinking about their identity. A more clearly conceived self-image can help to resolve doubts about the profession's proper role in the building industry. In turn, the resolution of uncertainty in this area should enable the architectural community to choose an effective strategy for dealing with other building professions. The American Institute of Architects (AIA) in particular exhibits repeated difficulties in handling this problem. It appears to be unable to develop a consistent policy for dealing with interior designers and engineers, and it is very unclear about the stance it should take with respect to the increasing power of building firms and developers. The inadequacies of the AIA necessarily must be a source of major concern. It is the principal official spokesman for the architectural community.

Consistent attitudes toward practice are hard to find among architects because of the variety of pressures influencing them, and the tremendous difficulties involved in finding a secure position in the marketplace. Even so, one is more likely to discover such a set of attitudes in individual firms, very much so in firms that have been able to attain, and then preserve, an influential position in the industry; but it certainly is not a characteristic of the whole architectural community. Commentators on the history of professions tell us that "faction is the distinctive feature of architectural politics," more conspicuous than in any other profession. However, there are many more such "cliques and coteries" within the community than in previous periods and the spectrum of attitudes displayed covers a wider range.[3] The variety is partly the result of the sheer volume of firms and architects who have a voice. However, it is also connected to the conditions of practice discussed in this book. These transformations have induced firms to pursue different strategies in order to survive and become renowned.

One reason prospects for the development of a consistent philosophy are much better in a single firm than in the profession is that the firm is a small group. It can force out members who resist a consensus; or it can break up and start over again with a more compatible group of principals. Within the architectural community, there is no authority structure comparable to the power that principals hold in a private firm. This is also true for the AIA. As an association of professionals, it must mirror the

---

[3] A. M. Carr-Saunders and P. A. Wilson, *The Professions*, p. 184.

concerns and preferences of its members, who in turn respond to the many different conditions under which practice is conducted. It is evident that the number of issues that are critical to the fate of architecture, on which the AIA can speak with one voice and take a stand, has been diminishing. The forced abandonment of its mandatory code of professional conduct for a period of almost a decade is a telling sign of the decline of the AIA's influence on architects. Furthermore, although the AIA is the major spokesman for the profession, it is facing increasing competition from other organizations which claim they represent the interests of architects. These include the schools that grant professional degrees, the registration boards that are now organized into a national council, and, of course, the many clubs and institutes set up by practitioners in major metropolitan centers, and the independent architectural press. Each of these types of organizations has its own ideas about what the profession is, and what the aims of practice ought to be, which often deviate from the approach favored by the average firm constituting the bulk of AIA membership. The resulting diversity and heterogeneity contribute many positive benefits to the creative life of architects, but they also act as another force encouraging the fragmentation of the architectural community. The extent of this fragmentation is undermining the power of the profession within the building industry, and in its relations with clients and the public.

### Challenge III. The Need to Maintain a Secure Hold on the Market for Services, in a Period When the Competition from Other Professions Is Increasing

This need is a long-standing concern of the profession, for the reason that so much architectural knowledge and skill overlaps the expertise of other professions and occupations, including engineers, interior designers, surveyors, construction workers and managers, materials experts, and real estate economists. As I have pointed out, with the growth in the complexity of building projects, the competition from other professions and occupations is increasing. The individual architect cannot possibly keep up with the progress of knowledge in all the fields involved in the design and construction of a modern building.

In response to this difficulty, several strategies are pursued. First, the profession has welcomed the development of specialized training programs *within* the schools, so that there are now architects who are reasonably competent as architects *and* as building technologists, managers, and programmers. Second, and to an increasing extent this is the dominant strategy now, the profession has been transformed from a collection of self-employed practitioners into a community of *firms* staffed by salaried architects. Firms can make use of some of the principles of the division of labor, thus exploiting the knowledge of specialist architects and other building professionals.

A third strategy is also being used more frequently. This is for firms to concentrate in the one area where architects have claimed a special competence historically, and in which there is relatively little competition from other professionals who hold power in the building industry: the artistic side of architecture.[4] The popularity of

---

[4]Gerald M. McCue and William R. Ewald Jr., *Creating the Human Environment*, p. 280. Also see Magali Sarfatti Larson, "Emblem and Exception: The Historical Definition of the Architect's Professional Role."

this strategy is exemplified by the tremendous emphasis on design, to the exclusion of other professional services. The success of the strategy is related to the expansion in the consumption of "culture" discussed earlier. It is represented by the emergence of a clientele who relies on fashionable design and distinctive imagery to advertise and sell its products and services. The emergence of firms that function mainly as design architects, and sometimes indicate as much to clients, is a significant shift in emphasis from the attitude that dominated practice between the two world wars. During the 1920s and 1930s, it was more common for the leaders of the profession to argue that the key to survival was the ability to provide comprehensive services. The separation of the design function is not confined to simple projects and is not engaged in only by small "boutique" offices but is now a feature of the entire industry.[5] It is a development made possible by the availability of other practices that are prepared to handle the preparation of working drawings and specifications, deal with contractors, and generally supervise construction. On projects in which a design firm is located in New York, Philadelphia, Chicago, or Los Angeles and the project is in another state or smaller city, the combination of firms often meets other goals as well, such as giving the associated local firm responsibility for the review of shop drawings and site supervision.

It should be said that the tradition of specializing in the art of design is not without its precedents in the history of architecture, when the conditions of the building industry, the prescribed social position of architects, or the characteristics of a building type made it feasible. The architects of the Gothic cathedrals and other medieval buildings may have been master builders who knew a great deal about fabrication and worked along with the masons and carpenters on the site. However, many Italian Renaissance architects, students at the Royal Academy in London at the end of the eighteenth century, and nineteenth-century graduates of the Ecole des Beaux Arts in Paris, distanced themselves from the building trades as a mark of upper class status. The French architects have maintained a separate design tradition almost continuously from the time of Colbert. Still today, an architect in Paris or Lyon expects that construction drawings and building supervision will be handled by an independent enterprise, known as a "bureau d'etudes." The conception of a practice concentrated on design to the exclusion of other services is a relatively recent development in the United States. It responds to a combination of conditions external to the profession, including the interests of clients in the art of architecture mentioned earlier, the greater specialization within the building professions, and a vastly more competent and technologically more advanced construction industry.

The problem of all strategies for coping with competition is that if architects know about them, they are also known by other groups in the industry. Civil engineers, builders, and contractors have tried to acquire a knowledge of the skills possessed by architects, and to use them. If they are not able to pick up the skills themselves, the large supply of architects turned out by the schools, and the desire of employees in

---

[5]The term "boutique economy" has become standard parlance in discussions of the changing structure of the American economy. It refers to the emergence of small specialized firms and companies that are said to be a source of technological innovation and are more adaptable and flexible in responding to market demand. "Boutiques" have become common in other producer service and professional businesses, too, including law and accountancy. Gail Appleson, "Boutique Firms Hold Their Own," p. 1. Also "Can America Compete?" p. 47.

big firms to moonlight, offer a ready supply of itinerant designers. Furthermore, interior designers are now conducting practices that handle the shell and facade of buildings along with the inside scenery.[6]

There is no single method for fending off competition from other building professions. The "professional project" is a movement that began in the nineteenth century among lawyers, doctors, engineers, and other learned occupations to control their respective domains by securing an exclusive license to practice from governments. Although successful for some professions, the movement failed in terms of achieving the objective for architects.[7] It is true that only registered architects can use the title of architect. However, persons with other training and experience who work in the building industry are allowed to perform many of the duties that architects would like to arrogate to themselves. Architecture remains, and is always likely to be, a highly vulnerable profession, although the various strategies that have been explored now and in previous periods will be able to improve the economic security and professional influence of some practitioners and some firms. Still, anyone who chooses architecture as a career must be prepared to endure a greater degree of risk in fulfilling his or her career aspirations than a person educated as a lawyer or a physician. In architecture, as in many other occupations, there are, of course, rewards other than a high income and job security that make the choice of the field attractive and worthwhile.

### Challenge IV. The Need to Find Ways to Maintain Profitability and Solvency When the Costs of Running a Design Firm Are Steadily Increasing

It has proven extremely difficult to manipulate all the variables that determine the final profit and loss in an enterprise: to keep wages, salaries, and benefits down while attracting and holding onto qualified architects; to reduce other overhead costs; to maintain the level of staffing appropriate to a constantly fluctuating work load; to charge higher fees; and to discover new sources of steady income.

Three solutions have been widely advocated in recent years. First, to generate income from other sources, including construction and development subsidiaries. The revision of the mandatory AIA code of professional conduct has legitimatized the use of this approach. A second strategy is to manage projects and the work in the office more efficiently. Indeed, the belief in the value of better management as the key to survival and also to the achievement of high-quality performance is almost an obsession in the profession. It is exemplified by the appointment in 1984 of Louis Marines, an architect who had a career as a management consultant, to be executive director of the AIA. Third, to maintain a steady flow of jobs through the office, relying on marketing and

---

[6]Since interior designers, even in the two or three states in which they are registered, are not licensed to design *buildings*, formal responsibility must rest with staff architects or professionals outside the firm. The practice of having third-party architects sign drawings for approval purposes is, of course, standard in many sectors of the building industry.

[7]The concept of the professional project has received its fullest treatment in the writings of Magali Sarfatti Larson, especially her book, *The Rise of Professionalism: A Sociological Analysis*. Larson points out that there were many other methods through which professionals attempted to secure their control over a domain, in addition to obtaining a grant of authority from the state, including the construction of a definition of reality that would be accepted and believed in by clients. It can be said that one of the problems of architects is that they and their clients and users do not necessarily regard buildings from the same perspective.

public relations programs. More so than other types of producer service or consulting firms, architectural offices are either in a trough or are suddenly overwhelmed by tasks that must be completed in a great hurry. The volatility of the resulting work load can have a devastating effect on the organizational integrity of firms, not to mention a deleterious effect on profitability.

In some firms, one or more of these alternative solutions has been tried with magnificent results. There are several offices with reputations for design excellence or high-quality service, that are supported by profits from land speculation or housing and office development. There are other well-known and respected firms that are very well managed, and as a result, their profitability has increased. These offices sometimes adopt a two-track method of operations. They produce dull, very efficient space for one group of clients, making use of their management expertise or their knowledge of how to make cheap buildings. However, the profit from these jobs is then used to underwrite losses on other projects in which superior design quality is the principal criterion. The surfacing of the marketing mentality in architecture, long after it became institutionalized in other producer service businesses, has been a boon to some offices whose principals once thought it was indecent to admit they hustled for work. Well-conceived, persistent marketing programs have fostered many newly affluent firms, and have probably contributed to the overall increase in the demand for architectural services. However, as I said in earlier chapters, there are some firms for which none of the strategies appear to work. They lead to heavier overhead expenditures and less profit, but do not result in additional business.

One difficulty in the application of conventional ideas about effective management to architectural practice has been the inability of firms and their consultants to understand the special characteristics of architectural work, and then to establish a philosophy of management appropriate to it. Architecture requires closer collaboration between personnel employed in advisory and management capacities and production workers than is necessary in almost any other service business. Most building jobs are unique undertakings, in which the organization of the design and building team must be defined *de novo* for each project. With their specialized technical knowledge, different team members are integral to drawing up the plan, just as they are essential for solving the problems that arise inevitably during construction. Because of the necessary collaboration between parties that in other industries are regarded as "management" and "labor," questions about the respective responsibilities of producer service personnel and production workers are often indistinguishable from substantive questions about the design of the project. The system of responsibilities and rights in the organization of the work group is therefore developed along with the building. This arrangement demands tremendous flexibility in patterns of authority, and a continuous process of exchange of power and influence between designers, managers, technical experts, contractors, and members of the building trades. It usually prevents the formation of the type of clearly established and acknowledged table of command that is standard elsewhere, and that management theorists so often advocate.[8]

---

[8]This discussion leans on a commentary by Piore and Sabel dealing with the conditions of work that encourage the preservation of fragmentation in the construction industry. I have revised and adapted it because it is such a good analogy to the operations of a design team. Michael Piore and Charles Sabel, *The Second Industrial Divide*, p. 117.

The problem with the marketing approach is that, often in opposition to the purported aims that supposedly distinguish a professional organization from other producer service businesses, it leads architectural offices to put success in getting jobs ahead of dedication to the interests of users and the general public. This is not the intention of the advisors who have urged architects to become more proficient marketers. These consultants emphasize repeatedly that good service is a condition for effectiveness in selling. Nevertheless, the prevailing manifestation of the marketing mentality has been troubling for firms who are confused by how to balance professional values with profitability. The uncertainty seems to be intensifying, given the continued growth in the number of firms and the greater reliance on competitive pricing for getting jobs. To the extent that clients become aware that their architects are confused, and begin to suspect the depth of a firm's commitment to ideals of professional service, the result could be greater harm to the future status of the profession than several years of faulty management and mistakes in marketing.

In a profession as competitive as architecture has become, there is constant pressure to invent new approaches in order to stay ahead of the pack. However, managing and marketing consultants have spread their advice so effectively that different firms now use identical approaches, thus undermining the advantage each office hoped to achieve in formulating a strategy. Theodore Hammer, the managing partner of Hanes, Lundberg and Waehler in New York City, has commented on the irony of the situation.

> Clients are pretty smart. They've seen all the folks who have taken the [marketing] seminars come in and say, "We are truly dedicated professionals, we provide quality service on the budget, and we have a team coming off a project next week identical to yours." Then the other firm comes in and says, "We are truly dedicated professionals . . ." Everybody sounds the same lately.[9]

It may turn out that the principal beneficiary of the doctrine that marketing programs can upgrade profitability is the business of the consultants themselves.

### Challenge V. The Need to Have a Competent Organization Exhibiting High Morale and Motivated to Produce Good Work

In all modern professions, a conflict develops between the demands of the organizations in which professionals work and the personal aspirations of individuals. This conflict is receiving increasing attention because so much professional work is done now in bureaucratic settings. These settings require routine work procedures and hierarchical structures for making decisions. Such demands clash with the expectation of professionals that they will be able to exercise the independent authority under which lawyers, physicians, and architects have traditionally operated. The sociologist Eliot Freidson has pointed out that even when they work in organizations, professionals are less subject to supervision and control than other employees. At the very least, professionals enjoy technical autonomy.[10] Nevertheless, it can be argued that the modern professional situation may be less conducive to responsible, creative work than when more professionals were self-employed and worked in smaller groups.

---

[9] Society for Marketing Professional Services, *Distinguished Paper Series*, 1986, p. 21.
[10] Eliot Freidson, *Professional Powers*, p. 166.

These conflicts may well be more common in architecture than in other types of professional practice. They emerge when architects work, as more of them do each year, as staff architects for corporations, real estate developers, and government agencies. Similar disputes also develop in professional offices run by architects. The tendency is accentuated by the special characteristics of architecture as a type of work activity. In the previous section I discussed the structure of the design and building team which demands close exchanges between service and production workers. These relationships inevitably challenge the standard formula according to which service businesses are managed. In addition, architecture attracts students who assume that practice permits an unusual degree of individualized, creative self-expression. These egotistic attitudes are encouraged in schools of architecture. The temperament of architects reinforced by educational experiences yields an employee population who probably are more prone than other professional workers to insist on autonomy. Furthermore, the determination of design quality depends on informed, but intuitive, judgments. This method of evaluation limits the power of senior managers and administrators to impose their own standards on lower level employees. It is just this fear, that judgment based on status, age, and experience will not carry the weight that it does in other professions, that leads older architects and principals to adopt a dictatorial manner. In turn, younger architects are inclined to demand greater autonomy than they are capable of exercising.

The clash between the personality of many architects and the modes of thinking and managing required in practice contributes to the intensity of the debate about how to produce a combination of good staff morale and quality performance in an office. There is reason to think that this debate is more prominent in architectural circles than in other professions. The argument is critical because the ideals of the profession tend to equate excellence in practice with design excellence, frequently ignoring the quality of professional service or the performance of the building. Design excellence often does result from providing relatively free reign to individual imaginations. As a consequence architects feel some kind of special responsibility to resolve the management dilemma in their firms.

There is no single correct method for attracting good designers, and motivating them to invest their work with significance and meaning. Staffing policies are determined by different objectives. These include the position principals wish to attain in the status hierarchy of the profession, the scale of projects on which they work, the range of services the firm provides, and the importance principals assign to maintaining continuity and reducing employee turnover. One might assume that firms celebrated for the design quality of their work would be good situations for architects with artistic talent, who would work happily and effectively in these settings. But it often turns out that such firms are dreadful employers for people who wish to exercise these skills, because the principals make all the interesting and important design decisions, while the majority of the staff is relegated to drawing up plans and details. Sometimes gifted designers do better in a firm with a strong commercial orientation, where their skills are more exceptional and therefore more highly valued. Firms that do small projects, including suburban housing developments, office buildings, and space planning, are frequently more stimulating environments for talented designers. The projects are simple enough so that principals are content to turn over major responsibility to junior staff. These offices may reproduce the atmosphere of old-fashioned ateliers, in which the principals serve as critics of schemes developed by young designers. However, because these offices are managed haphazardly, they often have a short life. If an atelier-type firm does survive and grow, it is usually as a result of having installed bureaucratic management techniques. However, as I have said, this is exactly the setting some of the best design architects abhor, and they then quit. In a comprehensive service firm, with its many projects concentrated on pragmatic issues, the morale of the designers may be poor, but architects proficient in technical skills may fare rather well. Principals usually allow employees more responsibility and authority than when design is the issue, because so many answers have become standardized and are routine. If the firm is large and successful, there may be opportunities for internal job mobility. We should not overlook the many architects who choose to work in technical areas for just this reason, even though their jobs do not receive the recognition from the profession or the public that is accorded to design tasks.

Various career paths and types of practice have been advocated as solutions to the motivational and organizational problems of principals and employed architects. There is no one method that can handle the diverse requirements of an increasingly complex, competitive, and fragmented profession with equal effectiveness. The combination of diversity and fragmentation are major factors that help to explain why architecture is populated by a higher proportion of alienated and disappointed men and women than any other major profession, why so many firms are badly managed, and why when offices are managed efficiently, they achieve work of dubious architectural quality. Of all the challenges facing the profession, the problems of motivating architects and sustaining office morale and performance may be the most difficult to address.

It would be nice to believe that the obstacles can be dealt with by a little tinkering: to find a more experienced marketing consultant, to separate the drafting room from the business office, to return to the model of the one-person or the two-partner office, to hire a new manager, to formulate an improved personnel plan, to recruit graduates from another school of architecture, or to acquire clients who are more appreciative and behave like patrons.[11] Each of these strategies has been tried by some firm and by individual architects. There are cases in which the adoption of such strategies has led to jobs which are more fulfilling and has improved the quality of work and the prospects of particular offices. But it is very unlikely that any one of these strategies can have a lasting impact, unless it is combined with significant shifts in career objectives, the patterns of office organization, or the backgrounds and experiences of the staff.

The problems of managing an architectural practice run too deep to be influenced by minor events and simple adjustments in the conduct of a career or a firm. Like the other challenges identified in this chapter they are rooted in the conditions discussed throughout the book. They are challenges that have arisen from great institutional changes that have occurred precipitously and intensively over the last fifty years, in the building industry, among different levels of government, in the corporate sector, and in environmental experiences of the entire American population. These changes have transformed the system of building production, the methods through which

---

[11]The suggestion about maintaining distance between drafting room (an older term for the place where design work is done) and the office is made by Morris Lapidus, *Architecture: A Profession and a Business*, p. 2.

clients choose architects, the roles assigned to architects in the building process, and the standards according to which the merits of buildings are judged and architects and their firms evaluated. To deal with the challenges in coming decades, the adoption of ingenious management techniques by individual offices or the use of clever public relations programs by the architectural community is unlikely to prove sufficient. Intensive research, thought, and policy initiatives focussing on these challenges are needed. To achieve these initiatives, the best minds and talents of the profession must be mobilized. But not only of the profession itself. Architecture is too important to the quality of American life for us to assume that the knowledge of architects working alone is adequate to address it. The issues and conditions this book has described demonstrate that architecture and building touch the interests of building owners, users, and the public at large. The production of architecture is achieved through the participation of clients, users, builders, manufacturers, other design professions, government officials, and financiers. These groups must be involved in the investigation along with the architectural community. Only if joint programs with this scope and on this scale are undertaken, is the profession likely to formulate persuasive policies that will assure the independence of architectural practice in future years.

## BIBLIOGRAPHY—BOOKS AND ARTICLES

Appleson, Gail. "Boutique Firms Hold Their Own." *National Law Journal,* March 7, 1983, pp. 1, 22, and 23.

"Can America Compete?" *Business Week,* April 20, 1987, pp. 45–69.

Carr-Saunders, A. M., and P. A. Wilson. *The Professions.* London: Oxford University Press, 1933.

Freeman, Richard B., and James L. Medoff. "New Estimates of Private Sector Unionism in the United States." *Industrial and Labor Relations Review,* 32 (1979), pp. 143–74.

Freidson, Eliot. *Professional Powers.* Chicago: University of Chicago Press, 1986.

Lapidus, Morris. *Architecture: A Profession and a Business.* New York: Reinhold Publishing, 1967.

Larson, Magali Sarfatti. *The Rise of Professionalism: A Sociological Analysis.* Berkeley: University of California Press, 1974.

McCue, Gerald M., and William R. Ewald Jr. *Creating the Human Environment.* Urbana: University of Illinois Press, 1970.

Piore, Michael, and Charles Sabel. *The Second Industrial Divide: Possibilities for Prosperity.* New York: Basic Books, 1984.

Schuman, Tony. "Professionalism and Social Goals of Architects, 1930–1980." In ed. Paul L. Knox. *The Design Professions and the Built Environment,* Beckenham, England: Croom Helm, 1987.

Society for Marketing Professional Services. *Distinguished Paper Series 1986.*

选读 34
"建筑设计中的一些转变以及它们对设计实践与管理的启示"
大卫·S. 哈维兰

哈维兰在这篇文章中调查了正在影响美国建筑实践和建设业的因素。他讲述了在塑造建成环境中影响建筑师角色转变的五类变化：从新建向重建和修复的转变；从设计作为个人追求向公共追求的转变；从设计与建造为先后过程向平行过程的转变；设计控制权从建筑师主导向专业销售商和制造商为主导的转变以及从建筑过程整体推进向各独立领域分担设计责任的转变。哈维兰用传统与当下建筑实践的定义来考量和分析这些因素，得出了对行业未来发展的重要启示。

READING 34

# Some Shifts in Building Design and Their Implications for Design Practices and Management

David S. Haviland

Haviland investigates in this article the factors that are affecting architectural practice and the construction industry in the United States. He describes five changes that impact the role of the architect in shaping the built environment: the shift from new building construction to reconstruction and restoration; the view of design as a communal rather than an individual pursuit; the collapse of design and construction into a parallel rather than a sequential process; the shift of design control from the architect to specialty vendors and manufacturers; and the disintegration of the architectural process into separate areas of design responsibilities. Haviland analyzes these factors in terms of traditional and contemporary definitions of architectural practice and draws important implications for the profession in the future.

## FIVE KEY SHIFTS

Change is constant. Some changes, however, have broad or deep implications. This paper looks at five specific changes in American building practice which are creating substantial pressures on design, designers, and design management.

*Source:* "Some Shifts in Building Design and Their Implications for Design Practices and Management" by David S. Haviland. Copyright © 1996 Locke Science Publishing Co., Inc. Reprinted with permission from the *Journal of Architectural and Planning Research*, Locke Science Publishing Co., Inc., 117 West Harrison, Suite 640-L221, Chicago, IL 60605. This article originally appeared in Volume 13, Number 1, pp. 50–62.

**Construction Is More Frequently Reconstruction**

As we move toward the next century, the majority of domestic American building activity is being directed to reconstructing existing facilities—refitting, renovating, rehabilitating, and reusing existing buildings and places to keep them vital and productive.

A fraction of U.S. nonresidential construction, building improvements, maintenance, and repair has been growing steadily, reaching over $110 billion and approximately one half of the 1991 U.S. nonresidential construction market.[1] According to a special 1986 Census Bureau study of nonresidential improvements and upkeep, reconstruction is especially prevalent—as much as 75 to 80 percent of total construction activity—in the institutional facilities sector that has long been a mainstay of architectural practice.[2] While these figures may be influenced by recession, it is safe to say that reconstruction has become and will remain a very substantial share of American building.

A shift to reconstruction influences a number of design and building practices. First, much reconstruction is, in fact, "refitting" a building's exterior cladding, interior partitioning, caseworks, mechanical and electrical subsystems. It has become necessary or at least desirable to refit buildings to accommodate changes in work and building uses as well as changes in technology and the marketplace. Continuous refitting has become a way of life in shopping facilities, restaurants, hotels, offices, and a variety of other buildings. As a result of the emphasis on refitting, we are seeing a growing division between building fit-up and shell (structure, roof, and civil services) subsystems, intensive innovation in some fit-up systems (as an example, workstations and personal environments), a tendency to defer building fit-up until the last moment to respond to changes in needs, to take advantage of innovation and the growing influence of interior and services designers.[3]

Reconstruction projects tend to be smaller than new construction projects. They are technically complicated, involve existing conditions, and are often pursued as programs of work (for example, as broad scale re-imaging, asbestos removal, accessibility for the disabled, or lighting retrofit programs). By their nature these projects are intimately involved with building users and occupants, involve complicated logistics, and require broad management skills. Often these projects are conceived and executed within larger facilities management programs.

**Design Is Communally Negotiated by Several Powerful Stakeholders**

Design has become an important aspect of American culture. There is ample evidence to suggest that individuals, organizations, and communities care about design.[4] At the same time resources and environmental concerns have heightened concerns about building. As a result, we have seen the emergence of four sets of powerful stakeholders: owners, financiers, occupants, and the citizenry at large. Each demands a seat at the table and each is prepared to exert muscle.

The growth of the strong owner as a key force in contemporary building has been well documented.[5] Project owners are now more prepared than ever to assert themselves as the primary beneficiaries of the design and building process. Increasingly, institutional owners are viewing facilities as portfolio assets which embody both costs and benefits and which must be managed strategically.[6] The shift to reconstruction has placed even smaller owners in the constant (re)building business, and they are developing or accessing the expertise to gain value for their investments in this activity.

Following the well-documented excesses of the 1980s, the owner of the 1990s is expected to be extremely careful in making the decision to build and be much more committed to retaining and managing the resulting facilities. These owners may have substantial facilities management expertise; dictate the terms of engagement for design professionals; set high expectations for design services; and assume a partner's role in design and construction. These owners may be as sophisticated as the professionals and contractors they engage, and even more knowledgeable on matters of special interest. Likewise, these owners may establish their own discipline for advancing the work and use their muscle to resolve differences and problems.[7]

Turning to financiers, we have always understood that those who supply capital for design and construction are key actors. Emerging from the easy money and ready incentives of the 1980s, American building is relearning this lesson. Reacting to overstimulation and overbuilding, traditional sources of debt and equity capital are now absorbing losses rather than making capital available. Pressured by a recessionary economy and structural problems, American businesses and institutions are not in a position to provide substantial internal financing. Competition for capital is severe and many otherwise qualified projects lie in wait. Moreover, all indications point to an era of very carefully considered and highly conservative facilities financing.

While owner and financier influences have been felt for some time, now we also see the rising influence of the public—as building users and as citizens at large. Fueled by growing societal interest in human capital and productivity as well as more specific concerns for occupational health and safety, building occupants and users are demanding seats at the table of project planning and design. The Americans with Disabilities Act, now being implemented, further raises the ante. As citizens at large, the public is increasingly heard in community zoning, planning, and design review processes.

Within this arena of new and often competing stakeholders, design becomes a broadly discussed and negotiated activity.[8] Project plans are described and debated in planning and zoning hearings, in complex financial negotiations, community forums, and in the media. As localities become even more strapped for cash, new projects are seen as resources for community development.[9] Owners and designers are required to make accommodations and compromises to secure approvals—often in the glare of public meetings or in the heat of bargaining sessions in hearing rooms, cloakrooms, and corridors.

## Design and Construction Are Proceeding as Parallel Enterprises

One of the consequences of protracted negotiation among powerful stakeholders is an extended and often difficult "birth" process for many projects. Months and even years are consumed in making plans, gaining organizational buy-ins, testing feasibility, securing buildable sites, committing capital, and navigating all this through the shoals of community land use and design review.

When finally the factors are assembled and approved, most owners are under great pressure to move their projects to completion as quickly as possible. The motivations for speedy action are many: owners may wish to take advantage of agreements among the stakeholders before they disintegrate. Owners may desire to reduce further interim financing costs, take advantage of market conditions in the construction industry, or simply gain the intended advantages of new or reconstructed facilities as soon as possible.

The design team may be carefully designed for success in this negotiation process. The owner may assemble combinations of firms with the array of talents needed to move the project to realization. This assembly process may add power to the project team in the form of a signature designer, a high-powered market or financial consultant, a firm with local political clout, a construction manager, or others with special skills. Design itself may be separated into concept and detail stages, with responsibility for the latter deferred and perhaps even shifted to other designers, vendors, purchasers, tenants, or even the owner himself.

In these circumstances, construction may begin before design is fully developed. This usually takes the form of overlapped or "fast-track" design and construction. In order to limit their financial risk, many owners seek guaranteed maximum price contracts from contractors or construction managers based on partially developed design and documents. Design may be divided into packages whose content and timing may follow market and procurement logic rather than design logic. Once packaging and scheduling decisions are made, design necessarily marches to its tune. Architects and engineers may produce dozens or even hundreds of sets of contract documents. A project or construction manager may be added to coordinate design and construction activities and to enforce schedules and budgets.

### Substantial Design Responsibility Is Shifting to Vendors

The twin pressures to get an early construction start and to defer building fit-up to a later date are not only spreading out the design process but also re-allocating design responsibility from project architects and engineers to suppliers and contractors.

As suggested in Figure 34-1, the dividing line between owner or consultant design on the one hand and contractor or vendor design on the other may occur at a number of different places. A common ingredient in all of these is substantial allocation of design responsibility to organizations who are essentially vendors:

- Even where both conceptual and detailed design are accomplished by a licensed design professional (the first option), the American approach to building shifts a good deal of detailed component and assembly design to manufacturers, suppliers, and specialist trade contractors through the mechanisms of required shop drawings, samples, and product submissions.[10]
- The three middle options move substantial design responsibility to contractors and suppliers. These may introduce considerable ambiguity into the procurement process as contractors commit to cost and time on incomplete design information. There may be considerable difference of opinion as to the intent of "scope" or "performance" documents.
- The final option results in full delegation of design to a vendor—a manufacturer, a design/build entity, or an organization that provides facilities on a build-operate-transfer basis. Design/build contracting is on the increase in the United Kingdom and appears to be so in the U.S.[11] This approach provides an owner with early construction cost and time commitments and with single-point responsibility for design

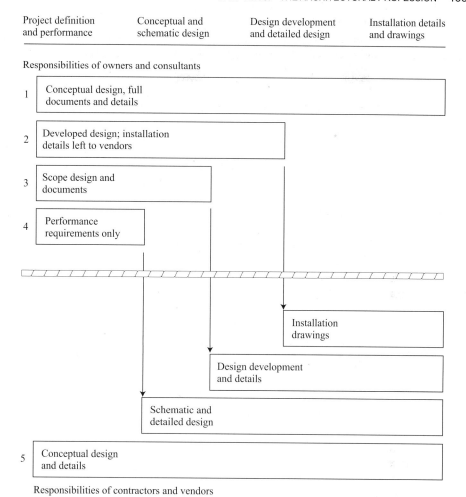

**FIGURE 34–1** Options for allocating design responsibility. [Adapted from *Faster Building for Commerce* (London: National Economic Office, 1988).]

and construction. The designer works within an organization committed to and involved with building. These organizations can develop the design and communications disciplines necessary for effective fast-track building; they can also institutionalize and apply learning curves from one project to the next. Design/build companies may further integrate vertically by offering site selection and land acquisition services, financing, and even ongoing facilities management services. Some are prepared to own and operate these facilities, transferring them to the owner after amortizing their capital investments and earning agreed-upon operating fees.

There are, of course, reasons for allocating design responsibility to vendors. These companies understand their products in detail. This has become especially so with building fit-up systems where there has been a good deal of innovation. It is difficult for design professionals to keep up and, in competitive bidding situations, it may be difficult for them to gain access to these vendors during design. This situation continues through construction where vendors typically have no contractual access to the project designers or, for that matter, to each other as they attempt to resolve interface problems involving their products.

### Design Is Becoming Disintegrated Even as It Becomes More Pervasive

It has always been convenient to view building design as a coherent and bounded activity that starts at a specific moment—traditionally after the owner has completed a brief, established a schedule and budget, and selected a designer—and ends with occupancy of the building. As convenient as this view of design may be, it is no longer entirely appropriate.

As suggested here, design often "begins" deep within a program of ongoing strategic planning and facilities management. It is nurtured and given shape in complex negotiations among powerful stakeholders, only some of whom may be design professionals or have construction expertise. It is hardened in the crucible of financial and regulatory review. It may be apportioned among partners on a design team, divided among design professionals and others who may be vendors, or assigned to vertically-integrated organizations which may also provide land acquisition, financing, construction, and facilities management expertise. Moreover, design may continue long into a building's life. Fit-up design may not occur until users or tenants are identified. Similarly, design decisions may be deferred as long as possible to accommodate rapidly developing technologies. The process of refitting a facility to meet changes in need as well as new strategic opportunities continues through its lifetime.

The net result of these forces is that even as design becomes important and pervasive, it also becomes fragmented and disintegrated.

## PRESSURE ON SOME OF OUR CENTRAL IDEAS ABOUT PRACTICE

These shifts in building practice are placing substantial pressure on some of our most clearly understood and widely cherished ideas of architectural practice, ideas that have served as cornerstones for both practice and teaching about practice.[12] This section examines pressures on five such ideas—design as something done principally by professionally-based designers; the architecture firm as the principal venue for design; the project as the building block of practice; design as a process that can be regulated by standard forms of agreement; and the professional standard of reasonable care—and extracts implications for contemporary design management.

### Design as Something Done Principally by Profession-Based Designers

Even as design has become more important and pervasive, it is done increasingly by people and groups which are not based in the design professions. Owners, occupants, financiers, regulators, and the citizenry at large have substantial roles in establishing design parameters and negotiating design concepts while manufacturers, suppliers, fabricators, and installers are increasingly responsible for design details and interfaces. As reconstructing and refitting existing buildings becomes more prominent, design development and documentation may shift to interior designers and mechanical contractors.

In these circumstances, design management spans the entire facilities cycle and involves coordinating and reconciling many diverse and competing issues, organizations, and ways of operating. Responsibility may inure to people with general management skills who know little about buildings. Emphasis shifts to the role of information systems as mechanisms for "tying together" a diverse and diffuse planning, design, construction, and facilities management process.

Similarly, design management involves adjudicating among many diverse and competing positions and values. While this is hardly new, architects and engineers can no longer assume that their values, long considered to be central to building, will prevail. In fact, those responsible for design may not feel the traditional professional's responsibility for the larger good.

Ultimately this phenomenon raises questions of identity for the design professions themselves. Should they expand their boundaries to be more inclusive of people actually doing what they have staked out as their work? How long can they do this without losing their rights to exist as professions apart from the laity? Or without losing the identity that binds a group of like-minded people together in common enterprise?[13] The current struggles in some states to demarcate architecture and engineering and nationally to define a profession of interior design can be cited as skirmishes in this larger war.

## The Architecture Firm as the Principal Venue for Design

As design and design responsibility spread, it follows that the independent private architecture, engineering, or combined A/E practice is no longer the only place where design is done.

Many such practices, of course, have "repositioned" themselves by offering a broader range of design services, especially those that can be considered as predesign and post-construction. Others have "partnered" with construction companies, transformed themselves into design/build entities, opened software development subsidiaries, or made similar moves to meet the marketplace.

Other steps are being taken by firms and organizations outside of professional practice. General contractors are transforming themselves into project or construction managers. Design/build enterprises are growing in influence. Owners are establishing substantial facilities design, construction, and management organizations, or, in some cases, "outsourcing" these services to private firms under long-term vendor agreements.

On the global scale, we are seeing the emergence of "giants" who have the capacity to design, construct, finance, and even sometimes manage the resulting facilities. As an example, a Japanese "big five" contractor in 1990 received building contract awards between $8.7 and $11.9 billion, employed between 9,900 and 14,900 staff, did 35% to 55% of its business on a design/build basis, and spent between $55 and $100 million on research and development. As the single European market comes into place, already large design and construction firms in Europe are both growing (through acquisition of smaller and even mid-size companies) and vertically integrating in order to become major actors on an enlarged playing field.[14]

These new and important design venues raise important issues for design management. An increasing fraction of designers are not in traditional agency relationships with project owners and, in fact, may not have access to them at all. Designers find themselves working in firm and organizational cultures that are not of their making. Design managers find themselves interacting with a larger set of corporate values, objectives, and operating styles. On the other hand, larger and more vertically-integrated organizations offer the possibility of closer collaboration between designers and those who plan and manage buildings on the one hand and those who construct them on the other. Converting this opportunity into reality presents yet another challenge to design management.

### The Project as the Building Block of Design Practice

Our tendency is to consider the individual project as the building block of practice. We see excellence as achievable on a project-by-project basis and practice as a body of projects.[15] Increasingly, however, design organizations find themselves conceiving, executing, and managing "programs" of projects.

Program and project management share many common ingredients, but they are not the same. The programmatic imperative brings a larger set of issues and probably a wider range of actors to the table. Program and project goals are interactive and reciprocal: programs parametrize their projects and, at the same time, a program is implemented one project at a time. These two sets of goals and structures exist within a third—those of the design organization, unit, or firm—requiring attention and adjudication at three levels.

For independent design firms, providing program services shifts the focus from acquiring projects—traditionally the grist of the marketing mill—to acquiring and especially retaining clients. Client management becomes a discernible objective and a set of skills; those who provide project services become the keys to maintaining and retaining clients. These firms may seek long-term agreements to provide program services, perhaps from institutions and corporations who are "outsourcing" these services, thus blurring the distinctions between designer and client, vendor and purchaser.

At the level of daily project activity, we are not well prepared to manage large numbers of small, complicated projects. Most of our design and construction project management theory and guidance is based originally on study and experimentation in "large systems" such as the space program and major defense projects. Only now are there efforts to look more carefully at small projects and design practices that are founded on the basis of doing lots of small projects all the time.[16]

### Design as Contractually Regulated by Standard Forms of Agreement

Building requires collaboration among many different entities who bring their individual talents, motivations, styles, and biases to a temporary multi-organization created to execute a project or a program of projects. Given the complexity and the integrated nature of the work to be done, the American building enterprise has relied on a series of standard forms of agreement to spell out, in contractual terms, a comprehensive pattern of responsibilities, relationships, risks, and rewards among the principal parties to design and construction.

Standard forms of agreement such as those promulgated by The American Institute of Architects and the Engineers Joint Contract Documents Committee traditionally have brought a level of industry-wide conception and negotiation of responsibilities for building in general as well as consistency and currency to individual projects. As they have enjoyed broad use, these forms of agreement and the practices they represent have become an integral part of the processes used to decide disputes, allocate liability, and insure risks accepted by the various parties involved.

The shifts described earlier are producing building procurement processes that are more diverse and more likely to entail a specific allocation of planning, design, and construction responsibilities suited to the project and its circumstances. At the same time, standard forms have had difficulty digesting changes in building practice, for example, projects intended to scope and define projects, the use of alternative dispute resolution mechanisms, new partnering arrangements, and the allocation of design responsibilities among owners, design professionals, and vendors.[17]

In response, we see a growing use of letters of intent, owner-developed contracts, and other project-specific forms. At the same time, there are efforts to make long-standing standard forms of agreement more flexible and adaptable to project circumstances.[18] Both of these tendencies raise important implications for design management.

Those responsible for structuring projects now have the added responsibility of designing specific procurement processes to suit the needs and circumstances of each project or program of projects. In the absence of a useful base of research in this area, these decisions are made largely on the basis of individual experience, bias, and salesmanship efforts by those providing design and construction services.[19]

In this environment, owners and financiers can be expected to play strong roles in establishing and dictating terms of engagement. At a general level, this can produce agreement forms that do not recognize the complexities of building.[20] At the project level, managers spend more time attempting to assess the implications of new language and requirements.

Customizing procurement approaches requires project and design managers to reason through expectations and capabilities among the parties at hand and also to assess the risk/reward consequences of various allocations of responsibility and liability.[21] This requires analytical capabilities on the one hand and a complex of interpersonal, negotiation, and communication skills on the other. These capabilities and skills lie quite outside the group of technical skills required to "do" project design and construction.

### The Professional Standard of Reasonable Care

Rising owner and public expectations of design coupled with strong marketing by architects and engineers ("just put your project in our capable hands . . . there will be no problems") is subtly repositioning design and designers: Design may be seen as commodity and designers as purveyors of that commodity. At the same time, there is a shift of design responsibility to design/builders, contractors, and suppliers—entities that are considered by the buying public and also the courts as vendors.

Traditionally, American law draws a clear distinction between the standards of performance expected of vendors and professionals. As suppliers to a consuming public, vendors are expected to provide products that are fit for their purpose; this is a standard of strict liability (if it leaks and it isn't supposed to, it must be redressed). Assuming that they do not establish contracts that set a higher standard, professionals are expected to perform (only) with reasonable care and skill (if it leaks, did the architect act as a reasonably prudent architect would do in the same community, in the same time frame, and given the same or similar facts and circumstances?). This standard of reasonable care is central to the contemporary definition of a profession; it is one of the key rights accorded to a body of professionals by the laity in recognition of the inexact nature of professional practice and a profession's obligation to act for the public good.[22]

Recalling that Anglo-American law reflects and ultimately, through its system of case-by-case decisions and precedents, incorporates principal cultural values and practices, we must assume that the standard of reasonable care will continue to be re-examined and redefined in current terms.[23] Given that such standards are evaluated in the crucible of actual experience, designers and design managers assume no little responsibility in developing and maintaining the standards that the courts will ultimately apply to their performance.

At the societal level, we can expect continuing thrusting and parrying on these questions. In the UK, for example, the growth of design/build contracting has sparked substantial discourse among construction lawyers on standards of professional performance.[24] Closer to home, the State Education Department which regulates professional conduct in New York State has recently notified architects and engineers that they may be violating the law by obtaining design details for contractors who fabricate and install structural steel.[25]

Even as owners demand perfection or certainty, designers are challenged to assert the role and value of professional judgment in their work. For some firms, this suggests the need to reposition their marketing efforts—always difficult in recessionary times—and at the same time to work toward a practice of values which include both service and reflection.[26]

## LOOKING FORWARD

Some of the current changes in building practice are producing shifts in what we design, where design occurs, who is involved, what help we can expect from professional institutions, and the standards by which designers' performance is to be judged. These shifts are substantial and many of them are structural. Individually and collectively, they create new conditions for design management. They also create new challenges for those who seek to undertake these responsibilities and, especially, for those who seek to educate, train, and support these people in their daily work.

## NOTES

1 The Construction outlook for 1991 in *Construction Review* (1990) places the forecasted 1991 total at $62.8 billion in nonresidential building improvements construction and $35.6 billion in nonresidential buildings maintenance and repairs, all in 1987 dollars. Moreover, the 1990 to 1995 forecast projected a 2% growth in nonresidential reconstruction and a 2% reduction in new nonresidential construction.

2 U.S. Department of Commerce, Bureau of the Census, "Expenditures for Nonresidential Improvements and Upkeep, 1986," issued March 1989.

3 It is instructive to note that, according to "Means Square Foot Costs," R. S. Means Company, Inc., Kingston, MA, 1991 edition, the combined costs of the exterior clo-

sure, interior partitioning, mechanical and electrical subsystems for a typical range of nonresidential building types range from 45% to 65% of the total initial construction cost. For multifamily housing and nonresidential housekeeping facilities, this range is from 55%–65%. When you consider that a substantial proportion of building reconstruction involves upgrading or replacing these subsystems, they account for a dominant segment of the design and construction market.

4 See, for example, the recent work of Colin Clipson, "First Things First," *Journal of Product Innovation Management,* no. 7 (1990), and "Contradictions and Challenges in Managing Design," a paper presented at the Symposium on Emerging Forms of Architectural Practice, University of Cincinnati, April 19–21, 1991.

5 Robert Gutman addresses this issue in a number of his writings including *Architectural Practice: A Critical View* (Princeton Architectural Press, 1988). In the UK, Donald Bishop cogently describes the contemporary building owner and its "search for certainty" in *Professional Liability: Report of the Study Teams* (London: HMSO, 1989).

6 See, for example, Hugh Nourse, *Managerial Real Estate* (Englewood Cliffs, NJ: Prentice-Hall, 1990). The *Times* of London, September 26, 1990, reported that 38% of the top 50 companies in the UK had property officers with positions on the board of directors or reporting directly to the chairman. The same article quotes *Corporate Real Estate* as indicating that 27% of the property executives in the 350 leading U.S. companies reported directly to the CEO and that 44% held positions as vice presidents or above.

7 Bishop, *Professional Liability,* notes that these owners tend to value a completed project over a successful lawsuit and are prepared to use their muscle to resolve disputes and move the work forward. In the U.S. we see growing use of mediation, conciliation, and related alternate dispute resolution techniques in design and construction; see, for example, "Quiet Revolution Brews for Settling Disputes," *Engineering News-Record,* August 26, 1991, pp. 21–23.

8 Dana Cuff describes this negotiation and provides some telling vignettes in *Architecture: The Story of Practice* (Cambridge, MA: MIT Press, 1991).

9 According to a 1990 study reported in Mike E. Miles, Emil E. Malizia, Marc A. Weiss, Gayle L. Berens, and Ginger Travis, *Real Estate Development Principles and Process* (Washington: The Urban Land Institute, 1991), average national impact fees charged by localities ranged from $968 per 1,000 square feet of industrial space to $3,321 per 1,000 square feet of retail space.

10 A careful study of this practice in the UK is reported in Colin Grey and Roger Flanagan, *The Changing Role of Specialist and Trade Contractors* (Ascot: The Chartered Institution of Building, 1989). From comparative studies done by Flanagan and his colleagues, it appears that this phenomenon is even more widespread in the U.S. See Roger Flanagan, George Norman, Vernon Ireland, and Richard Ormerod, *A Fresh Look at the U.K. and U.S. Building Industries* (London: Building Employers Confederation, 1986).

11 In surveys of contract forms in use in the UK, the Royal Institute of Chartered Quantity Surveyors reports that design-and-build contracts represented 12.2% of construction contract value in 1989, up from 5.1% in 1984. The RICS indicates that these figures include only contracts involving a quantity surveyor and that they are underestimates. In its August 12, 1991, issue, p. 9, *Engineering News-Record* noted that design/build contracts account for "less than 5% of all construction in the U.S. but owner demand is driving its use upward." *ENR*'s top 400 contractors more than doubled their design/build billings from $18 billion in 1987 to $37 billion in 1990.

12 For example, consider some of the central concepts embodied in David Haviland, ed., *The Architect's Handbook of Professional Practice,* 11th ed. (Washington, DC: The AIA Press, 1988). These themes also recurred in faculty submissions considered for

inclusion in the *Instructor's Guide* developed for use with the *Handbook*.

13. Gutman explores this question quite thoroughly in *Architectural Practice: A Critical View.*

14. A careful analysis of this phenomenon is being prepared for a forthcoming book on global construction by Roger Flanagan, David Haviland, and Yo Hisatomi. The statistics describing the Japanese "big five" companies (Shimizu, Taisei, Kajima, Takenaka, and Obayshi) are provided by Hisatomi. For a progress report on transcontinental alliances in Europe, see "Builders Take Their Partners for Europe," *The Economist,* July 14, 1990.

15. See, for example, Dana Cuff, "The Origins of Excellent Buildings," *In Search of Design Excellence* (Washington, DC: The AIA Press, 1989), pp. 77–87. Several of the other papers in this volume address this issue as well. Cuff's case studies, along with a larger framework for considering them, are also presented in her book, *Architecture: The Story of Practice.*

16. A review of the writings in *Project Management Journal,* the conference proceedings of the Project Management Institute and even *Construction Management and Economics* reveal a large project focus. Sounding as "another voice" on this issue, James R. Franklin, resident fellow at The American Institute of Architects, has sparked a program of workshops and writings on small projects and practices. See, for example, "Thinking Small," *Architecture,* July 1990, pp. 107–9, and *Current Practices in Small Firm Management* (Washington, DC: The AIA Press, 1990).

17. Using the AIA documents as examples, the most commonly used form of owner-architect agreement (AIA Document B141) starts with schematic design services and ends with completion of construction; the AIA has found it difficult to get architects to use the more broadly-based designated services approach embodied in Documents B161 and B162. On the other issues noted, the AIA documents incorporate only arbitration as a dispute resolution mechanism. The standard form of joint venture agreement was last revised in 1979. There are no efforts to establish linkages between project designers and specific specialist trade contractors (as is reflected in the "nominated subcontractor" variations in the UK's Joint Contracts Tribunal forms of agreement).

18. At this writing, the AIA is developing an electronic documents program that should allow more flexible use of its standard forms.

19. As a step in the right direction, see Rashid Mohsini and Colin Davidson, "Determinants of Performance in the Traditional Building Process," *Construction Management and Economics* (forthcoming); and Rashid Mohsini and A.F. Botros, "Pascon: An Expert System to Evaluate Alternate Project Procurement Processes," *CIB 90, Building Economics and Construction Management,* Vol. 2: *Design Economics/Expert Systems* (Sydney, 1990), pp. 525–37.

20. As an example we can cite the ill-fated model design and construction documents proposed by the National Association of Attorneys-General in 1988. Distribution of these documents was suspended in December 1989 after extensive criticism by both design professionals and contractors. For a summary of the problems with these documents, see Victor O. Schinnerer and Company, "Guidelines for Improving Practice," General Information #51, 1989.

21. For an especially cogent approach to this see Ava J. Abramowitz, "Risk Management," in *The Architect's Handbook of Professional Practice,* 11th ed., ed. David Haviland (Washington, DC: The AIA Press, 1988).

22. Adherence to a standard of reasonable care and the use of "other reasonably prudent architects" as arbiters of that standard in given circumstances can be seen as an example of a profession's right—and obligation—to develop and enforce its own standards. See Dana Cuff, "The Architectural Profession," in Haviland, *The Architect's Hand-*

*book.*

23  As an example, consider the following comment in a 1978 Minnesota opinion, made after a ringing endorsement of architecture as an "inexact science" and the importance of applying a standard of reasonable care to architects' performance: "We have re-examined our case law on the subject of professional services and are not persuaded that the time 'has yet arrived' for the abrogation of the traditional rule." *City of Moundsview v. Walijarvi,* Supreme Court of Minnesota, 1978, 263 N.W.2d 420 [emphasis supplied].

24  For example, see David L. Cornes, *Design Liability in the Construction Industry.* 2nd ed. (London: Collins, 1985).

25  *Engineering News-Record,* October 14, 1991, p. 10. The "shop drawings" issue was a major aspect of the intense national debate over the American Society of Civil Engineers' Quality in the constructed project manual issued in November 1990.

26  For a description of such practice, see Donald A. Schön, *The Reflective Practitioner: How Professionals Think in Action* (New York: Basic Books, 1983). Schön begins by examining contemporary indictment of the professions by the public, citing the professions' own positioning as "applied scientists" as a significant contributor to lack of public confidence in their work.

## REFERENCES

American Institute of Architects. *Standard Form of Agreement between Owner and Architect,* 1987 edition, and other standard forms of agreement. Washington, DC: The American Institute of Architects, 1987.

Clipson, C. "Contradictions and Challenges in Managing Design." Paper presented at the Symposium on Emerging Forms of Architectural Practice, University of Cincinnati, April 19–21, 1991.

Cornes, D. L. *Design Liability in the Construction Industry.* 2nd ed. London: Collins, 1985.

Cuff, D. *Architecture: The Story of Practice.* Cambridge, MA: MIT Press, 1991.

*Faster Building for Commerce.* London: Her Majesty's Stationery Office, 1988.

Franklin, J. R. *Current Practices in Small Firm Management.* Washington, DC: The AIA Press, 1990.

Grey, C, and R. Flanagan. *The Changing Role of Specialist and Trade Contractors.* Ascot, UK: The Chartered Institution of Building, 1989.

*Guidelines for Improving Practice.* Various circulars. Chevy Chase, MD: Victor O. Schinnerer and Company, n.d.

Gutman, R. *Architectural Practice: A Critical View.* Princeton: Princeton Architectural Press, 1988.

Haviland, D., ed. *The Architect's Handbook of Professional Practice.* Washington, DC: The AIA Press, 1988.

Miles, M. E.; E. E. Malizia; W. A. Weiss; G. L. Berens; and G. Travis. *Real Estate Development Principles and Process.* Washington, DC: Urban Land Institute, 1991.

*Professional Liability: Report of the Study Teams.* London: Her Majesty's Stationery Office, 1989.

U.S. Department of Commerce, Bureau of the Census. *Expenditures for Nonresidential Improvements and Upkeep 1986.* Washington, DC: Government Printing Office, 1989.

U.S. Department of Commerce. "Construction Outlook for 1991." *Construction Review,* November–December 1990, pp. v–vi.

选读 35

"建成环境评价：概念基础、收益和用途"

沃尔夫冈·F.E. 普赖泽尔

普赖泽尔研究了建筑过程应如何超越传统设计和建设阶段，延伸至囊括对正在使用的建成环境的评估。在这篇文章中，他提出了基于性能的建筑评价理论模型，并且对这一领域从 20 世纪 60 年代至今的演进情况进行了梳理。他概括出不同形式建筑物使用后评价的方法和裨益，并将"建筑性能"的概念认定为评价建成环境的重要基础。

## READING 35

## Built Environment Evaluation: Conceptual Basis, Benefits and Uses

Wolfgang F. E. Preiser

Preiser has researched the ways that the building process can be extended beyond the traditional design and construction phases to include the evaluation of built environments as they are being occupied. In this article he presents a theoretical model of performance-based building evaluations and traces the evolution of this field from the 1960s. He outlines the methods and benefits of various forms of post-occupancy building evaluations and defines the concept of "building performance" as the primary basis of assessing built environments.

## INTRODUCTION

Many parties participate in the use of buildings, including investors, owners, operators, maintenance staff, and perhaps most important of all, the end users, i.e., the actual persons occupying the building. The focus of this paper is on occupants and their needs as they are affected by building performance, and on occupant evaluations of buildings. The term evaluation contains the word "value" and thus, occupant evaluations must state explicitly whose values are referred to in a given case. An evaluation must also state whose values are used as the context within which performance will be tested. A meaningful evaluation focuses on the values behind the goals and objectives of those who wish their buildings to be evaluated, or those who carry out the evaluation. Shown in Figure 35–3 is the performance concept in the building process as a relativistic notion. This is the opposite of the idea of absolute objectivity in evaluations which does not exist, except in the case of physical measurements of building

*Source:* "Built Environment Evaluation: Conceptual Basis, Benefits and Uses" by Wolfgang F. E. Preiser. Copyright © 1994 Locke Science Publishing Co., Inc. All rights reserved. Reprinted with permission from the *Journal of Architectural and Planning Research,* Locke Science Publishing Co., Inc., 117 West Harrison, Suite 640-L221, Chicago, IL 60605. This article originally appeared in Volume 11, Number 2, pp. 92–107.

performance. "Performance criteria" used in evaluations are equated with "evaluation criteria" by this author. They are translated from goals and objectives that in themselves are derived from values held by individuals, groups, organizations, or entire socio-political systems.

This process of translation was described by Brauer and Preiser (1976), through a process of sub-setting whereby a hierarchy is established. Goals and missions of organizations imply functions, as expressed in organizational breakdowns, departments, etc. Functions in turn imply activities, carried out in designed environments. Finally, activities imply behaviors of occupants, who are affected by the quality of the built environment. Frequently, there are differences in values held by various groups or organizational units. Resolving such differences may help in achieving performance that results in a satisfactory building for all who interact with it. The establishment of a clear linkage between values and POE may also lead to a more enlightened, proactive stance by those who manage facilities with a desire to improve the quality of the building stock namely through a better understanding of various aspects of building performance.

There is a difference between the quantitative and qualitative aspects of building performance and the respective performance measures. Many aspects of building performance are in fact quantifiable, such as lighting, acoustics, temperature and humidity, durability of materials, amount and distribution of space, and so on. The evaluation of qualitative aspects of building performance, such as aesthetic beauty or visual compatibility with a building's surroundings, is somewhat more difficult and less reliable. In yet other cases, the expert evaluator will pass judgment. Examples are the expert ratings of scenic and architectural beauty awarded chateaux along the Loire River in France, as listed in the travel guide literature. The higher the apparent architectural quality and interest of a building, the more stars it will receive. Recent advances in the assessment methodology for visual aesthetic quality or scenic attractiveness are encouraging. Someday it is hoped, even this elusive domain will be treated in a more objective and quantifiable manner (Nasar, 1988).

In the following, the topic of post-occupancy evaluation and its theoretical base, the performance concept, are presented.

In the U.S.A., POEs derived their name from the "occupancy permit" which is issued when a building is completed, inspected, and deemed to be safe to occupy in accordance with building codes and regulations.

In the U.S.A., Canada, Australia and New Zealand, several government agencies have established ongoing POE programs in order to evaluate the performance of their facilities. The Department of the Army (1976), for example, created the so-called *Design Guide* series which was based on POEs of some 20 building types that are being built on a recurring basis. POE results are also the basis for technical manuals, criteria literature and standards of government agencies.

There is an increased acceptance of the post-occupancy evaluation concept and process in the private sector. More and more facility managers are conducting POEs upon completion of projects, and some are conducting continuing evaluation programs routinely in order to solicit user feedback on existing buildings. POEs are carried out by various groups, including in-house staff of organizations, design firms, as well as POE consultants.

While some design firms are fearful that POE results may be used against them, and others are unable to convince clients to pay for POEs, POE today constitutes an important contribution in the quest to provide "quality assurance." Furthermore, over the past few years, there has been an increasing concern for liability, e.g., litigation for negligence, budget overruns, building failure, or inappropriate design decisions made during the planning and development of a facility. As a consequence, and as part of the cost of doing business in architecture, liability insurance costs have risen astronomically to the point where they are unaffordable for small design firms.

## THE POE CONTEXT

POE is not the end phase of a building project, but rather, it is an integral part of the entire building delivery process. It is also part of a process in which a POE expert draws on available knowledge, techniques, and instruments in order to predict a building's likely performance over a period of time.

At the most fundamental level, the purpose of a building is to provide shelter for activities that could not be carried out as effectively, if at all, in the natural environment. A building's performance is its ability to accomplish this. POE is the process of the actual evaluation of a building's performance once in use by human occupants.

A POE necessarily takes into account the owners', operators', and occupants' needs, perceptions, and expectations. From this perspective, a building's performance indicates how well it works to satisfy the client organization's goals and objectives as well as the needs of the individuals in that organization. A POE can answer, among others, these questions: Does the facility support or inhibit the ability of the institution to carry out its mission? Are the materials selected safe (at least from a short term perspective) and appropriate to the use of the building? In the case of a new facility, does the building achieve the intent of the program that guided its design?

## THE PROCESS OF POE

Several types of evaluations are made during the planning, programming, design, construction, and occupancy phases of a building project. They are often technical evaluations related to questions about the materials, engineering, or construction of a facility. Examples of these evaluations would include structural tests, reviews of load-bearing elements, soil testing, and mechanical systems performance checks, as well as post-construction evaluation prior to building occupancy.

Technical tests usually evaluate some physical system against relevant engineering or performance criteria. While technical tests indirectly address them by providing a better and safer building, they do not evaluate it from the point of view of occupant needs and goals, or performance and functionality as it relates to occupancy. The client may have a technologically superior building, but it may provide a dysfunctional environment for people.

Other types of evaluations are conducted that address issues related to operations and management of a facility. Examples are energy audits, maintenance and operation reviews, security inspections, and programs which have been developed by professional building managers. While not POEs, these evaluations are relevant to questions similar to those described above.

The process of POE differs from these and technical evaluations in several ways:

- A POE addresses questions related to the needs, activities, and goals of the people and organization using a facility, including maintenance, building operations, and design-related questions. Other tests assess the building and its operation, regardless of its occupants.
- The performance criteria established for POEs are based on the stated design intent and criteria contained in or inferred from a functional program. POE evaluation criteria are not based on merely technical performance specifications.
- Measures used in POEs include indices related to organizational and occupant performance, such as worker satisfaction and productivity, as well as measures of building performance referred to above (e.g., acoustic and lighting levels, adequacy of space and spatial relationships, etc.).
- POEs are usually "softer" than most technical evaluations. POEs often involve assessing psychological needs, attitudes, organizational goals and changes, and human perceptions.
- POEs measure both successes and failures inherent in building performance.

In computing the cost of production over the useful life of an average building, about ninety percent of the costs are employee costs (Wineman, 1986; Thorne, 1980). Providing better buildings to meet the needs of their occupants has a significant payback for sponsoring agencies. A well designed facility can improve organizational and individual performance and can enhance the organizational climate.

## EVOLUTION OF POE

Historically, buildings were evaluated informally, but largely in an unsystematic and subjective manner. Around 1750 B.C., in Mesopotamia, for example, King Hammurabi is reported to have the builders of failing buildings put to their deaths, a rather drastic approach to quality assurance! Only recently, more systematic POEs have been conducted in which objective measures of performance and explicitly stated evaluation criteria were used.

Table 35–1 summarizes the most important conceptual contributions and milestones in the evolution of the field of POE which is of fairly recent origin and has been practiced primarily in English-speaking countries (Preiser, Rabinowitz and White, 1988).

Early efforts of POE in the U.S.A. were heavily biased toward dormitory evaluations because they offered a population of available and cooperative subjects. In the latter half of the 1960s, the first environmental evaluations and profiles were developed. Then, in the 1970s, several multi-method evaluation approaches were used on hospitals, public housing, schools, military facilities, and federal offices. Models of occupant satisfaction focused POEs were devised, design guides and standards were created, and some early POE programs were instituted in government facilities planning programs.

In the 1980s in the private sector, a number of POEs were conducted on occupants' satisfaction with hospital, school, and office design. In addition to advances in environmental design research in general, several contributions in POE practice were made. Some researchers attempted to link objective, environmental features with

**TABLE 35–1  MILESTONE IN THE EVOLUTION OF POE**

| Year | Author(s) | Building type(s) | Contribution to the field |
|---|---|---|---|
| 1967 | Van der Ryn/Silverstein | Student Dormitory | Environmental Analysis Concept and Methods. |
| 1969 | Preiser | Student Dormitories | Environmental Performance Profiles: Correlation of subjective and objective performance measures. |
| 1971 | Field | Hospital | Multi-method approach to data collection. |
| 1972 | Markus, et al. | Any facility type | Cost-Based Building Performance Evaluation Model. |
| 1974 | Becker | Public Housing | Cross-sectional, comparative approach to data collection and analysis. |
| 1979 | Francescato, et al. | Public Housing | Evaluation models of "resident satisfaction," allowing physical managerial intervention. |
| 1975 | General Services Administration | Office Buildings | Office Systems Performance Standards. |
| 1976 | Department of the Army | Military Facilities | Design Guide Series with updatable, state-of-the-art criteria. |
| 1976 | Rabinowitz | Elementary Schools | Comprehensive, full-scale evaluation on technical, functional, and behavioral factors. |
| 1979 | Public Works Canada | Government Facilities | POE incorporated into Project Delivery System. |
| 1980–81 | Daish, et al. | Military Facilities | POE as routine staff activity in Government Building Process. |
| 1981 | Marans and Spreckelmeyer | Offices | Evaluation model linking perceptual and objective attributes. |
| 1982 | Parshall and Pena | Any Facility Type | Simplified and standardized evaluation methodology. |
| 1983 | Orbit 1 | Offices | Office research linking buildings and information technology. |
| 1984 | Brill, et al. | Offices | Linking Worker Productivity and office design. |
| 1985 | White | Any Facility Type | Linking programming and POE in graduate architectural education. |
| 1986 | Kantrowitz, et al. | Architecture School | POE analysis of entire building process and documentation. |
| 1986 | Preiser, Pugh | Any Facility Type | POE Process Model and Levels of Effort. |
| 1989 | Vischer | Office Buildings | Office Environment Evaluation. |
| 1991 | Petronis, Preiser, et al. | VA Hospitals | Activation Process Analysis and Guide for VA Hospitals. |

subjective ratings and perceptions. Marans (1988) reported that many studies have found objective and subjective data to be poorly correlated.

Over the years, the conceptual base for POE has grown. Much of the work has focused on the built environment's effects on the perceptions and behavior of workers (Marans and Spreckelmeyer, 1981; Goodrich, 1976).

The influence of space on office productivity (Brill et al., 1984), and the relationship of POE results to building programming (Parshall and Pena, 1982) have been addressed. Preiser (1983) and Preiser, Rabinowitz and White (1988) described a POE process model, which is based on the performance concept and the evaluation research framework. The Department of Public Works in Canada (1979) and New Zealand (Daish et al., 1980, 1981) instituted POE programs to optimize space utilization and facility improvements.

As far as curricula and instructional programs in POE are concerned, at Florida A&M University, for instance, POE has been taught since the mid 1980s in conjunction with facility programming and design research which focuses on the entire building process and its documentation. The same is true for the schools of architecture at the University of New Mexico, the University of Cincinnati, and others in the U.S.A.

## PURPOSES AND TYPES OF POE

A POE can serve several purposes depending on a client organization's goals and objectives. POE can provide the necessary data for the following:

- To measure the functionality and appropriateness of design and to establish conformance with performance requirements as stated in the functional program. A facility represents policies, actions, and expenditures that call for evaluation. When POE is used to evaluate design the evaluation must be based on explicit and comprehensive performance requirements contained in the functional program statement referred to above.

- To fine-tune a facility. Some facilities incorporate the concept of "adaptability," such as office buildings where changes are frequently made. In that case, routinely recurring evaluations contribute to an ongoing process of adapting the facility to changing organizational needs.

- To adjust programs for repetitive facilities. Some organizations build what is essentially the identical facility on a recurring basis. POE identifies evolutionary improvements in programming and design criteria, and it also tests the validity of underlying premises that justify a repetitive design solution.

- To research the effect of buildings on their occupants. Architects, designers, environment behavior researchers, and facility managers can benefit from a better understanding of building-occupant interactions. This requires more rigorous scientific methods than design practitioners are normally able to use. POE research in this case involves thorough and precise measures, and more sophisticated levels of data analysis, including factor analysis and cross-sectional studies for greater generalizability of findings.

- To test the application of new ideas. Innovation involves risk. Tried and true ideas can lead to good practice but new ideas are necessary to make advances. POE can help determine how well a new concept works once applied.

- To justify actions and expenditures. Organizations have greater demands for accountability and POE helps generate the information to accomplish this objective.

## POE Process Models

General models of the POE process have been described by several authors in their writings (e.g., Daish et al., 1980; Marans and Spreckelmeyer, 1981). While there are variations in the process, depending on the nature and objectives of the respective POEs, three levels of effort can be generally distinguished in POE work. Preiser and Pugh (1986) developed this observation into the "POE Process Model" and used it to outline the levels of effort that can be found in a typical POE, as well as the 3 Phases and 9 Steps that are involved in the process.

Levels of effort refer to the amount of time, resources, and personnel, the depth and breadth of investigation, and the implicit cost involved in conducting a POE. The three levels are: (1) indicative, (2) investigative, and (3) diagnostic. Each higher level requires more data gathering and is more comprehensive than the previous level, as depicted in Figure 35-1.

- Indicative POEs give an indication of major strengths and weaknesses of a particular building's performance. They usually consist of selected interviews with knowledgeable informants, as well as a subsequent walk-through of the facility.
- Investigative POEs go into more depth. Objective evaluation criteria are explicitly stated either in the functional program of a facility or they have to be compiled from guidelines, performance standards, and published literature on a given building type.
- Diagnostic POEs correlate physical environmental measures with subjective occupant response measures. Case study examples of POEs at these 3 levels of effort

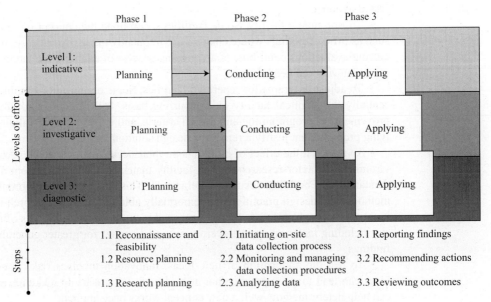

**FIGURE 35–1** POE Process Model. [Used by permission of Wolfgang F. E. Preiser]

can be found in Preiser, Rabinowitz, and White (1988).

## BENEFITS AND LIMITATIONS OF CURRENT POE PRACTICE

Each of these POEs can result in several benefits and uses. Recommendations can be brought back to the client, and remodeling can be done to correct problems. Lessons learned can influence design criteria for future buildings, as well as provide information about buildings in use to the building industry. This is especially relevant to the public sector which designs buildings for its own use on a repetitive basis.

## POE USES AND BENEFITS

The many uses and benefits which result from conducting POEs are listed below. These benefits provide the motivation and rationale for committing to POE as a concept and for developing POE programs (Kantrowitz et al., 1986).

### Short-Term Benefits

- Identification and solutions to problems in facilities.
- Pro-active facility management responsive to building user values.
- Improved space utilization and feedback on building performance.
- Improved attitude of building occupants through active involvement in the evaluation process.
- Understanding of the performance implications of changes dictated by budget cuts.
- Better informed design decision making and understanding of the consequences of design.

### Medium-Term Benefits

- Built-in capability for facility adaptation to organizational change and growth over time, including recycling of facilities into new uses.
- Significant cost savings in the building process and throughout the life-cycle of a building.
- Accountability for building performance by design professionals and owners.

### Long-Term Benefits

- Long-term improvements in building performance.
- Improvement of design databases, standards, criteria and guidance literature.
- Improved measurement of building performance through quantification.

There are three time frames for which POEs yield results as far as the performance concept in the building process is concerned. They are short-term, medium-term, and long-term outcomes of POEs. These time frames refer to immediate action, the three to five year intermediate time frame (which is necessary for the development of new construction projects), and the long-term time frame, respectively, which is necessary for strategic planning, budgeting, and master planning of facilities, i.e., ranging from ten to twenty-five years.

The most important benefit of a POE is its positive influence upon the delivery of humane and appropriate environments for people through improvements of the programming and planning of buildings (Figure 35–2). POE is a form of product research that helps designers develop a better design in order to support changing requirements of individuals and organizations alike.

POE provides the means to monitor and maintain a good fit between facilities and organizations, people, and activities that they support. POE can also be used as an integral part of a pro-active facilities management program.

## THE PERFORMANCE CONCEPT

The Performance Concept is based on the assumption that a building is designed and built to support, even enhance, the activities and goals of its occupants.

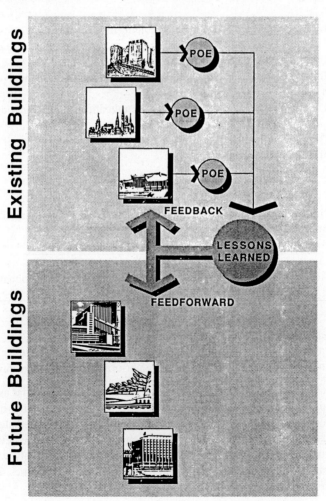

**FIGURE 35–2**  Feed forward from POE improves future buildings. [Used by permission of Wolfgang F. E. Preiser]

The early work on building performance was conducted by Ezra Ehrenkrantz and his associates on the School Construction Systems Development Project in California (Educational Facilities Laboratories, 1967). This work was further advanced at the Institute of Advanced Technology at the National Bureau of Standards. At the latter, under the direction of John Eberhard, the "Performance Concept" (Eberhard, 1965) was generated. Subsequent projects executed by the National Bureau of Standards (Wright, 1971) for the Department of Housing and Urban Development and the General Services Administration built upon these initial efforts.

Performance evaluation and feedback as shown in Figure 35–3 relates client goals and performance criteria to the objectively and subjectively measurable effects of buildings on people. The concept embodies two features. The first feature is that everything shown in the shaded area is dependent upon the relativity of person/environment relationships, i.e., the same building and its physical attributes (which can be objectively measured and described) may be perceived by the same people differently at different times, or differently by different people at the same time.

A second and important feature is that the evaluator is inside the shaded area, implying again the relativity of perceived building performance. The evaluator is the driving force of the evaluation system, and thus, he/she introduces biases, he/she sets the scope of the evaluation, and he/she presents the findings to the client.

Performance criteria used in evaluations are developed from goals and objectives which in themselves are derived from values held by individuals, groups, organizations, or entire socio-political systems. Frequently, there are differences among various groups or organizational units. Resolving such conflicts helps in satisfying building performance.

**FIGURE 35–3**  The performance concept. [Used by permission of Wolfgang F. E. Preiser]

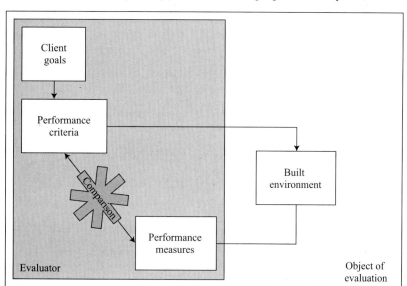

## ELEMENTS OF BUILDING PERFORMANCE

The third aspect of POE in the building process pertains to the elements of performance (Figure 35-4) that are measured, evaluated, and used to improve buildings.

The three major categories of elements in building evaluation relate to technical, functional, and behavioral performance. There are other factors that relate to buildings, such as location, economics, etc., but in terms of physical performance implications affecting owners, organizations, and building occupants, the three elements listed are the most important. The three categories of elements are part of a larger context, i.e., the scale of the environment in which it is measured, and the salient characteristics of its users and/or occupants.

One of the horizontal axes in Figure 35-4 indicates the building scale at which the performance in the three categories is measured and evaluated, as each type of performance element needs to be evaluated at an appropriate scale.

The elements of performance comprise the basic attributes of buildings. The elements of performance in this framework consist of explicitly stated evaluation criteria. Based on objective tests and performance measures, in reality many measures act together in terms of how the overall environment is experienced.

The first horizontal axis in Figure 35-4 differentiates among settings and places in hierarchical order of site or scale, whereby aggregates of a given category make up the next higher level category, e.g., rooms constitute buildings, aggregates of work stations make up rooms, etc.

The second horizontal axis in Figure 35-4 differentiates among users/occupants of buildings in terms of their numbers, and descriptors such as age group, sex, subculture group, race, or status within an organizational structure, to name just a few. An important aspect is also whether user/occupants of buildings are individuals, groups, or entire organizations.

**FIGURE 35-4** Elements of building performance. [Used by permission of Wolfgang F. E. Preiser]

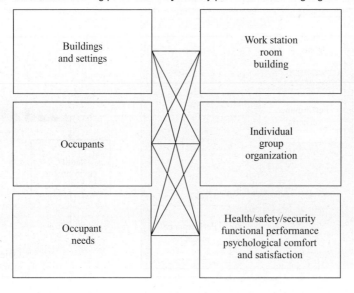

The vertical axis shown in Figure 35–4 contains performance criteria which are considered in buildings, broken down into the categories of technical, functional, and behavioral elements of building performance.

## THE PERFORMANCE EVALUATION RESEARCH FRAMEWORK

The performance evaluation research framework (Figure 35–5) connects the evaluation of buildings with: (1) measurement technology, (2) data bases and information systems (including clearinghouses), and (3) the development of performance criteria for buildings.

As shown in Figure 35–5, the cyclic framework contains three important features:

### Measurement Technology

Measurement technology employs all those techniques and technological aids that are used in the data collection and analysis of POE. They include interviews, questionnaire surveys, direct observation, mechanical recording of human behavior, measurement of light, indoor air quality, and acoustic levels, recording with video and time lapse film cameras, mapping of behavior, still photography, and so on.

To date, there is little or no standardization of measurement technology and methods used in POE, which poses problems regarding the generalizability of POE findings. As the field of POE matures, this issue will have to be addressed.

**FIGURE 35–5**  The performance evaluation research framework. [Used by permission of Wolfgang F. E. Preiser]

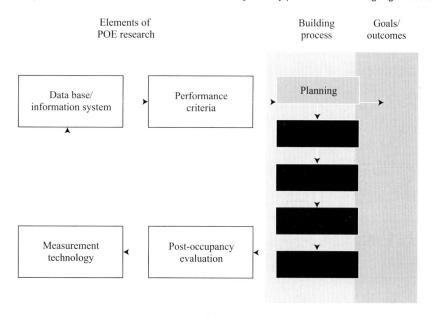

## Data Bases, Information Systems, and Clearinghouses

Data collected with appropriate measurement technology is fed into data bases (National Academy Press, 1987), information systems or clearinghouses which contain the results of POEs. They provide a much needed focus for sharing the POE results (see Figure 35-6). This activity is guided by organizations and associations concerned with specific building types, such as offices, schools, housing, or laboratories (Preiser and Krishnan, 1991).

At this time, only one clearinghouse for the dissemination of POE research exists, i.e., AEPIC, The Architecture and Engineering Performance Information Center.[1] It was created to collect and disseminate information concerning technical failures in buildings. Its core of information comes from the files of a major insurance company which donated them to the center.

## Performance Criteria

Performance criteria and guidelines are being developed from data bases/information systems for a given agency and/or building type. They are usually documented in either technical manuals, design guides, or in specialized data bases (Department of the Army, 1976).

**FIGURE 35-6**   Generic database. [Used by permission of Wolfgang F. E. Preiser]

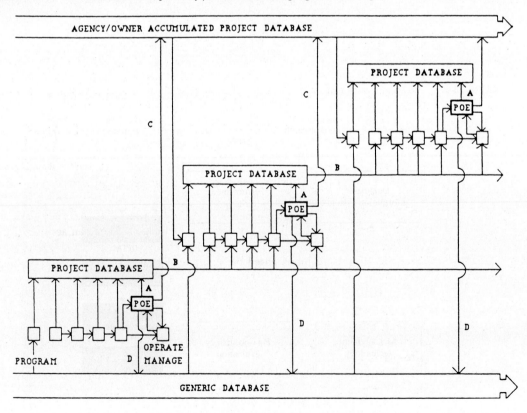

The criteria are building-specific and address particular sets of occupants and building functions. As such they are an evolving and improving set of performance "benchmarks" for a given building type.

Performance criteria and design guidelines feed the entire building delivery process, and thereby the cycle of continuously improving building performance is completed.

## CONCLUSIONS

A number of approaches to Post-Occupancy Evaluations which have been developed in recent years are documented in the book *Building Evaluation* (Preiser, 1989) and form the basis for some of the conclusions drawn below. This paper is the result of more than twenty years of theoretical and applied research in the topic area of Post-Occupancy Evaluation. A body of knowledge has been generated through the collective contributions of researchers in the field whereby the trend has been toward evermore systematic and sophisticated evaluation studies. Today, system-wide evaluations are carried out by such agencies as the Department of Veterans Affairs, which oversees 172 major hospitals and countless clinics and other facilities throughout the United States. For example, the U.S. Postal Service issued what must have been the largest request for proposals in the field of Post-Occupancy Evaluation ever for a system-wide evaluation of postal facilities, valued at approximately $.5 million in fees for services rendered. (Farbstein et al., 1989). This indicates that there are a number of benefits that may be derived by applying Post-Occupancy Evaluation techniques to the various phases of the building delivery process, especially in the large corporate and institutional sectors. These benefits may range from quick fixes to existing building-related problems, all the way to litigation and the establishment of planning and design guidance for an agency or corporation. Thus, Post-Occupancy Evaluation is being seriously considered by major players in the building industry and service sectors of society.

In advancing the state-of-the-art of Post-Occupancy Evaluation endeavors in the United States and elsewhere the following conclusions can be drawn:

• The one-off case study POE is no longer valued and is being replaced by cross-sectional and longitudinal evaluations of facility types such as schools, offices, etc., in order to increase the validity of the obtained results.

• The "performance concept" and "performance criteria," made explicit and scrutinized through Post-Occupancy Evaluations, have now become an accepted part of good design, by moving from primarily subjective experience based evaluations to more objective evaluations based on explicitly stated performance requirements in buildings.

• Critical in the notion of performance criteria is the focus on the quality of the built environment as perceived by its users/occupants. In other words, building performance is seen to be critical beyond aspects of energy conservation, life cycle costing and the functionality of buildings, but it focuses on the meaning and users' perceptions of buildings.

• Data gathering techniques for Post-Occupancy Evaluation studies need to be standardized and the results of POE need to become replicable.

- POEs have become more cost-effective due to the fact that short-cut methods have been devised which allow the researcher/evaluator to obtain valid and useful information in a much shorter time-frame than was previously possible. Thus, the cost of staffing and other expenses have been considerably reduced, especially for so-called "indicative" type POEs.

- Architectural research firms specializing in facility programming, evaluation and topical research have sprung up and have been able to gain a foothold in the professional world by not competing with the standard architectural firms in the design arena, but rather, by complementing and in fact enhancing those firms' ability to deliver a high-quality product. Most recently, POE methodology has been applied to a process analysis of hospital activation for the Department of Veterans Affairs system of hospitals and health care facilities. This project used interviews, document research and archival research methods, focused on an analysis of the building delivery process and activation of hospitals from a management perspective, i.e., for a given milestone or decision point in the system the actors, the resources, time and necessary guidance were identified. The entire system of activation was modeled for the first time. The result was a guide to activation which will be used at different levels of decision making ranging from local Medical Centers to regional offices and all the way to the Central Office of the Department of Veterans Affairs. Guidance is provided on when and what needs to be planned for, or what resources need to be budgeted, what equipment needs to be ordered, as well as what training should be provided when moving into a new hospital (Petronis, Preiser, et al., 1991).

- POE has gained a certain degree of respectability in tandem with the topic area of facility programming. It became an area of focus for the National Academy of Sciences Building Research Board (National Academy Press, 1987). This is extremely important, as the recommendations of that Board may become policy guidance and agenda items for funding agencies in the future. Thus, we can expect to hear more about POE in the years to come.

## NOTE

1 Architecture and Engineering Performance Information Center (AEPIC), 3907 Metzerot Road, University of Maryland, College Park, Maryland 20742.

## REFERENCES

Architects and Space Planners. *Technology and Office Design.* London, UK: Duffy Eley Giffone Worthington, 1983.

Becker, F.D. *Design for Living—The Residents View of Multi-family Housing.* Ithaca, NY: Center for Urban Development Research, Cornell University, 1984.

Brauer, R.L., and W.F.E. Preiser. "Impact of Organizational Form on Identification of User Requirements in Building Delivery." *Proceedings of the CIB W-65 Symposium on Organization and Management of Construction.* Washington, DC: National Academy Press, May 1976.

Brill, M., et al. *Using Office Design to Increase Productivity.* Vol. 1. Buffalo, NY: Workplace Design and Productivity, Inc, 1984.

Daish, J.; J. Gray; and D. Kernohan. *Post-Occupancy Evaluation of Government Buildings.* Wellington, New Zealand: Victoria University of Wellington, School of Architecture, 1980.

———. *Post-Occupancy Evaluation Trial Studies.* Nos. 1–3. Wellington, New Zealand: Victoria University of Wellington, School of Architecture, 1981.

Department of the Army. *Design Guide DG 1110-3-106 US Army Service Schools.* Washington, DC: Engineering Division, Military Construction Directorate, Office of the Chief of Engineers, 1976.

Eberhard, J.P. "Horizons for the Performance Concept in Building." In *Proceedings of the Symposium on the Performance Concept in Building,* Building Research Board. Washington, DC: National Academy of Sciences, 1965, pp. 93–98.

Ehrenkrantz, E. *School Construction Systems Development Project.* New York: Educational Facilities Laboratories, 1965.

Farbstein, J., et al. "Post-Occupancy Evaluation and Organizational Development: The Experience of the United States Postal Service." In *Building Evaluation,* ed. W.F.E. Preiser. New York: Plenum, 1989.

Field, H., et al. *Evaluation of Hospital Design: A Holistic Approach.* Boston: Tufts-New England Medical Center, 1971.

Francescato, G.; S. Weidemann; J.R. Anderson; and R. Chenoweth. *Residents' Satisfaction in HUD-Assisted Housing: Design and Management Factors.* Washington, DC: U.S. Department of Housing and Urban Development, U.S. Government Printing Office, March 1979.

General Services Administration. *The PBS Building Systems Program and Performance Specifications for Office Buildings.* 3rd ed. Washington, DC, November 1975.

Goodrich, R. *Post-Design Evaluation of Centre Square Project.* Philadelphia, PA: Atlantic Richfield, Inc., 1976.

Kantrowitz, M., et al. "POE: Energy Past and Future." *Progressive Architecture,* April 1986, pp. 114–23.

Marans, R., and K. Spreckelmeyer. *Evaluating Built Environments: A Behavioral Approach.* Ann Arbor, MI: The University of Michigan, Institute for Social Research and College of Architecture and Urban Planning, 1981.

Markus, T., et al. *Building Performance.* New York: Halsted Press, 1972.

Nasar, J.L., ed. *The Visual Quality of the Environment: Theory, Research and Application.* Cambridge, UK: Cambridge University Press, 1988.

National Academy of Sciences Building Research Board. *Post-Occupancy Evaluation: Practices in the Building Process—Opportunities for Improvement.* Washington, DC: National Academy Press, 1987.

Parshall, S.A., and W.M. Pena. *Evaluating Facilities: A Practical Approach to Post-Occupancy Evaluation.* Houston, TX: CRS Sirrine, Inc., 1982.

Petronis, J.P.; W.F.E. Preiser; et al. *Activation Process Analysis and Guide for VA Hospitals.* Unpublished draft report. Albuquerque, NM: Architectural Research Consultants, Inc., 1991.

Preiser, W.F.E. *Behavioral Design Criteria in Student Housing: The Measurement of Verbalized Response to Physical Environment.* Blacksburg, VA: Virginia Polytechnic Institute and State University, 1969.

Preiser, W.F.E., ed. *Building Evaluation.* New York: Plenum, 1989.

Preiser, W.F.E., and R. Krishnan. *Computerized POE Database Development for Medical and Scientific Laboratory Facilities.* Unpublished manuscript, School of Architecture and Interior Design, University of Cincinnati, 1991.

Preiser, W.F.E., and R.R. Pugh. "Senior Centers: A Process Description of Literature Evaluation, Walkthrough Post-Occupancy Evaluations, A Generic Program and Design for the City of Albuquerque." In *The Costs of Not Knowing, EDRA 17 Proceedings,* ed. J. Wineman, R. Barnes, and C. Zimring. Washington, DC: Environmental Design

Research Association, 1986.

Preiser, W.F.E.; H.Z. Rabinowitz; and E.T. White. *Post-Occupancy Evaluation.* New York: Van Nostrand Reinhold, 1988.

Public Works Canada. *Project Delivery System, Stage 10: Level 1 Evaluation, Users Manual.* 1st ed. Ottawa, Canada: Department of Planning and Coordination Branch, June 1979.

Rabinowitz, H.Z. *Buildings-in-Use Study.* Milwaukee, WI: University of Wisconsin, School of Architecture and Urban Planning, 1975.

Thorne, R. *Building Appraisal.* Videotape, scripted from material supplied by the Building Performance Research Unit, University of Strathclyde, Glasgow. Sydney, Australia: University of Sydney, Department of Architecture, 1980.

Van der Ryn, S. and M. Silverstein. *Dorms at Berkeley.* Berkeley, CA: University of California, Center for Planning and Research, 1967.

Vischer, J.C. *Environmental Quality in Offices.* New York: Van Nostrand Reinhold, 1989.

White, E.T. *Building Evaluation in Professional Practice.* Tallahassee, FL: School of Architecture, Florida A&M University, 1985.

Wineman, J. *The Behavioral Basis of Office Design.* New York: Van Nostrand Reinhold, 1986.

## THE ARCHITECTURAL PROFESSION—SUGGESTED READINGS

Coxe, Weld. *Managing Architectural and Engineering Practice.* New York: Wiley, 1980.

Cuff, Dana. *Architecture: The Story of Practice.* Cambridge, MA: MIT Press, 1991.

Grant, Donald P. *The Small-Scale Master Builder.* San Luis Obispo, CA: The Small-Scale Master Builder, 1983.

Haviland, David. *Managing Architectural Projects.* Washington, DC: American Institute of Architects, 1984.

Kostof, Spiro. *The Architect: Chapters in the History of the Profession.* New York: Oxford, 1977.

Oliver, Paul, and Richard Hayward. *Architecture: An Invitation.* Cambridge: Blackwell, 1990.

Saint, Andrew. *The Image of the Architect.* New Haven: Yale University Press, 1983.

St. John Wilson, Colin. *Architectural Reflections.* Boston: Butterworth, 1992.

Torre, Susana, ed. *Women in American Architecture.* New York: Whitney Library of Design, 1977.

# EPILOGUE 跋

选读 36
"对建筑行业的思考"
欧内斯特·勒罗伊·博耶和李·D. 米特冈

博耶和米特冈研究了当下建筑教育的状况和未来前景,其研究成果《建筑社区:建筑教育与实践的新未来》一书面面俱到,备受推崇。这篇选读就是这本书的绪论部分。两位作者认为建筑教育和建筑行业正在快速变化的大背景下运营,这种快速变化的大背景为建筑教育和建筑行业的发展提供了宏大的希望和各种可能性,"但是同样伴随而来的可能是这一领域前所未有的巨大焦虑"。博耶和米特冈回顾了建筑教育和实践的历史传统,然后以七个相互独立但又相互关联的重点方面为基础提出了他们自己的革新框架:内容丰富的任务、相互尊重下的多样性、没有标准化的标准、相互关联的课程、相互支持的学习氛围、更为整齐划一的行业以及公民积极参与的生活——为国家服务。

## READING 36

## A Profession in Perspective
### Ernest LeRoy Boyer and Lee D. Mitgang

This selection is the introductory chapter to Boyer and Mitgang's comprehensive and highly respected study of current conditions and future prospects of architecture education. The authors believe that architecture education and the profession are operating in a rapidly changing context that offers enormous promise and possibilities, "but also anxiety perhaps unmatched in the history of the field." Boyer and Mitgang trace the tradition and history of architecture education and practice and then offer their own framework for renewal based on seven separate but interlocking priorities: an enriched mission; diversity with dignity; standards without standardization; a connected curriculum; a supportive climate for learning; a more unified profession; and lives of civic engagement—service to the nation.

The education of architects—the men and women who design our skyscrapers and plazas, churches and museums, our schools and our homes—rests on traditions as old as history. From one generation of architects to the next, there is a rich legacy of principles and personalities that creates a common bond among veterans and novices alike. During one campus visit, we heard an eighteen-year-old student and a seasoned professor discussing intently the trinity of core values composed more than two millennia ago by the Roman architectural writer Vitruvius: "Firmness, Commodity, and Delight." Often, we heard students and faculty swap remembrances of architectural deities like "H. H." (Richardson), "Corbu," (Le Corbusier) or "Mies" (van der Rohe)—as if reminiscing fondly about familiar friends. And students throughout this

Source: "A Profession in Perspective" in *Building Community: A New Future for Architecture Education and Practice* by Ernest L. Boyer and Lee D. Mitgang. Copyright © 1996 The Carnegie Foundation for the Advancement of Teaching. All rights reserved. pp. 3–28.

country soon learn to converse comfortably in the arcane but richly fascinating language of architecture that has developed over centuries. Alongside technical twentieth-century terms such as "CAD" (computer-aided design), or "HVAC" (heat, ventilation, and air conditioning systems), aspiring architects quickly become acquainted with eighteenth-century French Beaux Arts terms such as "en charette," still the common parlance for the grueling work marathons students and practitioners engage in to solve design problems.

This sense of kinship with centuries of traditions, thoughts, and personalities is, in fact, the true tie that binds those who practice architecture with those who teach it and study it. The nobility of architecture has always rested on the idea that it is a *social art*—whose purposes include, yet transcend, the building of buildings. Architects, in short, are engaged in designing the physical features and social spaces of our daily lives, which can shape how productive, healthy, and happy we are both individually and collectively. The profound and permanent impact of the architecture profession demands an education not only highly technical and practical, but broad and intellectually liberating as well. A nineteen-year-old student told us: "Architects must use their unique skills and talents to revitalize both our communities and the world. Our purpose must be noble, to create sustainable environments which are healthy for nature and man, environments that can be improved by our intervention."

An architect in Michigan put it this way: "Architects must seek a balance between economic construction, safety and welfare of the general public, long-term value of the products they help produce—all while remaining sensitive to the aesthetic concerns of the owner, neighbors and the architect him/herself."

The essential purpose of architecture education, then, is not only the basic training of beginning practitioners, but also the initiation of students into this common legacy of knowledge, skills, and language, while instilling a sense of connectedness to the human needs that architecture, as a profession, must continually address. Architecture education, if it is to fulfill those ends, must celebrate and support, and also challenge, the profession and society as a whole.

During the conduct of this study, we discovered that despite serious concerns about the education of architects, schools of architecture draw at least as much praise as criticism from students, educators, and practicing architects. In hundreds of conversations, we met only a handful who proposed abandoning altogether the methods and traditions which have shaped architecture education for much of this century. What we heard most often were calls for *reform,* not *revolution,* and we were, in fact, impressed by the number of schools who are engaged in candid review of their programs. Reflecting on all we heard and read, we concluded that the fascination of architecture education lies far more in its *possibilities* than in its problems.

This is not to suggest that changes, even bold changes, should not be carefully considered. Still, we are convinced that architecture education, at its best, is a model that holds valuable insights and lessons for all of higher education as a new century approaches. We agree with a faculty member at a southeastern campus who called architecture education "one of the best systems of learning and personal development that has been conceived." The professor went on to say: "I would hate to see us abandon a system with so many wonderful qualities and successes for the sake of some 'bold experiment in education.'"

For the nation's 35,524 full- and part-time students who attended 103 accredited degree programs in 1994–95, the study of architecture is among the most demanding and stressful on campus, but properly pursued it continues to offer unparalleled ways to combine creativity, practicality, and idealism. Coming from public school environments where aesthetic and three-dimensional learning experiences are frequently downplayed and even disparaged, it's little surprise that eighteen- and nineteen-year-old students seeking an artistic avenue with a practical bent soon become fiercely devoted to the design studio culture.

There are several available educational paths to a professional degree. Most students pursue either a five-year undergraduate program directly from high school or a six-year master's program that combines undergraduate education and two years of graduate professional study. Some, particularly those who decide on architecture later in their college years, pursue a three- or four-year graduate degree. With few exceptions, students in practically every state are required to earn a professional degree, spend at least three years as interns, and gain a specific set of professional experiences before they can take the Architecture Registration Examination that must be passed to become licensed as an architect. (It even took Philip Johnson three tries to pass all ten parts of that dreaded test.) The path to licensure, then, is not just demanding, but long: at least eight years, and often more.

How satisfied, in fact, are students with their experiences in architecture programs? When we surveyed students and alumni from fifteen representative campuses whether they would attend their school of architecture again if they had it to do over, about eight out of ten said they would (Table 36-1).

And even with the demands of school and the often dire forecasts about job prospects, we found morale among architecture students to be remarkably high. Nearly

**TABLE 36-1**  IF YOU HAD IT TO DO OVER, WOULD YOU STILL ATTEND THIS SCHOOL OF ARCHITECTURE?

|  | Yes | No | Not Sure |
|---|---|---|---|
| Students | 82% | 11% | 6% |
| Alumni | 79 | 15 | 6 |

*Source*: The Carnegie Foundation for the Advancement of Teaching, Survey on the Education of Architects, 1994.

**TABLE 36-2**  HOW WOULD YOU RATE MORALE AMONG ARCHITECTURE STUDENTS? (SURVEY OF STUDENTS)

|  | Percentage |
|---|---|
| Very high | 10% |
| Fairly high | 64 |
| Fairly low | 17 |
| Very low | 3 |
| Not sure | 6 |

*Source*: The Carnegie Foundation for the Advancement of Teaching, Survey on the Education of Architects, 1994.

three out of four students we surveyed rated morale at their schools as either "very high" or "fairly high" (Table 36-2).

Revealingly, more than 90 percent of administrators, students, and alumni and more than 80 percent of faculty surveyed believe that graduates of architecture school leave "well prepared" as problem solvers—a quality that leading practitioners repeatedly told us was among the most prized in new employees (Table 36-3). A twenty-two-year-old student at a midwestern campus told us her school "does an excellent job of creating problem solvers. With this talent and skill, architecture graduates can go on to be leaders and innovators in an unlimited number of areas, as well as in the profession of architecture."

Despite the evident stresses of student life, nearly 4,500 architecture students earned professional bachelor's or master's degrees in 1994–95, and most seem quite aware of the gloom-and-doom job scenarios that may await them. When we asked students what their salary expectations were five years after earning their professional degree, 24 percent said less than $30,000 and another 44 percent said in the $30,000 range (Table 36-4). As one student put it, freshmen are routinely greeted at the portals of architecture school with the message: "I won't be rich, I won't be famous, and I have to work a lot."

We then asked students what motivated them to enter the architecture field. The most common response, given by 44 percent of those surveyed, was "putting their creative abilities to use." Intriguingly, the other most frequent responses—given by a combined 39 percent—were a desire to improve the quality of life in communities or improve the built environment as a whole. Hardly anyone cited the prestige of the profession or good salary prospects (Table 36-5).

**TABLE 36-3** ARCHITECTURE STUDENTS LEAVE THIS SCHOOL WELL PREPARED AS PROBLEM SOLVERS

|  | Agree | Disagree | Don't know |
|---|---|---|---|
| Administrators | 96% | 3% | 1% |
| Faculty | 82 | 14 | 4 |
| Students | 91 | 7 | 2 |
| Alumni | 96 | 4 | 0 |

Source: The Carnegie Foundation for the Advancement of Teaching, Survey on the Education of Architects, 1994.

**TABLE 36-4** LOOKING AHEAD, WHAT KIND OF SALARY DO YOU EXPECT FIVE YEARS AFTER EARNING YOUR PROFESSIONAL DEGREE IN ARCHITECTURE? (SURVEY OF STUDENTS)

|  | Percentage |
|---|---|
| $20,000–$29,999 | 24% |
| $30,000–39,999 | 44 |
| $40,000–49,999 | 22 |
| $50,000 or more | 10 |

Source: The Carnegie Foundation for the Advancement of Teaching, Survey on the Education of Architects, 1994.

**TABLE 36-5** PLEASE RANK THE THREE MOST IMPORTANT REASONS FOR ENTERING THE ARCHITECTURE PROFESSION (SURVEY OF STUDENTS)

|  | Percentage ranking item first |
| --- | --- |
| Putting creative abilities to practical use | 44% |
| Improving quality of life in communities | 22 |
| Improving the built environment | 17 |
| The prestige of the profession | 2 |
| Good salary prospects | 1 |
| Other | 14 |

*Source* : The Carnegie Foundation for the Advancement of Teaching, Survey on the Education of Architects, 1994.

**TABLE 36-6** OVERALL, I FEEL THIS IS AN EXCELLENT TIME TO BE ENTERING THE ARCHITECTURE FIELD

|  | Agree | Disagree | Don't know |
| --- | --- | --- | --- |
| Administrators | 67% | 31% | 3% |
| Faculty | 47 | 47 | 6 |
| Students | 71 | 29 | 0 |

*Source* : The Carnegie Foundation for the Advancement of Teaching, Survey on the Education of Architects, 1994.

For most, then, the intrinsic appeal of architecture education seems to transcend its practical uncertainties, and many students cling, against the odds, to ideal visions of their potential. For many, architecture school is an opportunity to be part of a tight-knit community on campus that is defiantly proud of its distinctive methods and its reputation for long hours and hard work. "We all have a dream," said one thirty-four-year-old student. "We all want to be recognized architects. You have to pay your dues. When we enter this program they tell us that it is not high paying. But I assume money will follow if you gain recognition."

In our survey, 71 percent of students agreed that "this is an excellent time to be entering the architecture field"—a level of optimism exceeding that of both faculty and administrators (Table 36-6).

Again, this is not to suggest that all is well with architecture education. Repeatedly in our travels, we witnessed the estrangement of the academy and the profession, the isolation and stress of student life, the disconnection of architecture from other disciplines, and the inflexibility of the curriculum on many campuses. We also saw the great unevenness of experiences in design studio, and the autocratic, one-way communication that often marks "design juries"—the tension-packed ritual during which invited critics, including faculty, practicing architects, and more rarely clients, review and critique student work.

Further, we frequently heard that many architecture schools fit uncomfortably within the campus culture. In 1900, almost all architects served apprenticeships in offices.[1] Over the last century, professional education was brought fully into academic life, but the values and rewards of these two cultures have never been fully reconciled. Compounding those difficulties, architecture programs are generally small compared with other academic units, expensive in terms of per-pupil staffing and physical space, and for the most part, produce little research revenue for the university.

As W. Cecil Steward, dean of the University of Nebraska's College of Architecture, has noted, many university administrators, especially those on research-driven campuses, tend to see the architecture field as splintered and disputatious, and the design orientation of architecture faculty places architecture among "the 'soft,' 'fuzzy,' and undervalued disciplines in the comprehensive universities."[2]

And in his memorable Walter Gropius Lecture in 1985, Henry N. Cobb, FAIA, then-chairman of Harvard's Department of Architecture, said that his own architecture school:

> . . . with its curious studio-based teaching methods, with its paucity of scholarly research, and its dedication to serving the highly "contaminated" professions of architecture, landscape architecture and urban planning, must appear, to borrow the language of "Peanuts," as a kind of "Pig-Pen" character in the university family—that is to say disreputable and more or less useless, but to be tolerated with appropriate condescension and frequent expressions of dismay.[3]

Repeatedly, administrators of schools of architecture also told us that lack of financial resources was their biggest problem. The dean of one private, East Coast school of architecture told us his program has been hit with a 25 percent budget cut over the last five years, with further reductions likely. Thomas R. Wood, director of Montana State University's School of Architecture, put it bluntly: "The biggest threats I see are financial survival and, related to that issue, student access to higher education. The past five years have been disastrous for higher education. The budget reductions damage program quality, faculty morale, and limit the opportunities of people seeking an architectural education."

In our surveys, architecture administrators and faculty agreed overwhelmingly that their schools have been living with more than their fair share of budget restraints over the past several years (Table 36-7).

At one school we visited, not a single wall clock worked, windows were broken, and paint was peeling off the walls. There were no electrical outlets in the studios—a costly problem if and when the school attempts to move decisively into the computer age. At present, 250 students at this school share a dozen outdated computer terminals—at a time when knowledge of computers has become practically a prerequisite for employment.

**TABLE 36-7** THIS SCHOOL/DEPARTMENT HAS HAD TO LIVE WITH MORE THAN ITS FAIR SHARE OF BUDGET RESTRAINTS OVER THE PAST SEVERAL YEARS

|  | Agree | Disagree | Don't know |
|---|---|---|---|
| Administratiors | 73% | 27% | 0% |
| Faculty | 77 | 13 | 10 |

*Source*: The Carnegie Foundation for the Advancement of Teaching, Survey on the Education of Architects, 1994.

Such problems are by no means universal. We found, in fact, quite a number of schools of architecture which have accommodated themselves quite comfortably within the university culture, and in our surveys, more than eight out of ten architecture school administrators and alumni and nearly seven out of ten faculty disagreed strongly that their programs "might be better off if they were not part of university campuses at all." More than 75 percent of all groups surveyed also considered their schools well regarded on campus (Table 36-8).

Along with the challenges posed by the institutional context of many architecture programs, the profession of architecture itself has become many professions, some connected only tenuously, if at all, to the work of designing buildings. Architectural practice, furthermore, is rapidly becoming more global. In 1994, nearly one-third of U.S. firms with at least twenty employees, and 15 percent of firms with ten to nineteen employees were engaged in international work, according to the American Institute of Architects.[4] With virtually all architectural firms now computerized, markets halfway around the globe are quickly becoming as accessible as those around the corner. These emerging trends—globalization and computerization—have implications for architecture education that many schools are only beginning to confront.

A profession that for much of its history prized the new and unique now finds more and more of its members engaged in preservation and renovation. A profession once populated almost exclusively by wealthy, white males now must struggle to connect to a more diverse range of clientele and communities. A profession long regarded as the orchestrator of the building process now finds itself increasingly competing with general contractors, developers, interior designers, construction firms, and others. Practitioners and students who once revered that lonest of lone architectural wolves, the fictional Howard Roark from *The Fountainhead,* now talk of replacing him with a new icon: the "team player." The field has become increasingly varied, yet there remains fixed in the minds of many a single image of the architect that may well be an antique.

"We are operating a 1900-year-old education program directed toward delivering a 500-year-old model architect as we head into the 21st Century," Professor Gregory Palermo, FAIA, of Iowa State University, wrote recently.[5]

We found, in short, a profession struggling both to fit in, and if possible, to *lead,* within a social and economic context that in a number of crucial respects has been dramatically altered. We also found a profession whose faith in its own future has been shaken. What seems missing, we believe, is a sense of common purpose con-

TABLE 36-8 HOW WELL REGARDED IS THIS SCHOOL OF ARCHITECTURE ON CAMPUS?

|  | Well regarded | Not well regarded | Don't know Not applicable |
|---|---|---|---|
| Administrators | 89% | 9% | 2% |
| Faculty | 77 | 14 | 10 |
| Students | 85 | 11 | 4 |
| Alumni | 78 | 7 | 15 |

*Source*: The Carnegie Foundation for the Advancement of Teaching, Survey on the Education of Architects, 1994.

necting the practice of architecture to the most consequential issues of society—and that same sense of unease permeates architecture education as well.

Urs Peter Gauchat, dean of the New Jersey Institute of Technology's School of Architecture, summarized the situation this way: "The rapidly changing context for architectural services has created a host of problems for architectural education. One problem stands out: the seed of self-doubt and the lack of a clear vision of what the architect can and should do. Many schools are permeated by a considerable lack of conviction about the future of architecture. This is often accompanied by the lack of a clear agenda about how to prepare students for a fulfilling professional life."

The atmosphere surrounding architecture education and the profession, then, is one of enormous promise and possibilities, but also anxiety perhaps unmatched in the history of the field. The mood resembles, in a fascinating way, that of public education in this country. Everyone agrees that elementary and secondary schools have serious problems. Yet at times of national self-doubt, there is a tendency to overlook the strengths of education. We expect the world of schools—demanding that they solve all of our complex social, economic and spiritual ills. When they inevitably fall short, we condemn them for failing to meet our high-minded expectations. The point is this: In the search for educational renewal at *all* levels, including the professional, care should be taken not to ascribe to schools alone what must, in the end, be a *shared* responsibility.

**THE MOVE TO THE ACADEMY**

In searching for common purpose and a more promising future, we begin by looking back—at the traditions, principles, and paradoxes that have, for so long, shaped the education and the professional lives of architects in the United States, and to a large degree, continue to do so.

Early in the nineteenth century, Thomas Jefferson and others with an interest in the future of architecture in a young and expanding nation sensed that more formal methods were needed to address the acute shortage of trained architects. Technical schools filled that need to some extent during the first half of the century, and a few university-based engineering schools began offering courses in architectural drawing.[6] But the dominant model of training architects was apprenticeship, at best a hit-or-miss proposition educationally.

It wasn't until the mid-nineteenth century that the momentous relocation of architecture education from office to campus began to take shape. In 1865, three years after Congress passed the Morrill Land Grant Act and eight years after the founding of the American Institute of Architects, the first formal, campus-based architecture courses in the United States were offered at MIT. The University of Illinois began its program in 1868, Cornell University in 1871, Syracuse University in 1873, and Columbia University in 1881.

Thus began a century-long process in architecture that paralleled, in many respects, the growing professionalism shaping many other fields. As Dana Cuff related in her recent study of architectural practice, training and the profession were largely unregulated during the nineteenth century. It wasn't until 1897 that Illinois became the first state to require a license to practice architecture. Most who called themselves architects were white, well-off and male, and got their training not at school but in apprenticeships of widely varying quality.[7] The rise of the city, the advent of new building materials and techniques, the birth of skyscrapers and new transportation systems—all made even more urgent the need for professionalization in architecture.

What was emerging in those early days of campus-based education, then, was a growing impulse to standardize architecture's expertise through a more specialized, organized, and regulated educational program, and a related need to assure the public that practitioners could assume legal responsibility for the quality and safety of their work.[8] These needs were decisively addressed by the arrival in America of a philosophy of education and practice developed at the Ecole des Beaux Arts, the leading center of architecture education in France. William Robert Ware, founder of both MIT's and Columbia University's architecture programs, was instrumental in adapting that French philosophy to American schools, and many of Ware's curricular precepts remain influential to this day.

Ware's principles were:

- that details of a practical nature that can be learned in the office should be postponed until after formal education;
- that courses in construction and history can be taught by means of "cooperative student investigation . . .";
- that architectural design should be conducted by a competitive method, with judgments by jury;
- that the study of design should be continuous through school, and design problems should not be overly practical, but rather should stimulate the imagination through the study of great masters;
- that the study of construction should be stressed; and
- that architectural curriculum should include as broad a cultural background as time permitted.[9]

The Beaux Arts philosophy has remained, for a century, perhaps the single greatest influence on how architecture education is conducted and thought about. As Professor Kathryn Anthony, of the University of Illinois at Champaign–Urbana, states: "The values behind the accreditation process, the way in which the curricula are structured, the jury system, and other key aspects of architectural education continue to primarily reflect an old Ecole des Beaux Arts model."[10]

The Beaux Arts school established, first of all, the hallowed place of European and classical traditions in design. While that had the advantage of distinguishing architecture from more technically driven approaches used by engineers and others, it also placed the profession more firmly in an aesthetic realm that made it seem, to many in the public, less vital to community concerns than some other professions. The Beaux Arts philosophy also elevated design above technical aspects of the curriculum, emphasized one-on-one teaching methods, created the present-day system of design juries using outside evaluators to judge student projects, and stressed the learning of past architecture to inform present design.[11] At the same time, the ascendance of design studio under the Beaux Arts system brought with it an atmosphere of individualism, criticism, and competition among faculty and students.

By the turn of the twentieth century, there were still only nine professional schools of architecture in the United States, enrolling slightly under four hundred students.[12] For the most part, those early schools offered similar four-year programs which attempted to strike a balance between design and structures, but given the weighty contents, schools gave only "meager treatment" to either.[13] By 1912, the number of schools had grown to thirty-two, enrollments had tripled, and the Association of Collegiate Schools of Architecture (ACSA) was founded in order to foster communication among schools and establish minimal standards. By 1932, there were fifty-two institutions of collegiate rank offering professional courses in architecture.[14] Teaching, meanwhile, was carried on by "technicians" rather than scholars, and thus research, in the academic sense, was almost totally lacking.[15]

An early history of architecture education by Arthur Clason Weatherhead noted that the typical curriculum in the early twentieth century lacked cohesiveness, the business side of architecture was neglected, and there was little effort to aid the transition of students between the academy and the office. Graduates gained a strong sense of design, were resourceful in solving problems in artificial settings, and well versed in Beaux Arts logic, but their thinking tended to be two-dimensional.[16]

Following World War I, the competitive precepts of Beaux Arts were increasingly questioned. By the 1920s and 1930s, the coming of modernism, the growing rejection of Neoclassicism, and the emergence of urbanism led to a fundamental reconsideration of architecture education. Modernism was further refined by the Bauhaus philosophy imported from Germany in the 1930s by Walter Gropius at Harvard, and Ludwig Mies van der Rohe at the Illinois Institute of Technology. Under their influence, greater stress was placed in U.S. schools on draftsmanship, structural logic, appreciation of the properties of materials, and an aesthetic that derived from the exploration of geometric forms.[17] The period was marked by a shift to a more uniform, well-defined training system in which the Association of Collegiate Schools of Architecture also acted as a unifying influence.

Modernism powerfully affected the curricula at many U.S. schools. Stiffer course requirements were introduced, the study of materials and construction was emphasized, more schools related themselves to allied professional fields, and schools began to explore the relationship of architecture to human and community needs. The University of Oregon rejected the atelier, "master-pupil" model of Beaux Arts studio education and permitted students more freedom to choose their projects according to their interests.[18] The University of Cincinnati began its practice of having students alternate school studies and field work, which has continued into the 1990s.[19]

National organizations concerned with architecture education, meanwhile, were considering ways to assure more complete preparation for practice.[20] Along with the rise of professionalism and the growth of architectural knowledge came a drive for monitoring and accountability. At first, the Association of Collegiate Schools of Architecture itself performed the accreditation function by establishing minimum standards for membership. Those early standards were abandoned in 1932, however, amid criticism that they were stultifying curricula. The result was a regulatory vacuum. In 1940, a coalition of education and professional organizations founded the National Architectural Accrediting Board (NAAB). The establishment of NAAB marked a milestone in efforts to create clearer standards and regulatory safeguards in professional degree programs while recognizing the differing resources, geographic circumstances, and missions of individual schools.[21]

By the end of World War II, the nearly century-long shift away from apprenticeship to a university-based system of architecture education was virtually complete. In

1953, architecture enrollments nationwide had reached nearly ten thousand, a 150 percent increase over 1930.[22]

All along, the rationale behind the move to a campus-based architectural education system was that certain kinds of intellectual growth and learning needed for professional education was more likely to occur on campus than in an office. Still, the shift from the old apprenticeship system planted the seeds of separation between education and practice. Because education, organizationally, has been divorced from the practice world, schools have had to make extraordinary and costly efforts to help students gain real-life experiences, such as field trips, undergraduate internships, or preceptorships. Graduates, for their part, have faced the challenge of acquiring certain skills needed to make the transition from school to office not readily available in most standard curricula.

In 1956, the American Institute of Architecture Students was founded, an outgrowth of the student chapters of the AIA that had already spread to many campuses. In 1962, the membership of the National Council of Architectural Registration Boards (NCARB) voted to establish a single, national registration examination, replacing the many different exams given by states.[23]

During the 1960s, powerful new forces buffeted architecture education and campus life. Many schools sought to include social and political issues in their curricula, and a number also tried to connect students to underserved urban and rural communities by opening, for example, "Community Design Centers." The 1960s and 1970s, however, also saw increased questioning of key precepts of modernism. Changing economic and political tides led architects and architecture educators to be less optimistic about their prospects to be leaders in shaping the built environment. The resulting "postmodern" movement of the 1970s and 1980s was not simply a stylistic revolt against the alleged "glass box" sameness of modernism, but also reflected a feeling that modernist architecture had led both practitioners and educators away from some of the profession's more interesting and vital historical precepts. However, as the name suggests, postmodernism seemed more a move *away* from modernism than *toward* any specific unified approach.

By the 1990s, enthusiasm for postmodernism seemed to be evolving into an eclecticism in which many now seem inclined to disavow allegiance to any single style or philosophy. Indeed, the very notion of "style" has fallen into disfavor in many circles. Philosophically, however, a sense of modernist engagement in political, social, economic, and environmental issues appears to be reviving among at least some architects, educators, and students.

What has emerged, then, over the past century, is an ordered, university-based educational system marked by a high degree of professionalization and regulation. Together, the 103 accredited schools of architecture, the five national architecture organizations, their publications, and leading architectural journals act as the guardians and arbiters of the language and traditions of the profession, the awards and recognition programs, the educational performance objectives, and the legal terms for admission to practice.

"All of which is to say," writes Professor Palermo, "architecture and architects have perhaps never been so systematically ordered, internally defined, and—well, professional."[24]

## A TRADITION OF SELF-EXAMINATION

As much as any profession, architecture has been engaged in continuous self-reflection, and over the last sixty-five years, as architecture education shifted from offices to campuses, at least half a dozen comprehensive studies have been written examining current conditions and future directions of education and practice. Those reports were, for the most part, *internal* studies, sponsored by groups within the profession itself and conducted by investigators drawn from the ranks of the profession or from schools of architecture.[25]

In reviewing those studies, what is most striking is the continued force of so many of their themes and observations, even those written decades ago. Viewed another way, however, the reports are also powerful testimony to the depth and durability of the dilemmas that continue to confront architecture education and practice.

What are the most critical issues identified in those past studies? Which still resonate today?

*A Study of Architectural Schools, 1929–1932,* sponsored by the Association of Collegiate Schools of Architecture (ACSA) and funded by the Carnegie Corporation, criticized the dominance of design faculty over those specializing in "construction." Design projects at many schools, it said, resulted in "paper architecture" whose real purposes and functions are often unclear.[26] The report noted the scarcity of "real research" in architectural schools,[27] and described the difficulty architecture schools often have fitting into the university culture: "The insistence of the architect on the problem method in design, with its emphasis on accomplishment rather than time spent, on student freedom rather than regimentation, completely upsets the machinery of marks, semesters, and quantitative measures. . . ."[28]

The report's description of school life more than sixty years ago also rings familiar. "Go through, of an evening, any university campus containing an architectural school. That school can be spotted without fail. It is the one brilliantly lighted attic. It is always an attic, usually in the oldest and least desirable building."[29]

Finally, architecture school was very much a man's world at the time of the Bosworth and Jones report. In 1930, just 271 female architecture students were enrolled in 34 U.S. schools of architecture.[30]

Almost twenty-five years later, in 1954, an American Institute of Architects commission produced a two-volume report, *The Architect at Mid-Century,* which presented forty-three recommendations, including an aptitude test for prospective architects, more concerted support for architectural research, establishment of "study institutes" for architecture faculty, uniform registration laws, and a single registration exam.[31]

Looking ahead to the second half of the twentieth century, the report admonished educators and practitioners to close the growing gap between them—advice we heard repeated in our own travels. Schools, the AIA report said:

> will do well to maintain the closest liaison with the profession in order to adjust content and method to the changing needs of practice. And, by the same token, the profession, too, must apply its highest wisdom, most sympathetic understanding, and most penetrating vision to the problems of education. The very term "professional education" reveals by its compound form, the necessity of enlightened and harmonious cooperation.[32]

Perhaps of all major studies of architecture education, the most widely cited yet frequently misconstrued was the 1967 *Study of Education for Environmental Design*—sponsored by the American Institute of Architects and widely referred to as the "Princeton Report." Ironically, the Princeton Report owes much of its renown to a recommendation it never explicitly made. The report is often credited with proposing the "four-plus-two" master's program—four years of undergraduate study, followed by two years of graduate professional study. While the report did, in fact, propose a dramatic restructuring of architecture education that lent support to the idea, the spread of the four-plus-two programs owes more directly to two earlier reports—*Report of the Special Committee on Education, AIA,* and *"Blueprint '65: Architectural Education and Practice."*

The more visionary aspects of the Princeton Report lay elsewhere. Most essentially, it hypothesized three goals for design education: first, helping students develop the competence to work within the realities of actual practice; second, preparing graduates to be adaptable enough to grasp, and work within, "the continuing changes in the social, economic, scientific and technological setting of our society"; and finally, preparing students to develop their own analytical framework in which to envision a better society and built environment, "beyond present day constraints. . . ."[33]

To realize those objectives, the study stressed the importance of ending the isolation of the architectural discipline. It called for making connections—"building ladders and bridges," as co-author Robert Geddes put it recently—between all professions engaged in environmental design: "We needed to make working connections with others—engineers, planners, landscape architects, and an array of non-designers—who participate in the making of the built environment."[34]

To build those bridges, the report called for a flexible architecture curriculum, a wide range of teaching methods, and diverse architecture programs. Rather than proposing a "core curriculum" for all schools, the report suggested an intricate "modular, jointed framework for environmental design education" aimed at allowing students to tailor their studies to prepare them for more than nine hundred possible design-related careers.

The Princeton Report stressed the need for better understanding of architecture and the built environment among *nonarchitects*.[35] And it emphasized the importance of *continuing education*—an issue that has moved in the 1990s to the top of the academic and professional agenda:

> The idea is still accepted implicitly at many schools that what is learned in a period of four years to six years must carry a student through his entire professional career. To the extent that we are capable of developing more powerful planning and design concepts and methods through research and experimentation in the field, this idea of a termination point for education becomes less and less tenable.[36]

The Princeton Report was not immediately welcomed by the architecture education community. As Professor Geddes recounted recently, the report was issued at a time of growing disenchantment between the profession and the academy, and the failure by the AIA to invite more active participation from the collegiate schools association and other architectural and nonarchitectural organizations made the report instantly suspect to many educators.[37] Still, the wisdom and foresight of many of its observations survived the initial cool reception, and the document is today referred to by many with renewed admiration.

In 1981, a two-volume, 1,448-page analysis of the design studio was produced by a consortium of eight East Coast schools of architecture. The first volume of the *Architecture Education Study,* often called the "MIT Study," consisted of papers by six eminent scholars, each focused mainly on the methods and culture of the design studio. The second volume offered detailed case studies of design studios and student experiences.

The study introduced a topic of enduring relevance: how design education can shape attitudes about clients, users of buildings, and even fellow architects. MIT architecture professor Julian Beinart wrote that design students often held a pejorative view of architects—considering them egotists, elitists, insecure and indecisive.[38] And he discussed the disturbing attitudes architecture students harbor about clients. Some students advocated cooperation. Others suggested "educating" the client, while still others wished they could "eliminate" the client altogether—in effect, designing without compromise or constraint. Finally, some students spoke of "beguiling the client," with various kinds of "sneaky behavior" to deceive clients into thinking their wishes were being fulfilled.[39]

Lee Bolman, a professor of education at Harvard University, wrote that many practicing architects felt their schools shortchanged them in nondesign topics: 43 percent of those Bolman interviewed said they hadn't learned enough about how buildings actually get built, 39 percent wished they had been taught management skills, and 22 percent regretted not having learned how to deal better with other people. No one thought that school had provided too little training in design.[40] And Bolman argued that these curricular imbalances were not correctable simply by adding a few courses, but related instead to the more fundamental question of whether the methods and climate at most schools might be contributing to a disdain for technical and practice-oriented topics.

Asked to weigh the prospects of the profession itself, architects told Bolman in 1981 that they were fond of their work, but some believed, nonetheless, that the architecture field was "a dinosaur doomed to extinction."

> One in four felt the profession was clinging to an obsolete model of the architect's role.... A similar number felt that the profession did not know how to convince others of the architect's value. One-fifth of the architects suggested that the profession was dying, and an equal number felt that society undervalued architects.... There were few optimistic predictions although some predicted that architects would broaden their functions by involving themselves in planning, land use, development, and energy-efficient design....[41]

Bolman concluded with this challenge: "Many practitioners blame the schools for failure to address such issues. They ask, in effect, 'Shouldn't the schools be doing the research and theory-building to help in charting the profession's future? If they do not engage in that inquiry, who will?'"[42]

In the fifteen years since the MIT study, other notable books have been published on aspects of architectural practice or education. Two deserve special mention. *Architectural Practice: A Critical View,* published in 1988 by Robert Gutman, a sociologist who has taught at Princeton's architecture program for more than twenty-five years, identified critical issues for the profession generally:

- how to bring the supply of architects more in line with demand for services;
- how to develop, industry-wide, a consistent "philosophy of practice" and a clearer self-image that conforms to the realities of the building industry;
- how architecture could pursue both educational and professional strategies to cope with increased competition from other building professions;
- how to maintain the profitability of design businesses at a time of rising costs and growing competition; and
- how to balance the desires and aspirations of employees with the need for effective management and teamwork within design firms.[43]

Those challenges, aimed principally at practitioners, raised disturbing questions for educators as well. How effectively, for example, do teaching methods and curricular content at most architecture schools prepare students for a professional climate in which cooperation, management acumen, and specialization of services are increasingly valued? Have the changes in professional climate described by Gutman dangerously widened the gap between education and practice, and if so, what should be done?

University of Southern California Professor Dana Cuff's 1991 study, *Architecture: The Story of Practice,* employs the language of social science to describe the role of schools as "socializers" of young architects into the "culture" of the profession. Tracing the development of this culture, Cuff probed how schools of architecture help create in students, through their curricula and methods, a professional "ethos": what is valued, what is devalued.[44]

Cuff's telling portrait of life in architecture school adds modern insights to the picture presented sixty years earlier in the first ACSA-sponsored report. Students, she wrote, "stay up late, are never home, spend all their time in studio, and belong to a clique of other architecture students. . . . Here, in this earliest phase of becoming an architect, we see kernels of architects' later values, such as the principle of peer review and a developing segregation from the general public."[45]

Cuff especially notes the difficulties women face in entering academies where the curriculum and classroom culture reflect a long history of male dominance. Whatever their virtues, the design jury, the desk-side critique of student design work known as the "desk crit," and the grueling string of all-nighters known as the "charette"—reinforce a macho, boot camp atmosphere. Adding to the discomfort of female students is the fact that role models in studio and in the curriculum remain predominantly male.[46] Most essentially, Cuff argued, schools should do more to extend education "beyond context-free design," and she urged both the academy and the profession to pay more attention to "the social aspects of architecture."[47]

To summarize, our review of the history of American architectural education and of past efforts to assess it reveals as many possibilities as problems and as much continuity as change. A rich legacy of traditions continues to shape the methods and content of architecture education as well as professional practice, and an understanding of the historical roots of those practices is surely a necessary prelude to considering current conditions or future directions.

The distinguished scholar and practitioner Gerald McCue, of Harvard University, notes that the common pitfalls facing *any* study of architecture education include a tendency to exaggerate "newness and change," while downgrading that which is traditional, fundamental, and likely to continue. There is also, McCue says, a tendency to assume that problems, once identified, will necessarily be solved rationally. "It's like having a toothache forever, and yet we did. We're still dealing with some of these toothaches thirty years later."

With those wise precautions in mind, we begin our examination of current conditions and future prospects of architecture education by recalling, once more, the two-thousand-year-old Vitruvian trinity: Firmness, Commodity, and Delight. Throughout history, what has distinguished "architecture" from the mere building of buildings is the insight and skill to blend the useful with the timeless, the technically sound, with the beautiful. The challenge that has always faced both the academy and the profession has been discovering the right balance of those three ancient ideals, each so indispensable to successful architecture. That challenge continues today.

## A NEW VISION

In the chapters [of *Building Community*] that follow, we recommend a new framework for renewing architecture education and practice based on seven separate but interlocking priorities. The new vision builds on the traditions, history, and critiques already described, as well as best practices we discovered in campuses and offices around this country. If widely embraced, these "designs for renewal" would, we believe, help promote a more fruitful partnership between educators and practitioners that would not only enhance the competence of future architects but also lead the profession into more constructive engagement with the most pressing problems of our communities, our nation, and our planet.

We propose, as the first and most essential goal, *an enriched mission*—connecting schools and the profession more effectively to the changing social context. Specifically, we recommend that schools of architecture should embrace, as their primary objectives, the education of future practitioners trained and dedicated to promoting the value of beauty in our society; the rebirth and preservation of our cities; the need to build for human needs and happiness; and the creation of a healthier, more environmentally sustainable architecture that respects precious resources.

Second, we propose a more inclusive institutional context for the scholarly life of architects based on the principle of *diversity with dignity*. We imagine a landscape of architecture programs in which the multiple missions of schools are celebrated, and the varied talents of architecture faculty are supported and rewarded in a scholarly climate that encourages excellence in research, teaching, the application of knowledge, and the integration of learning.

To support the priority of diversity with dignity, we propose, as a third goal, *standards without standardization*. Such standards would affirm the rich diversity among architecture programs, establish a more coherent set of expectations at all schools that would support professional preparation, and bring into closer harmony the scholarly activities of students and faculty.

Fourth, the architecture curriculum at all programs should be better connected. A connected curriculum would encourage the integration, application, and discovery of knowledge within and outside the architecture discipline, while effectively making the connections between architectural knowledge and the changing needs of the profession, clients, communities, and society as a whole.

Fifth, each school of architecture should actively seek to establish a supportive *climate for learning*—where faculty, administrators, and students understand and share common learning goals in a school environment that is open, just, communicative, celebrative, and caring.

Sixth, educators and practitioners should establish *a more unified profession* based on a new, more productive partnership between schools and the profession. The priorities for sustained action between the academy and the profession should include strengthening the educational experience of students during school, creating a more satisfying system of internship after graduation, and extending learning throughout professional life.

Finally, we urge schools to prepare future architects for lives of civic engagement, of *service to the nation*. To realize this last goal for renewal, schools should help increase the storehouse of new knowledge to build spaces that enrich communities, prepare architects to communicate more effectively the value of their knowledge and their craft to society, and practice their profession at all times with the highest ethical standards.

Taken together, we propose an enriched educational climate in the academy and the profession—dedicated, with equal intensity, to promoting professional competence, and to placing architecture more firmly behind the goal of building not only great buildings but more wholesome communities.

## NOTES

1 Joseph Bilello, "Interaction and Interdependence: University-Based Architectural Education and the Architecture Profession," in *Architectural Education Report,* Special Education Task Force, American Institute of Architects (Washington, DC: American Institute of Architects, 1991), p. 18.
2 W. Cecil Steward, "Influences from Within the Academy upon Architectural Education," July 1988 (white paper prepared for the NAAB Board and distributed at the National Architectural Accrediting Board Validation Conference, Woodstock, Vermont, September 9–10, 1993), p. 14.
3 Henry N. Cobb, "Architectural Education: Architecture and the University," Walter Gropius Lecture, Harvard University, 1985; reprinted in *Architectural Record,* September 1985, p. 47.
4 American Institute of Architects, *Architecture Fact Book* (Washington, DC: American Institute of Architects, 1994), p. 34.
5 Gregory Palermo, "An Architect's Architect(s): Liberating Possibilities in Architectural Education," July 1995; draft of paper for *Dichotomy,* the student journal of the University of Detroit Mercy School of Architecture (projected publication date August 1996).
6 Turpin C. Bannister, ed., *The Architect at Mid-Century: Evolution and Achievement,* volume one of the Report of the Commission for the Survey of Education and Registration of the American Institute of Architects (New York: Reinhold Publishing Corp., 1954), p. 94.
7 Dana Cuff, *Architecture: The Story of Practice* (Cambridge, MA: MIT Press, 1991), pp. 23–26.
8 *Ibid.,* p. 26.
9 Arthur Clason Weatherhead, "The History of Collegiate Education in Architecture in the United States" (Ph.D. diss., Columbia University, 1941), pp. 27–29.
10 Kathryn H. Anthony, "ACSA Task Force on 'Schools of Thought in Architectural Edu-

cation'" (paper prepared for the Association of Collegiate Schools of Architecture, March 1, 1992), p. 13.
11. William L. Porter and Maurice Kilbridge, *Architecture Education Study, Vol. I:* The Papers (New York: Andrew W. Mellon Foundation, 1981), p. x.
12. Weatherhead, "The History of Collegiate Education," p. 63.
13. Bannister, *The Architect at Mid-Century,* p. 99.
14. F. H. Bosworth Jr. and Roy Childs Jones, *A Study of Architectural Schools* (New York: Charles Scribner's Sons, 1932), p. 3.
15. Weatherhead, "The History of Collegiate Education," p. 164.
16. *Ibid.,* pp. 173–74.
17. Bannister, *The Architect at Mid-Century,* p. 107.
18. Weatherhead, "The History of Collegiate Education," p. 193.
19. *Ibid.,* p. 195.
20. *Ibid.,* p. 244.
21. John M. Amundson, "An Historic Perspective of Accreditation in Architecture" (collected papers from the National Architectural Accrediting Board accreditation conference, New Orleans, March 18, 1976), p. 24.
22. Bannister, *The Architect at Mid-Century,* pp. 103, 109.
23. National Council of Architectural Registration Boards, "1993 Circular of Information No. 2: Architect Registration Examination" (Washington, DC: NCARB, 1993), p. 2.
24. Palermo, "An Architect's Architect(s)," p. 4.
25. Porter and Kilbridge, *Architecture Education Study,* pp. 842–45.
26. Bosworth and Jones, *A Study of Architectural Schools,* pp. 30–31, 43.
27. *Ibid.,* p. 36.
28. *Ibid.,* p. 129.
29. *Ibid.,* p. 110.
30. *Ibid.,* p. 104.
31. Bannister, *The Architect at Mid-Century,* pp. 442–49.
32. *Ibid.,* p. 109.
33. Robert L. Geddes and Bernard P. Spring, *A Study of Education for Environmental Design: The "Princeton Report"* (Washington, DC: American Institute of Architects, December 1967; reprinted June 1981), p. 4.
34. Robert L. Geddes, remarks to the AIA/*Architectural Record* Walter Wagner Education Forum, 1995 AIA National Convention, Atlanta, Georgia, May 6, 1995.
35. Geddes and Spring, *A Study of Education for Environmental Design,* p. 22.
36. *Ibid.,* p. 22.
37. Geddes, remarks.
38. Porter and Kilbridge, *Architecture Education Study,* pp. 293–94, 326.
39. *Ibid.,* pp. 300–307.
40. *Ibid.,* p. 683.
41. *Ibid.,* pp. 733–34.
42. *Ibid.,* p. 737.
43. Robert Gutman, *Architectural Practice: A Critical View* (Princeton, NJ: Princeton Architectural Press, 1988), pp. 97–108.
44. Cuff, *Architecture,* p. 43.
45. *Ibid.,* pp. 118.
46. *Ibid.,* p. 121.
47. *Ibid.,* p. 108.